FLUORINATED SURFACTANTS

SURFACTANT SCIENCE SERIES

1. Nonionic Surfactants, *edited by Martin J. Schick* (see also Volumes 19 and 23)
2. Solvent Properties of Surfactant Solutions, *edited by Kozo Shinoda* (out of print)
3. Surfactant Biodegradation, *by R. D. Swisher* (see Volume 18)
4. Cationic Surfactants, *edited by Eric Jungermann* (see also Volumes 34 and 37)
5. Detergency: Theory and Test Methods (in three parts), *edited by W. G. Cutler and R. C. Davis* (see also Volume 20)
6. Emulsions and Emulsion Technology (in three parts), *edited by Kenneth J. Lissant*
7. Anionic Surfactants (in two parts), *edited by Warner M. Linfield* (out of print)
8. Anionic Surfactants: Chemical Analysis, *edited by John Cross* (out of print)
9. Stabilization of Colloidal Dispersions by Polymer Adsorption, *by Tatsuo Sato and Richard Ruch*
10. Anionic Surfactants: Biochemistry, Toxicology, Dermatology, *edited by Christian Gloxhuber* (see Volume 43)
11. Anionic Surfactants: Physical Chemistry of Surfactant Action, *edited by E. H. Lucassen-Reynders* (out of print)
12. Amphoteric Surfactants, *edited by B. R. Bluestein and Clifford L. Hilton* (out of print)
13. Demulsification: Industrial Applications, *by Kenneth J. Lissant*
14. Surfactants in Textile Processing, *by Arved Datyner*
15. Electrical Phenomena at Interfaces: Fundamentals, Measurements, and Applications, *edited by Ayao Kitahara and Akira Watanabe*
16. Surfactants in Cosmetics, *edited by Martin M. Rieger*
17. Interfacial Phenomena: Equilibrium and Dynamic Effects, *by Clarence A. Miller and P. Neogi*
18. Surfactant Biodegradation, Second Edition, Revised and Expanded, *by R. D. Swisher*

19. Nonionic Surfactants: Chemical Analysis, *edited by John Cross*
20. Detergency: Theory and Technology, *edited by W. Gale Cutler and Erik Kissa*
21. Interfacial Phenomena in Apolar Media, *edited by Hans-Friedrich Eicke and Geoffrey D. Parfitt*
22. Surfactant Solutions: New Methods of Investigation, *edited by Raoul Zana*
23. Nonionic Surfactants: Physical Chemistry, *edited by Martin J. Schick*
24. Microemulsion Systems, *edited by Henri L. Rosano and Marc Clausse*
25. Biosurfactants and Biotechnology, *edited by Naim Kosaric, W. L. Cairns, and Neil C. C. Gray*
26. Surfactants in Emerging Technologies, *edited by Milton J. Rosen*
27. Reagents in Mineral Technology, *edited by P. Somasundaran and Brij M. Moudgil*
28. Surfactants in Chemical/Process Engineering, *edited by Darsh T. Wasan, Martin E. Ginn, and Dinesh O. Shah*
29. Thin Liquid Films, *edited by I. B. Ivanov*
30. Microemulsions and Related Systems: Formulation, Solvency, and Physical Properties, *edited by Maurice Bourrel and Robert S. Schecter*
31. Crystallization and Polymorphism of Fats and Fatty Acids, *edited by Nissim Garti and Kiyotaka Sato*
32. Interfacial Phenomena in Coal Technology, *edited by Gregory D. Botsaris and Yuli M. Glazman*
33. Surfactant-Based Separation Processes, *edited by John F. Scamehorn and Jeffrey H. Harwell*
34. Cationic Surfactants: Organic Chemistry, *edited by James M. Richmond*
35. Alkylene Oxides and Their Polymers, *by F. E. Bailey, Jr., and Joseph V. Koleske*
36. Interfacial Phenomena in Petroleum Recovery, *edited by Norman R. Morrow*
37. Cationic Surfactants: Physical Chemistry, *edited by Donn N. Rubingh and Paul M. Holland*
38. Kinetics and Catalysis in Microheterogeneous Systems, *edited by M. Grätzel and K. Kalyanasundaram*
39. Interfacial Phenomena in Biological Systems, *edited by Max Bender*
40. Analysis of Surfactants, *by Thomas M. Schmitt*
41. Light Scattering by Liquid Surfaces and Complementary Techniques, *edited by Dominique Langevin*
42. Polymeric Surfactants, *by Irja Piirma*
43. Anionic Surfactants: Biochemistry, Toxicology, Dermatology. Second Edition, Revised and Expanded, *edited by Christian Gloxhuber and Klaus Künstler*
44. Organized Solutions: Surfactants in Science and Technology, *edited by Stig E. Friberg and Björn Lindman*
45. Defoaming: Theory and Industrial Applications, *edited by P. R. Garrett*

46. Mixed Surfactant Systems, *edited by Keizo Ogino and Masahiko Abe*
47. Coagulation and Flocculation: Theory and Applications, *edited by Bohuslav Dobiáš*
48. Biosurfactants: Production · Properties · Applications, *edited by Naim Kosaric*
49. Wettability, *edited by John C. Berg*
50. Fluorinated Surfactants: Synthesis · Properties · Applications, *Erik Kissa*
51. Surface and Colloid Chemistry in Ceramic Processing, *edited by Robert J. Pugh and Lennart Bergström*

ADDITIONAL VOLUMES IN PREPARATION

Cationic Surfactants: Analytical and Biological Evaluation, *edited by John Cross and Edward J. Singer*

Industrial/Technological Applications of Dispersions, *edited by Robert B. McKay*

FLUORINATED SURFACTANTS

Synthesis • Properties • Applications

Erik Kissa

E. I. du Pont de Nemours & Co., Inc.
Deepwater, New Jersey

Marcel Dekker, Inc. New York • Basel • Hong Kong

Library of Congress Cataloging-in-Publication Data

Kissa, Erik
Fluorinated surfactants: synthesis • properties • applications / Erik Kissa.
p. cm. — (Surfactant science series; v. 50)
Includes bibliographical references and index.
ISBN 0-8247-9011-1 (acid-free)
1. Surface active agents. 2. Organofluorine compounds. I. Title. II. Series.
TP994.K57 1993
668'.1–dc20 93-9058
CIP

Atsurf, Fluorad, Fluosol, Fluowet, Forafac, Lodyne, Monflor, Oxydent, Pluronic, Scotchban, Surflon, and Zonyl are registered trademarks.

The publisher offers discounts on this book when ordered in bulk quantities. For more information, write to Special Sales/Professional Marketing at the address below.

This book is printed on acid-free paper.

MARCEL DEKKER, INC.
270 Madison Avenue, New York, New York 10016

Current printing (last digit):
10 9 8 7 6 5 4 3 2 1

PRINTED IN THE UNITED STATES OF AMERICA

Preface

Fluorinated surfactants are truly the super surfactants. Fluorinated surfactants can decrease the surface tension of water below the lower limit reached by hydrocarbon-type surfactants. The perfluorinated hydrophobe is extremely resistant to chemical attack, and fluorinated surfactants can be used in media where conventional surfactants do not survive. Since a perfluoroalkyl chain is not only hydrophobic but oleophobic, fluorinated surfactants can serve as oil and fat repellents. Compounds consisting of a fluorinated chain and a hydrocarbon group can function as surfactants in hydrocarbon media. Because of their unique properties, fluorinated surfactants are indispensable in certain practical applications and of great theoretical interest for the study of surfactants and micellar systems.

Chapter 1 presents an overview of fluorinated surfactants. The synthesis of fluorinated surfactants is discussed in Chapter 2. Since the space limitations precluded a detailed description of processes, patent citations are augmented by references to *Chemical Abstracts*. Physical and chemical properties are reviewed in Chapter 3. Chapters 4–7 are devoted to the theory of fluorinated surfactants: liquid-vapor and liquid-liquid interface (Chapter 4), solid-liquid interface (Chapter 5), solutions of fluorinated surfactants (Chapter 6), and the structure of micelles and mesophases, including mixed surfactant systems, in Chapter 7. The practical application of fluorinated surfactants is the subject of Chapter 8. Various applications are listed in alphabetical order for easy access to information. Chapter 9 reviews analytical and physical methods for the investigation of fluorinated surfactants. Chapter 10 examines the environmental and toxicological aspects, including the use of fluorinated surfactants in biological systems.

Because of my intention to write a stand-alone book, material dealt with in other monographs has been included. Related theories and principles are presented along with references to the literature for those who wish to study the fundamental theories in depth. Some discussion of hydrocarbon-type surfactants is given so they can be compared with fluorinated surfactants.

Several computer-aided literature searches were conducted. Ongoing research on fluorinated surfactants is very active, and while the book was being written new material had to be constantly reviewed. An effort was made to keep all chapters up to date.

Since the book was written at home after regular working hours, I am immensely grateful to my wife, Selma, whose support and patience made this book possible.

I am indebted for valuable comments and suggestions to Du Pont chemists who read the chapters in which they have expertise: Drs. J. E. Dowd, T. A. Liss, and J. F. Neumer (synthesis), K. S. Prowse (applications), M. W. Duch (ESCA), J. T. Cronin (IR), A. Foris (NMR), J. R. Valentine (MS), B. E. Baker (toxicology), R. C. Bergman and S. Raynolds (intravascular oxygen carriers). I am also grateful to my son Erik H. Kisa, M.D., for reviewing the chapter on blood substitutes.

Last but not least, I am indebted to Joseph Stubenrauch, Marilyn Ludzki, and Sandra Beberman of Marcel Dekker, Inc., for their assistance in preparing this volume.

Erik Kissa

Contents

FLUORINATED SURFACTANTS

1

Structure of Fluorinated Surfactants

1.1 INTRODUCTION

Surfactants have a very important role in our everyday life. Surfactants are essential in biological systems and industrial processes. Our food, cosmetics, medicine, and household items, such as soap and detergents, contain surfactants. The wide variety of surfactant applications has required different types of surfactants and a large number of surfactant structures is available for the specific need. The literature on surfactants is voluminous.

The word *surfactant* is an abbreviation of the more descriptive term *surface-active agent*. A surfactant is a substance which, even at low concentrations, effectively lowers the surface tension of its medium by selective adsorption on the interface. A surfactant can be a pure chemical compound or a mixture of homologues or different chemical compounds. The characteristic feature of surfactants is their efficiency in lowering surface tension. The surface tension of a liquid can be lowered by mixing it with another liquid of lower surface tension. For example, one part of ethanol added to four parts of water decreases the surface tension of water from 73 mN/m to below 40 mN/m. However, only 0.1% of a typical surfactant is needed for the same surface tension reduction. The efficiency of surfactants in lowering surface tension is related to selective adsorption of the surfactant at the interface. The adsorption, in turn, is a result of the amphiphilic nature of the surfactant.

The term amphiphilic or amphipathic, as it is sometimes called, implies attraction to two different kinds of media. The surfactant structure can be described as consisting of two parts with vastly different solution characteristics: a "solvent-soluble" lyophilic part and a "solvent-insoluble" lyophobic part.

Conventional surfactants consist of a water-soluble hydrophilic part and a water-insoluble hydrophobic part which is lipophilic, compatible with fats and hydrocarbons. The hydrophobe is usually a hydrocarbon group, but surfactants containing oxygen, nitrogen, sulfur, silicon, and/or halogens are also used.

In fluorinated surfactants the hydrophobic part of the surfactant molecule contains fluorine. At least one hydrogen atom in the hydrophobic segment of a surfactant has been replaced by fluorine. Both the extent of fluorination and the position of fluorine atoms in the surfactant molecule affect the characteristics of the surfactant. Hence fluorinated surfactants can be classified as perfluorinated surfactants or partially fluorinated surfactants. In perfluorinated surfactants all hydrogens in the hydrophobic segment have been replaced by fluorine. In partially fluorinated surfactants the hydrophobic part of the surfactant molecule contains both fluorine and hydrogen atoms. The location and the number of fluorine atoms in the partially fluorinated hydrophobe are important. Partially fluorinated surfactants with a terminal CF_3-group differ in their characteristics from partially fluorinated surfactants with a hydrogen-containing terminal group.

Substitution of fluorine for hydrogen changes the properties of a surfactant drastically [1–12]. The hydrophobic part of the fluorinated surfactant not only repels water but repels oil and fat as well. Hence fluorinated surfactants exhibit both water and oil repellency when adsorbed on substrates such as textiles or paper. Fluorinated surfactants are much more surface active than their hydrocarbon counterparts. Fluorinated surfactants can lower surface tension of aqueous systems to below 20 mN/m and are effective at very low concentration. Only 10 ppm of a fluorinated surfactant may be needed to lower the surface tension of water to 40 mN/m. Fluorinated surfactants exhibit surface activity in organic systems and are stable to heat, acids, and bases, as well as reducing and oxidizing agents. On the negative side is the higher price of fluorinated surfactants, but this is at least partially offset by the small quantities usually needed. Because of their unique properties, fluorinated surfactants are irreplaceable in many applications.

The term *fluorinated surfactant*, although widely used, can be misleading, since it implies that the hydrocarbon segment of a surfactant has been fluorinated. This, of course, is not the real synthetic route to surfactants with a fluorine-containing hydrophobe. The author therefore prefers the shorter term *fluorosurfactant*, in analogy to the frequently used terms *fluorochemical* and *fluorocarbon*. However, the term fluorinated surfactants is conventional and, consequently, the title of this book.

Some surfactants have counterions which contain fluorine but do not have fluorine in their hydrophobic part. Although such surfactants do not really belong to the class of fluorinated surfactants proper, the presence of fluorine in the counterion affects the behavior of the surfactant. Such surfactants have therefore been included in this book.

1.2 STRUCTURAL FEATURES OF FLUORINATED SURFACTANTS. THE HYDROPHOBE

To understand how surfactants function and to select a surfactant for a specific purpose it is necessary to classify surfactants according to their structural features.

Like all surfactants, fluorinated surfactants are either ionic or nonionic. Ionic surfactants can, unlike nonionic surfactants, dissociate into ions in an aqueous medium. The hydrophobic part can belong to a negative or positive ion. Some surfactants have negatively and positively charged functional groups on the same backbone. The surfactants can therefore be classified into four types:

Anionic surfactants—the hydrophobic part is an anion, for example, $R_fCOO^-Na^+$, where R_f is a fluorine-containing hydrophobe.

Cationic surfactants—the hydrophobic part is a cation, for example, $C_7F_{15}CONH(CH_2)_3N^+(CH_3)_3I^-$.

Amphoteric surfactants—have at least one anionic and one cationic group at their isoelectric point.

Nonionic surfactants—do not dissociate into ions, for example, $C_7F_{15}CH_2CH_2O(CH_2CH_2O)_nH$. A special class of nonionic fluorinated surfactants are compounds which do not have a hydrophile but consist of an oleophobic (fluorinated) segment and a oleophilic segment (see Sec. 1.8).

The structure of the hydrophobe of an anionic fluorinated surfactant can be varied more extensively than the structure of the hydrophile. The hydrophobe can be a fully or partially fluorinated alkyl group having a linear or a branched alkyl chain. The hydrophobe can have an aromatic group or contain other elements (O, N, Cl, S, and Si) as well, as shown with the following examples:

C, H, F

$C_nF_{2n+1}-$

$C_nF_{2n+1}CH_2CH_2-$

C, H, F, O

$C_nF_{2n+1}OCF_2CF_2-$

$C_nF_{2n+1}OC_6H_4-$

C ,H, F, O, N

$C_nF_{2n+1}CONH(CH_2)_3N<$

C ,H, F, O, S

$C_nF_{2n+1}SO_2NH(CH_2)_3N<$

C, F, Cl

$CF_3Cl_2(CF_2CFCl)_{n-1}CF_2-$

C, F, Si

$C_8F_{17}CH_2CH_2Si(CH_3)_2-$

The hydrophobes of partially fluorinated surfactants contain both fluorine and hydrogen atoms. Unlike the hydrophobe of hydrocarbon surfactants, the partially fluorinated hydrophobe consists of two mutually phobic parts which are not compatible. Partially fluorinated surfactants therefore exhibit anomalies in macroscopic characteristics, such as the critical micelle concentration (cmc), and in microscopic phenomena as well.

Partially fluorinated surfactants have several advantages over perfluorinated surfactants. The hydrocarbon segment provides solubility in more commonly used solvents, lowers the melting point of the surfactant, reduces volatility, and decreases the acid strength of fluorinated acids [13].

Hydrocarbon-type surfactants with fluorinated counterions are not truly fluorinated surfactants, because the surface-active ion is not fluorinated. However, fluorination of the counterion affects the solution characteristics of the surfactant and has been the subject of several investigations. Moss and coworkers [14,15] used the CF_3SO_3- anion as a counterion for a sulfonium methylating agent and a hydroperoxy surfactant. Hoffmann et al. [16,17] investigated surfactant association in solutions of dodecylammonium and tetradecylammonium trifluoroacetates and tetradecylpyridinium perfluorobutyrate. Sugihara et al. [18] studied the solubility and cmc of dodecylammonium perfluorocarboxylates in water. The effect of the counterion (trifluoroacetate, pentafluoropropionate, and heptafluorobutyrate) hydrophobicity on solubility, cmc, and the Krafft point were determined.

1.3 ANIONIC FLUORINATED SURFACTANTS

Ionic surfactants dissociate in water and form a surface-active ion with an oppositely charged counterion. The surface-active ions of anionic surfactants bear a negative charge. Anionic fluorinated surfactants can form water-insoluble ion pairs with cationic species and are usually not compatible with most cationic surfactants.

Anionic surfactants are the most important class of fluorinated surfactants. Based on the hydrophile structure, anionic fluorinated surfactants can be divided into four main categories:

Carboxylates	$R_fCOO^-M^+$
Sulfonates	$R_fSO_3^-M^+$
Sulfates	$R_fOSO_3^-M^+$
Phosphates	$R_fOP(O)O_2^{--}M_2^+$

where R_f is a fluorine-containing hydrophobe and M^+ an inorganic or an organic cation. Some anionic surfactants contain nonionic hydrophilic polyoxyethylene segments which increase their solubility and compatibility with cationic or amphoteric surfactants.

The fluorinated hydrophobe can be a perfluoroalkyl group or a partially fluorinated alkyl group. Some fluorinated surfactants have hydrophobes containing aryl groups as well. The hydrophobe may be attached via a sulfide, carbonamido, or sulfonamido linkage to the rest of the surfactant molecule.

Examples of anionic fluorinated surfactants with a carboxylic acid or carboxylate hydrophile are given in Table 1.1. Although fluorinated carboxylic acids are stronger acids than alkanoic acids, fluorinated carboxylates have some disadvantages characteristic of their hydrocarbon-type counterparts, such as insolubility in strong acids and in water containing di- or trivalent metal ions.

Examples of fluorinated surfactants with a sulfonic acid group (or its salt) as the hydrophile are shown in Table 1.2. Fluorinated sulfonic acids are less sensitive to low pH, electrolytes, and calcium ions than their carboxylic acid analogues.

Some examples of fluorinated surfactants with a sulfate group are given in Table 1.3. A sulfate group is a stronger hydrophile than the sulfonate group. Sulfated fluorinated surfactants are readily available from fluorinated alcohols but their lower hydrolytic stability limits their use.

Table 1.4 shows examples of fluorinated surfactants featuring a phosphate hydrophile. In general, the phosphates are less prone to cause foaming than other anionic fluorinated surfactants and some phosphate ester salts function as antifoaming agents.

1.4 CATIONIC SURFACTANTS

In cationic fluorinated surfactants the fluorinated hydrophobe is attached directly or indirectly to a quaternary ammonium group, a protonated amino group, or a heterocyclic base. Cationic surfactants dissociate in water, forming a surface-active positively charged ion and a negatively charged counterion. Like anionic surfactants, cationic surfactants are sensitive to electrolytes and the pH of the medium. Cationic surfactants are usually incompatible with anionic surfactants. Cationic fluorinated surfactants adsorb on negatively charged surfaces. Since most surfaces and particles are negatively charged, the adsorption can be advantageous or disadvantageous, depending on the intended use of the surfactant. For example, adsorption on clay and sludge in wastewater cleaning systems simplifies the removal of cationic fluorinated surfactants from effluents.

Some cationic fluorinated surfactants contain both quaternary and secondary amino groups and carbonamido or sulfonamido linkages. Examples of cationic fluorinated surfactants are given in Table 1.5.

Amine oxides, or more correctly amine oxides of tertiary amines, are nonionic in an alkaline or neutral solution but cationic in acid solutions. The syn-

(*Text continues on page 9.*)

TABLE 1.1 Carboxylic Acids and Salts

Structural features	References
Perfluoroalkanoic acid	
R_fCOOH or R_fCOOM	1, 2, 10, 19
Fluorinated alkanoic acid	
$R_f(CH_2CF_2)_nCH_2COOH$	20
Alkanoic acids with terminal R_f group	
$R_f(CH_2)_mCOONa$	21
Alkenoic acids with terminal R_f group	
$R_fCH{=}CH(CH_2)_nCOONa$	21
Perfluoroalkoxyalkanoic acids	
$R_fO(CH_2)_nCOOH$	22
Perfluoroalkylethyleneoxyalkanoic acids	
$C_8F_{17}CH_2CH_2OCH_2CH_2COOH$	23
Perfluoroalkoxybenzoic acid	
$R_fOC_6H_4COOH(Na)$	6, 7
Sulfide linkage	
$R_fCH_2CH_2SCH_2CH_2COOLi$	24
Bis(polyfluoroalkoxyalkylcarboxylic acid)sulfide	
or $-S(CH_2)_mS-$ linkage ($m = 2$ or 6)	
$[(CF_3)_2CFO(CF_2)_n(CH_2)_9CH(COOH)]S$	25
Perfluoroacylaminoalkanoic acids (Na or NH_4 salts)	
$C_8F_{17}CONH(CH_2)_5COONa$	26
Perfluoroalkanesulfonamido group	
$R_fSO_2NH(CH_2)_3N(CH_2COONa)_2$	27
$R_fSO_2N(CH_2CH(OH)CH_2COOH)(CH_2)_3N(CH_3)CONHC_2H_5$	28
Hydroxy and amine function	
$C_9F_{19}CH_2CH(OH)CH_2N(CH_3)CH_2COOK$	29
Perfluoroalkylated aminocarboxylates with oxy and hydroxy groups	
$R_fCH_2CH(OH)CH_2NH(CH_2)_3O[(CH_2)_4O]_8(CH_2)_3NHCH_2COONa$	30
Perfluoropolyether carboxylic acids	
$CF_3[OCF_2CF(CF_3)]_nOCF_2COOH$ ($n = 1,2,3$)	
$HOOCCF_2(OCF_2CF_2)_n(OCF_2)_mOCF_2COOH$ (n, $m = 1,2,3$)	31
Oxyethylated segment	
$F_3C(CF_2)_mSO_2N(C_2H_5)(CH_2CH_2O)_n(CH_2)_3COOK$ (or Na, Li)	
($m = 3$–25, $n = 2$–50)	32
$C_9F_{19}CH_2CH(OH)CH_2NH(CH_2)_3O(CH_2CH_2O)_8(CH_2)_3NHCH_2COONa$	30
$R_fCH_2CH_2SCH(COOH)CH_2COO(CH_2CH_2O)_{22}H$	
$R_fCH_2CH_2SCH(CH_2COOH)COO(CH_2CH_2O)_{22}H$	33
Hexafluoropropene oxide as the hydrophobe	
$C_3F_7O[CF(CF_3)CF_2O]_{n-2}CF(CF_3)COONa$ ($n = 2$–6)	34

TABLE 1.2 Fluorinated Sulfonic Acids and Salts

Structural features	References
Perfluoroalkanesulfonic acids	
$C_nF_{2n+1}SO_3H$ (or salts)	8, 35–37
Tetraethylammonium perfluorooctanesulfonate	
$C_8F_{17}SO_3N(C_2H_5)_4$	8, 38
Ammonium salts of fluorinatedsulfonic acids	
$R_fSO_3^- R_2N^+(CH_2CH_2OH)_2$, R = alkyl	39, 40
Perfluoroalkylethanesulfonates	
$C_nF_{2n+1}CH_2CH_2SO_3NH_4$	41
Perfluoropropylalkanesulfonates	
$C_3F_7(CH_2)_nSO_3Na$ (n = 5,7,9)	42
Perfluoroalkylbenzenesulfonates	
$C_nF_{2n+1}C_6H_4SO_3H$ (or salts)	43
Perfluoroalkoxybenzenesulfonic acid	
$R_fOC_6H_4SO_3H$ (or salts)	6, 7
Perfluoroacylbenzenesulfonates	
$R_fCOC_6H_4SO_3H$ or salts	
$HC_nF_{2n}COC_6H_4SO_3H$ or salts	44
Tri(perfluoroalkyl)methoxyalkanesulfonates	
$(C_2F_5)_3CO(CH_2)_3SO_3K$	45
Neos "Ftergent NF"	
$(CF_3)_2C=C(CF_3)OC_6H_4SO_3Na$	
$C_{3n}F_{6n-1}C_6H_4SO_3H$,	46, 47
Fluorinated hydrophobes with sulfide and carbonamide linkages	
$CF_3(CF_2)_nCH_2CH_2SCH_2CH_2CONHC(CH_3)_2CH_2SO_3H$ (or salts)	48, 49
Perfluoroalkanesulfonamido and carbonamido functions	
$R_fSO_2NH(CH_2)_3NMeCONH(CH_2)_2SO_3Na$	50
Perfluoroacylcarbonamido group	
$R_fCONR(CH_2)_3SO_3Na$	51
Oligo(hexafluoropropene oxide) hydrophobe	
$CF_3CF_2CF_2O-(\underset{\displaystyle CF_3}{\underset{\vert}{C}}FCF_2O)_n-Ar-SO_3H$ where Ar is an arylene group	52
Oxafluoroalkanesulfonates	
$CF_3CF_2[CF_2CF(CF_3)O]_nCF_2CF_2SO_3M$, M = K, Na	53
Oxyethylenesulfonates	
$CF_3C_6H_{12}CH_2O(C_2H_4O)_5SO_3Na$	54, 55
Perfluoroalkyl ether amides	
$F[CF(CF_3)CF_2O]_nCF(CF_3)CONHCH_2CH_2SO_3Na$	56
Fluorinated sulfosuccinates	
$R_fC_2H_4OOCCH(SO_3Na)CH_2COOC_2H_4R_f$	
$H(CF_2)_nC_2H_4OOCCH(SO_3Na)CH_2COOC_2H_4(CF_2)_nH$	10, 57
Perfluoroalkylethersulfonate	
$CF_3CF_2(CF_2CF_2)_nOCF_2CF_2SO_3M$, M = alkali metal	58
Perfluoroalkylsulfopropionates and sulfobutyrates	
$CF_3(CF_2)_nCH_2OOCCH(CH_3)CH_2SO_3Na$	59

TABLE 1.3 Fluorinated Alkyl Sulfates

Structure	References
Perfluoroalkylated methyl sulfate	
$C_7F_{15}CH_2OSO_3Na$	10
Fluorinated sulfatopoly(oxyethylene)	
$H(CF_2CF_2)_nCH_2(OCH_2CH_2)_mOSO_3NH_4$	
$CF_3(CF_2CF_2)_nCH_2(OCH_2CH_2)_mOSO_3NH_4$	60
Perfluoropropoxylated sulfate	
$(F_3C)_2CFO(CH_2)_6OSO_3Na$	61
Fluorinated aminosulfate	
$C_8F_{17}SO_2NH(CH_2)_3NH(CH_2)_3NHCH_2CH_2OSO_3Na$	62

TABLE 1.4 Fluorinated Phosphates

Structure	References
Perfluoroalkylethyl phosphates	63, 64
$CF_3(CF_2)_nCH_2CH_2OP(O)(OH)_2$	
$[CF_3(CF_2)_nCH_2CH_2O]_2P(O)(OH)$	
$[CF_3(CF_2)_nCH_2CH_2O]_2P(O)(ONH_4)$	
Mono- and bis(fluoroalkyl)phosphate amine salts	65
$(C_8F_{17}CH_2CH_2O)_{1.5}PO[(OH)NH(C_2H_4OH)_2]_{1.5}$	
Perfluoroalkyl phosphates	66
$(CF_3)_2CF(CF_2)_6FCH_2CH_2OP(O)(OH)_2$	
$[(CF_3)_2CF(CF_2)_6FCH_2CH_2O]_2P(O)(OH)$	
$[(CF_3)_2CF(CF_2)_6FCH_2CH_2O]_3P(O)$	
Oligomer phosphate	6, 7
$C_6F_{13}CH=C(CF_3)OPO(OH)_2$	
(Perfluoroalkyl)glycol monophosphates	
$F(CF_2)_nCH(OH)CH_2OP(O)(OH)_2$	67
$F(CF_2)_nCH(CH_2OH)OP(O)(OH)_2$	
$C_6F_{13}CH_2CH_2S(CH_2)_3P(O)(OC_2H_5)_2$	68
Perfluoroalkanesulfonamide derivatives	69
$CF_3C(CF_2)_7SO_2N(C_2H_5)CH_2CH_2OP(O)(OH)_2$	
$CF_3C(CF_2)_7SO_2N(C_2H_5)CH_2CH_2OP(O)(ONa)_2$	
$[CF_3C(CF_2)_7SO_2N(C_2H_5)CH_2CH_2O]_2P(O)ONa$	
Polymerizable phosphate esters	70
$CH_2=CRCOOCH_2CH(OH)CH_2OP(O)(OR')ONa$, where R = H or CH_3, R′= alkyl, fluoroalkyl	

TABLE 1.5 Cationic Fluorinated Surfactants

Structural feature	References
(Perfluoroalkyl)alkyl-	
$C_nF_{2n+1}CH_2CH_2N^+(CH_3)_2C_2H_5\,I^-$ n = 6 or 8	71
$C_7F_{15}CH_2NH(CH_2)_2N^+(CH_3)_3\,Cl^-$	
$H(CF_2)_nCH_2N^+R(CH_3)_2\,I^-$, n = 2,4,6; R = CH_3, C_2H_5, C_3H_7, C_4H_9	72
$(CF_3)_2CF(CF_2)_6CH_2CH(OH)CH_2N^+(CH_3)_3\,I^-$	73
Perfluoroalkaneamido-	
$C_7F_{15}CONH(CH_2)_3N^+(CH_3)\,I^-$	74
$F[C(CF_3)CF_2O]_nCF(CF_3)CONH(CH_2)_3N^+(C_2H_5)_2CH_3\,I^-$	75
$R_fO(C_3F_6O)CF(CF_3)CONH(CH_2)_3N^+(CH_3)_3\,Cl^-$	76
$CF_3CF_2CF_2CONHCH_2CH_2CH_2N^+(OH)(CH_3)_2\,HOCH_2COO^-$	77
Perfluorooctanesulfonamido-	
$C_8F_{17}SO_2NH(CH_2)_3N^+(CH_3)_3\,I^-$	78
Sulfide linkage	
$R_f(CH_2)_nS(CH_2)_mN^+(CH_3)_2CH_2COOH\,Cl^-$	79
$C_6F_{13}CH_2CH_2SCH_2N^+(CH_3)_2CH_2CH_2OH\,Br^-$	80
$(CF_3)_2CF(CF_2)_4(CH_2)_2SCH_2CH(CH_3)_2COOCH_2CH_2N^+(CH_3)_3\,I^-$	81
$R_fCH(OH)CH_2SCH_2CH_2N^+(CH_3)_2CH_2C_6H_5\,Br^-$, $R_f = C_6F_{13}$ or C_8F_{17}	82
Ester linkage	
$C_6F_{13}SO_2O(CH_2)_3N^+(CH_3)_3\,I^-$	83
Amine salts	
$[F(CF_2)_8CH(OH)CH_2]_2NCH_2CH_2NH_2 \cdot H_2SO_4$	84
Heterocyclic nitrogen	
N-(Perfluorooctanesulfonyl)piperazine derivatives	85
Terminal perfluoroalkyl (CF_3-)	
$CF_3(CH_2)_nN^+(CH_3)_3Br^-$	86

thetic routes to amine oxides and cationic surfactants are similar. Although amine oxides are intrinsically nonionic, they may be considered to belong to the class of cationic surfactants.

1.5 AMPHOTERIC SURFACTANTS

Amphoteric fluorinated surfactants are bifunctional compounds which can function both as anionic and as cationic surfactants depending on the pH of the medium [87]. Amphoteric surfactants have at least one cationic group and at least one anionic group and are electrically neutral around their isoelectric points. Outside the isoelectric range, amphoteric surfactants function as anionic or cationic surfactants depending on the pH of their medium. Ideally, typical amphoteric fluorinated surfactants function in an alkaline medium as anionic surfactants, while in an acid medium they assume a cationic character:

$$R_fN^+H_2CH_2CH_2COOH \underset{H^+}{\overset{OH^-}{\rightleftharpoons}} R_fN^+H_2CH_2CH_2COO^- \underset{H^+}{\overset{OH^-}{\rightleftharpoons}} R_fNHCH_2CH_2COO^-\ M^+$$

In reality, surfactants having cationic and anionic functional groups are not always truly amphoteric. The dissociation constants of anionic and cationic functional groups are rarely equal. As a consequence, either the cationic or the anionic character may dominate the behavior of the surfactant in solution.

Amphoteric fluorinated surfactants usually have a (a) carboxybetaine (I), (b) sulfobetaine (II), or (c) sulfatobetaine (III) structure:

$$R_fXYN^+(CH_3)_2(CH_2)_nCOO^- \qquad \text{(I)}$$
$$R_fXYN^+(CH_3)_2(CH_2)_nSO_3^- \qquad \text{(II)}$$
$$R_fXYN^+(CH_3)_2(CH_2)_nOSO_2O^- \qquad \text{(III)}$$

where $n = 1, 2,$ or 3, (d) an amino acid function, or (e) a dialkylated heterocyclic nitrogen, for example, *n*-dialkylpiperazine [88] or *n*-dialkyl-1,4-oxazine [89].

Amphoteric fluorinated surfactants are compatible with other types of surfactants and adsorb on either negatively or positively charged surfaces. Amphoteric fluorinated surfactants are used in fire-extinguishing agents, foam stabilizers, wetting agents, spreading agents on hydrocarbon surfaces, emulsifiers for manufacturing fluoropolymers, repellents for paper and textiles, cleaning agents for degreasing metal surfaces, and body or hair shampoos. Their main disadvantage is a higher price.

Examples of amphoteric fluorinated surfactants are given in Table 1.6.

1.6 NONIONIC SURFACTANTS

Nonionic surfactants do not dissociate into ions in water. As a consequence, nonionics are less sensitive to electrolytes and pH changes. Nonionic fluorinated surfactants are soluble in an acid or an alkaline medium and are compatible with ionic and amphoteric species. Unlike ionic surfactants, nonionic fluorinated surfactants are not preferentially adsorbed on charged surfaces.

The hydrophile of nonionic surfactants is usually a polyoxyethylene chain or consists of polyoxyethylene and polyoxypropylene segments. The length of the hydrophilic chain can be conveniently varied to modify the hydrophile-lipophile balance (HLB) of the surfactant, a property which affects interfacial behavior and the stabilization of emulsions.

The solubility of nonionic surfactants decreases with increasing temperature and at the cloud point (see Sec. 6.4) the aqueous solution becomes turbid. In

TABLE 1.6 Amphoteric Fluorinated Surfactants

Structure	References
Carboxybetaines	
$R_fCH_2CH(OOCCH_3)CH_2N^+(CH_3)_2CH_2COO^-$	90
$R_fCH_2CH(OH)CH_2N^+(CH_3)_2CH_2COO^-$	89
$R_fCH_2CH(OCOCH_3)CH_2N^+(CH_3)_2CH_2COO^-$	24
$C_9F_{19}CONH(CH_2)_3O(CH_2)_2N^+(CH_3)_2CH_2COO^-$	91
$CF_3(CF_2)_nCH_2CH_2CONH(CH_2)_mN^+(CH_3)_2CH_2CH_2COO^-$ ($n = 7$ or 5, $m = 3$ or 2)	92
$R_fCH_2CH_2SCH_2CH_2N^+(CH_3)_2CH_2COO^-$	93
$R_fCH_2CH_2SCH_2CH(OH)CH_2N^+H(CH_3)CH_2COO^-$	94
$CF_3(CF_2)_nCH_2CH_2SO_2NHCH_2CH_2N^+(CH_3)_2CH_2CH_2COO^-$ ($n = 5$, 7, or 9)	95
$p-C_9F_{17}OC_6H_4SO_2NH(CH_2)_3N^+(CH_3)_2CH_2COO^-$	96
Perfluoroalkyletheramidoalkyl betaines	
$C_3F_7OCF(CF_3)CF_2OCF(CF_3)CONH(CH_2)_3N^+(CH_3)_2CH_2COO^-$	97
Sulfobetaines	
$C_8F_{17}CH_2CH_2CONH(CH_2)_3N^+(CH_3)_2CH_2CH_2CH_2SO_3^-$	98
$C_9F_{19}OC_6H_4CONH(CH_2)_6N^+(CH_3)_2CH_2CH_2CH_2SO_3^-$	99
Sulfatobetaines	
$CF_3(CF_2)_6CF{=}CHCH_2N^+(CH_3)_2CH_2CH_2OSO_3^-$	100
$R_fCH_2CH_2SCH_2CH(OSO_3^-)CH_2N^+(CH_3)_3$	101
Trianion-type amphoteric fluorinated surfactants	
$C_8F_{17}CONH(CH_2)_3N^+(C_2H_4OH)(CH_2COOH)_2CH_2COO^-$	102
$C_8F_{17}SO_2NHCH_2CH(OH)CH_2N^+CH_3[CH_2CH(OH)CH_2SO_3Na]_2\,Cl^-$	103
$C_6F_{13}SO_2NHCH_2CH_2CH_2N^+CH_3(CH_2CH_2CH_2SO_3Na)_2\,I^-$	103
$n-C_6F_{13}SO_2NH(CH_2)_3N^+(CH_2COONa)[CH_2CH(OH)CH_2SO_3Na]_2\,Cl^-$	104
$4-n-C_8F_{17}C_6H_4SO_2NH(CH_2)_6N^+(CH_2CH_2OSO_3Na)_3\,Cl^-$	104
$n-C_8F_{17}SO_2NH(CH_2)_3N^+(CH_2COONa)_3\,Cl^-$	105

general, nonionic fluorinated surfactants are more soluble in organic solvents than ionic fluorinated surfactants.

The polyoxyethylene hydrophile is less stable chemically than carboxylate or sulfonate hydrophiles. Hence nonionic fluorinated surfactants cannot be used in strongly oxidizing media.

Examples of nonionic fluorinated surfactants are given in Table 1.7. Typical nonionic fluorinated surfactants are oxyethylated alcohols, amines, or thiols (mercaptans).

TABLE 1.7 Nonionic Fluorinated Surfactants

Structure	References
Oxyethylated alcohols	
$CF_3(CF_2)_nCH_2O(CH_2CH_2O)_mH$	10, 106–109
$CF_3CF_2(CF_2CF_2)_mCH_2CH_2O(CH_2CH_2O)_nH$	24
$F(CF_2)_q(CH_2)_m(OCH_2CH_2)_nOH$ ($q = 6,7$; $m = 1,3$; $n = 4,5,6,8$)	110
$H(CF_2CF_2)_nCH_2O(CH_2CH_2O)_mH$	106, 111–113
$(CF_3)_2CFO(CH_2)_6O(CH_2CH_2O)_nH$	114
Propylene oxide segments	
$CF_3CHFCF_2CH_2O[CH(CH_3)CH_2O]_mH$	114
$CF_3CHFCF_2CH_2O[CH(CH_3)CH_2O]_m(CH_2CH_2O)_nH$	114
Hydroxyl hydrophile	
$C_9F_{19}CH_2CH(OH)CH_2OC_2H_5$	115
Fluorinated polyhydric alcohols	
$C_8F_{17}C_2H_4O[CH_2CH(CH_2OH)O]_nH$ ($n = 1.7$ average)	116
$C_8F_{17}C_2H_4SO_2NH[CH_2CH(CH_2OH)O]_nH$ ($n = 1.7$ average)	116
Oxyethylated perfluorophenol	
$C_6F_5(OCH_2CH_2)_{10}OH$	117
Perfluoroalkyl-2-ethanethiol derivatives	
$C_6F_{13}CH_2CH_2S(CH_2CH_2O)_3H$	118, 119
$R_fCH_2CH_2SCH_2CH(OH)CH_2O(CH_2CH_2O)_nCH_3$ (avg. $n = 7$ $R_f = C_{4-12}$)	120
$R_fC_2H_4S(CH_2CH_2O)_mC_2H_4S(CH_2CH_2O)_nH$	121
Carbonylamido function	
$C_nF_{2n+1}CONH(CH_2CH_2O)_mH$ ($n = 6$–9, $m = 2$–4)	122
$C_7F_{15}CONH(CH_2)_3N(CH_2CH_2OH)_2$	10
$C_nF_{n+1}CH_2CON[(CH_2CH_2O)_mCH_3]_2$	123
Sulfonamido function	
$C_8F_{17}CH_2CH_2SO_2N(CH_3)CH_2CH_2OH$	124
$F_3C(CF_2)_7CHFCF_2SO_2N[(CH_2CH_2O)_nH]_2$	125
$R_fSO_2N(CH_3)CO(OCH_2CH_2)_nOC_4H_9$	8
$C_{10}F_{19}OC_6H_4SO_2N(C_2H_5)(CH_2CH_2O)_nH$	6

1.7 SILICONE-CONTAINING FLUORINATED SURFACTANTS

The surface tensions of silicone surfactants [126–131] are in general below those of hydrocarbon surfactants. For example, the surface tension of dimethicones (polydimethylsiloxanes) in water is about 20-21 mN/m. Since perfluoroalkyl substituents usually decrease the surface tension of the parent compound, fluorination of silicone surfactants can be expected to yield surfactants with the lowest surface tension. The reduction of this concept to practice is not straightforward, however.

Fluorinated substituents in the α and β position to silicon weaken the Si–C bond and make it vulnerable especially to a nucleophilic attack [126]. Fluorination of a silane such as CH_3SiCl_3 or $(CH_3)_2SiCl_2$ yields compounds of considerable thermal stability, but the exceptionally high electronegativity of fluorine and the consequent strong inductive effect make the Si–C bond hydrolytically unstable [132,133]. Hydrolytic cleavage of the Si–F bond in compounds CF_3SiCl_3 and $(CF_3)_2SiCl_2$ liberates the fluorinated hydrocarbon CF_3H.

To reduce the inductive effect on the Si–C bond the perfluorinated group has to be isolated from the silicon atom. Fluorination of an alkylsilane even in the β position yields hydrolytically unstable compounds which undergo thermal rearrangements and liberate $CF_2{=}CH_2$. Fluorination of an alkyl group in the γ position, $CF_3CH_2CH_2-$, does not affect the Si–F bond significantly and the hydrolytic stability is adequate for practical use [132]. The dichlorosilane $CF_3CH_2CH_2Si(CH_3)Cl_2$ hydrolyzes to form a polysiloxane. However, the ethylene link increases the bulkiness of the side chains, increases the hydrocarbon/fluorocarbon ratio, and causes fluorocarbon-hydrocarbon and fluorine-silicon interactions and orientation effects which may reduce the surface activity of the fluorinated compound. The surface tension of poly(3,3,3-trifluoropropylmethylsiloxane) is higher than that of the lower-cost dimethicones [poly(dimethylsiloxane)] [134,135]. Because of the high affinity of fluorine to silicon [136], fluorine atoms may be inclined to coordinate with silicon atoms. The distorted orientation may partially expose the hydrocarbon link of the pendant side chain.

TABLE 1.8 Surface Tension of Fluorosilicones

Structure	Surface tension (nM/m)
Surface tension of liquid	
$(Me_3CO)_2[CF_3(CF_2)_6CH_2O]_2Si^a$	18.4
$CF_3(CF_2)_6(CH_2)_2Si(OMe)_3$	18.1
$CF_3(CF_2)_3(CH_2)_2Me_2SiOSiMe_2H$	16.4
Aqueous solution (1.0% conc.)	
$[CF_3(CF_2)_6CH_2O_2CCH_2CH(SO_3^-Na^+)CO_2(CH_2)_3Me_2Si]_2O$	12.5
$CF_3(CF_2)_6CH_2O_2CCH_2CH(SO_3^-Na^+)CO_2(CH_2)_3Me_2SiOMe_2Si(CH_2)_3OH$	13.0

[a] Me = CH_3-.
Source: Ref. 135.

To increase the surface activity, Owen and Groh [135] increased the fluorocarbon content of the side chains while maintaining the ethylene link between the fluorocarbon group and the silicon atom. The surface tension of the fluorosilicones is shown in Table 1.8.

Owen and Growh [135] found fluorosilicone copolymers to be more effective antifoam agents than conventional silicone antifoams at defoaming silicone surfactants. However, the fluorosilicone copolymers are ineffective against foaming of fluorinated surfactants having a very low surface tension (see also Foams and Defoaming, Chapter 8).

Siloxanes have been designed by connecting the perfluoroalkyl group to silicon with a link other than ethylene. Polysiloxanes featuring perfluoroalkylcarbonamide groups have been described in patent literature [137–139]:

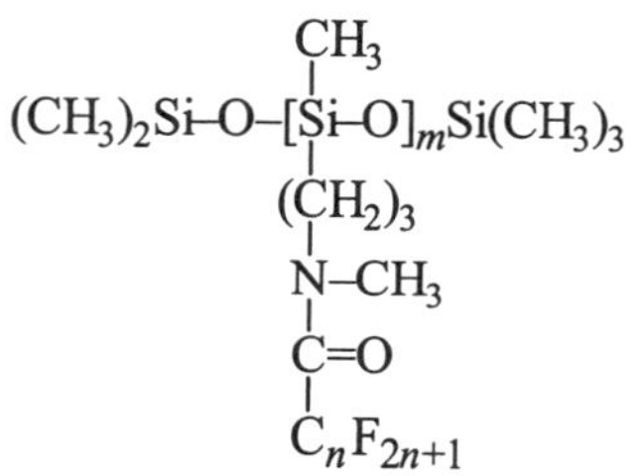

1.8 FLUORINATED SURFACTANTS WITHOUT A HYDROPHILE

Surfactants are used most frequently in aqueous systems. Their amphiphilic character arises from the presence of a hydrophilic group and a hydrophobic group in the same molecule. A perfluoroalkyl group of fluorinated surfactants is not only hydrophobic but oleophobic as well. In a hydrocarbon medium a perfluoroalkyl group is a lyophobe (oleophobe) while a unfluorinated alkyl group is a lyophile (oleophile). Hence a compound featuring a perfluoroalkane segment and a hydrocarbon segment is amphiphilic in a hydrocarbon medium.

Semifluorinated alkanes are low-molecular-weight block copolymers of normal perfluorocarbons and hydrocarbons [140–144]. Their structural formula $F(CF_2)_m(CH_2)_nH$ is usually abbreviated to F_mH_n [144] or F_nH_m [143]. The semifluorinated alkanes so far reported have 6 to 32 carbon atoms $(m + n)$. The compounds $m = 12$ and $n = 8$, 14, or 18 lower the surface tension of *n*-dodecane [144]. Semifluorinated alkanes $F_{10}H_{12}$,$F_{12}H_{14}$, and $F_{12}H_{18}$ form a gel with *n*-dodecane at higher concentrations (10^{-2} M), when allowed to cool to ambient temperature from above their melting points [141,144]. Light scattering, nuclear magnetic resonance (NMR), and fluorescence techniques have shown that semifluorinated alkanes F_8H_{16} and $F_{10}H_{16}$ form micelles in toluene. F_8H_{12} and F_8H_{16} form reversed micelles in perfluorotributylamine and perfluorooctane, respectively [145]. The estimated aggregation numbers of the micelles are small, about 4 to 6. The compound $F_{12}H_{10}$ forms a gel in fluorocarbon solvents (perfluorodecalin) [146]. $F_{10}H_{10}$ forms a liquid crystal phase [147]. When dissolved in a hydrocarbon solvent and placed on a water surface, semifluorinated alkanes, e.g., $F_{12}H_8$, $F_{10}H_{12}$, and $F_{12}H_{18}$, spread and form monolayers [144]. The observed area occupied by a molecule suggests that the perfluoroalkyl segment extends out from water, while the hydrocarbon segment is immersed in water [144].

The polymeric surfactants $(HFPO)_nAr$, where $(HFPO)_n$ is a hexafluoropropylene oligomer and Ar an aryl group [52], lower the surface tension of hydrocarbons such as *m*-xylene and therefore belong to the class of hydrophileless surfactants (see Sec. 1.9).

1.9 POLYMERIC FLUORINATED SURFACTANTS

Perfluorocarbon chains, especially linear ones, are less flexible than hydrocarbon chains and the melting points of fluorinated surfactants are usually higher than those of their hydrocarbon analogues. Fluorinated polyethers, such as oligomers of hexafluoropropene oxide (HFPO), are more flexible and have a lower melting point than perfluoroalkanes. Fluorinated polyethers are therefore more suitable as long-chain hydrophobes or oleophobes for polymeric surfactants [148–152]. Some examples of anionic fluorinated surfactants with an oligo (hexafluoropropene oxide) chain (HFPO) are [52] shown below (see also Chapter 4, Figs. 4.21 and 4.22):

$$CF_3CF_2CF_2O(\underset{\displaystyle CF_3}{\underset{|}{C}}FCF_2O)_n\underset{\displaystyle CF_3}{\underset{|}{C}}FCOONa$$

and

$$CF_3CF_2CF_2O(\underset{\displaystyle CF_3}{\underset{|}{C}}FCF_2O)_n\text{–}ArSO_2Na$$

where Ar is an aryl group.

Nonionic fluorinated surfactants $(HFPO)_n$-Ar are soluble in aromatic solvents, e.g., toluene or xylene, and lower the surface tension of these solvents [52] (see Table 4.17 in Chapter 4 and Sec. 1.8 of this chapter). Ishikawa and Sasabe [52] found the effectiveness of these fluoroalkylated benzenes to increase with the increasing HFPO chain length. This is in apparent discord with the results obtained by Abe et al. [153] for low-molecular-weight fluoroalkylated benzenes with a short HFPO chain. Abe et al. found that surfactants with only one or two HFPO linkages lowered the surface tension of *m*-xylene further than the surface tension decrease reported by Ishikawa and

Sasabe for the long-chain analogues. Although different techniques were used for surface tension measurement, the apparent incongruity remains to be resolved.

Monduzzi et al. [154] studied three- or four component water-oil microemulsions containing perfluoropolyether (PFPE) oils and a anionic surfactant, molecular weight 723, with a PFPE oleophobic chain

$$R(OCF_2CF)_n(OCF_2)_mCF_2COONH_4$$

where R is a perfluoroalkyl group. The surfactant formed microemulsions even without a cosurfactant, behaving in some ways similar to a double-chained ionic surfactant, such as didodecyldimethylammonium bromide.

Myamoto et al. [155,156] prepared poly[(acylimino)alkylene] block copolymers featuring alkyl and perfluoroalkyl hydrophobes. Short perfluoroalkyl groups (C_2F_5 or C_3F_7) of the copolymers were sufficiently hydrophobic to lower the surface tension of water below 20 mN/m.

Polymeric fluorinated surfactants $R_fO(CH_2CH_2O)_nR_f$ (n = 7, 13, 23, or 90) featuring a long-chain hydrophile have been prepared from polyethylene glycols. Houghton [157] reacted tetrafluoroethylene, $CF_2{=}CF_2$, pentamer with polyethylene glycol (molecular weight 1000) to produce the surfactant $C_{10}F_{19}O(CH_2CH_2O)_{23}C_{10}F_{19}$. The sulfonyl chloride $C_{10}F_{19}OC_6H_4SO_2Cl$ yielded $C_{10}F_{19}OC_6H_4SO_2O(CH_2CH_2O)_{23}O_2SC_6H_4OC_{10}F_{19}$. The polymer $C_7F_{15}COO(CH_2CH_2O)_{23}OCC_7F_{15}$ was prepared by reacting polyethylene glycol with $C_7F_{15}COCl$. These surfactants improve the gloss and leveling of coatings, polishes, and paints.

1*H*,1*H*,2*H*,2*H*,3*H*-Perfluoro-1,2-nonylene oxide and *t*-butyl glycidyl ether gave a surfactant of the statistical composition $C_{10}F_{21}CH_2CH_2O\{[CH_2CH(CH_2C_6F_{13})O]_5[CH_2CH(CH_2OH)O]_{18}\}H$ [158], used as a surfactant in shampoos.

Mueller [159] prepared perfluoroalkyl-substituted half esters and amides featuring an $R_f{-}R{-}S{-}$ function, a carboxylic group, and a polyoxyethylene chain. As an example, maleic anhydride was reacted in methyl ethyl ketone with polyoxyethylene of molecular weight 1000. The reaction product was reacted with perfluoroalkylethylmercaptan (R_f consisted of C_6, C_8, and C_{10} isomers) in the presence of triethylamine as a catalyst to form the fluorinated surfactant

$$\begin{array}{l} R_fCH_2CH_2S\underset{|}{C}HCOOH \\ \qquad\qquad\quad CH_2COO(CH_2CH_2O)_{22}H \end{array}$$

Oligomer-type surfactants, $R_fCH_2CH_2OOCCH_2S[CH_2C(CONH_2)H]_{10}H$, have been prepared from perfluoroalkylethanol, mercaptoacetic acid, and acrylamide [160]. The surfactant (0.01% w/w) lowered the surface tension of a protein-containing fire-extinguishing foam composition to 23.4 mN/m.

REFERENCES

1. H. G. Bryce, in "Fluorine Chemistry," J. H. Simons, ed., Vol. 5, p. 370, Academic Press, New York (1950).
2. J. O. Hendricks, Ind. Eng. Chem. 45, 99 (1953).
3. W. A. Zisman, in "Contact Angle, Wettability, and Adhesion," R. F. Gould, ed., ACS Adv. Chem. Ser., p. 43, American Chemical Society, Washington, DC (1964).
4. W. Holzapfel, Fette Seifen Anstrichm. 68, 837 (1966).
5. R. Gritzky, Fette Seifen Anstrichm. 71, 971 (1969).
6. J. Hutchinson, Fette Seifen Anstrichm. 76, 158 (1974).
7. R. Jones, Fette Seifen Anstrichm. 77, 455 (1975).
8. H. G. Klein, J. N. Meuβdoerffer, and H. Niederprüm, Metalloberfläche 29, 559 (1975).
9. H. Liebig and K. Ulm, Chem. Zeit. 99, 477 (1975).
10. E. Schuierer, Tenside 13, 1 (1976).
11. J. N. Meuβdoerffer and H. Niederprüm, Chem. Zeit. 104, 45 (1980).
12. H. C. Fielding, in "Organofluorine Chemicals and Their Industrial Applications," R. E. Banks, ed., p. 214. E. Horwood, Chichester, England (1979).
13. E. G. Shafrin and W. A. Zisman, J. Phys. Chem. 66, 740 (1962).
14. R. A. Moss and W. J. Sanders, Tetrahedron Lett. 19, 1669 (1979).
15. R. A. Moss and K. W. Alvis, Tetrahedron Lett. 21, 1303 (1980).
16. H. Hoffmann, G. Platz, and W. Ulbricht, J. Phys. Chem. 85, 1418 (1981).
17. M. Angel and H. Hoffmann, Z. Phys. Chem. (München) 139, 153 (1984); CA 101, 178036.
18. G. Sugihara, S. Nagadome, T. Yamashita, N. Kawachi, H. Takagi, and Y. Moroi, Colloids Surf. 61, 111 (1991).
19. C. M. Hu, Z. Q. Xu, and W. Y. Huang, J. Fluorine Chem. 49, 433 (1990).
20. M. Hauptschein and M. Braid J. Am. Chem. Soc. 83, 2500 (1961).
21. N. O. Brace, J. Org. Chem. 27, 4491 (1962).
22. N. Nakayama, H. Shimizu, H. Tomota, and F. Nemoto (Neos), Jpn. Kokai Tokkyo Koho JP 62103041 (1985).
23. G. Billentefaniak (Hoechst), Ger. Offen. 2353392 (1975).
24. Zonyl® FSA, Du Pont Performance Products.
25. K. B. Gilleo, E. S. Jones, and E. G. Tajkowski (Allied), U.S. Patent 3,828,098 (1974).
26. T. Hayashi and Y. Osai (Asahi), Jpn. Kokai Tokkyo Koho JP 54044610 (1979).
27. Dainippon Ink, Jpn. Kokai Tokkyo Koho JP 59046252 (1984).
28. Dainippon Ink, Jpn. Kokai Tokkyo Koho JP 57209259 (1982).
29. I. Hisamoto, C. Maeda, and M. Hirai (Daikin), Eur. Patent Appl. EP 83,077 (1983).
30. I. Hisamoto and M. Yamana (Daikin Kogyo), Eur.Patent Appl. EP 144,844 (1985).
31. G. Caporiccio, F. Burzio, G. Garniselli, and V. Biancardi, J. Colloid Interface Sci. 98, 202 (1984).

32. K. C. Liu (GAF), U.S. Patent 4,720,578 (1988).
33. K. F. Mueller and T. W. Cooke (Ciba-Geigy), Ger. Offen. 2852593 (1979).
34. K. Ogino, H. Murakami, N. Ishikawa, and M. Sasabe, Yukagaku 32(2), 96 (1983).
35. J. Burdon, I. Farazmand, M. Stacey, and J. C. Tatlow, J. Chem. Soc. 2574 (1957).
36. T. Gramstad and R. N. Haszeldine, J. Chem. Soc. 2640 (1957).
37. R. N. Haszeldine and A. E. Tipping, Ger. Offen. 2238458 (1973).
38. H. Niederpruem, P. Voss, and V. Beyl, Ger. Offen. 1966931 (1975).
39. I. A. Muggli (3M), U.S. Patent 4,873,020 (1989).
40. M. Wechsberg, Tenside 15, 2 (1978).
41. R. H. Goldbaum and W. R. Remington (Du Pont), U.S. Patent 4,784,809 (1988).
42. T. Kuwamura, M. Ohshima, and E. Kameyama, Nippon Kagaku Kaishi (3) 545 (1974).
43. D. Prescher, B. Dreher, W. Dmowski, and R. Wozniacki, Ger. Patent (East) DD 239789 (1986).
44. B. Dreher, D. Prescher, W. Dmowski, and R. Wozniacki, Ger. Patent (East) 239788 (1986).
45. B. Tittle (ICI), Brit. Patent 1,153,854 (1969).
46. N. Funasaki and S. Hada, J. Colloid Interface Sci. 73, 425 (1980).
47. Neos, Jpn. Tokai Tokkyo Koho, JP 57164198 (1982).
48. R. E. A. Dear and E. K. Kleiner (Ciba-Geigy), U.S. Patent 4,014,926 (1977).
49. R. A. Falk (Ciba-Geigy), U.S. Patent 4,420,434 (1983).
50. Dainippon Ink, Jpn. Kokai Tokkyo Koho, JP 5948449 (1984).
51. C. Kimura, K. Kashiwaya, M. Kobayashi, and T. Nishiyama, J. Am. Oil Chem. Soc. 61, 105 (1984).
52. N. Ishikawa and M. Sasabe, J. Fluorine Chem. 25, 241 (1984)
53. L. Knunyants, A. L. Belferman, Yu. A. Pazderskii, L. S. German, and S. R. Sterlin, USSR Patent 1,019,798 (1986).
54. Y. Nakamura, M. Umemura, M. Yonekura (Asahi Denka), Japan Kokai JP 50121223 (1975).
55. Y. Nakamura, M. Umemura, M. Yonekura (Asahi Denka), Japan Kokai JP 52035185 (1977).
56. Neos Co., Ltd., Jpn. Kokai Tokkyo Koho JP 57059843 (1982).
57. N. Yoshino, N. Komine, J. Suzuki, Y. Arima, and H. Hirai, Bull. Chem. Soc. Jpn. 64, 3262 (1991).
58. G. A. Bargigia, G. Caporiccio, and M. Pianca, J. Fluorine Chem. 19, 403 (1982).
59. G. S. Goldin, K. O. Averbakh, and L. A. Nekrasova, Zh. Prikl. Khim. 49, 1341 (1976).
60. A. Greiner, B.Krüger, and M. Herbst, Z. Chem. 18, 383 (1978).
61. A. J. Szur (Diamond Shamrock), U.S. Patent 4,140,709 (1979).
62. Dainippon Ink, Jpn. Kokai Tokkyo Koho JP 59051252 (1959).
63. N. O. Brace and A. K. Mackenzie (Du Pont), U.S. Patent 3,083,224 (1963); CA 59, 5023.
64. A. L. Lore (Du Pont), U.S. Patent 4,064,067 (1977).

65. Asahi Glass, Jpn. Kokai Tokkyo Koho JP 60064990 (1985).
66. Daikin Kogyo Co., Jpn. Kokai Tokkyo Koho JP 56025190 (1981)
67. H. Huber-Emden (Ciba-Geigy), Ger. Offen. 2429401 (1975).
68. B. Boutevin, Y. Hervaud, and Y. Pietrasanta (Atochem), Fr. Demande FR 2616150 (1988).
69. R. F. Heine (3M), Fr. Patent 1,317,427 (1963); CA 59, 9799.
70. J. Wakatsuki, T. Kato, and T. Kurosaki (Kao), Ger. Offen. 3609491 (1985).
71. L. Foulletier and J. P. Lalu, Ger. Offen. 2244297 (1973).
72. K. Matsui, Y. Kikuchi, K. Sugimoto, and K. Suzuki, Jpn. Kokai Tokkyo Koho JP 61207362 (1986).
73. C. Kimura, K. Murai, K. Kashiwaya, T. Kanno, and T. Nishiyama, Yukagaku 31, 464 (1982).
74. I. Hisamoto, T. Maeda, and M. Nishiwaki (Daikin Ind.) Jpn. Kokai Tokkyo Koho JP 6187648 (1986).
75. L. Qian, L. Chen, Y. Zeng, J. Wang, Y. Liang, C. Hu, and G. Zhao, Huaxue Xuebao 40, 945 (1982).
76. P. L. Bartlett (Du Pont), U.S. Patent 3,555,089 (1971).
77. A. J. Szur (Diamond Shamrock) U.S. Patent 4,049,668 (1977).
78. C. Kimura, K. Murai, K. Kashiwaya, H. Takahashi, and T. Nishiyama, Yukagaku 31, 448 (1982).
79. S. S. Toukan and M. Hauptschein (Pennwalt), U.S. Patent 4,126,633 (1978).
80. F. Marty, E. Rouvier, and A. Cambon (Atochem), French Patent 2,592,648 (1987).
81. R. B. Hager and G. J. Walter (Pennwalt), Ger. Offen. 2219642 (1972).
82. S. Szonyi and A. Cambon, French Patent 2,530,623 (1984).
83. Sanyo Chem.Ind., Jpn. Kokai Tokkyo Koho JP 55157691 (1980).
84. H. Huber-Emden and A. Deflorin (Ciba-Geigy), Ger. Offen. 2453774 (1975).
85. A. R. Katritzky, T. L. Davis, G. W. Rewcastle, G. O. Rubel, and M. T. Pike, Langmuir 4, 732 (1988).
86. N. Muller, J. H. Pellerin, and W. W. Chen, J. Phys. Chem. 76, 3012 (1972).
87. B. R. Bluestein and C. L. Hilton, eds., "Amphoteric Surfactants," Surfactant Science Ser. Vol. 12, Marcel Dekker, New York (1982).
88. T. Hayashi and Y. Otoshi (Asahi Glass), Jpn. Kokai Tokkyo Koho JP 54062990 (1979).
89. J. W. Richter (Du Pont), U.S. Patent 4,283,533 (1981).
90. Du Pont, Jpn. Kokai Tokkyo Koho JP 56122336 (1981).
91. T. Hayashi and Y. Otoshi (Asahi Glass), Jpn. Tokkyo Koho JP 54147184 (1979).
92. R. Bertocchio and L. Foulletier (Ugine Kuhlmann), Ger. Offen. DE 2119302 (1971).
93. D. P. Cords (Du Pont), Ger. Offen. DE 2357916 (1974).
94. R. A. Falk (Ciba-Geigy), Eur. Patent Appl. EP 115251 (1984).
95. R. Bertocchio and L. Foulletier (Ugine Kuhlmann), Ger. Offen. DE 2120868 (1971).
96. Neos Co, Jpn. Kokai Tokkyo Koho JP 57164198 (1982).
97. P. L. Bartlett (Du Pont), U.S. Patent 3,839,425 (1974).
98. R. Bertocchio, L. Foulletier, and A. Lantz (Ugine Kuhlmann), Eur. Patent Appl. EP 17568 (1980).

99. Dainippon Ink & Kawamura Phys. Chem. Inst., Jpn. Kokai Tokkyo Koho JP 59048457 (1984).
100. S. Billenstein and W. Ehri, Ger. Offen. 2749329 (1979).
101. R. A. Falk (Ciba-Geigy), U.S. Patent 4,435,330 (1984).
102. T. Hayashi, Y. Otoshi and T. Seki (Asahi Glass), Japan Kokai JP 53045683 (1978).
103. Dainippon Ink/Kawamura Inst., Jpn. Kokai Tokkyo Koho JP 58179300 (1983).
104. Dainippon Ink Chem., Jpn. Kokai Tokkyo Koho JP 58203959 (1983).
105. Dainippon Ink/Kawamura Inst., Jpn. Kokai Tokkyo Koho JP 58206769 (1983).
106. R. Tatematsu, S. Ozaki, S. Nagase, and K. Inukai, Yukagaku 25, 287 (1976).
107. L. L. Gervits, K. N. Makarov, L. F. Komarova, N. V. Konovalova, V. F. Zabolotskikh, and Yu. A. Cheburkov, Izv. Akad. Nauk. SSSR, Ser. Khim. (10), 2256 (1974).
108. R. Tatematsu, S. Ozaki, S. Nagase, and K. Inukai, Yukagaku 26, 367 (1977).
109. P. Leempoel, C. Selve, G. Mathis, B. Castro, and J. J. Delpuech, Eur. Patent Appl. EP 51527 (1982).
110. C. Selve and B. Castro, Tetrahedron 39, 1313 (1981).
111. A. R. Naik (Unilever), Ger. Offen. 2261681 (1973).
112. A. Greiner and M. Herbst, Z. Chem. 16, 229 (1976).
113. A. Greiner and W. Gerhardt, Ger. (East) DD 111522 (1975).
114. A. J. Szur (Diamond Shamrock), U.S. Patent 3,980,715 (1976).
115. I. Hisamoto, C. Maeda, and M. Nishiwaki (Daikin), Eur. Patent Appl. EP 57443 (1982).
116. F. Wehowsky (Hoechst), Ger. Offen. 3611302 (1987).
117. J. Schulze, G. G. Yakobson, K. V. Dornikova, I. Endler, O. I. Osina, V. E. Platonov, and D. Prescher, Ger. (East) DD 249384 (1987).
118. J. P. Lampin, A. Cambon, F. Szoni, J. J. Delpuech, G. Serratrice, G. Thiollet, and L. Lafosse, Eur. Patent Appl. EP 165 853 (1985).
119. A. Cambon, J. J. Delpuech, L. Matos, G. Serratrice, and F. Szoni, Bull. Soc. Chim. Fr. 965 (1986).
120. T. W. Cooke (Ciba-Geigy), Eur. Patent EP 10523 (1980).
121. F. Szönyi and A. Cambon, J. Fluorine Chem. 36, 195 (1987).
122. J. Afzai, B. M. Fung, and E. A. O'Rear, J. Fluorine Chem. 34, 385 (1987).
123. C. Selve, E. M. Moumni, and J. J. Delpuech, J. Chem. Soc., Chem. Commun. 1437 (1987).
124. P. Bouvet and J. P. Lalu (Ugine Kuhlmann), Ger. Offen. 2004962 (1970).
125. U. Gross and H. R. Holzbauer, Ger. (East) DD 265398 (1989)
126. W. Noll, "Chemie und Technologie der Silicone," Chemie GmbH, Weinheim (1968).
127. B. Grüning and G. Körner, Tenside 26, 312 (1989).
128. D. Schäfer, Tenside 27, 154 (1990).
129. G. Schmidt, Tenside 27, 324 (1990).
130. H. F. Fink, Tenside 28, 306 (1991).
131. M. Gradzielski, H. Hoffmann, P. Robisch, and W. Ulbricht, Tenside 27, 366 (1990).
132. O. R. Pierce, Appl. Polym. Symp. 14, 7 (1970).

133. M. J. Owen, in "Silicon Based Polymer Science," J. M. Zeigler and F. W. Gordon Fearon, eds., Advances in Chemistry Series 224, American Chemical Society, Washington, DC (1990).
134. M . J. Owen, J. Appl. Polym. Sci. 35, 895 (1988).
135. M. J. Owen and J. L. Groh, J. Appl. Polym. Sci. 40, 789 (1990).
136. I. A. Lavygin, I. I. Skorokhodov, M. V. Sobolevskii, D. V. Nazarova, M. B. Lotarev. O. M. Kudinova, and G. V. Vorapayeva, Vysokomol. Soedin. A 18, 90 (1976).
137. Dow Corning, U.S. Patent 3,012,006 (1958).
138. ICI, Brit. Patent 1,188,212 (1967).
139. W. Noll, W. Büchner, H. J. Lücking, and C. Sucker, Kolloidz. Z. Polym. 250, 836 (1972).
140. R. J. Twieg and J. F. Rabolt, J. Polym. Sci., Polym. Lett. Ed. 21, 901 (1983).
141. J. F. Rabolt, T. P. Russell, and R. J. Twieg, Macromolecules 17, 2786 (1984).
142. R. J. Twieg, T. P. Russell, R. L. Siemens, and J. F. Rabolt, Macromolecules 18, 1361 (1985).
143. T. P. Russell, J. F. Rabolt, R. J. Twieg, R. L. Siemens, and B. L. Farmer, Macromolecules 19, 1135 (1986).
144. G. L. Gaines, Langmuir 7, 3054 (1991).
145. M. P. Truborg and J. E. Brady, J. Am. Chem. Soc. 110, 7797 (1988).
146. J. Höpken, C. Pugh, W. Richtering, and M. Möller, Makromol. Chem. 189, 911 (1988).
147. W. Mahler, D. Guillon, and A. Skoulios, Mol. Cryst. Liq. Cryst., Lett. Sect. 2, 111 (1985).
148. H. S. Eleuterio, A. S. Milian, Jr., and E. P. Moore, Jr. (Du Pont), U.S. Patent 3,250,808 (1966).
149. E. P. Moore (Du Pont), U.S. Patent 3,322,826 (1967).
150. F. L. Arbogast (Du Pont) U.S. Patent 3,412,148 (1968).
151. H. S. Eleuterio, J. Macromol. Sci., Chem. A6 (6), 1072 (1972).
152. L. A. Shits, L. V. Dikhtievskaya, S. P. Krukovskii, L. V. Cherendnichenko, and V. A. Pomarenko, Kolloidn. Zh. 38, 1130 (1976).
153. M. Abe, K. Morikawa, K. Ogino, H. Sawada, T. Matsumoto, and M. Nakayama, Langmuir 8, 763 (1992).
154. M. Monduzzi, A. Chittofrati, and M. Visca, Langmuir 8, 1278 (1992).
155. M. Myamoto, K. Ago, and T. Saegusa, Macromolecules 21, 1880 (1988).
156. M. Myamoto, K. Ago, and T. Saegusa, Macromolecules 22, 3540 (1989).
157. L. E. Houghton (ICI), Ger. Offen. 2215388 (1972).
158. G. Vanierberghe and H. Sebag, Ger. Offen. 3021447 (1980).
159. K. F. Mueller (Ciba-Geigy), U.S. Patent 4,171,282 (1979).
160. Asahi Glass, Jpn. Kokai Tokkyo Koho JP 60118228 (1985); CA 104, 132007.

2
Synthesis

2.1 STARTING MATERIALS

Direct fluorination with elemental fluorine is not practical for commercial synthesis of fluorinated surfactants. Elemental fluorine is extremely reactive and difficult to handle. The heat of formation of the C – F bond (about 460 kJ/mol or 110 kcal/mol) and the H – F bond (566 kJ/mol or 135 kcal/mol) exceeds that of the C – C bond (about 348 kJ/mol or 83 kcal/mol) [1]. Hence the fluorination with elemental fluorine leads to a violent fragmentation of the substrate unless the reaction is carefully controlled and the reaction heat effectively dissipated [2,3]. Commercially important pathways to fluorinated surfactants are electrochemical fluorination, telomerization, and oligomerization of tetrafluoroethylene [4–6].

Electrochemical fluorination utilizes anhydrous hydrofluoric acid as the fluorine source. Hydrofluoric acid is produced by a reaction of calcium fluoride with sulfuric acid [7]:

$$CaF_2 + H_2SO_4 \longrightarrow 2HF + CaSO_4$$

Industrially produced hydrofluoric acid contains impurities, mainly fluorosulfonic acid, silicon tetrafluoride, sulfur dioxide, sulfuric acid, and water. Purification of hydrogen fluoride by simple physical means is difficult and various chemical methods have been developed for the removal of impurities. Anhydrous hydrogen fluoride has been prepared by thermal decomposition of anhydrous sodium hydrogen fluoride or potassium hydrogen fluoride. The salt, KHF_2, is difficult to dry and electrolysis of the salt has been employed to remove traces of water and other impurities [8–10]. Highly pure hydrogen fluoride can be manufactured by electrolytically oxidizing impurities remaining in hydrogen fluoride [11].

The starting materials of the *telomerization* process are tetrafluoroethylene and a perfluoroalkyl iodide. Of the various perfluoroalkyl iodides used in telomerization processes, pentafluoroethyl iodide is the most important (see Sec. 2.3). The one-step process for manufacturing pentafluoroethyl iodide is based on the addition of iodine fluoride to tetrafluoroethylene. Iodine fluoride, IF, is too unstable to be isolated and is therefore formed from iodine pentafluoride and iodine in situ. Iodine pentafluoride is synthesized from iodine and elemental fluorine [12]:

$$I_2 + 5F_2 \longrightarrow 2IF_5$$

Iodine pentafluoride and iodine react with tetrafluoroethylene in the presence of a catalyst, such as antimony trifluoride, under pressure [13,14]:

$$5CF_2{=}CF_2 + IF_5 + 2I_2 \xrightarrow{SbF_3} 5CF_3CF_2I$$

The reaction of hexafluoropropylene with IF, formed in situ from IF_5 and I_2, gives heptafluoroisopropyl iodide [15,16]:

$$5CF_3CF{=}CF_2 + IF_5 + 2I_2 \longrightarrow 5(CF_3)_2CFI$$

Heptafluoroisopropyl iodide as a telomerization agent (telogen) produces telomers featuring a branched fluorocarbon chain end.

The disadvantage of the IF addition process for making perfluoroalkyl iodides is that it requires elemental fluorine. Fluorine in the form of fluorides is abundant in nature. However, liberation of fluorine from fluorides requires vigorous conditions to break the strong bond between fluorine and alkali and alkaline earth metals. Oxidation of hydrogen fluoride or fluorides to elemental fluorine is difficult, although such reactions have been reported in the literature [17]. Elemental fluorine is produced by electrolysis of anhydrous hydrogen fluoride containing dissolved potassium hydrogen fluoride [17–19]. Handling of elemental fluorine, because of its high reactivity and corrosiveness, aggravates the difficulty. Hence numerous attempts have been made to synthesize perfluoroalkyl iodides without using elemental fluorine [20–26].

Scherer and Futterer [20] prepared pentafluoroethyl iodide by passing pentafluoroethyl bromide through a layer of KI at 500°C.

Millauer [21–23] prepared pentafluoroethyl iodide and heptafluoropropyl iodide by a reaction of 1,2-diiodotetrafluoroethane, CF_2ICF_2I, with HF or of $CF_2CF{=}CF_2$ with iodine and HF, respectively. 1,2-Diiodotetrafluoroethane, obtained by the addition of iodine to tetrafluoroethylene, is heated in the presence of HF and a catalyst or an oxidant to form pentafluoroethyl iodide [21–24]:

$$CF_2ICF_2I + HF \longrightarrow CF_3CF_2I + HI$$

Chlorine, sulfonyl chloride, and antimony pentafluoride have been used as catalysts with HF to facilitate the conversion of the diiodide to the monoiodide [21,22].

A Montecatini Edison patent [25] describes a reaction of 1,2-diiodotetrafluoroethane with HF and PbO_2 to form pentafluoroethyl iodide. However, oxidizing acids, such as HNO_3, $HClO_3$, HIO_3, and H_5IO_6, form only volatile by-products and therefore have an advantage over metal oxides, such as HgO and PbO_2 [23]. The reaction with H_5IO_6 is carried out in an excess of HF at 150°C:

$$7CF_2ICF_2I + 7HF + H_5IO_6 \longrightarrow 7CF_3CF_2I + 4I_2 + 6H_2O$$

Electrochemical synthesis of pentafluoroethyl iodide from CF_2ICF_2I in the presence of anhydrous hydrogen fluoride has been claimed in Hoechst patents [26].

2.2 ELECTROCHEMICAL FLUORINATION

Electrochemical fluorination of organic compounds in anhydrous hydrogen fluoride [27–34] was invented by Simons [27–30]. The organic substance to be fluorinated is dissolved or dispersed in liquid hydrogen fluoride. A direct electric current of a voltage below 8 V, usually between 5 and 7 V, is passed through hydrogen fluoride containing the substance. Hydrogen is evolved at the cathode and the organic substance is fluorinated. The voltage applied is insufficient for fluorine evolution but adequate for fluorination at the anode. All hydrogen atoms in the molecule are replaced by fluorine but some functional groups such as carboxylic halides and sulfonyl halides are retained.

A diagram of a electrochemical fluorination plant is shown in Fig. 2.1. The electrolytic cell is built of a metal, resistant to corrosion by hydrogen fluoride, like copper, nickel, Monel, or iron. The anode plates are made of nickel and the cathode plates are usually made of iron or nickel. The cell is equipped with a reflux condenser, cooled to −10 to −40°C, to retain hydrogen fluoride. Alternatively, solid sodium fluoride is used to trap hydrogen fluoride escaping through the condenser. The temperature of the cell is usually kept at 0°C, although a temperature range of −80° to +75°C has been claimed in the patent literature [30,36]. The cell temperature is selected as a compromise between two opposing factors. On one hand, the conductivity of solutions in hydrogen fluoride increases with increasing temperature. On the other hand, anhydrous hydrogen fluoride boils at 19°C and lower fluorination temperatures are therefore more convenient. The process can be operated batchwise or continuously.

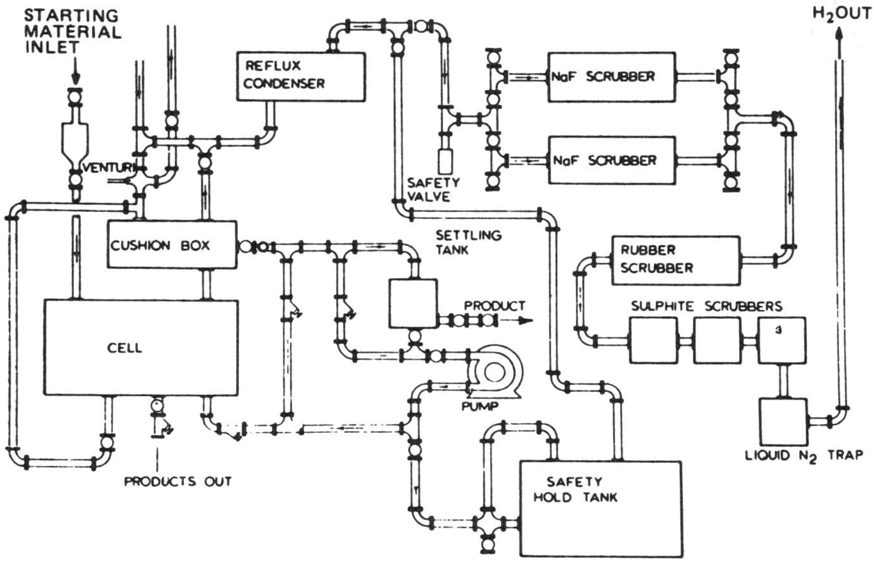

FIG. 2.1 Schematic outline of an electrochemical fluorination plant. (From Ref. 35. Reproduced by permission of Chapman & Hall Ltd.)

The conductivity of pure anhydrous hydrogen fluoride is very low. If the substrate dissolved in hydrogen fluoride does not adequately increase the conductivity, organic or inorganic additives are used to increase conductivity. Some additives are consumed in the process, some are inert. Water increases conductivity of HF. However, the amount of water in hydrogen fluoride has to be kept below 1%. Water in amounts above 1% lowers the yield [30,31,37], and above 10% w/w water produces dangerously large amounts of explosive oxygen difluoride. Alkali or alkaline earth metal fluorides used as conductivity additives do not react during electrolysis. They do not need to be replenished during fluorination but can corrode the anode.

The mechanism of electrochemical fluorination is still incompletely understood [6,31,38–41]. Fluorination is believed to occur by fluorine adsorbed on the nickel fluoride layer formed on the anode surface, while hydrogen is liberated on the cathode:

Cathode: $2H^+ + 2e^- \longrightarrow H_2$

Anode: $2F^- \longrightarrow 2F + 2e^-$

A free radical mechanism assumes the following fluorination reaction sequence:

$$F^- \xrightarrow[\text{Ni}]{>2.8\ \text{V}} F^{\cdot}_{ads} + e^-$$

$$RH + F^{\cdot}_{ads} \longrightarrow R^{\cdot}_{ads} + HF$$

$$R^{\cdot}_{ads} + F^{\cdot}_{ads} \longrightarrow RF$$

According to an ionic fluorination mechanism, adsorption of organic molecules on the NiF_2 layer is the first step, followed by their anodic oxidation to carbonium ions which react with fluoride ions [42]:

$$CH_4 \xrightarrow{-e^-,\,-H^+} CH^{\cdot}_{3ads} \xrightarrow{-e^-} CH^{+}_{3ads} \xrightarrow{F^-}$$

$$CH_3F_{ads} \xrightarrow{-e^-,\,-H^+} CH_2F^{\cdot}_{ads} \xrightarrow{-e^-} CH_2F^{+}_{ads} \xrightarrow{F^-}$$

$$CH_2F_{2ads} \xrightarrow{-2e^-,\,-H^+,\,F^-} CHF_3 \xrightarrow{-2e^-,\,-H^+,\,F^-} CF_4$$

Electrochemical fluorination of an alkanoic acid yields a perfluoroalkanoic acid fluoride [43–47]:

$$C_nH_{2n+1}COOH + (2n + 2)HF \rightarrow C_nF_{2n+1}COF + \text{by-products}$$

Electrochemical fluorination of carboxylic acids produces perfluoroalkanoic acid fluorides in a low yield (about 10–20%) [31] and water as a by-product. Water forms explosive oxygen difluoride and causes oxidative degradation of the carboxylic acid. For acids with six or more carbon atoms, cyclic perfluoroethers are among the by-products formed. The anhydrides of carboxylic acids give perfluorocarboxylic acid fluorides in a higher yield than the parent carboxylic acids. The electrochemical fluorination of carboxylic acids or their anhydrides is now obsolete.

The yield of a perfluorocarboxylic acid fluoride is higher when a carboxylic acid chloride or fluoride is fluorinated instead of the carboxylic acid or its anhydride [48,49]. For example, acetic acid, acetic acid anhydride, and acetyl fluoride give trifluoroacetyl fluoride in 17, 32, and 76% yields, respectively [32,49]:

$$C_nH_{2n+1}COCl + (2n + 2)HF \rightarrow C_nF_{2n+1}COF + HCl + \text{by-products}$$

The carboxylic acid fluorides are soluble in HF and the solutions of alkanoic acid fluorides with more than four carbon atoms are conductive. The yield of the perfluoroalkanoic acid fluoride decreases with increasing chain length of the carboxylic acid fluoride fluorinated, from as much as about 80% for acetyl fluoride to 10% for perfluorooctanoyl fluoride [34]. The yield is lower when an acyl chloride is fluorinated instead of an acyl fluoride, but acyl chlorides are more readily available than acyl fluorides [48].

Electrochemical fluorination of dicarboxylic acid fluorides also gives a higher yield than fluorination of the parent acids. The lower yield given by the parent carboxylic acids has been attributed to decarboxylation, similar to the Kolbe reaction [31].

The perfluorocarboxylic acid fluorides can be converted to perfluorocarboxylic acids, esters, amides, or other intermediates for surfactants (Fig. 2.2). Hydrolysis of the acid fluoride yields the carboxylic acid (I) or its salt (II). Alcohols form esters (III), which can be hydrogenated to an alcohol (IV), an intermediate for nonionic surfactants. The reaction of the acid fluoride with an amine, for example $NH_2C_3H_6N(CH_3)_2$, yields an intermediate (V) for an amine salt (VI), a cationic surfactant (VII), or an amphoteric surfactant (VIII). The surface activity of surfactants derived from perfluorocarbonyl fluoride is shown in Fig. 2.3.

Electrochemical fluorination of alkanesulfonyl chlorides or fluorides yields the corresponding perfluorosulfonyl chloride or fluoride in a 12–79% yield, depending on the alkane chain length [51]:

$$C_nH_{2n+1}SO_2Cl + (2n+2)HF \rightarrow C_nF_{2n+1}SO_2F + HCl + \text{by-products}$$

Fluorination of ethanesulfonyl chloride yields pentafluoroethanesulfonyl fluoride in 79% yield [52]. Fluorination of octanesulfonyl chloride yields perfluorooctanesulfonyl fluoride in a 25% yield [52], but perfluorodecanesulfonyl fluoride is obtained in only 12% yield [53].

The by-products formed are shorter-chain sulfonyl fluorides or chlorides and fluorocarbons [51–54]. The structures of the by-products suggest a cleavage of carbon-sulfur and carbon-carbon bonds and some oxidative degradation as well.

Fluorination of alkanesulfonyl fluorides or chlorides gives higher yields of perfluoroalkanesulfonic acids than fluorination of alkanesulfonic acids [49,51,52,54]. Another disadvantage associated with the fluorination of free

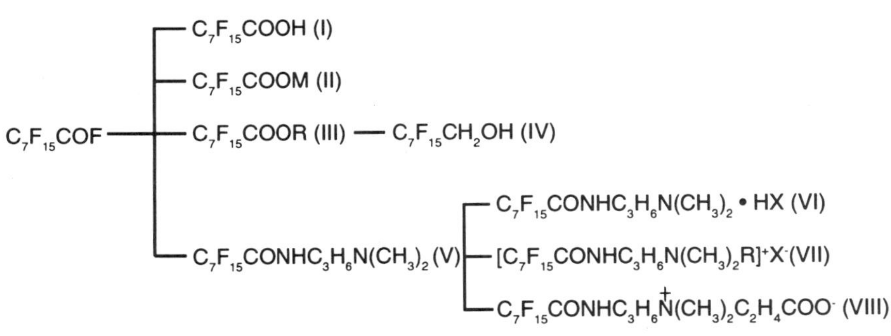

FIG. 2.2 Fluorinated surfactants derived from perfluorocarbonyl fluoride. (From Ref. 50.)

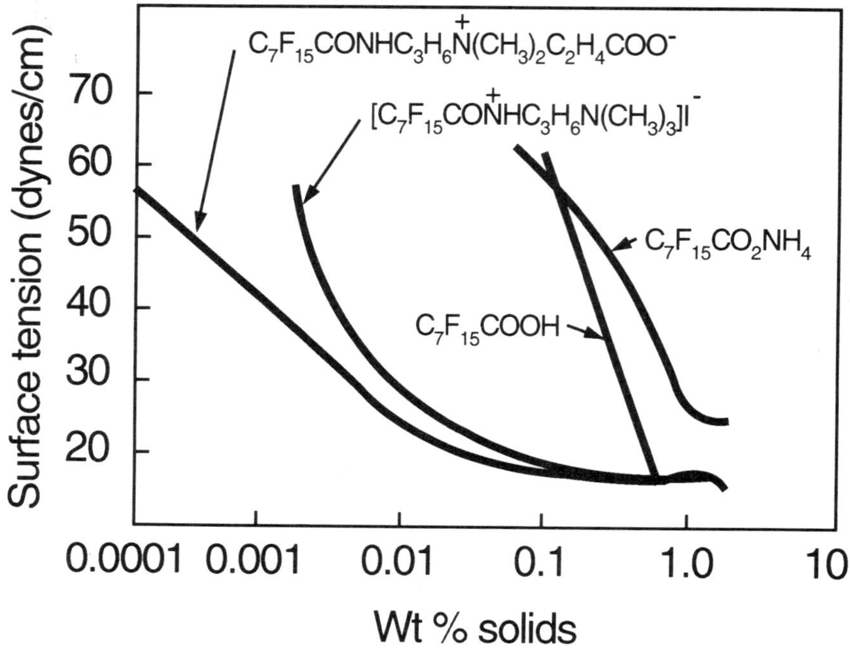

FIG. 2.3 Surface activity of aqueous solutions of fluorinated surfactants derived from perfluorocarbonyl fluoride. (From Ref. 50.)

alkanesulfonic acids is the potential explosion hazard caused by excessive amounts of oxygen difluoride and hypofluorites formed as by-products [55].

The perfluorosulfonyl fluorides obtained by the electrochemical fluorination process can be readily converted into the corresponding acids and salts used as surfactants (Fig. 2.4). Hydrolysis yields the perfluoroalkane sulfonic acid (I) or its salt (II). Primary or secondary amines yield sulfonamides (III), which can be converted to a carboxylic acid (IV) or an alcohol (V). The alcohol is an intermediate for nonionic surfactants (VI), phosphates (VII), or sulfates (VIII). The tertiary aminosulfonamide (IX) can form a cationic surfactant (X), a amphoteric surfactant (XI), or an amine salt (XII).

The surface activity of fluorinated surfactants derived from perfluorosulfonyl fluoride is shown in Fig. 2.5.

2.3 TELOMERIZATION

Telomerization was initially developed by the Du Pont Company for free radical polymerization of ethylene [56–61] and defined as a process of reacting a

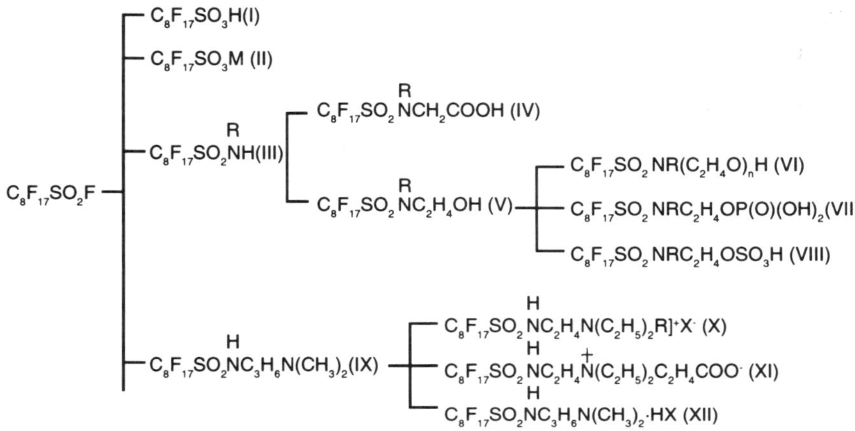

FIG. 2.4 Fluorinated surfactants derived from perfluorosulfonyl fluoride. (From Ref. 50.)

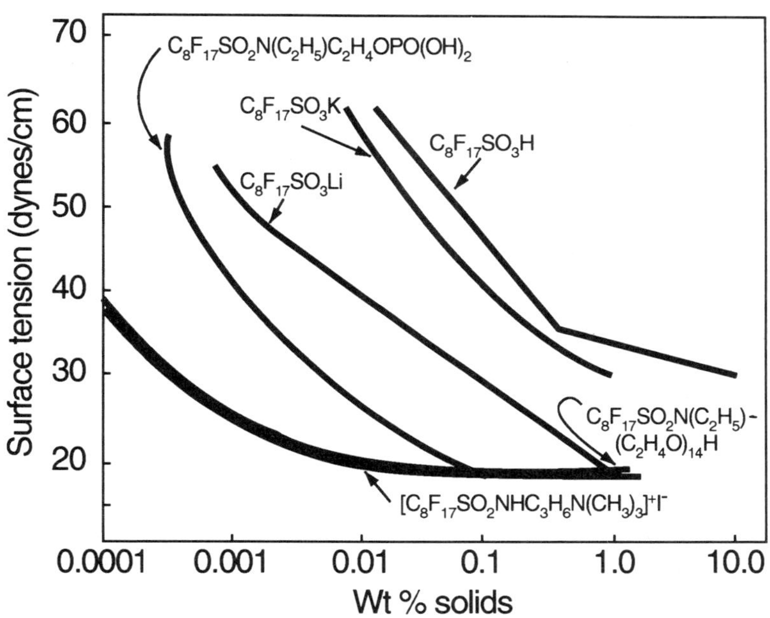

FIG. 2.5 Surface activity of aqueous solutions of fluorinated surfactants derived from perfluorosulfonyl fluoride. (From Ref. 50.)

molecule, called telogen, with two or more ethylenically unsaturated molecules, called taxogens:

$$\underset{\text{Telogen}}{YZ} + \underset{\text{Taxogen}}{nA} \longrightarrow \underset{\text{Telomer}}{Y-(A)_n-Z}$$

Haszeldine [62,63] allowed trifluoromethyl iodide to react with ethylene and obtained oligomers of the type $CF_3[CH_2CH_2]_nI$ (n = 1,2, and 3). The reaction of trifluoromethyl iodide with tetrafluoroethylene, catalyzed by UV irradiation, yielded short-chain polymers of the general formula $CF_3[CF_2CF_2]_nI$, where n is 1 to 10. Some of the members of the telomeric series were isolated.

The photochemically catalyzed reaction of trifluoromethyl iodide with tetrafluoroethylene involves a radical chain [64]:

$$CF_3I + h\upsilon \rightarrow CF_3\cdot + I\cdot$$

$$CF_3\cdot + CF_2{=}CF_2 \rightarrow CF_3CF_2CF_2\cdot \qquad \text{(initiation)}$$

$$CF_3CF_2CF_2\cdot + CF_2{=}CF_2 \rightarrow CF_3CF_2CF_2CF_2CF_2\cdot \qquad \text{(chain propagation)}$$

$$CF_3CF_2CF_2\cdot + CF_3I \rightarrow CF_3CF_2CF_2I + CF_3\cdot \qquad \text{(transfer)}$$

$$CF_3(CF_2CF_2)_n\cdot + CF_3I \rightarrow CF_3(CF_2CF_2)_nI + CF_3\cdot$$

$$2CF_3\cdot \rightarrow CF_3CF_3 \qquad \text{(chain termination)}$$

The polymerization mechanisms have been extended from the original free radical reaction to anionic, cationic, and transition metal–catalyzed telomerization.

The value of n in $CF_3(CF_2CF_2)_nI$ depends on the relative concentration of the chain transfer agent (the telogen), the monomer (tetrafluoroethylene), and the length of the radical chain. Telomerization in vapor phase favors the formation of longer-chain perfluoroalkyl iodides with $n > 2$ because the relative concentration of the telogen is lower. The probability is therefore higher that heptafluoropropyl iodide undergoes a subsequent homolytic fission and generates a $C_3F_7\cdot$ radical, which can react with tetrafluoroethylene.

Pentafluoroethyl iodide (1-iodopentafluoroethane) reacts with tetrafluoroethylene under conditions similar to those for iodotrifluoromethane and yields perfluoroalkyl iodides, $CF_3CF_2[CF_2CF_2]_nI$, with an even number of carbon atoms.

Commercial telomerization of tetrafluoroethylene with pentafluoroethyl iodide was developed by the Du Pont Company [65,66]. The telogen, pentafluoroethyl iodide, is prepared by reacting iodine pentafluoride and iodine with tetrafluoroethylene [13,14] (see Sec. 2.1):

$$5CF_2{=}CF_2 + IF_5 + 2I_2 \xrightarrow{SbF_3} 5CF_3CF_2I$$

Telomerization of tetrafluoroethylene with pentafluoroethyl iodide produces a mixture of even-carbon-numbered telomers differing in their overall carbon chain length:

$$C_2F_5I + nC_2F_4 \rightarrow C_2F_5(C_2F_4)_nI$$

Numerous modifications of the original Du Pont telomerization process have been patented [67–79]. Fluorides, such as HF/SF_4, SbF_3/SbF_5, or $IF_5/AlCl_3$, used in older telomerization processes as catalysts [4], are corrosive and have been replaced by selected metal salts dissolved in amines or solid metal salt–amine complexes. Hauptschein [67] has prepared telomers by reacting tetrafluoroethylene with a secondary iodide, having two same or different perfluoroalkyl groups, at 150–220°C under pressure. A Ciba patent [68] discloses telomerization of perfluoroalkyl iodides with perfluoroethylene or perfluoropropylene, catalyzed by a system consisting of an amine and a salt of a metal belonging to the group IB or IIB [68] or group IIIA, IIIB, or VIIIB [69]. The preferred compound is $ZrCl_4$ [69]. Kali-Chemie patents [70,71] disclose metal salt–amine telomerization catalysts, prepared from ethanolamine or butylamine and chlorides of Zn, Cu, Ti, or Sb. Pentafluoroethyl iodide was telomerized with tetrafluoroethylene at 180°C in the presence of a catalyst, on silica or alumina, prepared from CuCl, $TiCl_4$, $SbCl_3$, and ethanolamine [71].

Organic peroxides are used as telomerization catalysts, in spite their potential instability and handling hazards. Telomerization processes patented by Hoechst use bis(trichloroacryloyl)peroxide [72], $(C_nF_{2n+1}COO)_2$ [73], or bis(4-*t*-butylcyclohexyl)peroxydicarbonate [74] as a telomerization catalyst. Peroxyacids of the formula $C_nX_{2n+1}C(O)OOH$, where X = H, F, or Cl and n = 1 to 12, have been claimed in Kali-Chemie patents [75].

Both primary and secondary alkyl iodides are useful as telogens in telomerization processes. The telomerization of 1,2-diiodotetrafluoroethane is complicated by its thermal instability. The formation of ethylene iodide has been suppressed with tetrafluoroethylene to telomerize 1,2-diiodotetrafluoroethane with ethylene [76].

The perfluoroalkyl iodides do not react with nucleophiles, such as OH^- or NH_3, and cannot be converted directly to common intermediates for fluorinated surfactants. Perfluoroalkyl iodides are therefore reacted with ethylene

$$C_2F_5(C_2F_4)_nI + CH_2CH_2 \rightarrow C_2F_5(C_2F_4)_nCH_2CH_2I$$

A Ciba patent [77] describes a reaction of olefins with a perfluoroalkyl iodide of the formula $C_nF_{2n+1}I$, where n = 4 to 14. The olefins were bubbled through the liquid perfluoroalkyl iodides at 50 to 220°C in the presence of a free radical catalyst. Perfluoroalkyl iodides with 6 to 24 carbon atoms were reacted with an olefin in the presence of a metal-amine complex [78].

The perfluoroalkylethyl iodides can be readily converted to the corresponding alcohols, thiols, and sulfonyl chlorides used as intermediates for fluorinated surfactants (Fig. 2.6).

Perfluoroalkyl-2-ethanols are obtained by hydrolysis of 2-perfluoroalkylethyl iodides. A variety of processes have been described in the patent litera-

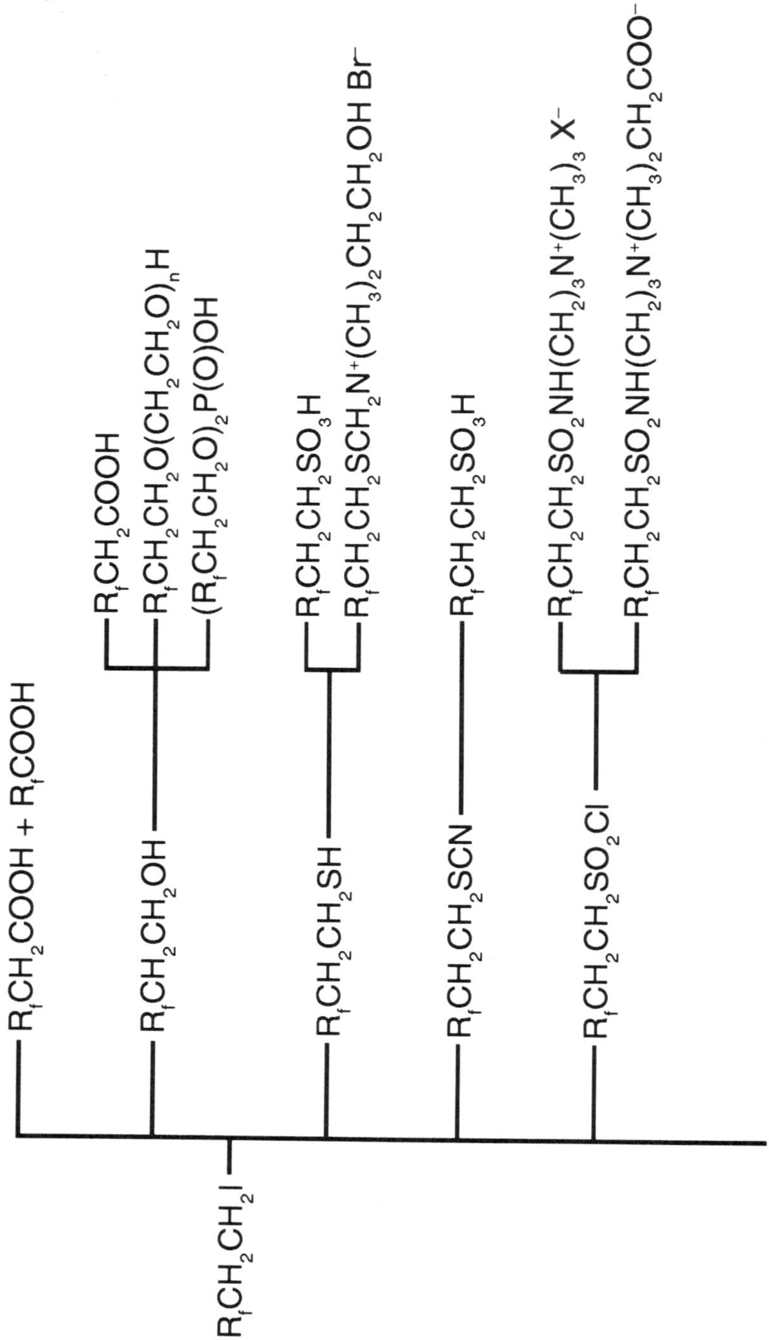

FIG. 2.6 Fluorinated surfactants derived from perfluoroalkylethyl iodides.

ture. A Du Pont patent describes a process of treating 2-perfluoroalkylethyl iodide with oleum [80] at 25°C, pouring the reaction mixture into water containing $NaHSO_3$ and distilling the organic layer. As an alternative to the Du Pont process, 2-perfluoroalkylethyl iodides have been hydrolyzed with nonoxidizing acids, like sulfuric acid or an arylsulfonic acid [81].

A Hoechst process hydrolyzes 2-perfluoroalkylethyl iodides in *N*-methylpyrrolidone and water by heating at 150°C for 13 h [82]. An Asahi process uses aqueous dimethylformamide (DMF) for hydrolysis at 140°C for 6 h, followed by a treatment with aqueous 10% KOH for 3 h at 120°C [83].

A two-stage process [84] for preparing 1,1,2,2-tetrahydroperfluoroalkanols converts perfluoroalkylethyl iodides, $R_fCH_2CH_2I$, to nitrates, $R_fCH_2CH_2ONO_2$, by a reaction with 70–98% HNO_3. In the second stage the nitrates are hydrogenated in the presence of a hydrogenation catalyst to the corresponding alcohols.

The hydrolysis of the perfluoroalkylethyl iodides is catalyzed by certain metals. Copper salts, for example, cupric sulfate, catalyze the hydrolysis of perfluoroalkylethyl iodides in aqueous acetonitrile heated at 160°C for 12 h, followed by heating in 10% NaOH at 70°C for 2 h [85]. A Hoechst patent [86] describes a process for preparing 2-perfluoroalkylethanol by a reaction of the iodide in water containing a metal and a phase transfer agent in an acid medium (pH 1 to 6). Another Hoechst patent [87] discloses a reaction of 2-perfluoroalkylethyl iodide with a peracid to yield 2-perfluoroalkylethanol.

Du Pont patents claim the preparation of fluorinated alcohols, $H(CF_2CF_2)_nCH_2OH$, by telomerization of methyl alcohol with tetrafluoroethylene in the presence of a peroxy or azo catalyst [88,89].

Perfluoroalkyl-2-ethanethiols are obtained by reacting a perfluoroalkyl-2-ethyl iodide with thiourea and hydrolyzing the thiouronium salt formed [90–94]. Perfluoroalkyl-2-ethanethiols can be converted to perfluoroalkylethylsulfonyl chlorides by a reaction with chlorine and water [92,93]:

$$R_fCH_2CH_2SH \xrightarrow{Cl_2/H_2O} R_fCH_2CH_2SO_2Cl$$

The perfluoroalkylethanesulfonyl chlorides are useful intermediates for producing cationic and amphoteric fluorinated surfactants.

2.4 OLIGOMERIZATION OF TETRAFLUOROETHYLENE

A process developed by ICI [95–97] is based on anionic polymerization of tetrafluoroethylene. Unlike high-molecular-weight poly(tetrafluoroethylene) produced by free radical polymerization, anionic polymerization catalyzed by a fluoride (e.g., cesium, potassium, or tetraalkylammonium fluoride) produces highly branched oligomers. The main products are a tetramer, a pentamer, and hexamers (Fig. 2.7).

The pentamer, the most abundant of the oligomers, is an unsaturated perfluorocarbon susceptible to a nucleophilic attack. The pentamer can react with a phenol. The product can be sulfonated to form a sulfonic acid, which after neutralization, functions as an anionic surfactant:

$$C_{10}F_{20} + C_6H_5OH \xrightarrow{\text{base}} C_{10}F_{19}OC_6H_5 \xrightarrow{SO_3} C_{10}F_{19}OC_6H_4SO_3H$$

Reaction of the pentamer with *p*-cresol yields an ether which can be oxidized to yield a fluorinated carboxylic acid. The acid can be neutralized to give an anionic fluorinated surfactant:

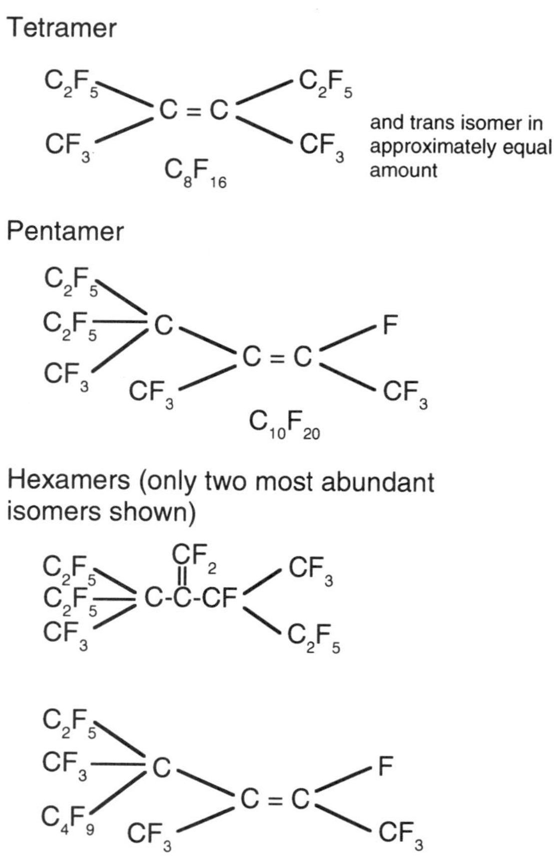

FIG. 2.7 Structures of tetrafluoroethylene oligomers. (From Ref. 96.)

$$C_{10}F_{20} + CH_3C_6H_4OH \xrightarrow{\text{base}} C_{10}F_{19}OC_6H_4CH_3 \xrightarrow{\text{oxidation}} C_{10}F_{19}OC_6H_4COOH \xrightarrow{\text{NaOH}} C_{10}F_{19}OC_6H_4COONa$$

The same fluorinated carboxylic acid can be prepared by reacting the pentamer with *p*-hydroxymethylbenzoate and hydrolyzing the ester formed:

$$C_{10}F_{20} + HOC_6H_4COOCH_3 \xrightarrow{\text{base}} C_{10}F_{19}OC_6H_4COOCH_3 \xrightarrow{\text{hydrolysis}} C_{10}F_{19}OC_6H_4COOH \xrightarrow{\text{NaOH}} C_{10}F_{19}OC_6H_4COONa$$

Alkali causes a stepwise degradation of the pentamer to a ketone which, like the pentamer itself, can react with strong alkali and yield an anionic fluorinated surfactant:

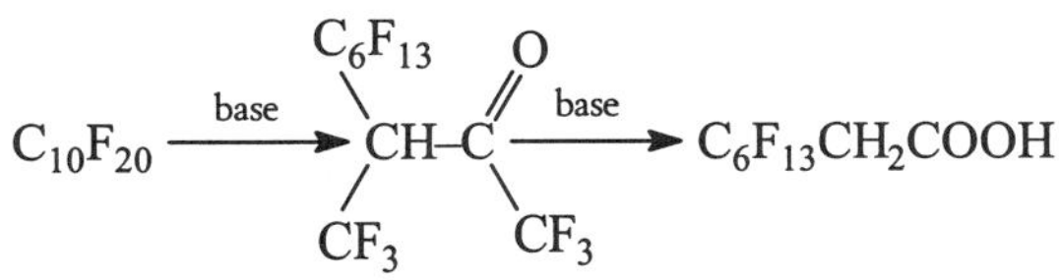

Cationic fluorinated surfactants can be prepared by reacting the pentamer with phenol and subsequently with chlorosulfonic acid. The sulfonyl chloride reacts with *N,N*-dimethylpropanediamine to form a tertiary amine sulfonamide, which can be quaternized to form a cationic surfactant:

$$C_{10}F_{20} + C_6H_5OH \xrightarrow{\text{base}} C_{10}F_{19}OC_6H_5 \xrightarrow{ClSO_3H} C_{10}F_{19}OC_6H_4SO_2Cl$$

$$\downarrow$$

$$C_{10}F_{19}OC_6H_4SO_2NHCH_2CH_2N(CH_3)_2 \xrightarrow{CH_3I} C_{10}F_{19}OC_6H_4SO_2NHCH_2CH_2N^+(CH_3)_3\ I^-$$

Reaction of the amine with β-propiolactone instead of methyl iodide yields the corresponding amphoteric fluorinated surfactant, of the structure $C_{10}F_{19}OC_6H_4SO_2NHCH_2CH_2N^+(CH_3)_2CH_2CH_2COO^-$.

Alternatively, the pentamer can be reacted with 3-hydroxypyridine. The pyridyl ether formed can be quaternized to give a cationic surfactant:

$$C_{10}F_{20} + HOC_5H_4N \xrightarrow{\text{base}} C_{10}F_{19}OC_5H_4N \xrightarrow{(CH_3)_2SO_4} C_{10}F_{19}OC_5H_4N^+CH_3\ CH_3SO_4^-$$

Nonionic surfactants can be prepared by a reaction of the pentamer with alcohols or with phenol. The subsequent sulfonation of the product gives a sulfonyl chloride which with *N*-ethylethanolamine yields an alcohol. An addition of ethylene oxide to the alcohol produces the nonionic surfactant:

$$C_{10}F_{19}OC_6H_4SO_2Cl \longrightarrow C_{10}F_{19}OC_6H_4SO_2N(C_2H_5)CH_2CH_2OH$$

$$\downarrow \text{ethylene oxide}$$

$$C_{10}F_{19}OC_6H_4SO_2N(C_2H_5)\,(CH_2CH_2O)_nH$$

The base-catalyzed reaction of the oligomers with oxyethylated alcohols gives ethers which can also be converted to nonionic surfactants:

$$C_{10}F_{20} + HO(CH_2CH_2O)_nR \longrightarrow C_{10}F_{19}O(CH_2CH_2O)_nR$$

$$R = H,\ \text{alkyl, aryl etc.}$$

If R is H, the reaction yields a nonionic fluorinated surfactant with a fluorocarbon group at both ends of the molecule:

$$C_{10}F_{20} + HO(CH_2CH_2O)_nH \longrightarrow C_{10}F_{19}O(CH_2CH_2O)_nC_{10}F_{19}$$

or

$$C_8F_{16} + HO(CH_2CH_2O)_nH \longrightarrow C_8F_{15}O(CH_2CH_2O)_nC_8F_{15}$$

Unlike the phenyl ethers, the alkyl ethers thus formed react readily with nucleophilic reagents. The reaction with another oxyethylated alcohol molecule gives a nonionic surfactant with two oxyethylene chains,

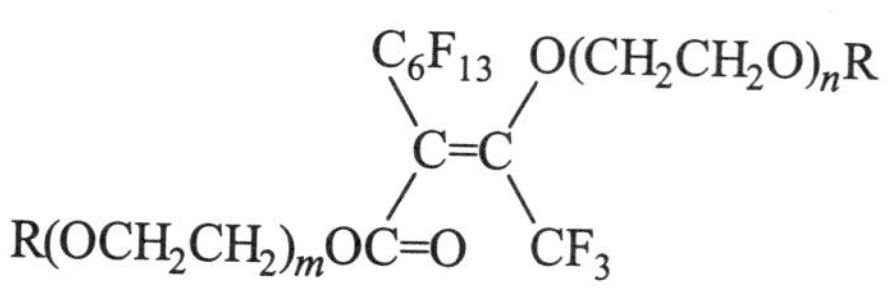

Another example of nucleophilic attack on the alkyl ethers is the reaction with an aqueous base which replaces a CF_3- group with hydrogen and yields a stable nonionic fluorinated surfactant $C_6F_{13}CH{=}C(CF_3)O(CH_2CH_2O)_nR$.

2.5 MISCELLANEOUS REACTIONS

Anionic fluorinated surfactants containing an oligo(hexafluoropropene oxide) chain have been prepared by anionic oligomerization of hexafluoropropene oxide (HFPO) [98,99] and hydrolysis of the oligomer, an acid fluoride [100]:

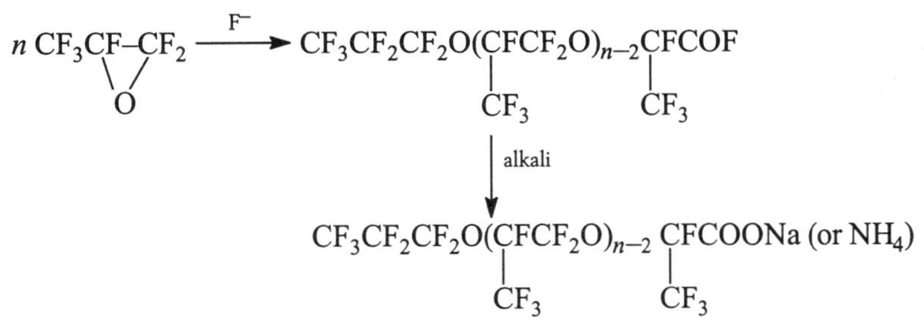

The HFPO oligomers are acid fluorides and acylate benzene or toluene in the presence of a Friedel-Crafts catalyst such as $AlCl_3$ [101,102]. The resulting oligomers with an aryl end group are soluble and exhibit surface activity in toluene and *m*-xylene [102]:

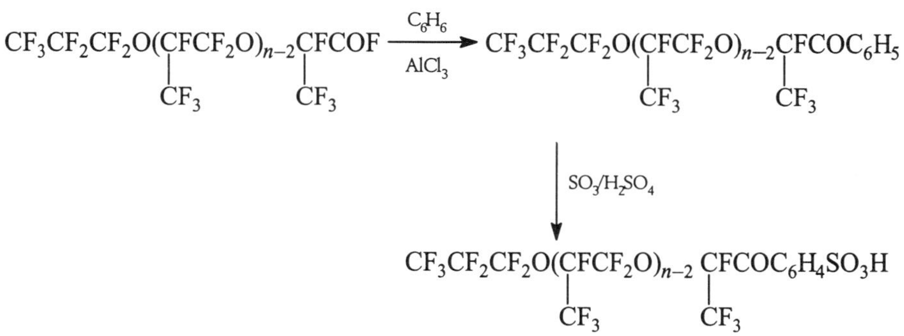

The acylated benzenes and toluenes were sulfonated with 30% oleum at 50–60°C. The sulfonic acid group was found to be in the meta position relative to the acyl $(HFPO)_n$ group. The sodium salts of the sulfonic acids, $(HFPO)_n-Ar-SO_3Na$ (where Ar is a phenyl or tolyl group), lowered the surface tension of water (see Sec. 4.3).

Pittman et al. [103,104] treated hexafluoroacetone with nucleophilic agents such as an alkoxide, cyanide, and an alkali metal fluoride to prepare fluoroalkyl acrylates and methacrylates with a perfluoroisopropyl group:

$$(CF_3)_2C{=}O + KF \longrightarrow (CF_3)_2CFO^-K^+$$

$$(CF_3)_2CFO^-K^+ + CH_2{=}CHCOCl \longrightarrow CH_2{=}CHCOOCF(CF_3)_2 + KCl$$

These fluoroacrylate monomers can be copolymerized with other acrylates to form polymeric surfactants.

The reaction of potassium fluoride with hexafluoroacetone has been employed to prepare nonionic fluorinated surfactants (see Sec. 2.9). Since hexafluoroacetone has been found to be highly toxic and a teratogen, it is no longer used for industrial preparation of fluorinated surfactants.

Kokelenberg and Pollet [105] attempted to introduce a few short perfluorinated groups into compounds featuring one or more hydrophilic functions. The addition of chlorotrifluoroethylene and hexafluoropropene to aliphatic or aromatic polyhydroxy compounds in the presence of a base and a dipolar aprotic solvent gave mixtures of products resulting from incomplete addition and partial dehydrofluorination.

2.6 ANIONIC FLUORINATED SURFACTANTS

The synthetic method selected for the preparation of an anionic fluorinated surfactant depends on the hydrophile of the surfactant: a carboxylate, sulfonate, sulfate, sulfite or phosphate group. Within each group the choice of the preparative method is determined by the availability of starting materials.

Carboxylates

From Carbonyl or Sulfonyl Fluorides or Chlorides

Perfluoroalkanecarbonyl fluorides produced by the electrochemical fluorination process are hydrolyzed to produce a perfluoroalkanoic acid or its salt:

$$C_7F_{15}COF + 2NaOH \rightarrow C_7F_{15}COONa + NaF + H_2O$$

Perfluoroalkanesulfonyl fluorides, prepared by the electrochemical fluorination process, yield fluorinated carboxylic acids via the sulfonamide as the intermediate. The N-substituted perfluorohexanesulfonamide is condensed with sodium monochloroacetate in aqueous isopropyl alcohol [106]:

$$C_6F_{13}SO_2F + NH_2(CH_2)_3NH_2 \longrightarrow C_6F_{13}SO_2NH(CH_2)_3NH_2$$

$$\xrightarrow{ClCH_2COONa} C_6F_{13}SO_2NH(CH_2)_3N(CH_2COONa)_2$$

A fluorinated sulfonyl chloride can be reacted with an amino acid to give a fluorinated anionic surfactant with a carboxylic group and a sulfonamido linkage [107]:

$$C_9F_{17}OC_6H_5 \xrightarrow{ClSO_3H} C_9F_{17}OC_6H_4SO_2Cl \xrightarrow[NaOH]{CH_3NHCH_2COOH}$$

$$C_9F_{17}OC_6H_4SO_2N(CH_3)CH_2COONa \xrightarrow{acid} C_9F_{17}OC_6H_4SO_2N(CH_3)CH_2COOH$$

Alkanoic acids with a terminal perfluoroalkyl segment have been prepared by the addition of perfluoroalkanesulfonyl chlorides to unsaturated acids [108].

From Telomer Iodides

Perfluoroalkyl iodides produced by the telomerization process can be oxidized to carboxylic acids, but drastic conditions are needed for the reaction to occur. Perfluoroalkyl iodide is oxidized with oleum (15–45% SO_3) at 100–180°C under pressure. The upper layer is treated with water and the perfluoroalkanoic acid formed is separated by distillation [109]:

$$CF_3(CF_2CF_2)_nI \xrightarrow{\text{oleum}} CF_3(CF_2CF_2)_{n-1}CF_2COOH + \tfrac{1}{2}I_2 + \text{by-products}$$

The liberated iodine is recovered by filtration.

Perfluoroalkanoic acids have also been obtained by carbonation of fluoroalkyl iodides in the presence of a copper-zinc catalyst [110] or group VIII transition metal complexes and subsequent hydrolysis [111]:

$$CF_3(CF_2)_7I + CO_2 \xrightarrow[(1)\,\text{DMF},\ (2)\,H_2O]{Pd[P(C_6H_5)_3]_4} CF_3(CF_2)_7COOH$$

The Grignard reaction has been employed to prepare fluorinated carboxylic acids available hitherto only by electrochemical fluorination [112]:

$$CF_3(CF_2)_5I + C_6H_5MgBr \longrightarrow C_6H_5I + CF_3(CF_2)_5MgBr$$

$$CF_3(CF_2)_5MgBr \xrightarrow[(2)\ \text{Hydrolysis}]{(1)\ (C_2H_5O)_2CO} CF_3(CF_2)_5COOH$$

The conversion of telomer iodides to chlorosulfates and the subsequent hydrolysis of the chlorosulfate yield perfluorinated carboxylic acids [113,114]:

$$C_mF_{2m+1}(CF_2CF_2)_nI \xrightarrow{ClSO_3H} C_mF_{2m+1}(CF_2CF_2)_nOSO_2Cl + HI$$

$$C_mF_{2m+1}(CF_2CF_2)_nOSO_2Cl + 6NaOH \longrightarrow C_mF_{2m+1}(CF_2CF_2)_{n-1}CF_2COONa + 2NaF + Na_2SO_4 + NaCl + 3H_2O$$

Long-chain alkanoic acids with terminal perfluoroalkyl segments can be prepared by a two-step synthesis [115,116]:

$$R_fI + CH_2{=}CH(CH_2)_{m-2}COOR \rightarrow R_fCH_2CHI(CH_2)_{m-2}COOR \xrightarrow[\text{alcohol}]{Zn} R_f(CH_2)_mCOOR$$

Hydrolysis of the ester gives a alkanoic acid with a terminal perfluoroalkyl segment.

Perfluoroalkyl iodides are converted with ethylene to perfluoroalkylethyl iodides, which can be oxidized more readily than the parent perfluoroalkyl iodides. Oxidation with chromic acid yields a perfluoroalkanoic acid and a fluorinated alkanoic acid with an α-methylene group [93,117]:

$$CF_3(CF_2CF_2)_nCH_2CH_2I \xrightarrow[H_2SO_4]{K_2Cr_2O_7} CF_3(CF_2CF_2)_nCH_2COOH + CF_3(CF_2CF_2)_nCOOH$$

From Perfluoroalkylethanol

Perfluoroalkylethanols, readily available by hydrolysis of perfluoroalkyl iodides, can be oxidized to fluorinated carboxylic acids. As an example, oxidation of $H(CF_2CF_2)_nCH_2OH$ with nitrogen oxides gave ω-hydroperfluorocarboxylic acids [118]:

$$HCF_2(CF_2CF_2)_nCH_2CH_2OH \xrightarrow{NO_x} HCF_2(CF_2CF_2)_nCH_2COOH$$

Photochemical chlorination of perfluoroalkylethanols gives perfluoroalkylacetyl chlorides [119,120], which are hydrolyzed to the carboxylic acid:

$$CF_3(CF_2)_nCH_2CH_2OH \xrightarrow{Cl_2/UV} CF_3(CF_2)_nCH_2COCl \longrightarrow CF_3(CF_2)_nCH_2COOH$$

The reaction of perfluoroalkylethanols with acrylonitrile and hydrolysis of the nitrile formed yields a fluorinated carboxylic acid [110]:

$$C_8F_{17}CH_2CH_2OH + CH_2{=}CHCN \xrightarrow[70\text{–}80^\circ C]{NaOH} C_8F_{17}CH_2CH_2OCH_2CH_2CN$$

$$\downarrow \text{conc. HCl at reflux}$$

$$C_8F_{17}CH_2CH_2OCH_2CH_2COOH$$

From Perfluoroalkylthiols

The reaction of perfluoroalkylthiols (see Sec. 2.3) with an alkyl acrylate in the presence of a free radical catalyst yields esters, which are hydrolyzed to obtain perfluoroalkylcarboxylates [90]:

$$R_fCH_2CH_2SH + CH_2{=}CHCOOCH_3 \longrightarrow R_fCH_2CH_2SCH_2CH_2COOCH_3 \xrightarrow{KOH}$$

$$R_fCH_2CH_2SCH_2CH_2COOK + CH_3OH$$

From Telomer Chlorides

Telomerization of $ClCF{=}CF_2$ with CF_3CCl_3 gives the telomer $CF_3CCl_2(CF_2CFCl)_nCl$, which can be converted to carboxylic acids. Halogen interchange with aluminum chloride and subsequent hydrolysis of the trichloromethyl end group yields a carboxylic acid containing fluorine and chlorine [121]:

$$CF_3CCl_2(CF_2CFCl)_nCl \xrightarrow{AlCl_3} CF_3CCl_2(CF_2CFCl)_{n-1}\,CF_2CCl_3$$

$$\downarrow \text{hydrolysis}$$

$$CF_3CCl_2(CF_2CFCl)_{n-1}CF_2COOH$$

From Esters

Esters of fluorinated acids can be hydrolyzed to obtain the parent acid [116,122]. Thus hydrolysis of hydrocarbon-segmented fluorinated carboxylic acid esters with KOH in 90% aqueous alcohol gives the potassium salt of the parent carboxylic acid [116]:

$$(CF_3)_2CF(CH_2)_{10}COOC_2H_5 + KOH \longrightarrow (CF_3)_2CF(CH_2)_{10}COOK + C_2H_5OH$$

Condensation of an aminocarboxylic acid with the isopropyl ester of a perfluorocarboxylic acid yields a fluorinated surfactant with a modified hydrophobic chain [123]:

$$H_2N(CH_2)_5COONa + C_8F_{17}COOCH(CH_3)_2 \longrightarrow$$

$$C_8F_{17}CONH(CH_2)_5COONa + HOCH(CH_3)_2$$

From Oligomers

Tetrafluoroethylene pentamer forms with *p*-hydroxymethylbenzoate an ester which can be hydrolyzed to prepare a salt of a fluorinated carboxylic acid [96,97]:

$$C_{10}F_{19}OC_6H_4COOCH_3 \xrightarrow{\text{hydrolysis}} C_{10}F_{19}OC_6H_4COONa$$

Oxidation of a methyl group on a cresyl ether, obtained by reacting *p*-cresol with the tetrafluoroethylene pentamer, yields a fluorinated carboxylic acid. The acid is neutralized to give a fluorinated surfactant [96,97]:

$$C_{10}F_{19}OC_6H_4CH_3 \xrightarrow{\text{oxidation}} C_{10}F_{19}OC_6H_4COOH \xrightarrow{NaOH} C_{10}F_{19}OC_6H_4COONa$$

From Fluorinated Epoxides

Fluorinated epoxides produce fluorinated surfactants with a hydroxyl in the hydrophobic chain [124]:

$$C_nF_{2n+1}\underbrace{CH{-}CH_2}_{O} + H_2NCH(CH_3)_2 \longrightarrow$$

$$C_nF_{2n+1}CH(OH)CH_2HNCH(CH_3)_2 \xrightarrow[C_2H_5ONa/C_2H_5OH]{ClCH_2COONa}$$

$$C_nF_{2n+1}CH(OH)CH_2N[CH(CH_3)_2]CH_2COONa$$

Fluorinated epoxides condensed with potassium sarcosine in aqueous isopropyl alcohol at 50°C yield a fluorinated aminocarboxylate [125]:

$$C_9F_{19}CH_2\text{–}\underset{\diagdown O \diagup}{CH\text{–}CH_2} + CH_3NHCH_2COOK \longrightarrow C_9F_{19}CH_2CH(OH)CH_2N(CH_3)CH_2COOK$$

By Photooxidation

Perfluoropolyether surfactants [126–129] have been prepared using perfluoroalkeneoxide as the hydrophobic group. Monocarboxylic acids were obtained starting from intermediates produced by photooxidation of hexafluoropropylene [129]:

$$CF_3(OCF_2\underset{\displaystyle CF_3}{\underset{|}{C}}F)_nOCF_2COOH \qquad (n = 1,2,3)$$

Dicarboxylic acids were prepared by reductive cleavage of perfluoropolyperoxide obtained by photooxidation of tetrafluoroethylene [129]:

$$HOOCCF_2[(OCF_2CF_2)_n(OCF_2)_m]_pOCF_2COOH \qquad (n = m = 1;\ p = 1,2,3)$$

By Ozonization

Perfluoroalkyl ethylene, $CF_3(CF_2CF_2)_nCH{=}CH_2$, obtained by dehalogenation of perfluoroalkylethyl iodide, is treated with ozone to give ozonides. Oxidative cleavage of the ozonides with hydrogen peroxide yields perfluoroalkanoic acids [130].

Sulfonates

By Electrochemical Fluorination

Perfluoroalkanesulfonyl fluorides or chlorides obtained by electrochemical fluorination are hydrolyzed to give the corresponding sulfonic acid or its salt [51–53]:

$$CF_3(CF_2)_nSO_2F + 2NaOH \longrightarrow CF_3(CF_2)_nSO_3Na + NaF + H_2O$$

The industrial process removes the fluoride liberated with calcium oxide [131]:

$$CF_3(CF_2)_nSO_2F + KOH + \tfrac{1}{2}CaO \longrightarrow CF_3(CF_2)_nSO_3K + \tfrac{1}{2}CaF_2 + H_2O$$

Quaternary ammonium salts of perfluoroalkanesulfonic acids can be prepared by neutralizing perfluoroalkanesulfonic acid with tetraalkylammonium hydroxide. This procedure is not economical, however, for industrial use. A discovery [131–133] that quaternary ammonium salts of perfluoroalkanesulfonic acids can be obtained directly by reacting the perfluoroalkanesulfofluoride with a tertiary amine and an alkoxysilane is the basis for a commercial process:

$$C_nF_{2n+1}SO_2F + N(C_mH_{2m+1})_3 + C_mH_{2m+1}OSiR_3 \longrightarrow C_nF_{2n+1}SO_3^-[N(C_mH_{2m+1})_4]^+ + FOSiR_3$$

The fluorinated surfactant Fluortensid FT 248 is made by reacting perfluorooctanesulfonyl fluoride with triethylamine and ethoxysilane in a anhydrous polar solvent, e.g., monochlorobenzene, diethyl ether, chloroform, acetonitrile, or tetrahydrofuran. The product precipitates and is separated by filtration or centrifugation.

The reaction mechanism was elucidated by reacting perfluorobutanesulfonyl fluoride with triethylamine and an amine:

$$C_4F_9SO_2F + H_2N(CH_2)_4Si(CH_3)_2OCH_3 + N(C_2H_5)_3 \longrightarrow C_4F_9SO_3^-[N(C_2H_5)_3CH_3]^+ + H_2N(CH_2)_4Si(CH_3)_2F$$

Contrary to the expected reaction path, the methoxy group reacted instead of the primary amine with the sulfonyl fluoride. The reaction mechanism is supported by the observation that in the absence of triethylamine the reaction of the amine with the perfluorobutanesulfonyl fluoride produces the compound

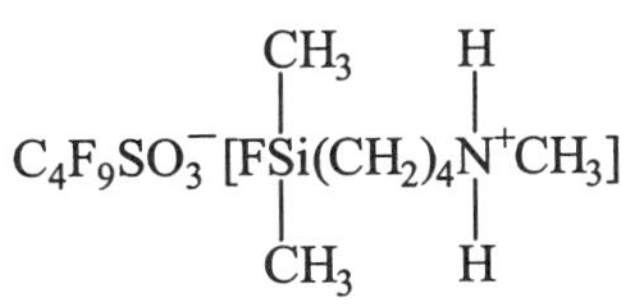

Perfluoroacylbenzenesulfonic acids $R_fCOC_6H_4SO_3H$ or their salts [134] have been prepared by acylating benzene with a perfluoroacyl halide in the presence of a Lewis acid and sulfonating the reaction product. Sodium (perfluorooctanoyl)benzenesulfonate was prepared by reacting perfluorooctanoyl chloride with benzene in the presence of $AlCl_3$.

$$C_7F_{15}COCl + C_6H_6 \xrightarrow{AlCl_3} C_7F_{15}COC_6H_5 \xrightarrow{\text{oleum}} C_7F_{15}COC_6H_4SO_3H$$

The sulfonic acid was neutralized with NaOH to give the sodium sulfonate $C_7F_{15}COC_6H_4SO_3Na$.

Perfluoroalkylether amides featuring a sodium sulfonate or sodium carboxylate group have been prepared by reacting an acid fluoride with an aromatic or aliphatic amino acid. For example, $F[CF(CF_3)CF_2O]_3CF(CF_3)COF$ has been reacted with sulfanilic acid and triethylamine, followed by NaOH, to give $F[CF(CF_3)CF_2O]_3CF(CF_3)CONHC_6H_4SO_3Na$ [135].

By Telomerization

Perfluoroalkylethyl iodides can be converted to sulfonic acids utilizing a thiol as an intermediate. The thiol formed with thiourea [92–94] is dispersed in acidified water and oxidized with chlorine:

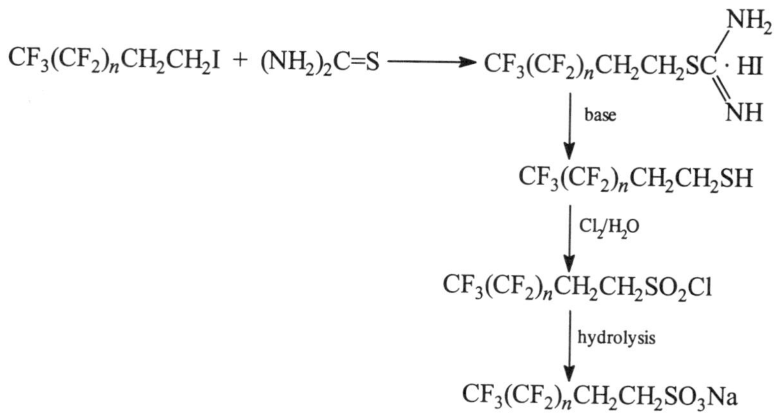

Alternatively, a perfluoroalkylethanethiol, $C_6F_{13}CH_2CH_2SH$, can be reacted in the presence of a free radical catalyst with 1-allyloxy-2,3-epoxypropane to form epoxides. A reaction of the epoxides with $CH_3NHCH_2CH_2SO_3Na$ gives the fluorinated surfactants

$$C_6F_{13}CH_2CH_2S(CH_2)_3OCH_2CH(OH)CH_2N(CH_3)CH_2CH_2SO_3Na$$

and $C_6F_{13}CH_2CH_2SCH(CH_3)CH_2OCH_2CH(OH)CH_2N(CH_3)CH_2CH_2SO_3Na$ [136].

A fluorinated surfactant $CF_3(CF_2)_nCH_2CH_2SO_3^-NH_4^+$ (NH_4^+ or H^+) is obtained by a reaction of the perfluoroalkylethyl iodide with KSCN. The thiocyanate formed is oxidized with a peroxycarboxylic acid, for example, peracetic acid [137]:

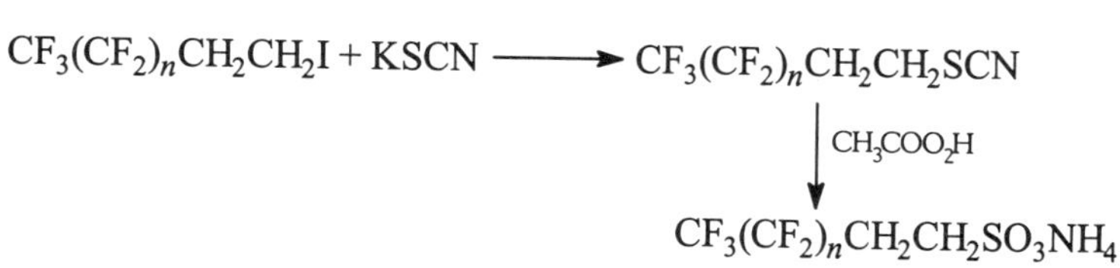

Potassium perfluoro-3-oxaalkanesulfonate,

$$CF_3CF_2(CF_2CF_2)_nOCF_2CF_2SO_3K,$$

can be prepared using sulfur trioxide and tetrafluoroethylene as the starting materials to form the tetrafluoroethane sultone as an intermediate [138]:

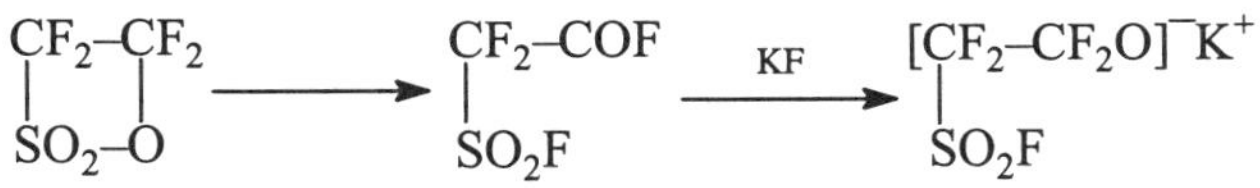

$$[FSO_2C_2F_4O]^-K^+ + C_2F_4 + I_2 \longrightarrow FSO_2C_2F_4OC_2F_4I + KI$$

$$FSO_2C_2F_4OC_2F_4I \xrightarrow{C_2F_4} FSO_2C_2F_4O(C_2F_4)_nC_2F_4I$$

The resulting iodide is converted to an anionic surfactant by substituting iodine with fluorine. A fluorine-nitrogen mixture (1:3) is bubbled into a solution of the telomers in 1,1,2-trichlorotrifluoroethane containing NaF and MgF_2. The solvent is removed by distillation and the residue is treated with KOH. The surfactant, $CF_3CF_2(CF_2CF_2)_nOCF_2CF_2SO_3K$, is isolated by neutralization and extraction with methanol.

By Oligomerization

The tetrafluoroethylene pentamer (see Sec. 2.4, Fig. 2.7) is allowed to react with phenol. The phenylether formed is treated with oleum to yield a sulfonic acid, which is neutralized with NaOH [96,97]:

$$C_{10}F_{19}OC_6H_5 + SO_3 \longrightarrow C_{10}F_{19}OC_6H_4SO_3H \xrightarrow{NaOH} C_{10}F_{19}OC_6H_4SO_3Na$$

By Sulfonation with a Sultone

Sulfopropylated *N*-alkylperfluorooctanamides [139] have been prepared by reacting ethyl perfluorooctanoate with an amine. The amide formed was reacted with metallic sodium and 1,3-propanesultone to give the fluorinated surfactant $R_fCONR(CH_2)_3SO_3Na$, where R = H or an alkyl in the C_1 to C_{12} range. The *n*-butyl group generally gave the best surfactant properties.

From Hexafluoropropene Oligomers

Fluorinated surfactants with an oligo(hexafluoropropene oxide) hydrophobe have been prepared by sulfonating $(HFPO)_nAr$, where $(HFPO)_n$ is an oligo(hexafluoropropene oxide) group and Ar an aryl group [102].

Sulfosuccinates

Fluorinated alkyl-2-sulfosuccinates are prepared by a reaction of a fluorinated alcohol with maleic anhydride. The reaction yields mono- or diesters, which are treated with Na_2SO_3 or $Na_2S_2O_5$. Two moles of the fluorinated alcohol $CF_3CF_2CH_2CH_2OH$ and one mole of maleic acid are heated at 140°C to give the diester, which is heated at reflux with $Na_2S_2O_5$ in aqueous isopropyl alcohol to obtain a fluorinated sulfosuccinate [140]:

$$2CF_3CF_2CH_2CH_2OH + HOOCCH{=}CHCOOH \longrightarrow$$

$$CF_3CF_2CH_2CH_2OOCCH{=}CHCOOCH_2CH_2CF_2CF_3$$

$$\downarrow Na_2SO_3$$

$$CF_3CF_2CH_2CH_2OOCCH(SO_3Na)CH_2COOCH_2CH_2CF_2CF_3$$

Yoshino et al. [141] esterified a fluorinated alkanol, such as *1H,1H,9H*-hexadecafluorononanol-1, with maleic anhydride in the presence of *p*-toluenesulfonic acid monohydrate. The resulting bis-maleate ester reacted with sodium hydrogen sulfite to yield the sodium salt of the fluoroalkyl-2-sulfosuccinate.

Sulfates

Perfluoroalkylmethyl sulfates, $X(CF_2)_nCH_2OSO_2Na$, where X is H or F and $n = 5$ to 12, are prepared by a reaction of the corresponding alkanols with concentrated H_2SO_4 or $ClSO_3H$ and neutralization [142].

Greiner et al. [143] prepared fluorinated ether sulfates, $H-(CF_2CF_2)_nCH_2[OCH_2CH_2]_mOSO_3NH_4$, $n = 2, 3, 4$ $m =$ average 3. The telomer alcohols were oxyethylated using BF_3 as the catalyst, sulfonated with chlorosulfonic acid, and neutralized with ammonia.

The fluorinated surfactants $(CF_3)_2CFO(CH_2)_6OSO_3Na$ (Na^+ or NH_4^+) have been prepared by sulfation of $(CF_3)_2CFO(CH_2)_6OH$, which is obtained by a reaction of hexafluoroacetone, CF_3COCF_3, with KF and $Cl(CH_2)_6OH$ [144]. The toxicity of hexafluoroacetone limits the usefulness of this process.

Dipropylene glycol 5,5,6,6,6-pentafluorohexyl ether sodium sulfate has been prepared by reacting dipropylene glycol 5,5,6,6,6-pentafluorohexyl ether with chlorosulfonic acid and neutralizing with NaOH [145].

Bunte salts

Fluorinated Bunte salts, featuring an $-S_2O_3M$ group, are prepared by reacting sodium thiosulfate with perfluoroalkylethyl iodide [146,147]:

$$C_nF_{2n+1}CH_2CH_2I + Na_2S_2O_3 \longrightarrow C_nF_{2n+1}CH_2CH_2S_2O_3Na$$

Fluorinated Bunte salts are effective dispersants and emulsifiers for fluorinated polymers [147].

Phosphates

The alkyl phosphates are mixtures of mono- and diesters formulated as ammonium salts, for example, $(R_fCH_2CH_2O)P(O)(O^-NH_4^+)_2$ and

$(R_fCH_2CH_2O)_2P(O)(O^-NH_4^+)$, or as free acids, for example, $(R_fCH_2CH_2O)P(O)(OH)_2$ and $(R_fCH_2CH_2O)_2P(O)(OH)$.

By Telomerization

Bis-(fluoroalkyl)phosphate esters are prepared by reacting $POCl_3$ with a perfluoroalkylethanol [148]:

$$2CF_3(CF_2)_nCH_2CH_2OH + POCl_3 \longrightarrow (CF_3(CF_2)_nCH_2CH_2O)_2P(O)Cl + 2HCl$$

$$\xrightarrow{H_2O} (CF_3(CF_2)_nCH_2CH_2O)_2P(O)OH$$

Reaction conditions can be selected to minimize the formation of the monoester and the triester.

Perfluorooctylethanol has been reacted with $P_2O_3Cl_4$ and diethanolamine to obtain a product with the triester content of 3.0% and the average composition $(CF_3(CF_2)_7CH_2CH_2O)_{1.5}PO[OH \cdot NH(C_2H_4OH)_2]_{1.5}$ [149].

A reaction of perfluoroalkyl iodides with yellow phosphorus produces fluorinated surfactants [93]:

$$6R_fI + P_4 \longrightarrow 2R_{f2}PI + 2R_fPI_2$$

$$R_{f2}PI + R_fPI_2 \xrightarrow[H_2O]{H_2O} R_{f2}PO_2H + R_fPO_3H_2$$

By Electrochemical Fluorination

Perfluoroalkane sulfonyl fluorides produced by electrochemical fluorination are reacted with ethylaminoethanol. The fluorinated alcohol formed is converted with $POCl_3$ to a fluorinated phosphate ester [150]:

$$2C_8F_{17}SO_2N(C_2H_5)CH_2CH_2OH \xrightarrow{POCl_3} [C_8F_{17}SO_2N(C_2H_5)CH_2CH_2O]_2P(O)Cl$$

$$\xrightarrow{NaOH} [C_8F_{17}SO_2N(C_2H_5)CH_2CH_2O]_2P(O)ONa$$

By Oligomerization

Reaction of the tetrafluoroethylene tetramer [96,97] with dilute aqueous sodium hydroxide and acidification of the reaction product give the ketone $C_6F_{13}CH_2COCF_3$. Bases convert the ketone to its enol form, which reacts with phosphorus oxychloride. Hydrolysis of the reaction product gives the fluorinated surfactant $C_6F_{13}CH{=}C(CF_3)OPO(OH)_2$.

2.7 CATIONIC SURFACTANTS

Carbonyl or sulfonyl fluorides obtained by electrochemical fluorination yield cationic fluorinated surfactants by a reaction with *N,N*-dimethyl-1,3-diaminopropane and quaternization with methyl iodide or dimethyl sulfate [5,151]:

$$C_7F_{15}COF \xrightarrow[\text{base}]{(CH_3)_2N(CH_2)_3NH_2} C_7F_{15}CONH(CH_2)_3N(CH_3)_2 \xrightarrow{CH_3I} C_7F_{15}CONH(CH_2)_3N^+(CH_3)_3I^-$$

$$C_8F_{17}SO_2F \xrightarrow[\text{base}]{(CH_3)_2N(CH_2)_3NH_2} C_8F_{17}SO_2NH(CH_2)_3N(CH_3)_2 \xrightarrow{CH_3I} C_8F_{17}SO_2NH(CH_2)_3N^+(CH_3)_3I^-$$

In a similar fashion, cationic fluorinated surfactants are prepared from telomer iodides by converting perfluoroalkylethyl iodides to sulfonyl chlorides and reacting these with *N,N*-dimethyl-1,3-diaminopropane [152].

Likewise, cationic surfactants are prepared by the oligomerization process. The tetraethylene tetramer is allowed to react with phenol and the phenyl ether obtained is chlorosulfonated [96,97]. The sulfonyl chloride is reacted with *N,N*-dimethyl-1,3-diaminopropane and quaternized:

$$C_{10}F_{20} + C_6H_5OH \longrightarrow C_{10}F_{19}OC_6H_5 \xrightarrow{ClSO_3H} C_{10}F_{19}OC_6H_4SO_2Cl$$

$$\xrightarrow{(CH_3)_2N(CH_2)_3NH_2} C_{10}F_{19}OC_6H_4SO_2NH(CH_2)_3N(CH_3)_2 \xrightarrow{CH_3I} C_{10}F_{19}OC_6H_4SO_2NH(CH_2)_3N^+(CH_3)_3I^-$$

Perfluoroalkyl-2-ethanethiols are prepared by reaction of a perfluoroalkyl-2-ethyl iodide with thiourea and subsequent alkaline hydrolysis of the thiouronium salt formed [90,94]. A reaction of the perfluoroalkyl-2-ethanethiol with $ClCH_2CH(OH)CH_2N^+(CH_3)_3Cl^-$ gives the cationic surfactant $R_fCH_2CH_2SCH_2CH(OH)CH_2N^+(CH_3)_3Cl^-$ [153].

In some cationic fluorinated surfactants the positively charged nitrogen atom is part of a heterocyclic ring [154]. Katritzky et al. [155] prepared cationic fluorinated surfactants derived from *N*-(perfluorooctanesulfonyl)piperazine:

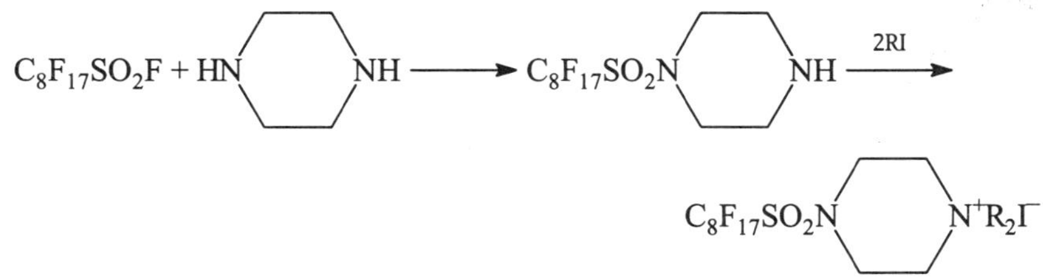

where R = C_nH_{2n+1}, n = 1, 2, 4, 6, or 8.

Cationic fluorinated surfactants have been prepared from perfluoroalkyl esters, obtained by converting an acid fluoride into an ester. Reaction of the ester with a diamine and alkylation with a halide or sulfonate gives a cationic surfactant, for example [156]:

$$CF_3CF_2[CF_2OCF(CF_3)]_2COOCH_3 + H_2N(CH_2)_3N(CH_3)_2 \longrightarrow$$

$$CF_3CF_2[CF_2OCF(CF_3)]_2COHN(CH_2)_3N(CH_3)_2 \xrightarrow{CH_3I}$$

$$CF_3CF_2[CF_2OCF(CF_3)]_2COHN(CH_2)_3N^+(CH_3)_3\ I^-$$

Matsui et al. [157] reacted the fluorinated ester $C_7F_{15}COOC_2H_5$ with $H_2N(CH_2)_2N(CH_3)_2$ and reduced the reaction product with $LiAlH_4$. Subsequent quaternization with CH_3I and ion exchange to replace iodide with chloride produced the cationic surfactant $C_7F_{15}CH_2NH(CH_2)_2N^+(CH_3)_3\ Cl^-$.

Using 3-(perfluoro-1,1-dimethyl-butyl)-1-propene as the starting material, cationic fluorinated surfactants have been prepared via the corresponding epoxypropane reacted with a secondary amine [158–160]:

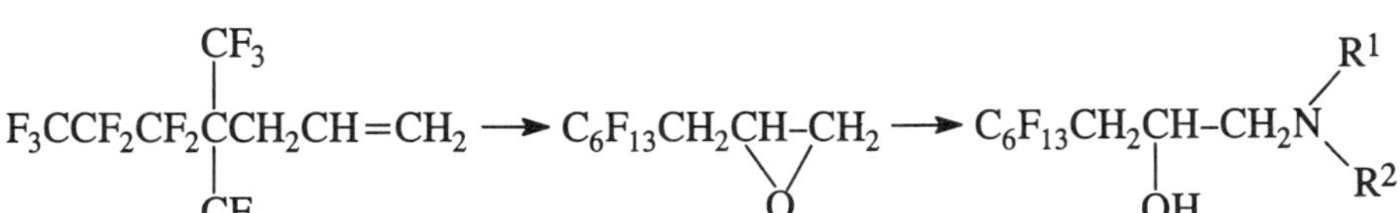

The alkylation with methyl iodide gave quaternary ammonium salts [160].

A reaction of the fluorinated epoxypropane with morpholine gave a cationic surfactant with a heterocyclic nitrogen:

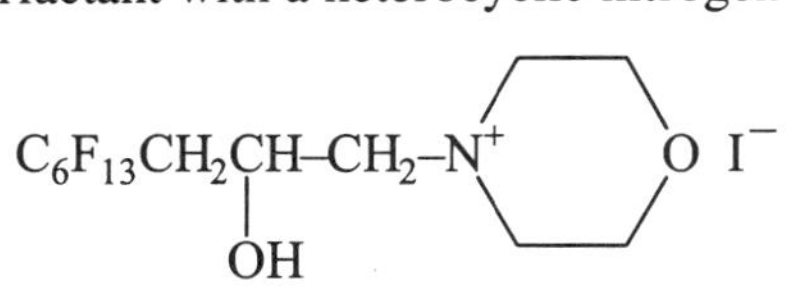

Amine oxides [161] or, more correctly, N-oxides of tertiary amines, are electrically neutral and belong formally to the class of nonionic surfactants. However, the chemistry involved in the production of amine oxides is related to that of cationic surfactants and is therefore described here.

The synthesis of amine oxides involves oxidation of a tertiary amine with hydrogen peroxide [160,162,163] or peracetic acid [164]:

$$C_8F_{17}CH_2CH(OOCCH_3)CH_2N(CH_3)_2 \xrightarrow{CH_3C(O)OOH} C_8F_{17}CH_2CH(OOCCH_3)CH_2\underset{\downarrow \atop O}{N}(CH_3)_2$$

Fluorinated amine oxides with a heterocyclic nitrogen have been prepared [160], for example:

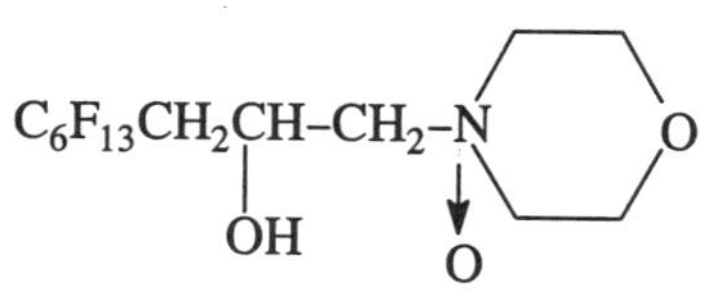

Amine oxides are effective foaming agents.

2.8 AMPHOTERIC SURFACTANTS

Amphoteric fluorinated surfactants have at least one anionic and at least one cationic group at about their isoelectric point [165]. The cationic group is usually a quaternary amine or a protonated tertiary or secondary amine. The anionic functionality is a carboxylate, sulfonate, sulfate, or phosphate group. Most widely used fluorinated surfactants are carboxybetaine-type amphoteric surfactants which have, like betaines, a cationic amine functionality and an anionic carboxylate group. Instead of a carboxylate group, the anionic component of sulfatobetaines is a sulfate group and that of sulfobetaines a sulfonate group.

Carboxybetaines have been prepared by reacting a fluorinated tertiary amine with β-propiolactone [166–170], chloroacetic acid or its sodium salt [169–178] or acrylic acid [166,169,170]. β-Propiolactone is a suspected carcinogen and is probably no longer used commercially as a quaternizing agent.

The synthetic methods utilized for the preparation of amphoteric fluorinated surfactants are similar to those used for cationic surfactants, except for the alkylation step. A tertiary amine can be prepared by reaction of a fluorinated ester or acid halide with a diamine which contains both a primary or a secondary amine and a tertiary amine [179]. The resulting amine is then treated with chloroacetic acid:

$$C_9F_{19}COOC_3H_7 + H_2N(CH_2)_3O(CH_2)_2N(CH_3)_2 \longrightarrow C_9F_{19}CONH(CH_2)_3O(CH_2)_2N(CH_3)_2$$

$$\downarrow ClCH_2COOH$$

$$C_9F_{19}CONH(CH_2)_3O(CH_2)_2N^+(CH_3)_2CH_2COO^-$$

Fluorinated tertiary amines with a sulfonamido linkage are obtained by a reaction with a fluorinated alkanesulfonyl chloride and a primary or a secondary amine. For example, 1-amino-3-dimethylaminopropane reacts with perfluoroalkylethanesulfonyl chloride to give a fluorinated tertiary amine which is quaternized to yield a betaine [180]:

$$C_6F_{13}CH_2CH_2SO_2Cl + H_2N(CH_2)_3N(CH_3)_2 \xrightarrow[2.\ CH_3Cl]{1.\ base}$$

$$C_6F_{13}CH_2CH_2SO_2HN(CH_2)_3N^+(CH_3)_3Cl^-$$

A reaction of perfluoroalkoxybenzenesulfonyl chloride with 1-amino-3-dimethylaminopropane yields a betaine with an aromatic sulfonamido linkage [178]:

$$C_9F_{19}OC_6H_4SO_2Cl + H_2N(CH_2)_3N(CH_3)_2 \xrightarrow{base} C_9F_{19}OC_6H_4SO_2NH(CH_2)_3N(CH_3)_2$$

$$\downarrow ClCH_2COOH$$

$$C_9F_{19}OC_6H_4SO_2NH(CH_2)_3N^+(CH_3)_2CH_2COO^-$$

Oligomerization of tetrafluoroethylene provides intermediates for amphoteric fluorinated surfactants. The tetrafluoroethylene pentamer reacts with phenol to form a phenyl ether [96]. Chlorosulfonation of the phenyl ether with chlorosulfonic acid yields a sulfonyl chloride, which is allowed to react with *N,N*-dimethylpropanediamine:

$$C_{10}F_{19}OC_6H_4SO_2Cl + NH_2CH_2CH_2CH_2N(CH_3)_2 \xrightarrow{base}$$

$$C_{10}F_{19}OC_6H_4SO_2NHCH_2CH_2CH_2N(CH_3)_2$$

A reaction of the resulting tertiary amine with β-propiolactone gives an amphoteric fluorinated surfactant:

$$C_{10}F_{19}OC_6H_4SO_2NHCH_2CH_2CH_2N(CH_3)_2 + \underbrace{CH_2CH_2CO}_{O} \longrightarrow$$

$$C_{10}F_{19}OC_6H_4SO_2NHCH_2CH_2CH_2N^+(CH_3)_2CH_2CH_2COO^-$$

Since β-propiolactone is a suspected carcinogen, sodium chloroacetate (or chloroacetic acid) has become the preferred alkylation agent:

$$C_{10}F_{19}OC_6H_4SO_2NHCH_2CH_2CH_2N(CH_3)_2 + ClCH_2COOH \longrightarrow$$

$$C_{10}F_{19}OC_6H_4SO_2NHCH_2CH_2CH_2N^+(CH_3)_2CH_2COO^-$$

Carboxybetaines with a sulfide linkage have been prepared by a reaction of perfluoroethyl iodide with (2-mercaptoethyl)dimethylammonium chloride and subsequent quaternization with chloroacetic acid [181,182]:

$$R_fCH_2CH_2I + HSCH_2CH_2N(CH_3)_2 \cdot HCl \xrightarrow{NaOH} R_fCH_2CH_2SCH_2CH_2N(CH_3)_2$$

$$\downarrow ClCH_2COOH$$

$$R_fCH_2CH_2SCH_2CH_2N^+(CH_3)_2CH_2COO^-$$

Carboxybetaines with a sulfide link and a hydroxyl group are obtained by reaction of perfluoroalkylethanethiol with epichlorohydrin. The resulting epoxide is allowed to react with sarcosine to give a betaine [183]:

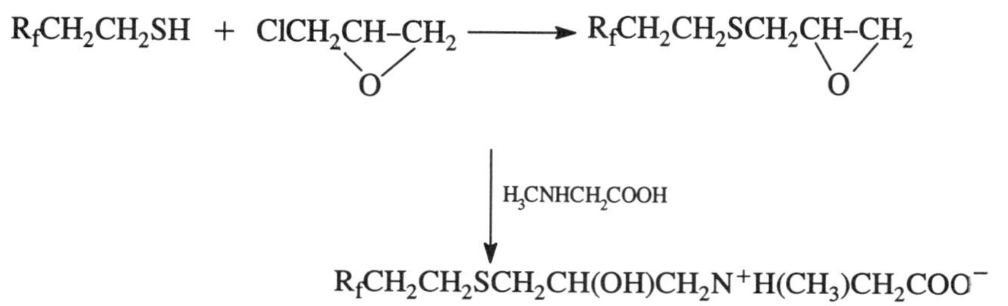

Amphoteric fluorinated surfactants with heterocyclic nitrogen have been prepared by treating a fluorinated ester with *N*-(2-hydroxyethyl)piperazine and subsequently with chloroacetic acid [179]:

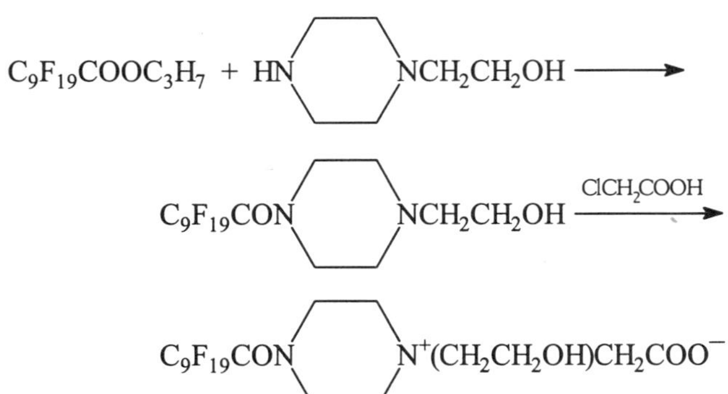

Perfluoroalkyl esters react with cyclic N-aminoquaternary salts to give an amphoteric fluorinated surfactant with a heterocyclic nitrogen [184].

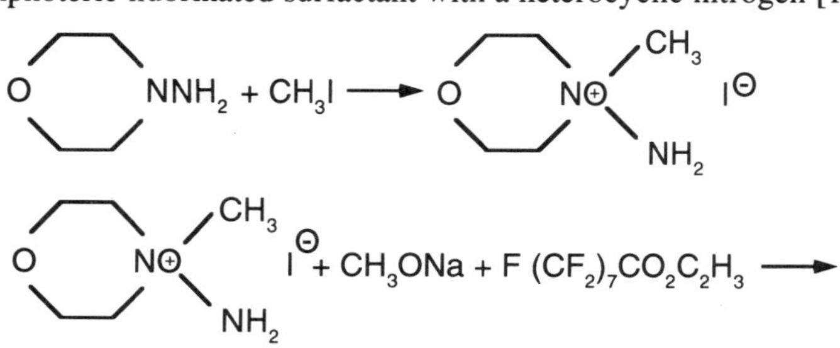

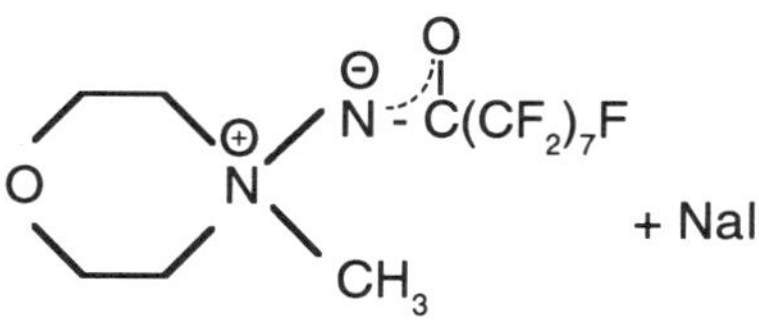

Amphoteric fluorinated surfactants with heterocyclic nitrogen have also been prepared by the Michael addition of 2-perfluoroalkylethanethiol to either a maleic monoamide or monoester [154,185–187]:

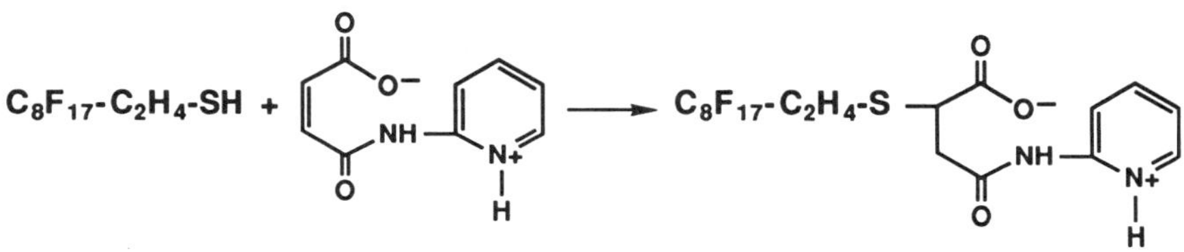

Sulfobetaines [171,188–191] were usually prepared by a reaction of a tertiary amine with propane sultone [188,189,192]. However, propane sultone has been found to be a carcinogen and is no longer used for this purpose. Instead of propane sultone, chlorosulfonic acid or its salt [193] and sodium 3-chloropropanesulfonate [190,191] are used as quaternizing agents to obtain fluorinated sulfobetaines.

Sulfatobetaines contain a sulfuric acid ester group, introduced by sulfating a hydroxyl group or an olefinic bond [193–196]. Sulfur trioxide, with dimethyl sulfate or as a trimethylamine complex, has been used as a sulfating agent to prepare sulfatobetaines. Billenstein and Ehrl [195] treated the compound $CF_3(CF_2)_6CF{=}CHCH_2N(CH_3)CH_2CH_2OH$ with a SO_3-air mixture and with dimethyl sulfate to give the betaine

$$CF_3(CF_2)_6CF{=}CHCH_2N^+(CH_3)_2CH_2CH_2OSO_3^-.$$

Sulfatobetaines $RCH_2CH(OSO_3^-)CH_2N^+(CH_3)_3$, where R = $C_6F_{13}-$, $C_6F_{13}CH_2CH_2S-$, $C_6F_{13}CH_2CH_2S(CH_2)_3O-$, $C_8H_{17}CH_2CH_2S-$, or $C_8H_{17}-$, have been prepared by reacting an epoxide $RCH_2\underset{\diagdown O \diagup}{CH{-}CH_2}$ with a $(CH_3)_3N \cdot SO_3$ complex [197].

Phosphatobetaines have been prepared from a fluorinated alkanol, $R_f(CH_2)_nOH$, where R_f = C_6F_{13} or C_8F_{17} and n = 2 or 11. Riess et al. [198,199] phosphorylated the fluorinated alcohol with phosphorus oxychloride in the presence of triethylamine:

$$R_f(CH_2)_nOH \xrightarrow{POCl_3\ +\ N(CH_3)_3} R_f(CH_2)_nOP(O)Cl_2$$

The reaction product, $R_f(CH_2)_nOP(O)Cl_2$, was subsequently reacted with bromoethanol. Hydrolysis of the phosphoryl chloride group and a reaction of the bromoethyl group with trimethylamine and silver carbonate yielded the perfluoroalkylated phosphotidylcholine:

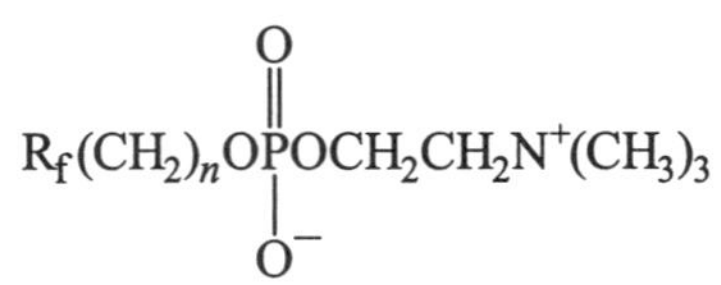

Alternatively, the phosphorylated alcohol, $R_f(CH_2)_nOP(O)Cl_2$, was treated with choline tosylate in pyridine. Subsequent hydrolysis gave the same perfluoroalkylated phosphotidylcholine:

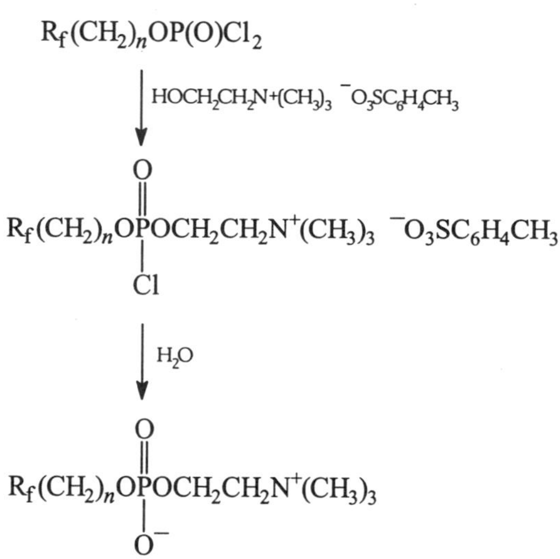

Chlorination of the fluorinated alcohol by phosphorus oxychloride was suppressed by using anhydrous diethyl ether as the solvent [199].

2.9 NONIONIC SURFACTANTS

The method selected for the preparation of a nonionic fluorinated surfactant depends on the hydrophile, which can be either an polyalkyl ether chain or a polyhydroxy group. Since oxyethylation results in a mixture of oligomers, special methods have been devised for the preparation of monodisperse surfactants. A unique group of nonionic surfactants are amphiphiles without a hydrophile. Semifluorinated alkanes with an oleophilic and an oleophobic segment function as nonionic surfactants in oleophilic solvents.

Nonionic Fluorinated Surfactants with a Polyoxyethylene Chain

From Fluorinated Alcohols:

The preparation of nonionic fluorinated surfactants from fluorinated alcohols is straightforward [200–211]. Perfluoroalkylethanol (see Sec. 2.3), obtained from telomer perfluoroalkylethyl iodides, is oxyethylated in the presence of a catalyst, for example BF_3:

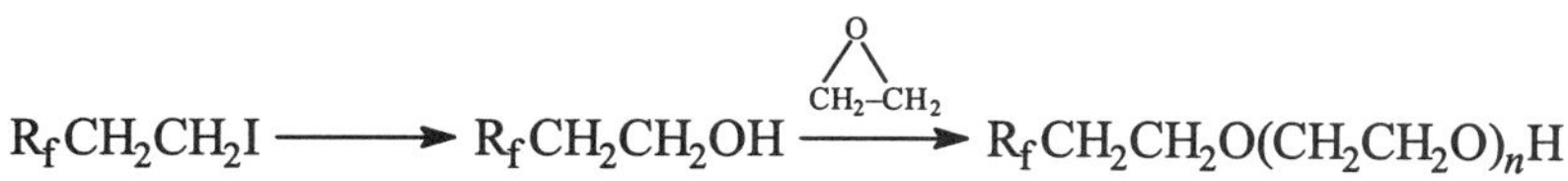

The oxyethylation conditions affect the formation of dioxane, a toxic by-product. Yang et al. [203] have found that a catalyst, consisting of an aluminum alkoxide (average $C_{10.7}$ alkoxide) and hydrogen fluoride, reduces the dioxane concentration to 9000 ppm, compared to 45,000 ppm for the surfactant prepared with BF_3 as the catalyst. Recent improvements of the oxyethylation process have reduced the dioxane content further.

Telomer perfluoroalkylmethanols have been oxyethylated to form nonionic surfactants having the structure $R_fCH_2O(CH_2CH_2O)_nH$ [204,205].

Oxyethylation of the telomeric alcohol $H(CF_2)_mCH_2OH$, where $m = 6, 8, 10$, in the presence of BF_3 yields the surfactant $H(CF_2)_mCH_2O(CH_2CH_2O)_nH$ [206,207]. The surfactant with the hydrophile length $n = 10$ was found to have the best surfactant properties.

The nonionic fluorinated surfactants $CF_3(CF_2)_mCH_2O(CH_2CH_2O)_nH$ and $H(CF_2)_pCH_2O(CH_2CH_2O)_nH$ have been prepared from the corresponding fluorinated alcohols in the presence of a base catalyst [208].

Perfluorooctylethanol reacted with epichlorohydrin gives an ester which, on hydrolysis with HCl, yields the surfactant of the structure $C_8F_{17}CH_2CH_2O[CH_2CH(CH_2OH)O]_nH$ (average $n = 1.7$) [209].

Leempoel et al.[210] and Selve and Castro [211] condensed $H(OCH_2CH_2)_4OH$ and $C_7F_{15}CH_2OH$ to form a nonionic fluorinated surfactant:

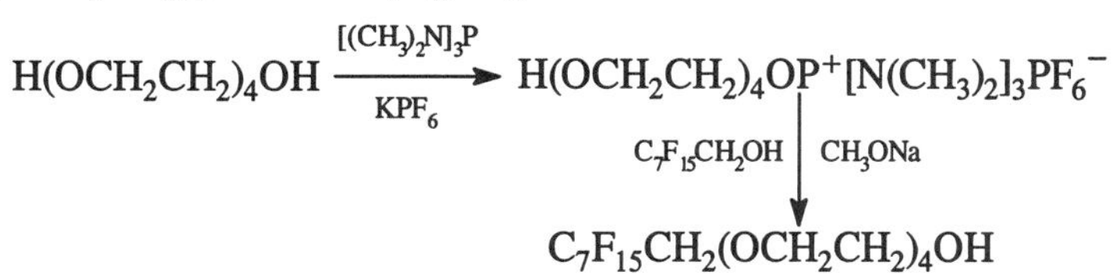

From Acids or Acid Halides:

Nonionic fluorinated surfactants are prepared from acid fluorides, obtained by electrochemical fluorination, usually with an amide as the intermediate. Sulfonyl fluorides are converted with an amine to the amide, which is reacted with ethylene chlorohydrin and oxyethylated:

$$C_8F_{17}SO_2F \xrightarrow{C_2H_5NH_2} C_8F_{17}SO_2NHC_2H_5 \xrightarrow{ClCH_2CH_2OH} C_8F_{17}SO_2N(C_2H_5)CH_2CH_2OH$$

$$\downarrow \overset{O}{CH_2-CH_2}$$

$$C_8F_{17}SO_2N(C_2H_5)CH_2CH_2O(CH_2CH_2O)_nH$$

The perfluorinated carbonyl fluorides, prepared by electrochemical fluorination, are converted to nonionic fluorinated surfactants via the amide by a similar process:

$$C_7F_{15}COF \xrightarrow{C_2H_5NH_2} C_7F_{15}CONHC_2H_5 \xrightarrow{ClCH_2CH_2OH} C_7F_{15}CON(C_2H_5)CH_2CH_2OH$$

$$\downarrow \overset{O}{CH_2-CH_2}$$

$$C_7F_{15}CON(C_2H_5)CH_2CH_2O(CH_2CH_2O)_nH$$

Meussdoerffer and co-workers [131,212] have developed the following process for preparing nonionic fluorinated surfactants:

$$C_8F_{17}SO_2F + 2H_2NCH_3 \longrightarrow C_8F_{17}SO_2NHCH_3 + H_2NCH_3 \cdot HF$$

$$C_4H_9OH + n\overset{O}{CH_2-CH_2} \longrightarrow H(OCH_2CH_2)_nOC_4H_9$$

$$COCl_2 + H(OCH_2CH_2)_nOC_4H_9 \longrightarrow Cl\underset{\overset{\|}{O}}{C}(OCH_2CH_2)_nOC_4H_9 + HCl$$

$$C_8F_{17}SO_2NHCH_3 + Cl\underset{\overset{\|}{O}}{C}(OCH_2CH_2)_nOC_4H_9 \longrightarrow$$

$$C_8F_{17}SO_2N(CH_3)\underset{\overset{\|}{O}}{C}(OCH_2CH_2)_nOC_4H_9 + HCl$$

Alkoxylated perfluoroalkane sulfonamides have been prepared by reacting aliphatic polyamines with less than the stoichiometric amount of a perfluoroalkanesulfonyl fluoride and alkoxylating the reaction product with ethylene oxide or propylene oxide [213]:

$$C_8F_{17}SO_2F + H_2N(CH_2CH_2NH)_3CH_2CH_2NH_2 \xrightarrow{(C_2H_5)_3N}$$

$$C_8F_{17}SO_2NH(CH_2CH_2NH)_3CH_2CH_2NH_2 \xrightarrow{(CH_2CH_2)O}$$

$$C_8F_{17}SO_2NH(CH_2CH_2NH)_3CH_2CH_2NH(OCH_2CH_2)_nH$$

Perfluoroalkanoic acids can be converted to nonionic fluorinated surfactants by a reaction with oxyethylated propylenediamine [93]:

$$C_7F_{15}COOH + H_2N(CH_2)_3N[CH_2CH_2O)_nH]_2 \longrightarrow C_7F_{15}CONH(CH_2)_3N[CH_2CH_2O)_nH]_2$$

Surfactants $R_fCONH(CH_2CH_2O)_{2\text{-}4}H$ have been prepared by condensing a perfluorocarboxylic acid derivative with an oxyethylated amine. A perfluorinated carbonyl halide is reacted with bromoethylamine and the product, $R_fCONHCH_2CH_2Br$, is condensed with $HO(CH_2CH_2O)_{2\text{-}4}H$. The same surfactant can be prepared by condensing the ester, R_fCOOCH_3, with $H_2N(CH_2CH_2O)_2H$ [214].

An esterification reaction has been utilized to introduce perfluoroalkyl groups into commercial nonionic hydrocarbon-type surfactants, Pluronic F68 and Butronic R1. Gangoda et al. [215] esterified Pluronic F68 (block copolymer of ethylene oxide and propylene oxide) with perfluorocarboxylic groups to increase the stability of fluorochemical (perfluorodecalin and perfluorotripropylamine) microemulsions.

Nonionic fluorinated surfactants having the structures
$R_fCH{=}CH(CH_2)_nOCH_2CH_2COO(CH_2CH_2O)_mH$,
$R_f(CH_2)_nOCH_2CH_2COO(CH_2CH_2O)_mH$,
$R_fCH{=}CH(CH_2)_nOCH_2CH_2COO(C_3H_6O)_mH$, or
$R_f(CH_2)_nOCH_2CH_2COO(C_3H_6O)_mH$ can be obtained by alkoxylation of the corresponding carboxylic acids [216].

From Tetrafluoroethylene Oligomers

For the oligomerization of tetrafluoroethylene, see Sec. 2.4, Fig. 2.7.

Nonionic surfactants can be prepared by a reaction of the tetrafluoroethylene pentamer with alcohols or with a phenol [96,97]. The reaction product obtained with phenol is sulfonated to produce a sulfonyl chloride. A reaction of the sulfonyl chloride with with *N*-ethylethanolamine yields an alcohol, which is oxyethylated to produce a nonionic surfactant:

$$C_{10}F_{19}OC_6H_4SO_2Cl \longrightarrow C_{10}F_{19}OC_6H_4SO_2N(C_2H_5)CH_2CH_2OH$$

$$\downarrow \text{ethylene oxide}$$

$$C_{10}F_{19}OC_6H_4SO_2N(C_2H_5)\,(CH_2CH_2O)_nH$$

Nonionic fluorinated surfactants with an aliphatic hydrophobe are produced by a base-catalyzed reaction of the oligomers with oxyethylated alcohols. The ethers formed are converted to nonionic surfactants:

$$C_{10}F_{20} + HO(CH_2CH_2O)_nR \longrightarrow C_{10}F_{19}O(CH_2CH_2O)_nR$$

$$R = \text{H, alkyl, aryl, etc.}$$

If R is H, both ends of the molecule react with the fluorocarbon [217]:

$$C_{10}F_{20} + HO(CH_2CH_2O)_nH \longrightarrow C_{10}F_{19}O(CH_2CH_2O)_nC_{10}F_{19},$$

or

$$C_8F_{16} + HO(CH_2CH_2O)_nH \longrightarrow C_8F_{15}O(CH_2CH_2O)_nC_8F_{15}$$

Unlike the phenyl ethers, the alkyl ethers formed react readily with nucleophilic reagents. The reaction with oxyethylated alcohol yields a nonionic surfactant with two oxyethylene chains,

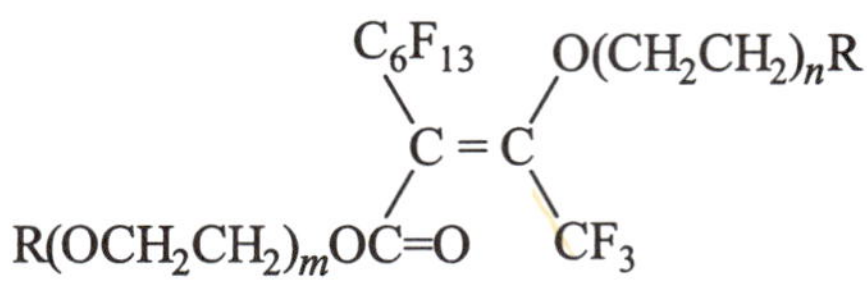

A stable nonionic fluorinated surfactant, of the structure $C_6F_{13}CH{=}C(CF_3)O(CH_2CH_2O)_nR$, is obtained by a reaction of the alkyl ether with an aqueous base which replaces a CF_3- group with hydrogen.

From Fluorinated Thiols

A fluorinated thiol is reacted with monochlorohydrin and the resulting alcohol is oxyethylated to yield a nonionic surfactant [218]. Perfluoroalkylmercaptan, $R_fCH_2CH_2SH$, epichlorohydrin, and Carbowax 350 were reacted to prepare the surfactant $R_fCH_2CH_2SCH_2CH(OH)CH_2O(CH_2CH_2O)_7CH_3$ [219].

From Hexafluoroacetone

Hexafluoroacetone has been converted with potassium fluoride to the alkoxide which is reacted with 1-chloro-6-hexanol:

$$(CF_3)_2C{=}O + KF \rightleftharpoons (CF_3)_2CFOK$$

$$(CF_3)_2CFOK + Cl(CH_2)_6OH \longrightarrow (CF_3)_2FCO(CH_2)_6OH + KCl$$

$$(CF_3)_2FCO(CH_2)_6OH + n\,\overset{O}{CH_2{-}CH_2} \longrightarrow (CF_3)_2FCO(CH_2)_6O(CH_2CH_2O)_nH$$

where n is from 1 to about 20 [220]. Hexafluoroacetone is highly toxic and a teratogen. It is therefore unlikely that hexafluoroacetone is used for the industrial production of fluorinated surfactants.

Monodisperse Nonionic Fluorinated Surfactants

Oxyethylation results in oligomeric mixtures, $R_f(CH_2)_{1\text{ or }2}O(CH_2CH_2O)_nH$, differing in the length, n, of their hydrophilic chain. Monodisperse nonionic surfactants can be prepared by grafting a perfluoroalkyl group to oligo(ethylene glycol) of defined molecular weight [221,222]. One end of the oligomer is protected by alkylation and the other end is activated by chlorination using thionyl chloride. The compound formed is condensed with a fluorinated alkanol and the end cap removed by hydrogenation in the presence of a catalyst.

Monodisperse bipodal nonionic fluorinated surfactants, perfluoroalkyl *N,N*-polyethoxylated amides, have been synthesized as dimethyl ether derivatives, $R_fCH_2C(O)N[(CH_2CH_2O)_nCH_3]_2$, to prevent equilibration with aminoesters [223].

A different method for preparing monodisperse nonionic fluorinated surfactants has been described by Szöny and Cambon [224]. The thiol $R_fCH_2CH_2SH$ is reacted with $Cl(CH_2CH_2O)_nH$, where $n = 1$–4, a known value. Thiolation of the compound, $n = 1$–3, and a second substitution reaction with $Cl(CH_2CH_2O)_nH$ yields the monodisperse nonionic surfactant $R_fCH_2CH_2S(CH_2CH_2O)_nCH_2CH_2S(CH_2CH_2O)_mH$ ($n = 0$–2, $m = 1$–4).

Nonionic Fluorinated Surfactants with a Polyhydroxy Hydrophile

Nonionic fluorinated surfactants with a tris(hydroxymethyl) hydrophilic group have been synthesized by Pavia et al. [225]. A perfluoroalkylthioethanol reacted with tris(hydroxymethyl)amidomethane in the presence of a radical initiator in boiling methanol:

$$CH_2{=}CHCONHC(CH_2OH)_3 + HSCH_2CH_2C_mF_{2m+1} \rightarrow C_mF_{2m+1}CH_2CH_2S(CH_2\underset{\displaystyle CONHC(CH_2OH)_3}{\underset{|}{C}H})_nH$$

Riess and co-workers [226–236] have prepared nonionic fluorinated surfactants with a polyhydroxy hydrophile by perfluoroalkylating natural products such as mono- and disaccharides, pentitols, hexitols, etc. Per-O-acetylglycopyranosyl bromides were reacted with 11-(F-alkyl)-10-undecenols, $R_fCH{=}CH(CH_2)_9OH$, $R_f = C_6F_{13}$ or C_8F_{17}, prepared by reacting the perfluoroalkyl iodide with undecenol using copper (I) chloride and ethanolamine catalysis [228]. The reaction product was deacetylated in a methanol-triethylamine-water mixture to give the fluorinated nonionic surfactant (see Sec. 10.4).

Perfluoroalkylated fatty acid 6-esters of sucrose and α,α-trehalose were prepared by a reaction of sucrose or α,α-trehalose with a perfluoroalkylated acid, $R_f(CH_2)_nCOOH$ ($R_f = C_4F_9$, C_6F_{13}, or C_8F_{17}, $n = 2$, 4, 10) in the presence of triphenylphosphine and diisopropyl azodicarboxylate in *N,N*-dimethylformamide [236]. The surfactants were used to prepare fluorochemical emulsions intended as intravascular oxygen carriers (see Sec. 10.4).

Fluorinated Surfactants Without a Hydrophile

Oleophilic/oleophobic fluorinated surfactants without a hydrophile, designed for use in hydrocarbon systems, are in a structural sense also nonionic fluorinated surfactants (see Sec. 1.7), for example, the semifluorinated alkanes [237–241], block polyethylene-polypropylene glycol ethers prepared with perfluoroalkene trimers [242], surfactants featuring an oligo(hexafluoropropene oxide) chain [243], and carboxamides and sulfonamides derived from *N*-(perfluorooctanesulfonyl)piperazine [244].

Semifluorinated alkanes are prepared by a free radical–initiated addition of perfluoroalkyl iodide to a terminal olefin [138,245,246]. Oil-soluble surfactants, $CF_3CF_2CF_2O[CF(CF_3)CF_2O]_{n-2}CF(CF_3)COR$ (R = phenyl or *p*-tolyl), are obtained by arylating hexafluoropropylene oxide oligomeric acid halides [243]. Carboxamides and sulfonamides are prepared by conventional reactions using *N*-(perfluorooctanesulfonyl)piperazine as the intermediate [244].

REFERENCES

1. G. Siegemund, W. Schwertfeger, A. Feiring, B. Smart, F. Behr, and H. Vogel, Fluorine Compounds, Organic, in "Ullmann's Encyclopedia of Industrial Chemistry," Vol. A11, p. 349, VCH Verlagsgesellschaft, Weinheim, Germany (1988).
2. F. R. Feher, P. W. Foerst, P. H. Liu, D. J. Kalota, and J. S. McConaghy (Monsanto), EP 332601 (1989).
3. K. V. Scherer, Jr., K. Yamanouchi, and T. Ono, J. Fluorine Chem. 50, 47 (1990).
4. (a) H. Liebig and K. Ulm, Chemiker Ztg. 99, 477 (1975). (b) E. Schuierer, Tenside 13, 1 (1976).
5. H. C. Fielding, in "Organofluorine Chemicals and Their Industrial Applications," R. E. Banks, ed., Ellis Horwood, Chichester, England (1979).
6. J. N. Meussdoerffer and H. Niederprün, Chem. Ztg. 104, 45 (1980).
7. J. Aigueperse, P. Mollard, D. Devilliers, M. Chemla, R. Faron, R. Romano, and J. P Cuer, Fluorine Compounds, Inorganic, in "Ullmann's Encyclopedia of Industrial Chemistry," Vol. A11, p. 307, VCH Verlagsgesellschaft, Weinheim, Germany (1988).
8. J. H. Simons, Chem. Rev. 8, 213 (1931).
9. J. H. Simons, Ind. Eng. Chem. 32, 178 (1940).
10. J. H. Simons, in "Fluorine Chemistry," Vol. I, p. 225, J. H. Simons, ed., Academic Press, New York (1950).
11. K. L. Motz, F. J. Freire, and E. G. Edwards (Du Pont), U.S. Patent 5,108,559 (1992).
12. W. K. R. Musgrave, in "Advances in Fluorine Chemistry," Vol. I, M. Stacey, J. C. Tatlow, and A. G. Sharpe, eds., Academic Press, London (1960).
13. R. E. Parsons (Du Pont), U.S. Patent 3,234,294 (1962).
14. R. E. Parsons (Du Pont), U.S. Patent 3,132,185 (1964). CA 61, 1755.

15. M. Hauptschein (Pennwalt), (a) Brit. Patent 996484 (1961). (b) Brit. Patent 1,067,063 (1963).
16. Daikin-Kogyo, (a) Brit. Patent 1,102,903 (1964). (b) Brit. Patent 1132141 (1965).
17. T. Moeller, "Inorganic Chemistry," John Wiley & Sons, New York (1952).
18. G. H. Cady, in "Fluorine Chemistry," Vol. I, p. 293, J. H. Simons, ed., Academic Press, New York (1950).
19. M. Jaccaud, R. Faron, D. Devilliers, and R. Romano, Fluorine, in "Ullmann's Encyclopedia of Industrial Chemistry," Vol. A11, p. 293. VCH Verlagsgesellschaft, Weinheim, Germany (1988).
20. O. Scherer and E. Futterer (Hoechst), Ger. Offen. DE 1805457 (1970). CA 73, 34765.
21. H. Millauer (Hoechst), Ger. Offen. DE 2139964 (1973). CA 78, 123993.
22. H. Millauer (Hoechst), Ger. Offen. DE 2154512 (1973). CA 79, 18052.
23. H. Millauer, Angew. Chem. 85, 988 (1973).
24. D. Bedford and K. Baum, J. Org. Chem. 45, 347 (1980).
25. Montecatini Edison, Brit. Patent 1210324 (1970).
26. H. J. Semmler and B. Felix (Hoechst), Ger. Offen. DE 2610148 (1972). CA 87, 183975.
27. J. H. Simons and co-workers, J. Electrochem. Soc. 95, 47 (1949).
28. J. H. Simons, ed., "Fluorine Chemistry," Vol. I, p. 414, Academic Press, New York (1950).
29. J. H. Simons and T. J. Brice, in "Fluorine Chemistry," J. H. Simons, ed., Vol. II, p. 340, Academic Press, New York (1954).
30. J. H. Simons (3M), U.S. Patent 2,519,983 (1950) CA 45, 510.
31. J. Burdon and J. C. Tatlow, in "Advances in Fluorine Chemistry," Vol. I., M. Stacey, J. C. Tatlow, and A. G. Sharpe, p. 129, Academic Press, London (1960).
32. S. Nagase, in "Fluorine Chemistry Reviews," P. Tarrant, ed., Vol. 1(1), p. 77, Marcel Dekker, New York (1967).
33. A. J. Rudge, in "Industrial Electrochemical Processes," A. T. Kuhn, ed., p. 71, Elsevier, Amsterdam (1971).
34. T. Abe and S. Nagase, in "Preparation, Properties, and Industrial Applications of Organofluorine Compounds," R. E. Banks, ed., Ellis Horwood, Chichester, England (1982).
35. F. G. Drakesmith and D. A. Hughes, J. Appl. Electrochem. 9, 685 (1979).
36. J. K. Wolfe, U.S. Patent 2,601,014 (1952).
37. E. Forche, in "Methoden der Organischen Chemie," Vol. V/3, p. 38, Georg Thieme Verlag, Stuttgart (1962).
38. J. Burdon, I. W. Parsons, and J. C. Tatlow, Tetrahedron 28, 43 (1972).
39. F. G. Drakesmith and D. A. Hughes, J. Appl. Electrochem. 6, 23 (1976).
40. J. S. Clarke and A. T. Kuhn, J. Electroanal. Chem. 85, 299 (1977).
41. D. Lines and H. Sutcliffe, J. Fluorine Chem. 17, 423 (1981).
42. F. Beck, "Organische Elektrochemie," p. 257, Verlag Chemie, Weinheim, Germany (1974).
43. E. A. Kauck and A. R. Diesslin, Ind. Eng. Chem. 43, 2332 (1951).
44. A. R. Diesslin, E. A. Kauck, and J. H. Simons (3M), U.S. Patent 2,567,011 (1951). CA 46, 1375.

45. A. R. Diesslin, E. A. Kauck, and J. H. Simons (3M), U.S. Patent 2,593,737 (1952). CA 46, 6016.
46. R. A. Guenthner (3M), U.S. Patent 2,606,206 (1952).
47. J. O. Hendrichs, Ind. Eng. Chem. 45, 99 (1953).
48. T. J. Brice, W. H. Pearlson, and H. M. Scholberg (3M), U.S. Patent 2,713,593 (1955).
49. H. M. Scholberg and H. G. Bryce (3M), U.S. Patent 2,717,871 (1955).
50. H. G. Bryce, in "Fluorine Chemistry," Vol. V, p. 370–440, J. H. Simons, ed., Academic Press, New York (1950).
51. T. Gramstad and R. N. Haszeldine, J. Chem. Soc. 173 (1956).
52. T. Gramstad and R. N. Haszeldine, J. Chem. Soc. 2640 (1957).
53. J. Burdon, I. Farazmand, M. Stacey, and J. C. Tatlow, J. Chem. Soc. 2574 (1957).
54. T. J. Brice and P. W. Trott, U.S. Patent 2,732,398 (to 3M Co.) (1956). CA 50, 13982.
55. E. Forche, in "Houben-Weyl, Methoden der Organischen Chemie," E. Müller, ed., Vol. V/3, p. 53, George Thieme, Verlag, Stuttgart (1962).
56. W. E. Hanford and R. M. Joyce, Jr. (Du Pont), U.S. Patent 2,440,800 (1942).
57. W. E. Hanford and J. Harmon (Du Pont), U.S. Patent 2,440,801 (1942).
58. W. E. Hanford and J. R. Roland (Du Pont), U.S. Patent 2,457,229 (1946).
59. R. M. Joyce, W. E. Hanford, and J. Harmon, J. Am. Chem. Soc. 70, 2529 (1948).
60. C. M. Starks, "Free Radical Telomerization," Academic Press, Orlando, FL (1974).
61. B. Gordon III and J. E. Loftus, in "Encyclopedia of Polymer Science and Engineering," Vol. 16, p. 533, John Wiley & Sons, New York (1989).
62. R. N. Haszeldine, J. Chem. Soc. 2856 (1949).
63. R. N. Haszeldine, J. Chem. Soc. 3037 (1950).
64. R. N. Haszeldine, J. Chem. Soc. 3761 (1953).
65. N. O. Brace (Du Pont), U.S. Patent 3,145,222 (1961).
66. W. A. Blanchard and J. C. Rhode (Du Pont), U.S. Patent 3,226,449, (1965) CA 64, 8031.
67. M. Hauptschein (Pennwalt), Brit. Patent 996484 (1965).
68. Ciba-Geigy, Brit. Patent 1189576 (1970).
69. Ciba-Geigy, Ger. Offen. DE 2001140 (1970).
70. W. Rudolph, J. Massonne, H. H. Jaeger, and H. Gress (Kali-Chemie), Ger. Offen. DE 2034472 (1972). CA 77, 6148.
71. W. Rudolph and S. Massonne (Kali-Chemie), Ger. Offen. DE 2132036 (1973). CA 78, 123990.
72. S. Rebsdat, E. Schuierer, and H. Hahn (Hoechst), Ger. Offen. DE 1915395 (1970). CA 73,76644.
73. B. Felix (Hoechst), Ger. Offen. DE 2164567 (1973). CA 79, 78077.
74. B. Felix and E. Weiss (Hoechst), Ger. Offen. DE 2542496 (1977). CA 86, 190648.
75. Kali-Chemie AG, Ger. Offen. DE 2162368 (1973).

76. N. O. Brace (Du Pont), U.S. Patent 3,016,407 (1962). CA 57, 11018e.
77. Ciba-Geigy, Brit. Patent 1210730 (1970).
78. Ciba-Geigy, Brit. Patent 1319898 (1973).
79. Montecatini Edison, Brit. Patent 1209982 (1970).
80. R. I. Day (Du Pont), U.S. Patent 3,283,012 (1966). CA 66, 18507.
81. H. Millauer (Hoechst), Ger. Offen. DE 2318677 (1974). CA 82, 72528.
82. U. Schwenk, I. Konig, and H. Streitberger (Hoechst), Ger. Offen. DE 2834795 (1980). CA 93, 7656.
83. T. Hayashi and M. Matsuo (Asahi), Ger. Offen. 2318941 (1973), U.S. Patent 4,001,309 (1977). CA 80, 14564.
84. S. Rebsdat, E. Schuierer, K. Ulm, and I. Wimmer (Hoechst), Ger. Offen. DE 2028459 (1971). CA 76, 58963.
85. G. Bargigia, C. Tonelli, and V. Tortelli (Ausimont), U.S. Patent 4,748,282, EP 247614 (1987). CA 109, 189819.
86. Hoechst, Ger. Offen. DE 3016571 (1981)
87. Hoechst, Ger. Offen. DE 3035641 (1982).
88. R. M. Joyce, Jr. (Du Pont), U.S. Patent 2,559,628 (1951). CA 46, 3063a.
89. K. L. Berry (Du Pont), U.S. Patent 2,559,629 (1951).
90. N. O. Brace (Du Pont), U.S. Patent 3,172,910 (1965). CA 63, 503c.
91. Hoechst, Ger. Offen. DE 2013103 (1971).
92. H. Stach (Hoechst), Ger. Offen. DE 2059257 (1971). CA 76, 13821.
93. E. Schuierer, Tenside 13, 1 (1976).
94. A. Lantz (Atochem), Ger. Offen. DE 3800392 (1988). CA 110, 57118.
95. W. R. Deem (ICI), Brit. Patent 1302350 (1973). CA 78, 111926.
96. J. Hutchinson, Fette Seifen Anstrichm. 76, 158 (1974).
97. R. Jones, Fette Seifen Anstrichm. 77, 455 (1975).
98. E. Moore, A. Milian, and E. Eleuterio (Du Pont), U.S. Patent 3,250,808 (1966).
99. E. Moore (Du Pont), U.S. Patent 3,322,826 (1967).
100. L. A. Shits, L. V. Dikhtievskaya, S. P. Krukovskii, L. V. Cherendichenko, and V. A. Ponomarenko, Kolloidn. Zh. 38, 1130 (1976).
101. K. C. Eapen, C. Tamborski, and T. Psarras, J. Fluorine Chem. 14, 243 (1979).
102. N. Ishikawa and M. Sasabe, J. Fluorine Chem. 25, 241 (1984).
103. A. G. Pittman, D. L. Sharp, and B. A. Ludwig, J. Polym. Sci. Pt. A-1, 6, 1729 (1968).
104. A. G. Pittman, D. L. Sharp, and R. E. Lundin, J. Polym. Sci., Pt. A-1, 4, 2637 (1966).
105. H. Kokelenberg and R. Pollet, Tenside 22, 1 (1985).
106. Dainippon Ink, Jpn. Kokai Tokkyo Koho, JP 59046252 (1984). CA 101, 9130d.
107. Neos Co. Jpn. Kokai Tokkyo Koho, JP 57,164,199 (1982).
108. G. V. D. Tiers, U.S. Patent 2,951,051 (1960).
109. K. v. Werner and A. Gisser (Hoechst), Eur. Patent Appl. EP49671; U.S. Patent 4,400,325 (1983). CA 97, 162372.
110. G. Billentefaniak (Hoechst), Ger. Offen. DE 2353392 (1975). CA 83, 58140.
111. M. Takahashi, H. Shuyama, and Y. Tsutsumi (Toyo Soda Mfg.), Jpn. Kokai Tokkyo Koho, JP 63010742 (1988).

112. H. Millauer (Hoechst), Ger. Offen. DE 1926942 (1970). CA 74, 41928.
113. M. Hauptschein and M. Braid, J. Am. Chem. Soc. 83, 2500 (1961).
114. M. Hauptschein and M. Braid, J. Am. Chem. Soc. 83, 2505 (1961).
115. N. O. Brace, J. Org. Chem. 27, 3027 (1962).
116. N. O. Brace, J. Org. Chem 27, 4491 (1962).
117. Hoechst, Ger. Offen. 1916669 (1970).
118. N. I. Galkina and D. Y. Kasaeva, USSR Patent 314748 (1971). CA 76, 13834.
119. H. Bathelt (Hoechst), Ger. Offen. 2558728 (1977). CA 87, 117597.
120. H. Bathelt (Hoechst), Ger. Offen. 2558711 (1977). CAS 87, 117598.
121. B. Boutevin, Y. Pietrasanta, and A. Sideris, C. R. Hebd. Seances Acad. Sci. Ser. C 281(11) 405 (1975).
122. Sagami Chem. Res. Ctr., Jpn. Kokai Tokkyo Koho, JP 57075046 (1982). CA 97, 181975.
123. T. Hayashi and Y. Osai (Asahi Glass), Jpn. Kokai Tokkyo Koho, JP 54044610 (1979). CA 91, 76024k.
124. S. Szöni, R. Vandamme, and A. Cambon, J. Fluorine Chem. 30, 37 (1985).
125. I. Hisamoto, C. Maeda, and M. Hirai (Daikin Kogyo), EP 83077 (1983). CA 99, 140395b.
126. D. Sianesi, A. Pasetti, R. Fontanelli, G. C. Bernardi, and G. Caporiccio, Chim. Ind. 55, 208 (1973).
127. R. J. Lagow, T. R. Bierschenk, T. J. Juhlke, and H. Kawa, in "Synthetic Fluorine Chemistry," G. A. Olah, R. D. Chambers, and G. K. Surya Prakash, eds., John Wiley & Sons, New York (1992).
128. K. Ogino, H. Murakami, N. Ishikawa, and M. Sasabe, Yukagaku 32(2), 96 (1983). CA 98, 145452.
129. G. Caporiccio, F. Burzio, G. Carniselli, and V. Biancardi, J. Colloid Interface Sci. 98, 202 (1984).
130. H. Ukihashi, T. Hayashi, and Y. Takasaki (Asahi Glass), Brit. Patent 1473807 (1977). CA 87, 133897 (1977).
131. H. G. Klein, J. N. Meussdoerffer, H. Niederprüm, Metalloberfläche 29, 559 (1975).
132. H. G. Klein, J. N. Meussdoerffer, H. Niederprüm, and M. Wechsberg, Tenside 15, 2 (1978).
133. V. Beyl, H. Niederprüm, and P. Voss, Liebigs Ann. Chem. 731, 58 (1970).
134. B. Dreher, DX. Prescher, W. Dmowski, and R. Wozniacki, Ger. (East) DD 239788 (1986). CA 46, 215901.
135. Neos Co., Jpn. Kokai Tokkyo Koho, JP 57059843 (1982).
136. R. A. Falk (Ciba-Geigy), U.S. Patent 4,577,036 (1986). CA 105, 42623.
137. R. H. Goldbaum and W. R. Remington (Du Pont), U.S. Patent 4,478,809 (1988).
138. G. A. Bargigia, G. Caporiccio, and M. Pianca. J. Fluorine Chem. 19, 403 (1982).
139. C. Kimura, K. Kashiwaya, and M. Kobayashi, J. Am. Oil Chem. Soc. 61, 105 (1984).
140. S. Billenstein, H. Brecht, and H. Hoffmann (Hoechst), Ger. Offen. DE 2329660 (1975). CA 82, 124798.
141. N. Yoshino, N. Komine, J. Suzuki, Y. Arima, and H. Hirai, Bull. Chem. Soc. Jpn. 64, 3262 (1991).

142. K. L. Berry and J. A. Bittles (Du Pont), U.S. Patent 2,559,751 (1951). CA 46, 3063.
143. A. Greiner, B. Krüger, and Margit Herbst, Z. Chem. 18, 383 (1978).
144. A. J. Szur (Diamond Shamrock), U.S. Patent 4,140,709. (1979) CA 90, 206258.
145. Y. Nakamura, H. Umemura, and M. Yonekura (Asahi Denka Kogyo), Jpn. Kokai, JP 50121179 (1975).
146. K. Waldmann (Hoechst), Ger. Offen. DE 2115140 (1972). CA 78, 59738.
147. K. Waldmann (Hoechst), Ger. Offen. DE 2115139 (1972). CA 78, 59737.
148. N. O. Brace and A. K. Mackenzie (Du Pont), U.S. Patent 3,083,224 (1963). CA 59, 5023f
149. Asahi Glass, Jpn. Kokai Tokkyo Koho, JP 60064990 (1985).
150. R. F. Heine (3M), Fr. Patent 1317427 (1963). CA 59, 9799. U.S. Patent 3,094,547 (1963).
151. C. Kimura, K. Murai, K. Kashiwaya, T. Kanno, and T. Nishiyama, Yukagaku 31, 464 (1982). CA 97, 129468.
152. H. Stach and D. Hoffmann (Hoechst), Ger. Offen. DE 2013104 (1971). CA 76, 13820.
153. T. W. Cooke, E. K. Kleiner, and R.E. A. Dear (Ciba-Geigy), Eur. Patent Appl. EP 2004 (1979). CA 91, 194960.
154. B. R. Bailey III, in "Cationic Surfactants. Organic Chemistry," J. M. Richmond, ed., Surfactant Science Series, Vol. 34, Marcel Dekker, New York (1990).
155. A. R. Katritzky, T. L. Davis, G. W. Rewcastle, G. O. Rubel, and M. T. Pike, Langmuir 4, 732 (1988).
156. P. Bartlett (Du Pont), U.S. Patent 3,555,089 (1971). CA 74, 140927k.
157. K. Matsui, Y. Kikuchi, K. Sugimoto and K. Suzuki, Jpn. Kokai Tokkyo Koho, JP 61207362 (1986). CA 106, 49601.
158. W. Dmowski, H. Plenkiewicz, and J. Porwisiak, J. Fluorine Chem. 41, 191 (1988).
159. H. Plenkiewicz and W. Dmowski, J. Fluorine Chem. 45, 389 (1989).
160. D. Prescher, J. Schulze, L. Richter, W. Dmowski, and H. Plenkiewicz, Tenside 29, 337 (1992).
161. J. D. Sauer, in "Cationic Surfactants. Organic Chemistry," J.M. Richmond, ed., Surfactant Science Series, Vol.34, Marcel Dekker, New York (1990).
162. R. Hager, W. Bonner, G. J. Walter (Pennwalt), Ger. Offen. DE 2219642 (1972).
163. A. J. Szur (Diamond Shamrock), U.S. Patent 4,049,668 (1977).
164. R. A. Clarke and D.P. Cords (Du Pont), Eur. Patent Appl. EP 149347 (1985).
165. B. R. Bluestein and C. L. Hilton, eds., "Amphoteric Surfactants," Surfactant Science Series, Vol. 12, Marcel Dekker, New York (1982).
166. R. Bertochio and L. Foulletier (Ugine Kuhlmann), Ger. Offen. DE 2119302 (1971). CA 76, 24720.
167. R. Bertochio and L. Foulletier (Ugine Kuhlmann), Ger. Offen. DE 2120868 (1971). CA 76, 33810m.
168. H. Stach, D. Hoffmann, and H. Brecht (Hoechst), Ger. Offen. DE 2127232 (1972). CA 78, 138225u.
169. T. Hayashi, Y. Otoshi, and T. Seki (Asahi Glass), Jpn. Kokai Tokkyo Koho, JP 53045683 (1978). CA 89, 113083k.

170. T. Hayashi, Y. Otoshi, and T. Seki (Asahi Glass), Jpn. Kokai Tokkyo Koho, JP 53045684 (1978). CA 89, 113084m.
171. P. L. Barlett (Du Pont), U.S. Patent 3,839,425 (1974). CA 82, 88056.
172. D. P. Cords (Du Pont), Ger. Offen. DE 2357916 (1974). CA 81, 123415.
173. T. Hayashi and Y. Otoshi (Asahi Glass), Jpn. Kokai Tokkyo Koho, JP 54062990 (1979). CA 91, 125254w.
174. T. Hayashi and Y. Otoshi (Asahi Glass), Jpn. Kokai Tokkyo Koho, JP 54147184 (1979). CA 92, 148920w.
175. T. Hayashi and Y. Otoshi (Asahi Glass), Jpn. Kokai Tokkyo Koho, JP 55007818 (1980) CA 93, 28184.
176. Du Pont, Jpn. Kokai Tokkyo Koho, JP 56122336 (1981). CA 96, 8518.
177. J. W. Richter (Du Pont), U.S. Patent 4,283,533 (1981). CA 95, 171468.
178. Neos, Jpn. Kokai Tokkyo Koho, JP 57164198 (1982). CA 98, 145460h.
179. T. Hayashi and Y. Otoshi (Asahi Glass), Jpn. Kokai Tokkyo Koho, JP 54062990 (1979). CA 91, 125254w.
180. H. Stach and D. Hoffmann (Hoechst), Ger. Offen. DE 2013104 (1971). CA 76, 13820.
181. D. P. Cords (Du Pont), U.S. Patent 2,357,916 (1974). CA 81, 123415.
182. S. S. Toukan and M. Hauptschein (Pennwalt), U.S. Patent 4,126,633 (1978). CA 90, 103406.
183. R. A. Falk (Ciba-Geigy), Eur. Patent Appl. EP 115251 (1984).
184. W. J. Middleton (Du Pont), U.S. Patent 3,963,776 (1976).
185. B. R. Bailey, in "Amphoteric Surfactants," B. R. Bluestein and C. L. Hilton, eds., Surfactant Science Series, Vol. 12, Marcel Dekker, New York (1982).
186. K. F. Mueller (Ciba-Geigy), U.S. Patent 4,069,244 (1978).
187. K. F. Mueller (Ciba-Geigy), U.S. Patent 4,242,516 (1980).
188. R. Bertochio, L. Foulletier, and A. Lantz (Ugine Kuhlmann), Eur. Patent Appl. EP 17568 (1980). CA 94, 49130.
189. Kawamura Phys. Chem. Res. Inst., Jpn. Kokai Tokkyo Koho, JP 56108747 (1981). CA 96, 103650.
190. Dainippon Ink, Jpn. Kokai Tokkyo Koho, JP 58179300 (1983). CA 100, 123184.
191. Dainippon Ink, Jpn. Kokai Tokkyo Koho, JP 59048457 (1984). CA 101, 56875.
192. W. Ehrl and S. Jung (Hoechst), Eur. Patent Appl. EP 57925 (1982). CA 98, 88811.
193. Dainippon Ink, Jpn. Kokai Tokkyo Koho, JP 59051252 (1984). CA 101, 90435.
194. H. Bathelt (Hoechst), Ger. Offen. DE 2749330 (1979). CA 91, 76023.
195. S. Billenstein and W. Ehrl (Hoechst), Ger. Offen. DE 2749329 (1979). CA 91, 56342.
196. A. R. Falk (Ciba-Geigy), U.S. Patent 4,435,330 (1984). CA 101, 9128.
197. Dainippon Ink, Jpn. Kokai Tokkyo Koho, JP 59053448 (1948). CA 101, 72426.
198. J. Riess, F. Jeanneaux, M. P. Krafft, C. Santaella, and P. Vierling, Int. Patent Appl. WO 90/15807 (1990).
199. M. P. Krafft, J. P. Rolland, P. Vierling, and J. G. Riess, New J. Chem. 14, 869 (1990).
200. A. R. Naik (Unilever), Ger. Offen. DE 2261681 (1973)
201. Neos, Jpn. Kokai Tokkyo Koho, JP 5925359 (1984). CA 101, 54698.

202. S. Rebsdat, E. Schuierer, K. Ulm, and I. Wimmer (Hoechst), Ger. Offen. DE 2028459 (1971). CA 76, 58963.
203. K. Yang, C. M. Starks, and O. C. Kerfoot (Conoco), U.S. Patent 4,490,561 (1984).
204. R. Tatematsu, S. Ozaki, S. Nagase, and K. Inukai, Yukagaku 26, 367 (1977).
205. R. Tatematsu, S. Ozaki, S. Nagase, and K. Inukai, Yukagaku 29, 23 (1980).
206. A. Greiner and W. Gerhardt, Ger. (East) DD111522 (1975). CA 84, 123793c.
207. A. Greiner and M. Herbst, Z. Chem. 16, 229 (1976).
208. R. Tatematsu, S. Ozaki, S. Nagase and K. Inukai, Yukagaku 25, 287 (1976). CA 85, 93732.
209. F. Wehowsky (Hoechst), Ger. Offen. DE 3611302 (1987). CA 108, 188957.
210. P. Leempoel, C. Selve, G. Mathis, B. Castro, and J. J. Delpuech, Eur. Patent Appl. EP 51527 (1982). CA 97, 12704.
211. C. Selve and B. Castro, Tetrahedron 39, 1313 (1983).
212. J. N. Meussdoerffer, H. Niederprüm, and M. Dahm, Ger Offen. 2238740 (1974).
213. G. Reitz and G. Boehmke (Bayer), Ger Offen. DE 2639473 (1978).
214. J. Afzai, B. M. Fung, and E. A. O'Rear, J. Fluorine Chem. 34, 385 (1987).
215. M. Gangoda, B. M Fung, and E. A. O'Rear, J. Colloid Interface Sci. 116, 230 (1987).
216. N. O. Brace (Du Pont), U.S. Patent 3,231,604 (1966). CA 64, 9963a.
217. L. E. Houghton (ICI), Ger. Offen. DE 2215388 (1972).
218. J. P. Lampin, A. Cambon, F. Szönyi, J. J. Delpuech, G. Serratrice, G. Thiollet, and L. Lafosse, Eur. Patent Appl. EP 165853 (1985). CA 105, 155095.
219. T. W. Cooke (Ciba-Geigy), Eur. Patent Appl. EP 10523 (1980). CA 93, 97302.
220. A. J. Szur (Diamond Shamrock), U.S. Patent 3,980,715 (1976). CA 85, 194514.
221. G. Mathis and J. J. Delpuech, Fr. 8022875 (1980).
222. T. Gartiser, C. Selve, L. Mansuy, A. Robert, and J. J. Delpuech, J. Chem. Res. (S) 292 (1984).
223. C. Selve, E. M. Moumni, and J. J. Delpuech, J. Chem. Soc. Chem. Commun. 1437 (1987).
224. F. Szönyi and A. Cambon, J. Fluorine Chem. 36, 195 (1987).
225. A. A. Pavia, B. Pucci, J. G. Riess, and L. Zarif, Bioorg. Med. Chem. Lett. 1(2), 103 (1991).
226. L. Zarif, A. Manfredi, C. Varescon, M. Le Blanc, and J. G. Riess, J. Am. Oil Chem. Soc. 66, 1515 (1989).
227. L. Zarif, J. Greiner, S. Pace, and J. G. Riess, J. Med. Chem. 33, 1262 (1990).
228. A. Milius, J. Greiner, and J. G. Riess, New J. Chem. 15, 337 (1991).
229. A. Milius, J. Greiner, and J. G. Riess, Colloids Surf. 63, 281 (1992).
230. J. G. Riess, C. Arlen, J. Greiner, M. Le Blanc, A. Manfredi, S. Pace, C. Varescon, and L. Zarif, Biomater. Artif. Cells Artif. Organs. 16, 421 (1988).
231. J. Greiner, A. Manfredi, and J. G. Riess, New J. Chem. 13, 247 (1989).
232. A. Manfredi, S. Abouhilale, J. Greiner, and J. G. Riess, Bull. Soc. Chim. Fr. 872 (1989).
233. L. Zarif, J. Greiner, and J. G. Riess, J. Fluorine Chem. 44, 73 (1989).
234. C. Varescon, A. Manfredi, M. Le Blanc, and J. G. Riess, J. Colloid Interface Sci. 137, 373 (1990).

235. S. J. Abouhilale, J. Greiner, and J. G. Riess, Carbohydr. Res. 212, 55 (1991).
236. S. Abouhilale, J. Greiner, and J. G. Riess, J. Am. Oil Chem. Soc. 69, 1 (1992).
237. R. J. Twieg and J. F. Rabolt, J. Polym. Sci. Polym. Lett. Ed. 21, 901 (1983).
238. J. F. Rabolt, T. P. Russell, and R. J. Twieg, Macromolecules 17, 2786 (1984).
239. R. J. Twieg, T. P. Russell, R. L. Siemens, and J. F. Rabolt, Macromolecules 18, 1361 (1985).
240. T. P. Russell, J. F. Rabolt, R. J. Twieg, R. L. Siemens, and B. L. Farmer, Macromolecules 19, 1135 (1986).
241. G. L. Gaines, Langmuir 7, 3054 (1991).
242. Neos Co., Jpn. Kokai Tokkyo Koho, JP 5952520 (1984). CA 101, 40225.
243. N. Ishikawa and M. Sasabe, J. Fluorine Chem. 25, 241 (1984).
244. A. R. Katritzky, T. L. Davis, G. W. Rewcastle, G. O. Rubel, and M. T. Pike, Langmuir 4, 732 (1988).
245. N. O. Brace, J. Org. Chem. 38, 3167 (1973).
246. N. O. Brace, J. Org. Chem. 44, 212 (1979).

3
Physical and Chemical Properties

3.1 CHEMICAL PROPERTIES

Perfluorinated surfactants are remarkably stable. Their outstanding thermal and chemical stability permits applications under conditions which would be too severe for conventional hydrocarbon-based surfactants. The very strong C–F bond is stable to acids, alkali, oxidation, and reduction, even at relatively high temperatures.

The unusual properties of fluorosurfactants arise from the unique properties of elemental fluorine [1]:

High oxidation potential $F_2 + 2e^- \rightarrow 2F^-$, E_0 2.65 volts (Table 3.1)
High ionization energy $F \rightarrow F^+ + e^-$ (Table 3.2)
High electron affinity $F + e^- \rightarrow F^-$ (Table 3.2)
High electronegativity of covalently bonded fluorine (Table 3.2). Fluorine is the most electronegative element.
Fluorine is very difficult to polarize [1].

The low dissociation energy of fluorine ($F_2 \rightarrow 2F$) (Table 3.3) provides a sufficient number of fluorine atoms for a reaction to occur. This is probably the main reason for the high reactivity of elemental fluorine [1].

The unusual chemical properties of fluorine as a substituent in organic compounds have been attributed to (1) the high electronegativity of fluorine, (2) the three nonbonding electron pairs on fluorine, and (3) the excellent match between the 2*s* and 2*p* orbitals of fluorine and the corresponding orbitals of other second period elements [3]. Fluorine can therefore form very strong covalent bonds with carbon and hydrogen. The carbon-fluorine bond is among

TABLE 3.1 Standard Electrode Potentials

Electrode reaction	E_o (V)
$Li^+ + e^- = Li$	−3.024
$Na^+ + e^- = Na$	−2.714
$2H^+ + 2e^- = H_2$	0.000
$I_2 + 2e^- = 2I^-$	0.536
$Br_2 + 2e^- = 2Br^-$	1.065
$Cl_2 + 2e^- = 2Cl^-$	1.358
$O_2 + 2H^+ + 2e^- = 2H_2O$	2.07
$F_2 + 2e^- = 2F^-$	2.65

Source: Ref. 33.

TABLE 3.2 Selected Ionization Energies, Electron Affinities, and Electronegativities

Element	Ionization energy[a]	Electron affinity[b]	Electronegativity[c]
F	401.5	83.5	4.0
Cl	300	87.3	3.0
Br	272.9	82.0	2.8
I	242.2	75.7	2.5
O	313.8	—	3.5
H	315.0	0	2.1
Li	125.8	0	1.0

[a] Values in kcal/mol for the first ionization energy of the reaction $X \rightarrow X^+ + e^-$ (Ref. 2, p. 57).
[b] Values in kcal for $X + e^- \rightarrow X^-$.
[c] Pauling scale (Ref. 2, p. 88).
Source: Ref. 2. Reproduced by permission of Cornell University Press.

TABLE 3.3 Dissociation Energies of Elements ($X_2 \rightarrow 2X$)

Element	Dissociation energy (kcal)
F_2	37
Cl_2	58
Br_2	46
N_2	225
O_2	118

Source: Ref. 1 and 9.

the strongest of known covalent bonds [4]. The bond strength increases further with increasing substitution at a carbon. The heat of formation of the carbon-fluorine bond increases in the order CH_3F 448 kJ/mol, CH_2F_2 459 kJ/mol, CHF_3 480 kJ/mol, and CF_4 486 kJ/mol.

The stability of fluorinated surfactants results from a strong C−F bond and effective shielding of carbon by fluorine atoms. The atomic radius of covalently bonded fluorine is only 0.72 Å. Because of their small size, fluorine atoms can shield a perfluorinated carbon atom without steric stress (Fig. 3.1.) [5]. Although the unique chemistry of fluorinated organic compounds is not completely understood, the stability of the C−F bond is very important in industrial applications.

The *thermal stability* of fluorinated surfactants is based on the outstanding stability of the fluorocarbon hydrophobe. Fluorocarbons are usually more stable than the corresponding hydrocarbons [6]. Other functional groups of the surfactant are usually less stable than the fluorocarbon group and may lower the thermal stability of the fluorinated surfactant. Hence not all fluorinated surfactants are thermally stable. Perfluoroalkanecarboxylic acids and perfluoroalkanesulfonic acids are the most stable fluorinated surfactants. Their salts decompose more readily, however, than the parent free acids.

Perfluoroalkanecarboxylic acids can be heated to 400°C in borosilicate glass without significant decomposition [7]. At higher temperatures (550°C) perfluoroalkanecarboxylic acids decompose, yielding an olefin, HF, and CO_2 [8] (Table 3.4) [1].

Salts of perfluoroalkanecarboxylic acids are less stable and are decarboxylated at a lower temperature than the parent acid, depending on the chemical nature of the cation [1,10,11]:

$$R_fCF_2CF_2COOM \rightarrow R_fCF{=}CF_2 + CO_2 + MF$$

The yields of perfluoroolefin and the decomposition temperature depend on the cation of the salt. The sodium salt of normal chain perfluoroalkanoic acids gives the terminally unsaturated perfluoroolefin in yields ranging from 65 to

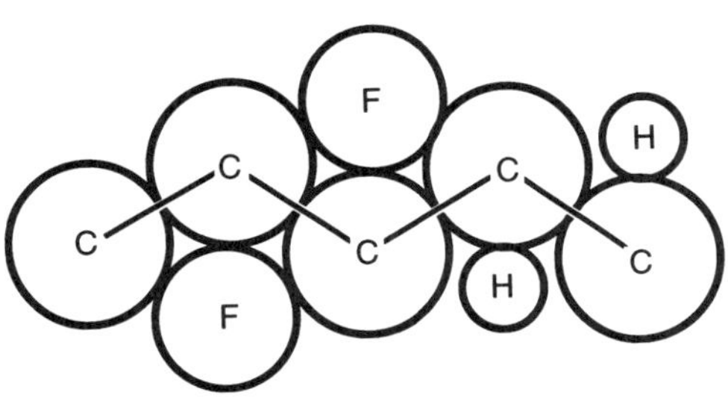

FIG. 3.1 Steric effects of fluorocarbons. (From Ref. 5.)

TABLE 3.4 Thermal Stability of Perfluoroalkanoic Acids and Alkanoic Acids

Conditions	Reactions
Acid heated	
Alkanoic acid	RCOOH → various decomposition reactions
Perfluoroalkanoic acid	$R_fCH_2COOH \xrightarrow{550°C} R_f{=}CF_2 + HF + CO_2$
Salt heated	
Alkanoic acid	$RCOONa \xrightarrow[CaO]{NaOH} RH + CO_2$ (poor yield)
Perfluoroalkanoic acid	$CF_3(CF_2)_3COOK \xrightarrow{165\text{–}200°C} CF_3CF_2CF{=}CF_2 + CF_3CF{=}CFCF_3$
	$CF_3(CF_2)_3COOK \xrightarrow[200°C]{HOCH_2CH_2OH} CF_3(CF_2)_2CHF_2$
	$2CF_3CF_2CF_2COOAg \xrightarrow{260\text{–}270°C} CF_3(CF_2)_4CF_3$

Source: Ref. 1.

100%, depending on the chain length of the parent acid [10], Lines and Sutcliffe [11] observed that the thermal stability of perfluorooctanoates decreases with increasing ionic radii of the cations within a group of the Periodic Table: $Li^+ > Ba^+ = Ca^+ > Na^+ > Pb^{2+} > Ag^+ > K^+ > Cs^+ > NH_4^+$ (Table 3.5). Barium appeared to be an exception in the correlation between thermal stability and ionic radii. However, Glöckner et al. [12,13]

TABLE 3.5 Decomposition of Perfluoroalkanoates

Salt	*20% decomposition temperature* (°C) Perfluoro-octanoates	Heptafluoro-butyrates [10]	Ionic radii	% Residue of *perfluorooctanoates* Found	Theoretical
Li	341	—	0.060	4	6
Ba	320	275	0.135	20	18
Ca	320	—	0.099	10	9
Na	298	235	0.095	9	10
Pb	290	—	0.121	15	24
Ag	287	295	0.126	19	20
K	244	200	0.135	14	13
Cs	237	—	0.169	30	28
NH4	167	185	0.148		

Source: Ref. 11. Reproduced by permission of Elsevier Sequoia.

found that the thermal stability of perfluoroalkanecarboxylates and ω-H-perfluoroalkanecarboxylates decrease in the order $Ca^{2+} > Li^+ > Na^+ > K^+ > NH_4^+$ (Figs. 3.2 and 3.3). Since this relationship is in accord with the ionization energies of the metals, the authors suggested that the primary step of the thermal decomposition reaction is dissociation of the salt and heterolysis of the carboxylate anion to a carbanion and CO_2. The ω-H-perfluoroalkane carboxylates are somewhat less stable than the perfluorocarboxylates, but the effect of the terminal hydrogen on thermal decomposition is minimal.

Anhydrous *perfluoroalkanesulfonic acids* are stable at 400°C in the absence of air but form hydrogen fluoride when moisture is present. Gramstad and

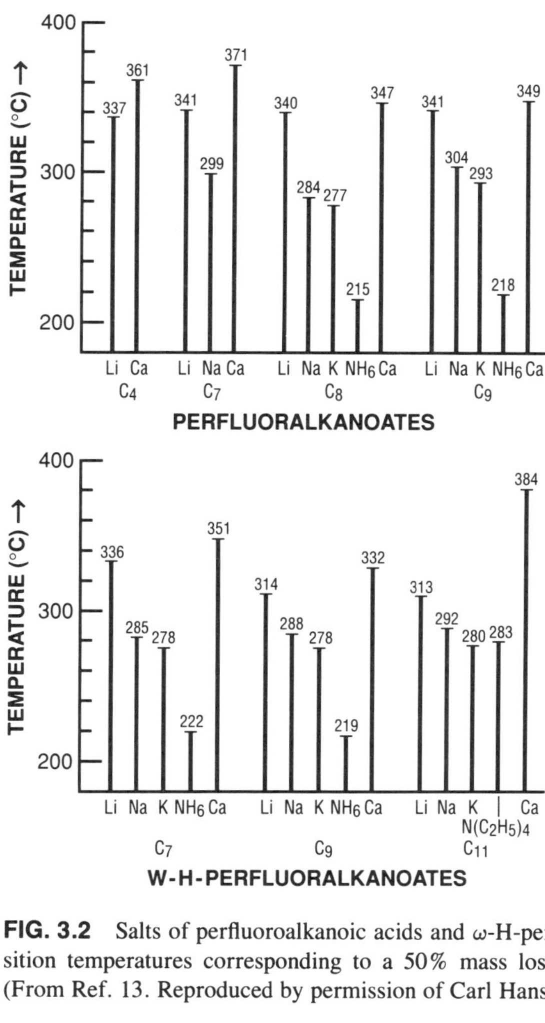

FIG. 3.2 Salts of perfluoroalkanoic acids and ω-H-perfluoroalkanoic acids. Decomposition temperatures corresponding to a 50% mass loss (10 °C/min; N_2 atmosphere). (From Ref. 13. Reproduced by permission of Carl Hanser Verlag.)

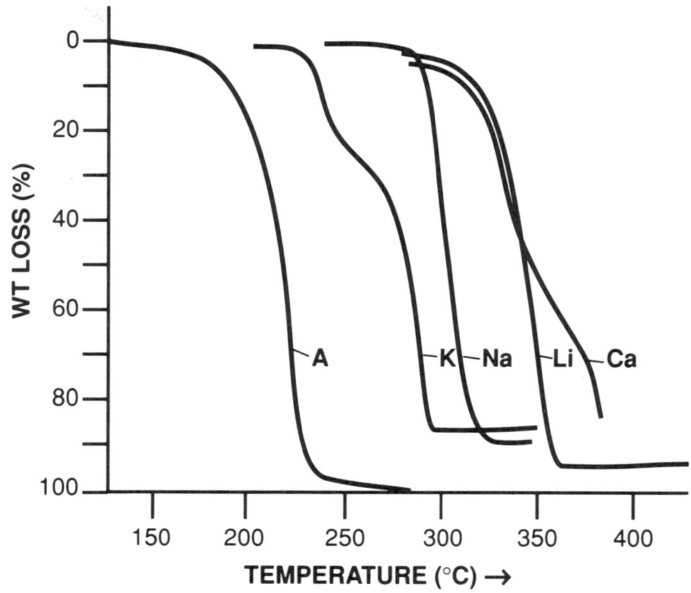

FIG. 3.3 Thermal decomposition of salts of perfluorooctanoic acid (10°C/min; N_2 atmosphere). (From Ref. 13. Reproduced by permission of Carl Hanser Verlag.)

Haszeldine [14] found that anhydrous perfluorooctanesulfonic acid, heated at 400°C for 3 h, liberates only a trace of hydrogen fluoride. Pyrolysis of anhydrous perfluoropropanesulfonic acid at 500°C gave perfluorohexane, pentafluoropropionyl fluoride, sulfur dioxide, carbonyl fluoride, sulfuryl fluoride, and hydrogen fluoride. The pyrolysis products indicate that (1) the C–S bond is the weakest link of the molecule and (2) the $C_3F_7\cdot$ radical is formed, which degrades further via the alkoxy radical during pyrolysis.

The *salts of perfluorosulfonic acids* are more thermally stable than the corresponding perfluoroalkanecarboxylates. Lithium, sodium, potassium, or barium salts of perfluorooctanesulfonic acid can be heated to 350–400°C without decomposition [14]. Fluoride is liberated at temperatures above 420°C. Gramstad and Haszeldine [14] found that anhydrous potassium perfluorooctanesulfonate was unaffected when heated in vacuo at 400°C for 5 h. Fluoride was liberated above 430°C. The thermal stability of tetraethylammonium perfluorooctane sulfonate is remarkable for a quaternary ammonium salt, although the tetramethylammonium salt is less stable than the corresponding alkali metal salt. Tetraethylammonium perfluorosulfonate can be heated to 350°C before decomposition begins (Fig. 3.4) [5,15,16]. Decomposition temperatures corresponding to a 50% weight loss have been reported for perfluoroalkane sulfonates by Glöckner and co-workers (Fig. 3.4) [13].

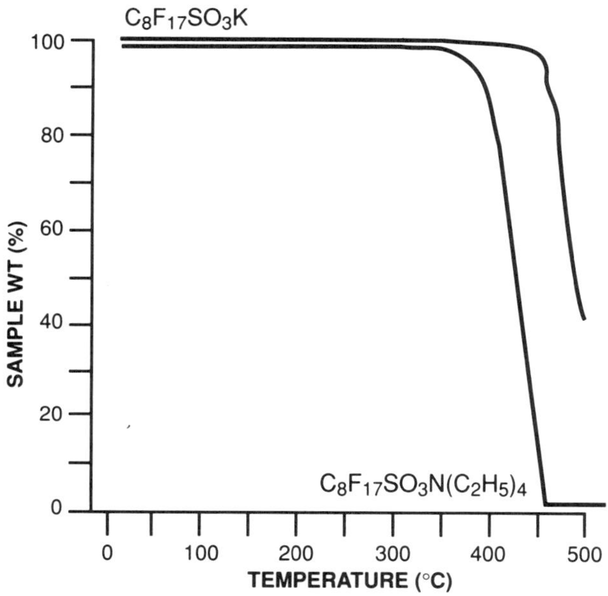

FIG. 3.4 Thermogravimetric analysis (TGA) curves of $C_8F_{17}SO_3K$ and $C_8F_{17}SO_3N(C_2H_5)_4$. (From Ref. 5.)

The pyrolysis of perfluoroalkanesulfonates proceeds by a different reaction mechanism than the pyrolysis of perfluoroalkanoates. Like its parent acid, potassium perfluorooctanesulfonate does not produce an olefin when heated but yields perfluoroalkanes of various chain lengths, SO_2, K_2SO_4, and soot [13]. The thermal stability of perfluoroalkanesulfonates decreases in the order of their cations: $Ca^{2+} > K^+ > Na^+ > Li^+ > NH_4^+$ (Fig. 3.5) [13]. Calcium perfluorooctanesulfonate is, like calcium perfluorooctanecarboxylate, more stable than the analogous alkali metal salts. However, the thermal stabilities of alkali metal perfluoroalkanesulfonates decrease with increasing ionic radii in the opposite order than the stabilities of perfluoroalkanecarboxylates.

The excellent *chemical stability* of perfluorinated alkanoic and alkanesulfonic acids to acids, oxidants, and alkali [5,13,17] allows their application under conditions too severe for hydrocarbon-based surfactants. Gramstad and Haszeldine [14] did not detect fluoride when they heated potassium perfluorooctanesulfonate (0.9 g) in water (5 mL) at 300°C for 8 h. However, when potassium perfluorooctanesulfonate was heated in 10% aqueous potassium hydroxide for 8 h at 270°C, 8% of the theoretical amount of fluoride was liberated. Perfluorooctanesulfonic acid, heated in water for 3 h at 400°C, produced 17% of the theoretical amount of hydrogen fluoride.

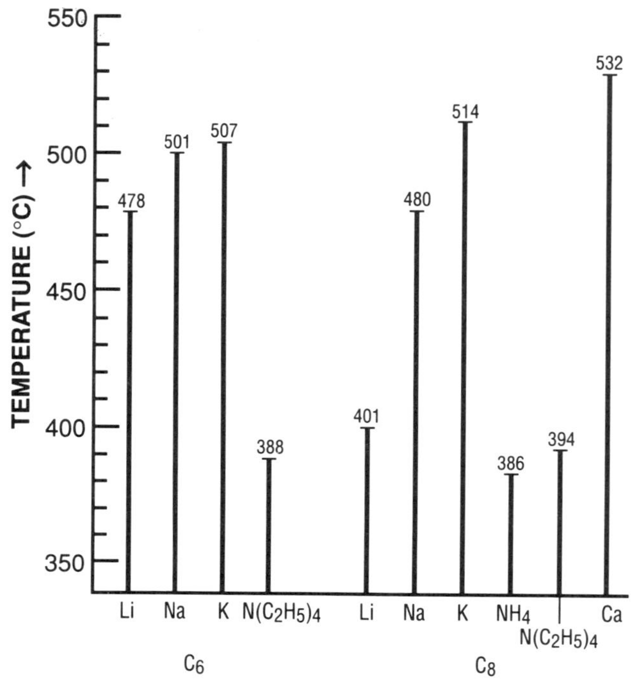

FIG. 3.5 Decomposition temperatures corresponding to a 50% mass loss of perfluoroalkane sulfonates. (From Ref. 13. Reproduced by permission of Carl Hanser Verlag.)

Perfluorooctanesulfonic acid is remarkably stable to oxidants: heating in concentrated nitric acid for 12 h at 160°C did not liberate hydrogen fluoride. Potassium perfluorooctanesulfonate is also unusually stable to oxidants (nitric acid, hydrogen peroxide) and reducing agents (hydrazine) [18].

Glöckner and co-workers [13] showed that anionic perfluorinated surfactants are stable even when stored for a long time in 60% HNO_3 or 98% sulfuric acid containing 10 g/L chromic oxide. After 15 days at ambient temperature, followed by 28 days at 90°C, the surface tension of the solutions did not change significantly (see also Sec. 4.4). However, nonionic surfactants having the structure $C_nF_{2n-1}(OCH_2CH_2)_mOCH_3$ were not stable in chromic-sulfuric acid. The solution turned green immediately.

The chemical properties of anionic fluorinated surfactants also differ from those of their hydrocarbon analogues by the increased *acid strength* of the parent acids. Replacement of hydrogen by fluorine increases the dissociation constant of carboxylic acids (Table 3.6) [19]. Perfluorinated alkanecarboxylic acids are strong acids, comparable to mineral acids [20]. Equivalent conduc-

TABLE 3.6 Acidity of Fluorinated Acids

Acid	CH_3COOH	CH_2FCOOH	CHF_2COOH	CF_3COOH
Dissociation constant, K	1.8×10^{-5}	2.2×10^{-3}	5.7×10^{-2}	5.5×10^{-1}

Source: Refs. 19 and 20.

tance measurements [7] have shown that perfluoroalkanecarboxylic acids are completely dissociated in water. The strength of perfluoroalkanoic acids has also been established by potentiometric titration of *n*-perfluorodecanoic acid, *n*-perfluoroheptanoic acid, and *n*-perfluoropropionic acid with NaOH. The similarity of the titration curves to the titration curve of nitric acid (Fig. 3.6) [23] indicated complete dissociation of the perfluoroalkanoic acids in water.

Perfluoroether surfactants featuring one or two carboxylic acid groups [24] are also strong acids. The pK_a values for the monocarboxylic acids are about 1.7, the $pK_{a1} \rightarrow pK_{a2}$ values for the dicarboxylic acids about 2 to 2.5.

Henne and Fox [20] observed that the shielding effect of one methylene group between the perfluorinated segment and the carboxylic acid group is

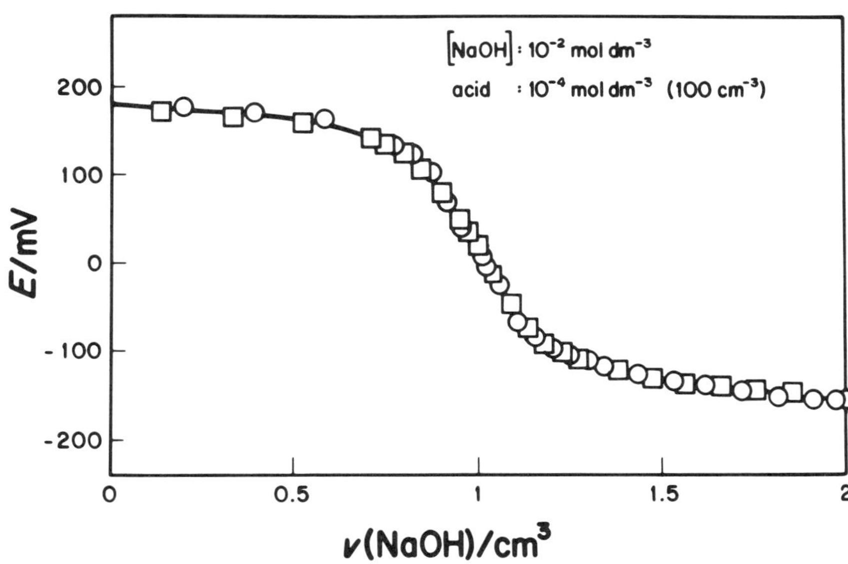

FIG. 3.6 Potentiometric titration curves of perfluorodecanoic acid ($C_9F_{19}COOH$) (○) and nitric acid (HNO_3) (□). Glass electrode, Ag/AgCl reference electrode at 25°C. (From Ref. 23. Reproduced by permission of Steinkopff.)

TABLE 3.7 Effect of Distance Between Perfluoroalkyl and Carboxyl Groups on Acidity of Fluorinated Acids

Fluorinated acid	Dissociation constant, K
CF_3COOH	5.5×10^{-1}
CF_3CH_2COOH	1×10^{-3}
$CF_3CH_2CH_2COOH$	7×10^{-5}
$CF_3CH_2CH_2CH_2COOH$	3.2×10^{-5}
$CH_3CH_2CH_2CH_2COOH$	1.56×10^{-5}

Source: Refs. 19–22. Macmillan Co.

considerable. Perfluoroacetic acid and perfluorobutyric acid are strong acids but CF_3CH_2COOH and $CF_3CH_2CH_2COOH$ are weak acids, although stronger than the corresponding nonfluorinated carboxylic acids. Hence even two methylene groups cannot completely shield the electronegative induction effect of the CF_3 group (Table 3.7).

Brace [25] determined dissociation constants of long-chain alkanoic and alkenoic acids featuring a terminal CF_3 group in 50% aqueous ethanol, because of limited solubility of the acids in water. Dissociation constants of perfluorooctanoic acid and the corresponding unfluorinated alkanoic acids were also measured (Table 3.8). The data revealed that the electron-withdrawing effect of the perfluoroalkyl group is transmitted through alkylene segments of two to five carbon atoms with decreasing force.

TABLE 3.8 Apparent Dissociation Constants for Perfluorooctanoic Acid, Perfluoroalkyl Segmented Alkanoic and Alkenoic Acids, and Alkanoic Acids in 50% Aqueous Ethanol

Compound	Concentration (M)	pK_a	$K \times 10^{-6}$
$CF_3(CF_2)_6COOH$	0.005	2.80 ± 0.03	159
$CF_3(CF_2)_6(CH_2)_2COOH$	0.002	5.12 ± 0.06	0.76
$CF_3(CF_2)_6(CH_2)_3COOH$	0.002	5.63 ± 0.03	0.22
$CF_3(CF_2)_6(CH_2)_4COOH$	0.002	5.81 ± 0.06	0.16
$CF_3(CF_2)_6(CH_2)_5COOH$	0.002	5.95 ± 0.005	0.11
$CF_3(CF_2)_6(CH_2)_6COOH$	0.002	6.26 ± 0.08	0.06
$CF_3(CF_2)_6CH{=}CH(CH_2)_2COOH$	0.004	5.40 ± 0.05	0.38
$CF_3(CF_2)_6CH{=}CH(CH_2)_3COOH$	0.003	5.83 ± 0.01	0.15
$CH_3(CH_2)_6COOH$	0.002, 0.005	6.17 ± 0.01	0.068
$CH_3(CH_2)_{10}COOH$	0.002, 0.005	6.13 ± 0.01	0.074

Source: Ref. 25. Reproduced by permission of the American Chemical Society.

3.2 MELTING POINTS

The melting point of a compound is defined as the temperature at which the solid and liquid phases are in equilibrium at the pressure of 1 atmosphere. The heat of fusion depends on the lattice energy of the solid, the type of the crystal lattice, and the entropy of fusion. Although the intermolecular attractive forces affect the heat of fusion and the melting point, a general correlation between the crystal energy and the melting point does not exist. Melting can be viewed as a transition from a highly ordered solid state to a liquid state of a low degree of order. Consequently, the entropy increases during melting. The numerical values of the melting points depend therefore not only on the heat of fusion but also on the entropy of fusion [26].

Perfluoroalkanes have higher melting points than the corresponding hydrocarbons, except for the first member, CF_4, of the homologous series (Fig. 3.7). However, the melting points of the perfluoroalkanes with less than five carbon atoms increase not continuously but alternately. In both series the homologue with three carbon atoms has the lowest melting point.

In analogy to perfluoroalkanes, the melting points of fluorinated surfactants are higher than those of the corresponding hydrocarbon-type surfactants. The melting points of fluorinated surfactants are affected by the stiffness of the fluorinated hydrophobe. Perfluorinated carbon chains of surfactant molecules, particularly the linear chains, are hard and inflexible. Perfluorinated surfactants therefore have a high melting point, a high Krafft point, and their solubility in

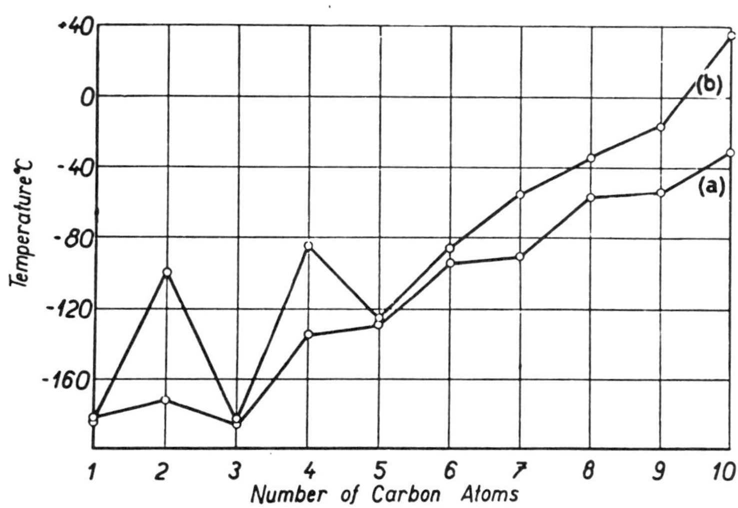

FIG. 3.7 Melting points of alkanes (a) and perfluoroalkanes (b). (From Ref. 19. Reproduced by permission of The Macmillan Co.)

solvents is low. Surfactants with a polyether segment are more flexible, have lower melting points, and are also more soluble [27].

Branching of the hydrophobe lowers the melting point of a fluorinated surfactant. Branched-chain perfluoroalkanecarboxylic acids have lower melting points than the normal-chain perfluoroalkanecarboxylic acids with the same number of carbon atoms (Fig. 3.8). The melting point differences between curves A, B, and C for equal-length fluorinated surfactants are larger than the difference between curves D and E for unfluorinated surfactants. Branching of a fluorinated chain has a larger disrupting effect on the melting point of a surfactant than a unfluorinated alkyl group. Fluoroalkyl groups are larger than unfluorinated alkyl groups and induce more steric hindrance than unfluorinated alkyl groups [28].

Brace [25] synthesized long-chain alkanoic and alkenoic acids with a terminal perfluorinated segment R_f (Table 3.9). The chain length and branching of the terminal R_f have a marked effect on the melting point of the segmented acids (Fig. 3.9). The shape of the curves depends on the interaction of the R_f and alkylene segments. As the hydrocarbon segment in the $CF_3(CF_2)_6(CH_2)_mCOOH$ series (curve A) is increased, the melting point increases first and then drops to a minimum. Further extension of the hydrocarbon segment gives a slope approaching roughly that of the curves for

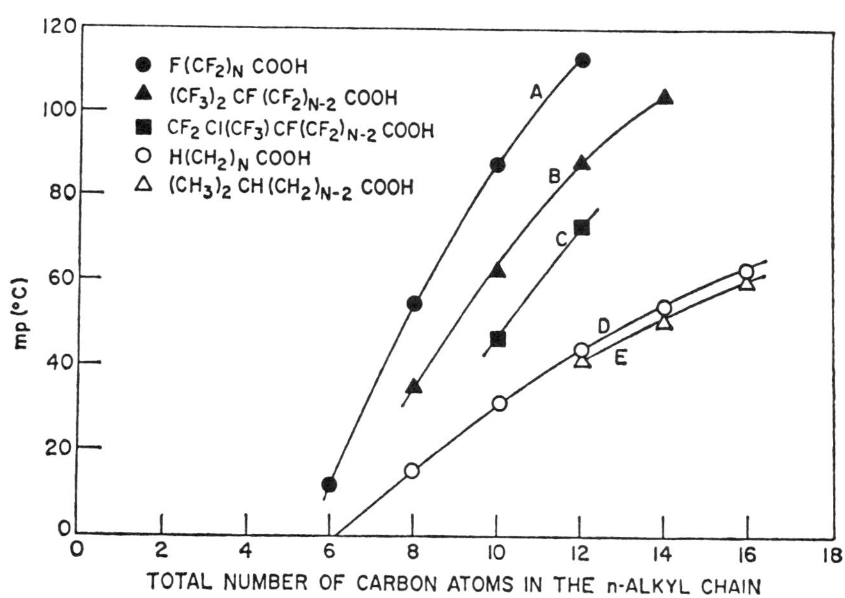

FIG. 3.8 Melting points of alkanoic and fluorinated alkanoic acids. (From Ref. 28. Reproduced by permission of American Chemical Society.)

TABLE 3.9 Melting Points of Terminal Perfluoroalkyl Segmented Alkanoic Acids

Compound	Melting point (°C)
$CF_3CF_2(CH_2)_{10}COOH$	48.5–50
$CF_3CF_2(CH_2)_{16}COOH$	70–71
$CF_3(CF_2)_2(CH_2)_{10}COOH$	49–50
$(CF_3)_2CF(CH_2)_{10}COOH$	36–37
$CF_3(CF_2)_2(CH_2)_{16}COOH$	73–74.5
$CF_3(CF_2)_3(CH_2)_5COOH$	41–42
$CF_3(CF_2)_3(CH_2)_{10}COOH$	55
$CF_3(CF_2)_4(CH_2)_{10}COOH$	58–59.8
$(CF_3)_2CF(CF_2)_2(CH_2)_{10}COOH$	43
$CF_3(CF_2)_4(CH_2)_{16}COOH$	77–79
$CF_3(CF_2)_5(CH_2)_{10}COOH$	67–69
$CF_3(CF_2)_6(CH_2)_2COOH$	86–88
$CF_3(CF_2)_6(CH_2)_3COOH$	73
$CF_3(CF_2)_6(CH_2)_4COOH$	79–80
$CF_3(CF_2)_6(CH_2)_5COOH$	64–65
$CF_3(CF_2)_6(CH_2)_6COOH$	79.5–80
$CF_3(CF_2)_6(CH_2)_{10}COOH$	82–83
$(CF_3)_2CF(CF_2)_4(CH_2)_{10}COOH$	63
$CF_3(CF_2)_6(CH_2)_{16}COOH$	91.5–92
$CF_3(CF_2)_7(CH_2)_{10}COOH$	89–90
$(CF_3)_2CF(CF_2)_6(CH_2)_{10}COOH$	88
$CF_3(CF_2)_9(CH_2)_{10}COOH$	111–112

Source: Ref. 25.

unfluorinated alkanoic acids The transition of the slope corresponds to a maximum interaction of the R_f and alkylene segments.

Long-chain carboxylic acids with a branched R_f segment have a lower melting point than the straight-chain analogues. The unusual melting behavior of the acids is caused by (1) the greater stiffness of the $(CF_2)_n$ chain than the $(CH_2)_n$ chain and (2) the larger cross-sectional area, 29 Å for $(CF_2)_n$ versus 20 Å for $(CH_2)_n$.

Melting points of perfluorooctanoates with inorganic cations are shown in Table 3.10 [11]. Their melting points do not increase regularly with increasing ionic radii, as may have been expected. The melting points of the salts are probably affected by the thermal stability of the salts, which decreases with increasing ionic radii.

Commercial fluorinated surfactants are usually mixtures and the relationship between structure and melting point is less clearly defined. For the tetraethylammonium salt of perfluorooctanesulfonic acid a broad melting

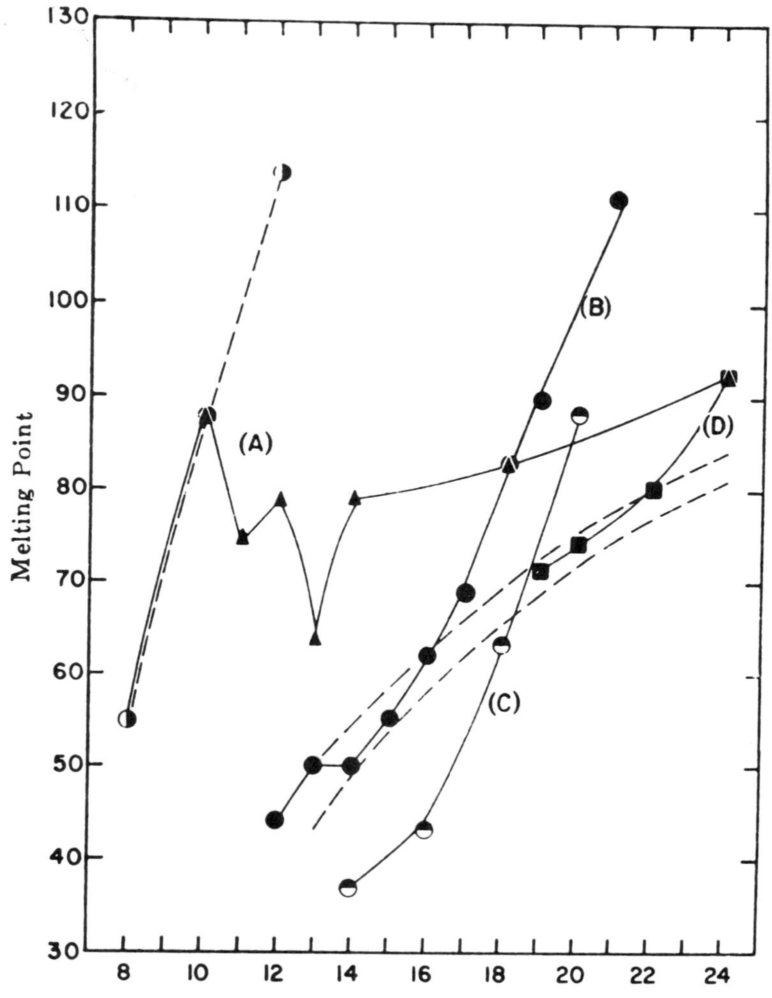

FIG. 3.9 Melting points of perfluoroalkanoic, segmented perfluoroalkylalkanoic, and alkanoic acids. $CF_3(CF_2)_{n-2}COOH$; $CF_3(CF_2)_6(CH_2)_mCOOH$ (A); $CF_3(CF_2)_n(CH_2)_{10}COOH$ (B), $n = 0$–10; $CF_3CF(CF_3)(CF_2)_n(CH_2)_{10}COOH$ (C), $n = 0$–6; $CF_3(CF_2)_n(CH_2)_{16}COOH$ (D), $n = 0$–6. The melting points of alkanoic acid are shown with dashed curves: even-carbon-number alkanoic acids (upper curve), odd-carbon-number alkanoic acids (lower curve). (From Ref. 25. Reproduced by permission of American Chemical Society.)

TABLE 3.10 Melting Points of Perfluorooctanoic Acid Salts

Salt	Melting point (°C)
Li	219–224
Na	273–275
K	222–235
Cs	230d
NH_4	157–165
Ca	145–155d
Ba	230–240
Ag	220–225
Pb	265–270

d = decomposition.
Source: Ref. 11. Reproduced by permission of Elsevier Sequoia.

range, about 170–190°C has been reported [15]. The parent acid of the salt was a mixture of isomers, formed by the electrofluorination process . The mixture consisted of the normal-chain isomer (70%) and branched isomers (30%).

3.3 BOILING POINTS

The transition of a liquid phase to its vapor phase involves the separation of molecules in the liquid and the removal of molecules from the surface of the liquid into the vapor phase. The energy absorbed when a definite quantity of a liquid is vaporized, the latent heat of vaporization, therefore depends on the intermolecular attractive forces which have to be overcome in order to separate molecules. According to Trouton's rule, the boiling points of nonassociated liquids, on the absolute temperature scale, are approximately proportional to their latent heat of vaporization. Hence the boiling point of a liquid depends on the relative strength of cohesive intermolecular forces.

The low polarizability of fluorine and the relatively weak attractive forces between fluorine atoms suggest that the boiling points of fluorinated compounds should be lower than those of other halocarbons. Indeed, the boiling points of alkyl halides increase in the order H < F < Cl < Br < J (Fig. 3.10) [19].

The unusual effect of fluorination on volatility can be illustrated with halomethanes. The boiling points of chlorinated and brominated methanes increase with increasing halogen substitution. The fluorinated methanes exhibit, however, a maximum boiling point for difluoromethane (Fig. 3.11) [1].

Perfluoroalkanes with more than four carbon atoms have lower boiling points than the corresponding hydrocarbons. However, the boiling points of

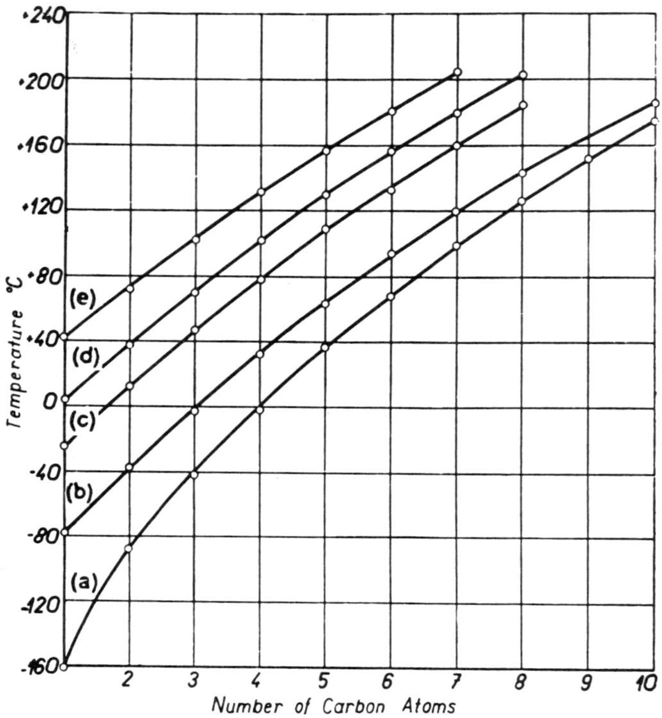

FIG. 3.10 Boiling points of alkanes (a), monofluoroalkanes (b), monochloroalkanes (c), monobromoalkanes (d), and monoiodoalkanes (e). (From Ref. 19. Reproduced by permission of The Macmillan Co.)

perfluoroalkanes with less than four carbons are higher than those of the corresponding hydrocarbons (Fig. 3.12).

Fluorination has a more pronounced effect on the boiling points of carboxylic acids than on the boiling points of hydrocarbons [30]. The boiling points of *perfluoroalkanoic acids* are about 45°C lower than those of the corresponding alkanoic acids (Fig. 3.13) [19,31]. The boiling points of perfluoroalkanoic acids increase with increasing chain length, like the boiling points of unfluorinated alkanoic acids.

Branching of the carbon chain decreases the boiling point of a perfluoroalkanoic acid. A terminal chlorine atom, increases the boiling points of branched perfluoroalkanoic acids $CF_2Cl(CF_3)CF(CF_2)_nCOOH$, (Table 3.11) [28].

In analogy to the effect of fluorination on the boiling points of perfluoroalkanoic acids, fluorination also lowers the boiling points of alkanesulfonic

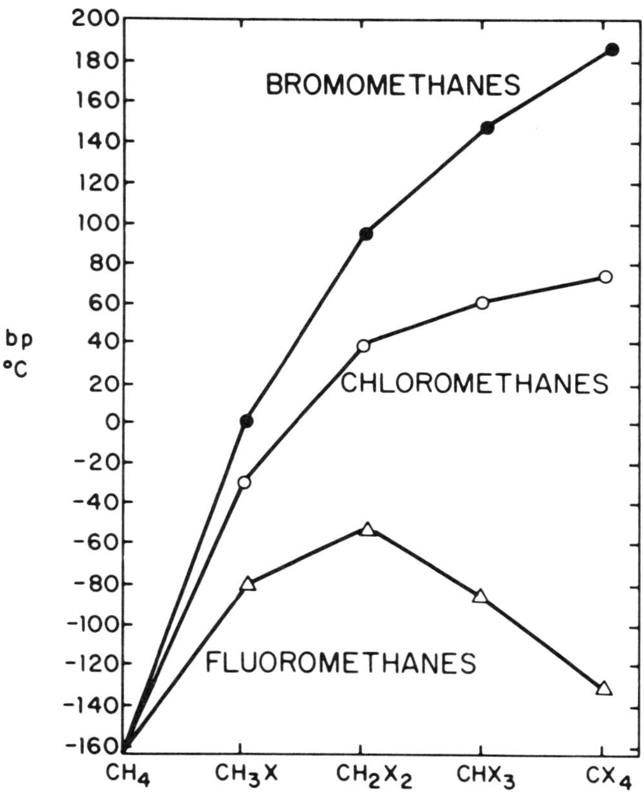

FIG. 3.11 Boiling points of halomethanes. (From Ref. 1.)

acids. Unlike alkanesulfonic acids, *perfluoroalkanesulfonic acids* are relatively volatile. The boiling points of perfluoroalkanesulfonic acids, like those of perfluorocarboxylic acids, increase with increasing carbon chain length (Table 3.12) [14]. Perfluorooctanesulfonic acid boils at 258–260°C at atmospheric pressure. However, the boiling points of perfluoroalkanesulfonic acids cannot be determined accurately at atmospheric pressure, because perfluoroalkanesulfonic acids form an anhydride when heated to the boiling point at atmospheric pressure. Like alkanesulfonic acids, perfluoroalkanesulfonic acids are hygroscopic and form a solid monohydrate.

Boiling points of *perfluoropolyether surfactants* (as well as their densities and refractive indexes) [24] increase with increasing length of the fluorocarbon chain and decrease in the order of their functional groups: carboxylic acid > alcohol > amine > nitrile (Table 3.13).

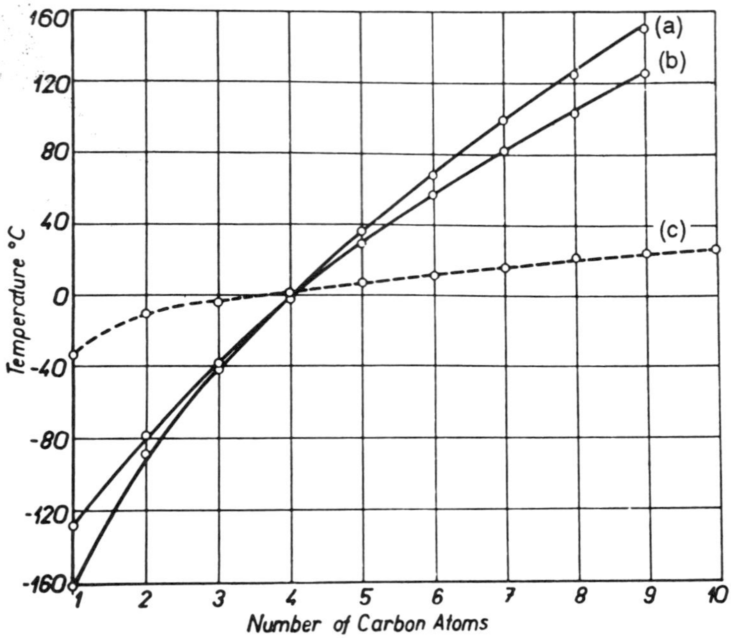

FIG. 3.12 Boiling points of alkanes (a), perfluoroalkanes (b), and the difference between the boiling points (boiling point of the alkane minus that of the perfluoroalkane). (From Ref. 19. Reproduced by permission of The Macmillan Co.)

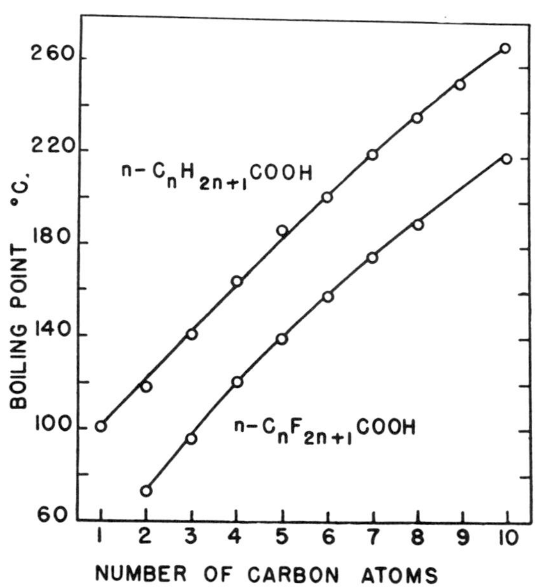

FIG. 3.13 Boiling points of alkanoic and perfluoroalkanoic acids. (From Ref. 7. Reproduced by permission of American Chemical Society.)

TABLE 3.11 Melting Points and Boiling Points of Branched Fluorinated Acids $CF_2X(CF_3)CF(CF_2)_nCOOH$

	X = F		X = Cl	
n	Melting point °C	Boiling point °C (mm)	Melting point (°C)	Boiling point °C (mm)
1	—	141 (760)	—	100 (50)
3	—	130 (150)	—	115 (28)
5	33–36	115 (20)	—	134 (24)
7	61–64	125 (10)	44–50	141 (8)
9	88–89	117 (0.5)	71–74.5	117 (0.3)
11	100–107	—		

Source: Refs. 28 and 29. Reproduced by permission of the American Chemical Society.

3.4 DENSITY

The density of organic compounds RX increases in the order X = H < F < Cl< I (Fig. 3.14). Substitution of fluorine for hydrogen increases the density of a hydrocarbon. The density increase resulting from substituting one fluorine for hydrogen at the same carbon atom decreases in the order of first > second > third > fourth [19].

TABLE 3.12 Boiling Points of Perfluoroalkanesulfonic Acids

Compound	Boiling point, °C (mm)
$C_2F_5SO_3H$	178° (760) 87° (25)
$C_3F_7SO_3H$	196° (760)
$C_4F_9SO_3H$	210–212° (760) 105° (22)
$C_5F_{11}SO_3H$	224°226° (760) 97° (4)
$C_6F_{13}SO_3H$	238–239° (760) 110° (5)
$C_7F_{15}SO_3H$	247–249° (760) 122° (5)
$C_8F_{17}SO_3H$	258–260° (760) 145° (10)

Source: Ref. 14.

TABLE 3.13 Physical Properties of Monofunctional Perfluoropolyether Surfactants

Structure	n	Boiling point (°C/760 mm Hg)	Density (d_4^{20})	Refractive index (μ_D^{20})
$CF_3(OCF_2-CF(CF_3))_n-OCF_2COOH$	1	158	1.7237	1.2989
	2	193	1.7794	1.2960
	3	220	1.8032	1.2994
$CF_3(OCF_2-CF(CF_3))_n-OCF_2-CH_2OH$	1	126	1.6835	
	2	163	1.7515	1.2950
	3	197	1.7910	1.2972
$CF_3(OCF_2-CF(CF_3))_n-OCF_2CH_2NH_2$	1	134		
	2	152		
	3	188		
$CF_3(OCF_2-CF(CF_3))_n-OCF_2CN$	1	115		
	2	124		
	3	160		

Source: Ref. 24. Reproduced by permission of Academic Press.

In accord with the densities of fluorinated organic chemicals in general, the densities of fluorinated surfactants are higher than those of their hydrocarbon-type analogues. Perfluoroalkanoic acids have higher densities than the corresponding unfluorinated alkanoic acids (Fig. 3.15) [7,32]. The densities of perfluoroalkanoic acids increase with increasing number of carbon atoms,

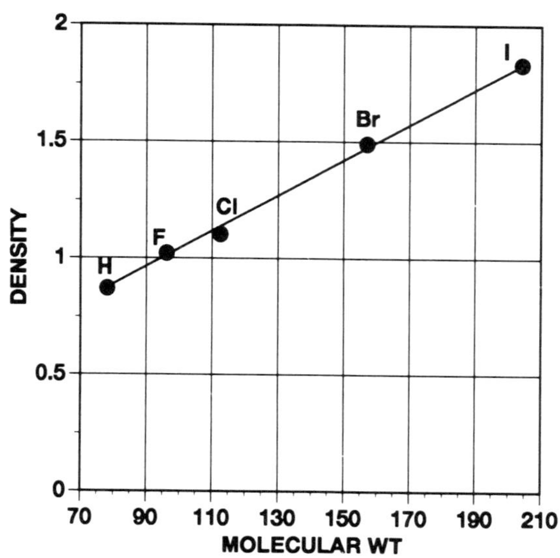

FIG. 3.14 Density of halobenzenes C_6H_5X. (Data from Ref. 33.)

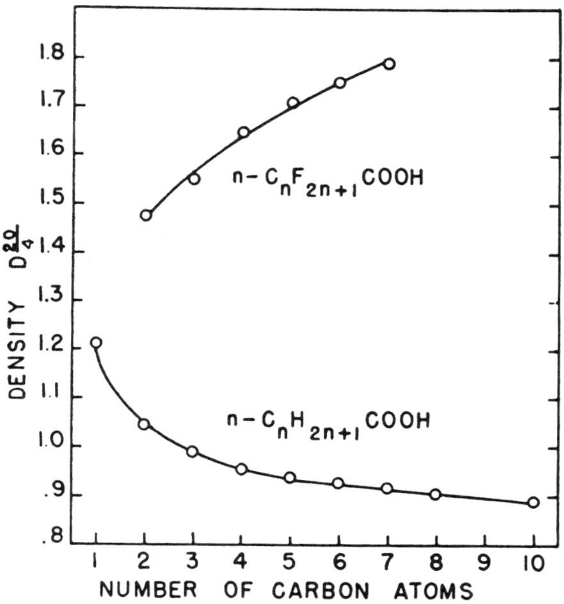

FIG. 3.15 Liquid densities of alkanoic and perfluoroalkanoic acids. (From Ref. 7. Reproduced by permission of American Chemical Society.)

whereas for the series of nonfluorinated alkanoic acids the opposite is true [7] (Fig. 3.15). Apparently, the difluoromethylene segment -CF_2- contributes more to the density of the acid than a carboxyl group. In contrast, the incremental effect of the methylene segment -CH_2- on density is smaller than the effect of a carboxyl group.

The density of carboxylic acids with a perfluoroether hydrophobe, derived from hexafluoropropylene, increases with increasing chain length (Table 3.13) [24].

3.5 REFRACTIVE INDEX

Fluoro compounds have a lower refractive index than their hydrocarbon and halocarbon analogues. The refractive indexes of monosubstituted halobenzenes increase in the order F < H < Cl < Br < I (Table 3.14). The low refractive index of fluorobenzene is related to the low polarizability of the fluorine atom. The polarizability, P_E, can be calculated from atomic contributions (Table 3.14) of which fluorine has the lowest value.

TABLE 3.14 Refractive Indexes and Surface Tensions of Benzene Derivatives C_6H_5X

X	Atomic refractive constant of X [33]	Refractive index [33] (n_D^{20})	Surface tension (γ) 20°C (vapor)
F	0.95	1.47	27.3
H	1.10	1.50	28.9
Cl	5.97	1.52	33.6
Br	8.87	1.56	35.8
I	13.90	1.57	39.3

The refractive index of perfluorinated compounds is exceedingly low [19]. Perfluoropentane has probably the lowest refractive index, (n_D^{20} 1.333), ever recorded.

Refractive indexes of some fluorinated surfactants are listed in Table 3.13.

REFERENCES

1. W. A. Sheppard and C. M. Sharts, "Organic Fluorine Chemistry," W. A. Benjamin, New York (1969).
2. L. Pauling, "The Nature of the Chemical Bond," 3rd ed., Cornell Univ. Press, Ithaca, NY (1960), p. 95.
3. B. E. Smart, "Fluorinated Organic Molecules," in "Molecular Structure and Energetics," Vol. 3. p. 141–191, (J. F. Liebman and A. Greenberg, eds.), VCH Publishers, Deerfield Beach, FL (1986).
4. N. V. Sidgwick, "The Chemical Elements and Their Compounds," Vol. II, p. 1099, Oxford University Press, Oxford (1951).
5. H. G. Klein, J. N. Meussdoerffer, and H. Niederprüm, Metalloberfläche 29, 559 (1975).
6. T. J. Brice, in "Fluorine Chemistry," J. H. Simons, ed., Vol. I, Academic Press, New York (1950).
7. E. A. Kauck and A.R. Diesslin, Ind. Eng. Chem. 43, 2332 (1951).
8. L. Hals, T. S. Reid, and G. H. Smith, J. Am. Chem. Soc. 73, 4054 (1951).
9. T. L. Cottrell, "The Strength of Chemical Bonds," 2nd ed., Butterworths Scientific Publications, London (1958).
10. J. D. LaZerte, L. J. Hals, T. S. Reid, and G. H. Smith, J. Am. Chem. Soc. 75, 4525 (1953).
11. D. Lines and H. Sutcliffe, J. Fluorine Chem. 25, 505 (1984).
12. V. Glöckner, K. Lunkwitz, and D. Prescher, Abh. Akad. Wiss. DDR, Abt. Math. Naturwiss., Tech. 1986, 539 (1987).

13. V. Glöckner, K. Lunkwitz, and D. Prescher, Tenside 26, 376 (1989).
14. T. Gramstad and R. N. Haszeldine, J. Chem. Soc. 2640 (1957).
15. H. G. Klein, J. N. Meussdoerffer, H. Niederprüm, and M. Wechsberg, Tenside 15, 2 (1978).
16. J. N. Meussdoerffer, and H. Niederprüm, Chemiker Zeit. 104, 45 (1980).
17. J. Burdon, I. Farzmand, M. Stacey, and J. C. Tatlow, J. Chem. Soc. 2574 (1957).
18. H. G. Bryce, in "Fluorine Chemistry," J. H. Simons, ed., Academic Press, New York (1964), Vol. V.
19. M. Hudlický, "Chemistry of Organic Fluorine Compounds," Macmillan, New York (1962).
20. A. L. Henne and C. J. Fox, J. Am. Chem. Soc. 73, 2323 (1951).
21. A. L. Henne and C. J. Fox, J. Am. Chem. Soc. 75, 5750 (1953).
22. E. T. MacBee, O. R. Pierce, and D. D. Smith, J. Am. Chem. Soc. 76, 3722 (1954).
23. X.-J. Fan, M. Ćolić, N. Kallay, and E. Matijević, Colloid Polymer Sci. 266, 380 (1988).
24. G. Caporiccio, F. Burzio, G. Carniselli, and V. Biancardi, J. Colloid Interface Sci. 98,202 (1984).
25. N. O. Brace, J. Org. Chem. 27, 4491 (1962).
26. A. R. Ubbelohde, Quart. Rev. (Chem. Soc. London) 4, 356 (1954).
27. N. Ishikawa and M. Sasabe, J. Fluorine Chem. 25, 241 (1984).
28. M. K. Bernett and W. A. Zisman, J. Phys. Chem. 71, 2075 (1967).
29. M. Hauptschein and M. E. Miville, Canadian Pat. 735,493 (1966).
30. A. V. Grosse and G. H. Cady, Ind. Eng. Chem. 39, 367 (1947).
31. J. O. Hendrichs, Ind. Eng. Chem. 45, 99 (1953).
32. C. H. Arrington, Jr., and G. D. Patterson, J. Phys. Chem. 57, 247 (1953).
33. Lange's Handbook of Chemistry, J. A. Dean, ed., 13th ed., McGraw-Hill, New York (1986).

4

Liquid-Vapor and Liquid-Liquid Boundaries. Surface Tension

4.1 THEORY OF SURFACE TENSION

An interface is the region at the boundary between two immiscible phases. Ideally, the interfacial boundary is considered to be the plane dividing the phases. In real systems the interface is not a plane but a region which has a composition different from that of the bulk phase (Fig 4.1). The boundary may be between solid-solid, solid-liquid, solid-vapor, liquid-liquid, or liquid-vapor immiscible phases. If one of the two phases in contact with each other is a gas or vapor, the term surface is used to denote the boundary.

Energetics predict that the free energy of interface formation has to be positive. If the free energy is negative, the interface is unstable or metastable. If the free energy is zero, the two phases are miscible and a boundary cannot exist.

Since the free energy of interface formation is positive, work has to be done in order to expand the interface. The work, W_{min}, required to expand the surface of a liquid reversibly and isothermally by a unit area is called the surface tension, γ,

$$W_{min} = \gamma \Delta A \tag{1}$$

where ΔA denotes the area created and γ the surface tension, in erg/cm^2, dyne/cm, or mN/m. The methods for the determination of surface tension are reviewed in Chapter 9.

As a consequence of surface tension, a liquid tends to contract its surface area. In the absence of other forces, a drop assumes a spherical shape.

The cause of surface tension is the difference in attractive forces acting on molecules at the surface. From the energetics point of view, molecules at or

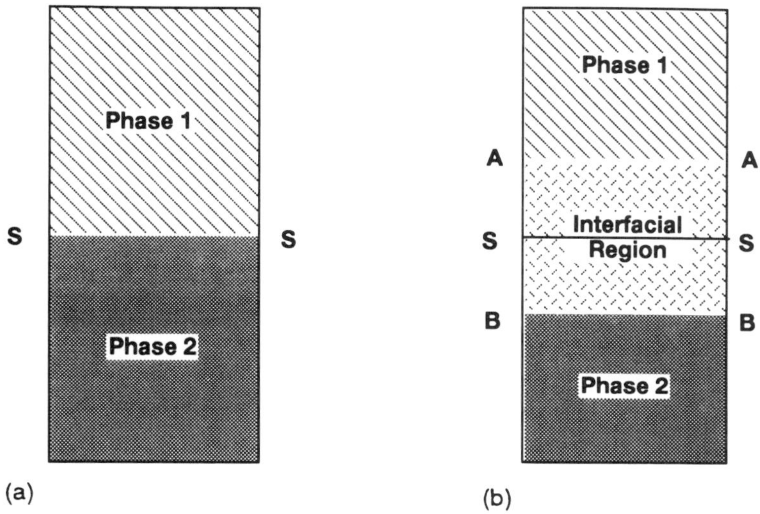

FIG. 4.1 Ideal interface (a) and real interface (b).

near the surface are in a different condition than the molecules in the bulk of the liquid. The molecules in the bulk phase are surrounded by other molecules. The force field around a molecule is uniform and the resulting net force acting on the molecule is zero. Unlike the molecules in the bulk, molecules at the surface are not surrounded by other molecules. The attractive forces on the inner side of the interface are not balanced because the interaction with the vapor phase is weak. The resulting net attractive force tends to pull the molecules out of the surface into the interior of the liquid. This migration of molecules out of the surface into the interior is in a dynamic equilibrium with the diffusion of molecules in the opposite direction, from the interior of the liquid to the surface [1]. The exchange rate, β, has been calculated from the equation [2]

$$\beta = \alpha(2\pi mkT)^{1/2}\rho_o \tag{2}$$

where α is a coefficient, ρ_o is the equilibrium vapor pressure of the liquid, m the mass of the molecule, T the abs. temperature, and k is Boltzmann's constant. The average residence time of a water molecule at the water surface at 25°C has been estimated to be 3 microseconds or less.

When a surface of a liquid is expanded, more molecules are brought from the interior of the liquid to the surface. The quantity of work required to bring more molecules to the surface depends on the strength of the attractive forces between the molecules and consequently on the energy of cohesion. Hence a correlation, albeit with many exceptions, exists between the boiling point and surface tension [3].

A surfactant molecule dissolved in water is surrounded by water molecules. The energy needed to transport the surfactant molecule to the surface depends on the effect the molecule has on water structure [4]. Water is believed to consist of regions of free unbound water and regions of hydrogen-bonded water with a structure similar to ice [5]. A surfactant molecule dissolved in water causes a distortion or disorientation of the water structure. Water molecules near the hydrophobic group(s) of the surfactant molecule are restructured into an even more extensively ordered structure [6]. The internal torsional vibrations of the hydrophobic chains are restricted by the highly structured water [7]. The surfactant molecule orients itself at the surface with the hydrophobic part directed away from water but the hydrophilic part in water. The resulting increase in entropy and decrease in free energy of the system decrease the work needed to transport the surfactant molecule to the surface of water. Since less work is needed to transport a surfactant molecule to the water surface than a water molecule, the surfactant reduces surface tension.

Because removal of the hydrophobic segment of the surfactant from water is entropically and energetically favorable, the surfactant molecules accumulate in the surface region of water. The equilibrium between the surfactant molecules in the interface or at the surface is given by the Gibbs equation

$$d\gamma = -\Sigma\Gamma_n d\mu_n \tag{3}$$

where $d\gamma$ is the change in the surface or interfacial tension, Γ_n is the surface excess concentration of the nth component, and $d\mu_n$ is the change in the chemical potential of the nth component. The surface excess concentration is the difference between the surfactant concentration in the surface region and the surfactant concentration in the bulk phase at equilibrium with the surface region. Surface excess concentration has been defined in general terms as the excess, per unit area of interface, of the amount of the component actually present in the system over the amount present in an ideal system of the same volume in which the bulk concentration remains constant up to the hypothetical dividing plane separating the two phases. The term *surface excess concentration* is unfortunately not very descriptive, especially when used simply as the *surface excess*. Attempts have been made to replace it, but the term *superficial density*, originally suggested by Gibbs, has not been a better choice.

At equilibrium between the interfacial and bulk phase concentrations

$$d\mu = -RT\,d\ln a_n \tag{4}$$

where a_n is the activity of the nth component in the bulk phase, R the gas constant, and T the absolute temperature. Hence

$$d\gamma = -RT\Sigma\,\Gamma_n\,d\ln a_n \tag{5}$$

For a dilute solution ($c \leqslant 0.01$M) containing only one nondissociating component the activity coefficients of the solute and solvent can be considered to be constant

$$d\gamma = -RT\Gamma\, d\ln c = -2.303RT\Gamma\, d\log c \tag{6}$$

where c is the molar concentration of the surfactant (or solute in general).

For a nondissociating surfactant in dilute solution the surface excess concentration can be calculated from the slope of a curve obtained by plotting the surface tension against log c

$$\Gamma = -\left(\frac{1}{2.303RT}\right)\frac{d\gamma}{d\log c} \tag{7}$$

For an ionic surfactant completely dissociated in a dilute solution

$$d\gamma = -RT(\Gamma_A\, d\ln a_A + \Gamma_B\, d\ln a_B) \tag{8}$$

In the absence of any other solute or electrolyte, assuming $\Gamma_A = \Gamma_B$ and $c_A = c_B$ to maintain electroneutrality,

$$d\gamma = -2RT\Gamma\, d\ln a = -4.606RT\Gamma\, d\ln a \tag{9}$$

For dilute solutions of a surfactant behaving like a strong electrolyte [8], the activity coefficient can be calculated using the Debye-Hückel relation and Eq. 9 becomes

$$d\gamma = -2RT\Gamma(1 - 0.58\sqrt{c})d\ln c \tag{10}$$

If the activity coefficient of a completely dissociated ionic surfactant in a dilute solution can be assumed to be unity,

$$d\gamma = -4.606RT\Gamma\, d\log c \tag{11}$$

The surface excess concentration Γ in a dilute solution of a completely dissociated surfactant in the absence of other electrolytes can be calculated from the slope of a surface tension plot against the logarithm of surfactant concentration:

$$\Gamma = -\left(\frac{1}{4.606RT}\right)\frac{d\gamma}{d\log c} \tag{12}$$

A typical plot of surface tension against the logarithm of concentration is shown in Fig. 4.2. The plot provides valuable practical information as well as theoretical insight. Initially, at low surfactant concentrations the surface tension decrease is gradual (section A–B). With increasing surfactant concentration the surface tension decreases more steeply and the curve becomes linear, as the liquid-vapor boundary becomes saturated with the adsorbed surfactant at the point (B). The plot is linear (section B–C) until an inflection point (C)is

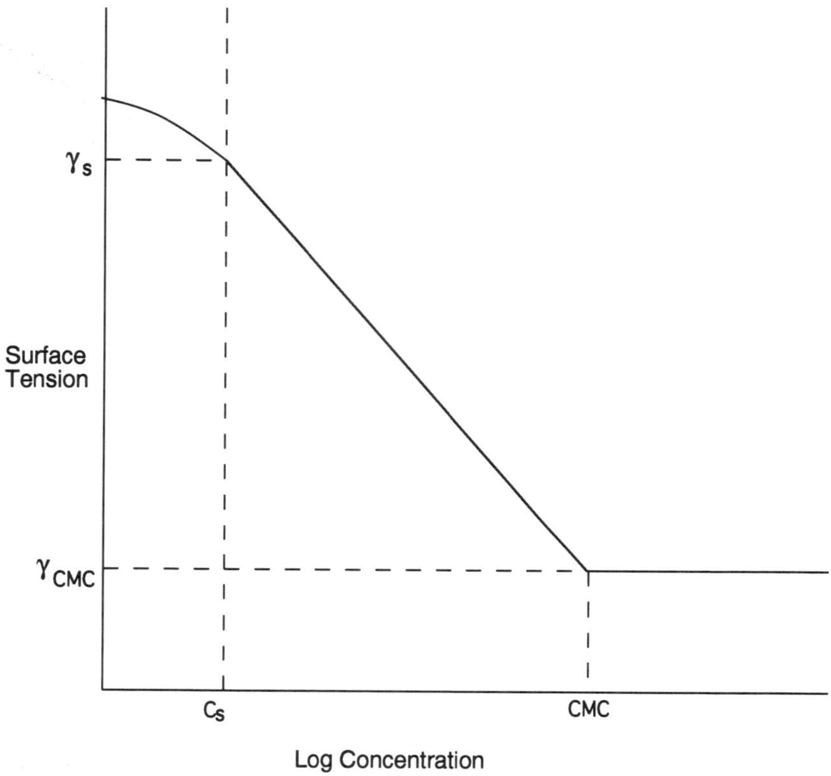

FIG. 4.2 Typical surface tension–log surfactant concentration curve.

reached. The inflection point usually corresponds to the critical micelle concentration (cmc) (see Chapter 6), although a limit of surfactant solubility, if reached below cmc, can also cause an inflection. Above cmc the surface tension is essentially constant, in the absence of complication factors.

The slope $d\gamma/d\log c$ is related to the surface excess concentration Γ (Eq. 12). Hence, the constant slope of the linear portion of the curve indicates that the surface excess concentration is constant and the surface is saturated. However, the surface tension continues to decrease with increasing surfactant concentration, in spite of $d\gamma/d\log c$ being constant. This apparent contradiction has been explained by surfactant concentration and activity changes in the bulk phase [9, 10].

The information needed for practical application of the surfactant is (1) the maximum surface tension decrease achievable with the surfactant, (2) the amount of surfactant needed to lower surface tension, and (3) the time needed to achieve the desired surface tension decrease (see Sec. 4.3).

Rosen [11] has introduced two terms to describe surfactant performance: efficiency and effectiveness. *Efficiency* is defined as the bulk liquid phase concentration of the surfactant at equilibrium required to depress the surface tension of the solvent by 20 dyne/cm. The definition of efficiency is a compromise because the surface saturation (Γ_m) is not achieved for all surfactants above the arbitrary surface tension value of 20 dyne/cm. *Effectiveness* is defined as the maximum surface tension reduction that can be obtained, regardless of bulk phase concentration of the surfactant. Instead of the term effectiveness the term surface tension at cmc, γ_{cmc}, is also being used.

Below the Krafft point the maximum surface tension decrease achievable is limited by the solubility of the surfactant. However, practical surfactant applications are usually above the Krafft point. Above the Krafft point (see Sec. 6.3), the minimum surface tension, γ_{cmc}, is indicated by the inflection point (C), corresponding to the cmc of the surfactant. Consequently, the surfactant concentration needed to achieve the minimum surface tension is the critical micelle concentration (cmc).

4.2 ADSORPTION AT LIQUID-VAPOR BOUNDARY

Surface tension curves, such as the one shown in Fig. 4.2, yield useful information, in addition to defining the minimum surface tension at cmc, γ_{cmc}. The surfactant concentration needed to attain the surface excess concentration, c_s, corresponds to surface saturation, Γ_m. The slope $d\gamma/d \log c$ of the linear portion of the curve is related to the surface excess concentration, Γ_m.

The excess surface concentration Γ_m can be determined from the linear portion of the surface tension–log surfactant concentration curve. Although direct determination of Γ is experimentally more difficult, several satisfactory methods have been developed [12]. McBain and co-workers [13,14] analyzed a slice of the surface removed with a microtome. Radioisotope tracer methods [15–21] are simpler and the agreement with calculated values has been quite satisfactory.

The adsorption of surfactant at the surface, and consequently Γ_m, depends on the surfactant structure. The Γ_m values listed by Rosen for nonfluorinated anionic, cationic, nonionic, and amphoteric surfactants [11] indicate that the hydrophilic group has a considerable effect. For ionic surfactants, Γ_m decreases with increasing cross-sectional area of the hydrated hydrophile. Carboxylates generally have higher Γ_m values than sulfonates or sulfates [10]. The Γ_m values for cationic surfactants decrease with increasing size of the quaternary hydrophile. The Γ_m values for nonionic surfactants decrease with increasing length of the oxyethylene chain. The length of the hydrophobic chain of ionic or nonionic surfactants beyond 10 carbon atoms, however, has very little effect below a chain length of 16 carbon atoms. A decrease in Γ_m for a chain length exceeding 18 carbon atoms has been explained by coiling of the carbon chain.

Branching of the hydrocarbon chain has very little effect. Similarly, a central placement of the hydrophile in the middle of the hydrocarbon chain, instead of a terminal position, has no significant effect on Γ_m. However, two terminal hydrophiles on opposite ends of the hydrophobe reduce Γ_m considerably. The hydrophobe is looped between both of its hydrophiles located in water (Fig. 4.3). [22]. In contrast, Γ_m values for nonionic surfactants with two oxyethylene chains attached to the *same* end of the hydrophobe are similar to the Γ_m values for surfactants with one oxyethylene chain and the same number of oxyethylene units [23].

Rosen [11] has suggested that fluorination of the hydrophobe has apparently only a small effect on Γ_m, in contrast to the large effect fluorination has on surface tension. If fluorination does not affect significantly the surface excess concentration Γ_m, the large effect of fluorination on surface tension must be related to the chemical potential of the fluorinated surfactant (Eq. 3). The chemical potential depends on the nature of the hydrophobe, as well as the hydrophile. The main factors in reducing surface tension are the decrease in the work needed to transport a surfactant molecule to the surface and the low surface energy of the monomolecular film formed by adsorbed surfactant molecules.

Under conditions of surface saturation, the Gibbs equation can be used to determine the maximum extent of surfactant adsorption on the surface. The area occupied by a molecule, A_s, expressed in nm^2, can be calculated from the surface excess concentration Γ_m of the saturated surface ($mole/cm^2$)

$$A_s = 10^{14}/N\Gamma_m \tag{13}$$

where N is Avogadro's number.

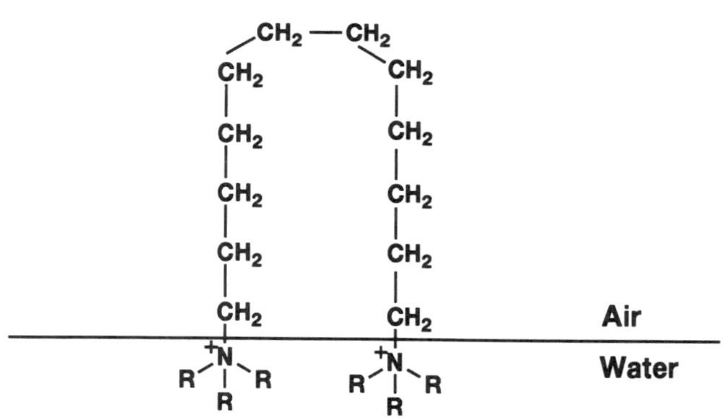

FIG. 4.3 Orientation of a surfactant with two hydrophilic groups on the air-water interface. (From Ref. 22. Reproduced by permission of American Chemical Society.)

The area occupied by a surfactant molecule provides valuable information on the orientation of the surfactant molecule at the surface. The adsorbed surfactant can be visualized to form a monomolecular film which exerts surface pressure, π,

$$\pi = \gamma_o - \gamma_s \tag{14}$$

where γ_o is the surface tension of the pure solvent and γ_s is the surface tension of the solution.

Shinoda and co-workers [24] calculated the cross-sectional area of the surfactants, using the equation [25,26]

$$-\left(\frac{\vartheta\gamma}{\vartheta \ln C_2}\right)_T = RT\Gamma_2\left(1 + \frac{K_g}{Z}\right)$$

where C_2 is the concentration of the surface-active ion, Γ_2 is the surface excess of surface-active ion, Z is the number of charges of counterion, and K_g is 0.52 for n-$C_7F_{15}COOK$ and is assumed to be 0.5 for other surfactants.

The cross-sectional areas of the salts of perfluorooctanesulfonic acid decrease in the order Li $\approx$ Na $>$ K $>$ NH_4 (Table 4.1). The sparse data available do not show a similar trend for perfluorocarboxylic acid salts.

Tadros [27] determined the area per molecule for Monflor surfactants from surface tension measurements. The area per molecule in the adsorbed monolayer was calculated from the surface excess concentration, Γ, determined from the linear portion of the γ–log C curve above cmc (Fig. 4.4). Equation 7 was used for nonionic and Eq. 12 for ionic surfactants (Table 4.2).

The area covered by the fluorinated surfactant Monflor 71 molecule, 0.26 nm^2, is much smaller than the area, 0.47 nm^2, covered by a similar hydrocarbon surfactant, cetyltrimethylammonium bromide (CTABr), molecule. The small area covered by the adsorbed fluorinated surfactant reveals very tight packing at the interface.

The amine(polyfluoroalkoxyacyl)imide surfactants

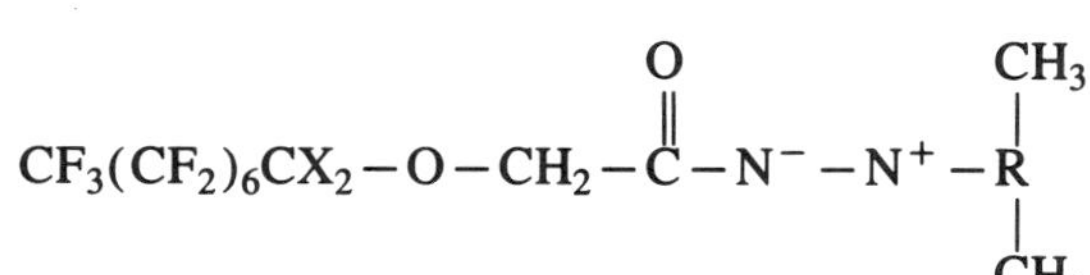

a. X = F R = $-CH_3$
b. X = H R = $-CH_3$
c. X = H R = $-CH_2CH_2OH$
d. X = H R = $-CH_2-CH(OH)-CH_2OH$

TABLE 4.1 Surface Tensions of Fluorinated Surfactants in Water at 25°C

Surfactant	γ_{min} (mN/m)	Γ_m (mol/cm^2 × 10^{10})	A (Å^2)	pC_{20}
n-$C_6F_{13}COOLi$	27.8			
n-$C_7F_{15}COOH$	15.2		41.5	
n-$C_7F_{15}COONa$	24.6	4.0	42.0	2.50
n-$C_7F_{15}COOK$	20.6	3.9	43.0	2.57
n-$C_8F_{17}COOLi$	24.6			
n-$C_8F_{17}COONa$	21.5			
n-$C_8F_{17}COOK$	20.6			
n-$C_8F_{17}COONH_4$	14.8			
n-$C_8F_{17}COONH_3C_2H_4OH$	15.9			
n-$C_8F_{17}COONH(C_2H_4OH)_3$	21.9			
n-$C_{10}F_{21}COOLi$	20.5			
n-$C_{10}F_{21}COONH_3C_2H_4OH$	13.8			
n-$C_{10}F_{21}COONH(C_2H_4OH)_3$	16.9			
$(CH_3)_2CF(CF_2)_4COOH$	≈15.5		48.0	
$(CH_3)_2CF(CF_2)_4COONa$	20.2	3.8	43.5	2.57
$(CH_3)_2CF(CF_2)_4COOK$	19.5		47.5	
$(CH_3)_2CF(CF_2)_4CHCH_2COOK$	19.4[a]			
n-$C_7F_{15}SO_3Na$	37.3	3.1	52.5	2.76
n-$C_8F_{17}SO_3Li$	29.8	3.0	55.2	3.20
n-$C_8F_{17}SO_3Na$	40.5	3.1	52.5	3.23
n-$C_8F_{17}SO_3K$	34.5	3.7	45.1	3.56
n-$C_8F_{17}SO_3NH_4$	27.8	4.1	41.0	3.40
n-$C_8F_{17}SO_3NH_3C_2H_4OH$	21.5	3.9	42.5	3.44
n-$C_9F_{19}SO_3\frac{1}{2}Mg \cdot 2\ H_2O$	22.0		45.2	

[a] In 0.02 N KOH.
Source: Refs. 24, 25, and 40.

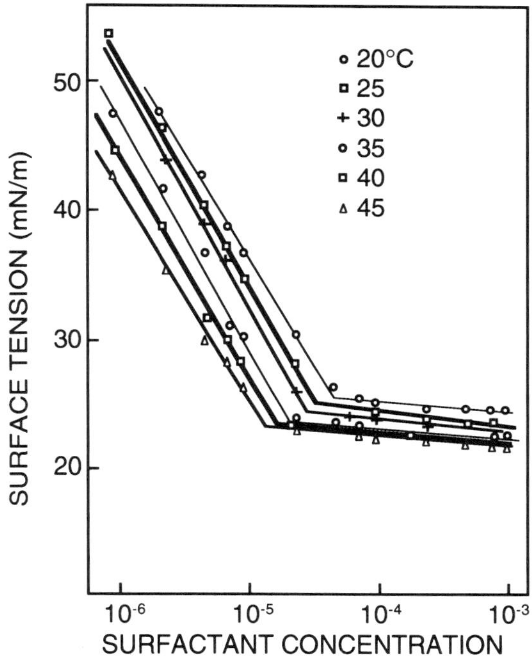

FIG. 4.4 Surface tension curves for Monflor 51 at various temperatures. (From Ref. 27. Reproduced by permission of Academic Press, Inc.)

TABLE 4.2 Area Covered by a Molecule in Adsorbed Monolayers of Monflor Surfactants

Trade name	Chemical formula	Area/molecule at 25°C (nm^2)
Monflor 51	$C_{10}F_{19}O(CH_2CH_2O)_nC_{10}F_{19}$, average $n \approx 23$	0.53
Monflor 31	$C_{10}F_{19}OC_6H_4SO_3Na$	0.36
Monflor 71	$C_{10}F_{19}OC_6H_4N(CH_3)_3^+CH_3SO_4^-$	0.26

Source: Ref. 27.

form aqueous solutions with extremely low surface tensions (16.7, 15, 16.2, and 15.8 mN/m^{-1} for a, b, c, and d, respectively) [28]. These low surface tension values are probably related to the very tight packing of the adsorbed surfactant. The surface areas occupied by the individual surfactant molecules, determined by using the Gibbs equation, Eq. 3, range from 32 to 39 $Å^2$.

For surface tension measurements the surfactant has to be adequately soluble in water. In contrast, surface pressure measurement using the Langmuir film balance requires that the surfactant be water immiscible and spread on the water surface.

Arrington and Patterson [29] spread the fluoroalcohol $H(CF_2)_{10}CH_2OH$ on water and determined the area occupied per molecule from the force-area curve of the spread monolayers (Fig. 4.5). By extrapolating the upper part of the curve to zero pressure, a close-packed area of 29 $Å^2$ was obtained. The fluorocarboxylic acid $H(CF_2)_{12}COOH$ was spread on water and on 0.01 N hydrochloric acid (Fig. 4.6). The force-area curve obtained for hydrochloric acid agrees with the curve shown for the fluoroalcohol in Fig. 4.5. When the fluorocarboxylic acid film on hydrochloric acid was recompressed, the curve was duplicated. The compression of the fluorocarboxylic acid film on water gave a smaller area per molecule than obtained on acid. A second compression gave even a smaller area (19 $Å^2$).

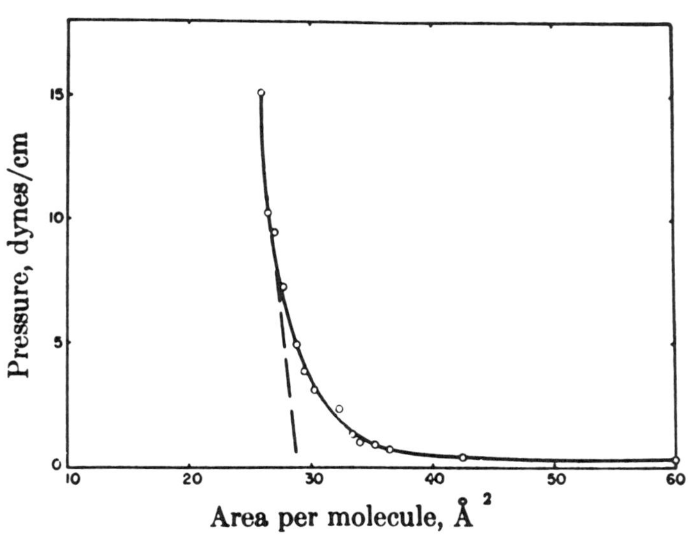

FIG. 4.5 Force-area curve for $H(CF_2)_{10}CH_2OH$. (From Ref. 29. Reproduced by permission of American Chemical Society.)

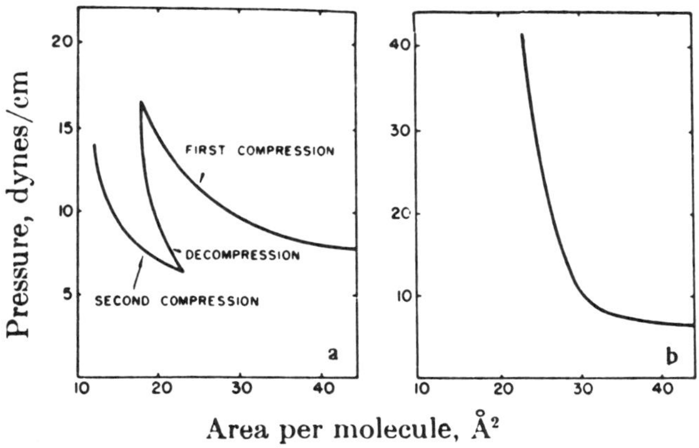

FIG. 4.6 Force-area curves for $C_{12}F_{25}COOH$ at (a) the water-air interface and (b) at a 0.01 N hydrochloric acid–air interface. (From Ref. 29. Reproduced by permission of American Chemical Society.)

The instability of the fluorocarboxylic acid film on water may have been caused by the significant, albeit very low, solubility of the acid in water. The fact that the fluorocarbon chain has a larger cross-sectional area than the carboxyl group may also have contributed to film instability.

Caporiccio et al.[30] prepared two series of surfactants:

Series A:

$$CF_3(O-CF_2-\underset{\displaystyle CF_3}{\underset{|}{CF}})_n-O-CF_2COOH \qquad (n = 1,2,3)$$

Series B:

$$HOOCCF_2-(O-CF_2CF_2)_n-(OCF_2)_m-OCF_2COOH \qquad (n = 1,2,3)\ (m = 1,2,3)$$

The area occupied by a surfactant molecule was calculated using Eq. 10 from linear plots of surface tension versus log surfactant concentration (Figs. 4.7 and 4.8). The area covered by a molecule in a monolayer was also determined from surface pressure–area curves (Figs. 4.9 and 4.10) obtained with the Langmuir film balance. Silver and zinc salts of (A) series and silver salts of (B) series were used for the measurement. Two sets of cross-sectional areas were calculated. One set represents molecules close together in a complete monolayer without strain (uncompressed monolayer). The other set is related to the limiting area per molecule below which the molecules in the monolayer begin to overlap (compressed monolayer). The values of the first set were obtained from surface pressure–area curves as the area per molecule below

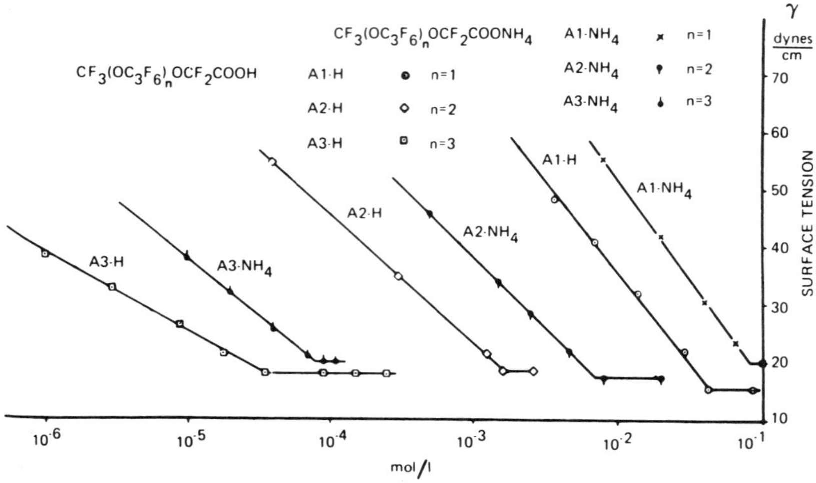

FIG. 4.7 Surface tension curves for monocarboxylic perfluoroether surfactants of the general formula $CF_3(OC_3F_6)_nOCF_2COOH$ or its ammonium salt. (From Ref. 30. Reproduced by permission of Academic Press, Inc.)

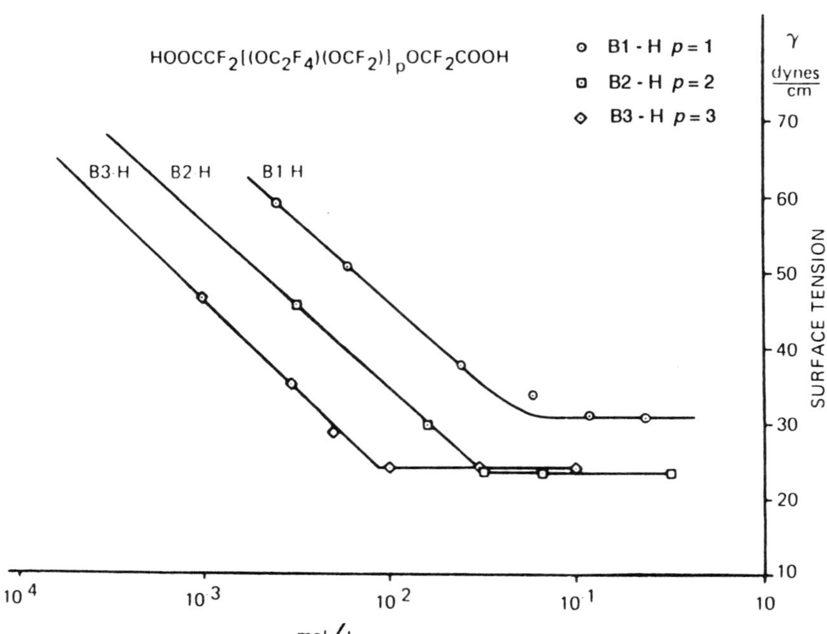

FIG. 4.8 Surface tension curves for dicarboxylic perfluoroether surfactants of the general formula $HOOCCF_2[(OC_2F_4)(OCF_2]_pOCF_2COOH$; B1-H, $p = 1$; B2-H, $p = 2$; BH-3, $p = 3$. (From Ref. 30. Reproduced by permission of Academic Press, Inc.)

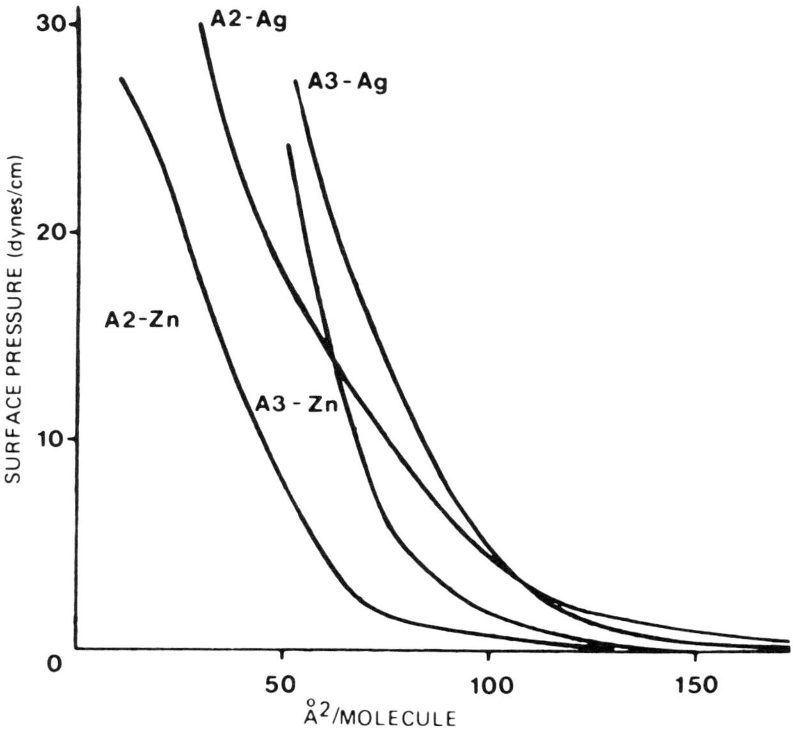

FIG. 4.9 Surface pressure–area per molecule curves for the silver and zinc salts of the monocarboxylic acid surfactants A2-Ag, $n = 2$; A2-Zn, $n = 2$; A3-Ag, $n = 3$; A3-Zn, $n = 3$. (From Ref. 30. Reproduced by permission of Academic Press, Inc.)

which surface pressure begins to increase rapidly. The limiting area values of the second set were calculated from the two-dimensional van der Waals equation of state [31].

$$\left[\frac{\pi + a}{A^2}\right](A - A_o) = kT \tag{15}$$

where A is the area per molecule at a surface pressure π, A_o is the limiting area per molecule, a is a constant, k is the Boltzmann constant, and T is the absolute temperature.

The cross-sectional areas are listed in Table 4.3, together with areas estimated from Stuart molecular models. The values calculated from surface excess concentration and values for uncompressed areas are in fairly good agreement. The areas of (A) series surfactants for the compressed monomolecular film agree reasonably well with the areas estimated using Stuart molecular

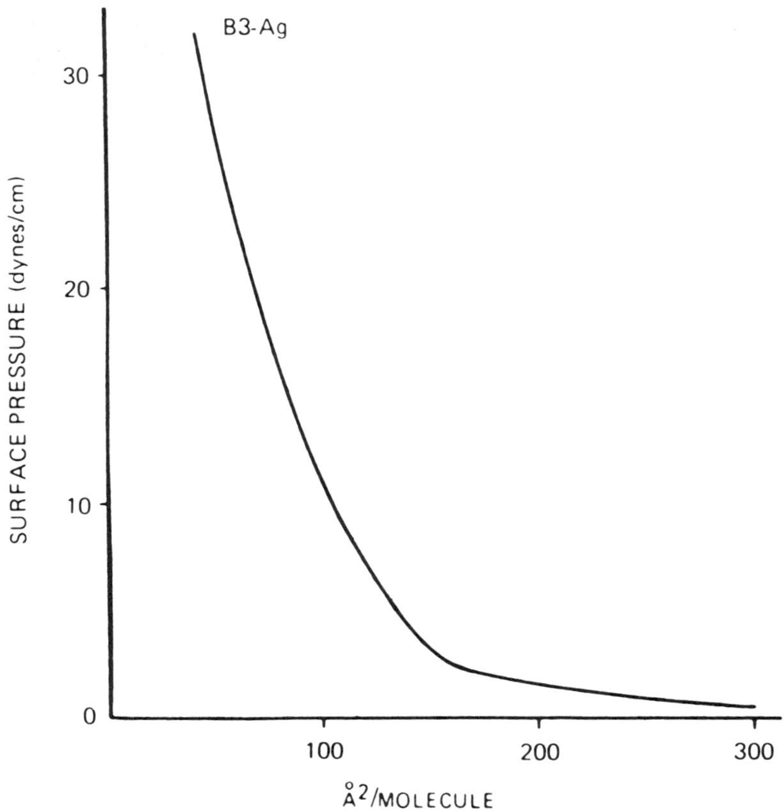

FIG. 4.10 Surface pressure–area per molecule curves for the silver salt of the dicarboxylic acid surfactant B-3-Ag, $p = 3$. (From Ref. 30. Reproduced by permission of Academic Press, Inc.)

models. However, the compressed monolayer areas of (B) series surfactants are twice as large as the Stuart areas. This suggests a folded structure of the polyether chain on the surface with both carboxylic acid groups oriented toward water [30].

Matos et al. [32] synthesized nonionic fluorinated surfactants which contained both oxyethylene and thioethylene groups:

$CF_3(CF_2)_mC_2H_4(SC_2H_4)(OC_2H_4)_nOH$ (Series I)

$CF_3(CF_2)_mC_2H_4(SC_2H_4)(OC_2H_4)_p(SC_2H_4)(OC_2H_4)_qOH$ (Series I)

The surface excess concentration and the area covered by the surfactant molecule adsorbed in the water-vapor boundary were calculated from the

TABLE 4.3 Limiting Area per Molecule (Å^2) for Monofunctional Silver and Zinc Salts and for a Difunctional Silver Salt

Code[a]	Area calculated from $d\gamma/dT$	Area determined with the Langmuir film balance		Stuart model area
		Uncompressed	Compressed	
A1	63			42
A2	81	72^b–70^c	47^b–47^c	42
A3	95	76^b–101^c	51^b–53^c	42
B1	136			28
B2	136			28
B3	128	112^c	56^c	28

[a] Surfactant structure code: see Figs. 4.7 and 4.8.
[b] Zn salt.
[c] Ag salt.
Source: Ref. 30.

slopes of the surface tension curves. The areas occupied by the adsorbed series I surfactant molecules with a short hydrophile ($n = 2$ or 3) were small (35 and 37 Å^2), indicating a dense packing and perpendicular orientation of the fluorocarbon chains. For the less hydrophilic surfactants in series II the areas were only a little larger, also revealing dense packing of the fluorocarbon chains (Fig 4.11). The areas occupied by the more hydrophilic surfactants of series I or II increased to values comparable to $A = 47$ Å^2 in the lamellar mesophases of $C_6F_{13}CH_2(OC_2H_4)_4OH$.

Selve at al. [23] synthesized fluorinated nonionic surfactants with a two-chain polyoxyethylene hydrophilic head linked to the hydrophobe via an amide bond, $F(CF_2)_l(CH_2)_mC(O)N[(C_2H_4O)_nCH_3]_2$. They calculated the area A per surfactant molecule adsorbed on the air-water interface from the slope of the surface tension curve using the Gibbs equation, Eq. 13. The area, A, increases with increasing number of oxyethylene units for both fluorinated and hydrocarbon surfactants. For nonionic surfactants with six oxyethylene units the area, A, occupied by linear fluorinated surfactants is smaller than those for the corresponding hydrocarbon-based nonionic surfactants. For nonionic surfactants with a larger number of oxyethylene units, the A values are similar for fluorinated or nonfluorinated surfactants.

In ideal systems, the surface area covered by mixtures of surfactants adsorbed at the liquid-vapor boundary is the sum of the areas covered by the components

$$A_{1,2} = X_1A_1 + X_2A_2 \tag{16}$$

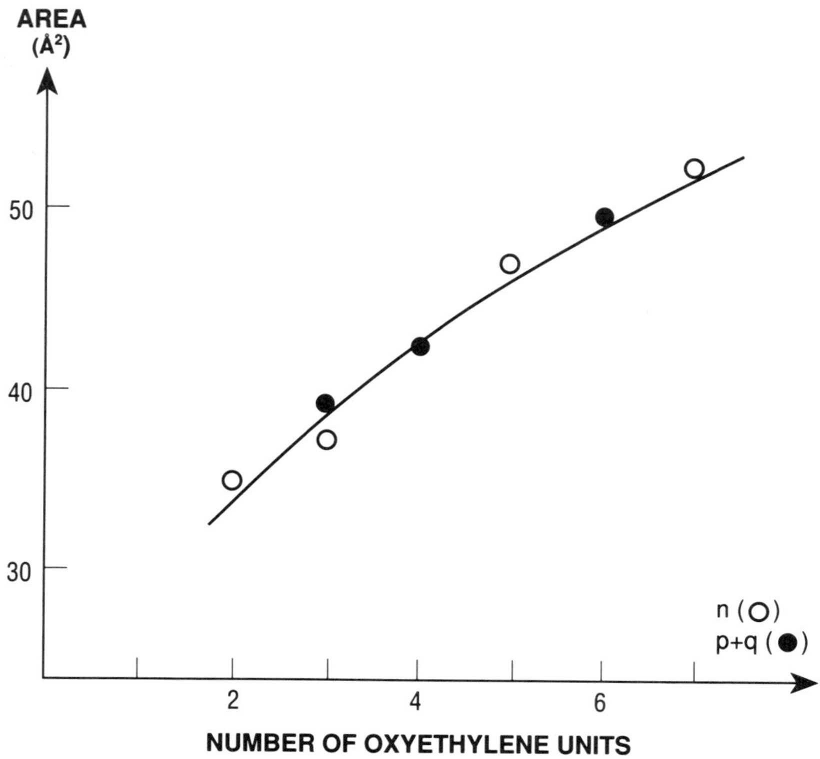

FIG. 4.11 Area A occupied by the adsorbed surfactant plotted against the number of oxyethylene units n (surfactant series I) and $p + q$ (surfactant series II). (From Ref. 32. Reproduced by permission of Academic Press, Inc.)

where X_1 and X_2 are the mole fractions and A_1 and A_2 are the molecular areas of pure components 1 and 2, respectively. When the mixed surfactants have similar molecular structures, the surface pressure–area isotherms obey the additivity rule. Thus the π-A curves of two cationic fluorinated surfactants,

$$C_3F_7OCF(CF_3)CF_2OCF(CF_3)CONH(CH_2)_3N^+(C_2H_5)_2CH_3I^-\ (FCN3)$$

and

$$C_3F_7OCF(CF_3)CF_2OCF(CF_3)CF_2OCF(CF_3)CONH(CH_2)_3N^+(C_2H_5)_2CH_3I^-\ (FCN4),$$

coincide with curves calculated by Zhang and co-workers using the additivity rule (Fig. 4.12) [33]. Surprisingly, the π-A curves of mixtures of perfluorononanoic acid and sodium octadecanesulfonate also obey the additivity rule. The experimental curves coincide with the calculated curves (Fig. 4.13), although the interaction between the hydrocarbon and fluorocarbon

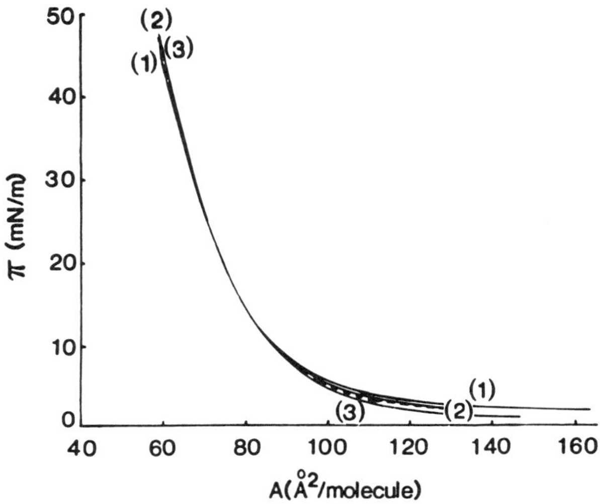

FIG. 4.12 Surface pressure–area per molecule isotherms for FCN3-FCN4 mixtures (20°C). The mole fractions of FCN3 are (1) 1.00; (2) 0.506; (3) 0.00. (From Ref. 33. Reproduced by permission of Academic Press, Inc.)

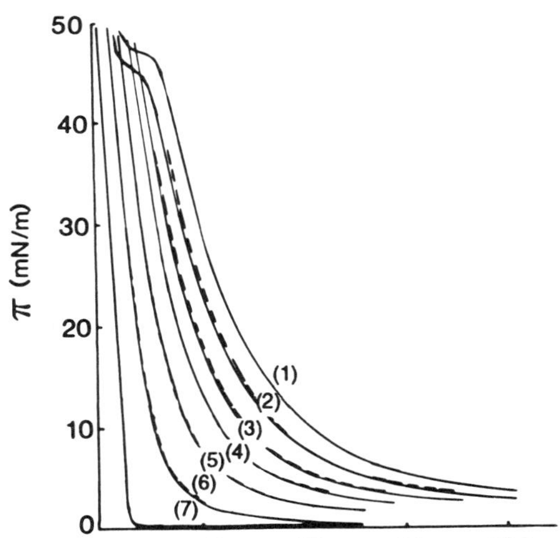

FIG. 4.13 Surface pressure–area per molecule isotherms for $C_9FH-C_{18}H_{37}SO_3Na$ mixtures (20°C). The mole fractions of C_9FH (perfluorononanoic acid) are (1) 1.00; (2) 0.832; (3) 0.665; (4) 0.512; (5) 0.335; (6) 0.167; (7) 0.00. (From Ref. 33. Reproduced by permission of Academic Press, Inc.)

chains is mutually phobic. Mixtures of perfluorooctanoic acid and sodium octadecylsulfate behave similarly. Hence the additivity rule is obeyed by two entirely different systems [12,33]: (1) the mixed surfactants are so similar that they can form a two-dimensional ideal solution, or (2) the surfactants are so different that they are incompatible and form separate monolayers. Similar results were obtained by Zhang et al. [33] using benzene-acetone-ethanol as the spreading solvent to form a monolayer.

A mixed monolayer can be considered an ideal monolayer when the surface pressure π is near zero. Zhang et al. [33] suggested that near zero surface pressure, mixtures of mutually phobic surfactants indeed form ideal monolayers. As the film is compressed the distance between molecules in the monolayer decreases and the mutual phobic interaction increases. Consequently, the mixture exhibits a positive deviation from the additivity rule. When the film is compressed further, the phobic interaction becomes so strong that the surfactants become incompatible and form separate monolayers. However, when the phobic interaction is very strong, the phase separation may occur at a low pressure where the deviations are experimentally insignificant and escape observation.

Mixtures of $CF_3C(C_2F_5)_2C(CF_3){=}C(CF_3)OC_6H_4SO_3Na$ (6201) and sodium octadecylsulfate deviated positively from the additivity rule at low pressures, but at higher pressures the π-A curves coincided with the calculated curves (Fig. 4.14). The apparent ideality at higher pressures is probably caused by the separation of the phobic components, in accord with the postulated mechanism. When the phobic interactions are reduced by a hydrocarbon segment in the fluorocarbon chain or branching of the hydrophobe chain, a positive deviation from the additivity rule may also be evidenced at higher pressures. As an example, the interaction between the FCN4 surfactant and octadecyltrimethylammonium bromide, both cationic surfactants, is not strong enough to form separate monolayers and the π-A curves deviate positively from the calculated curves (Fig. 4.15). When the hydrocarbon character of the hydrophobe is increased further, the deviation from ideal additivity decreases and may even become negative.

4.3 SURFACE TENSION IN WATER. SURFACTANT STRUCTURE

The surface tensions of Zonyl fluorinated surfactants [34] are shown in Table 4.4.

When selecting a surfactant to lower the surface tension of an aqueous solution it is useful to know

1. The concentration needed to lower surface tension to a given value
2. The maximum surface tension lowering, or the minimum surface tension achievable at any surfactant concentration

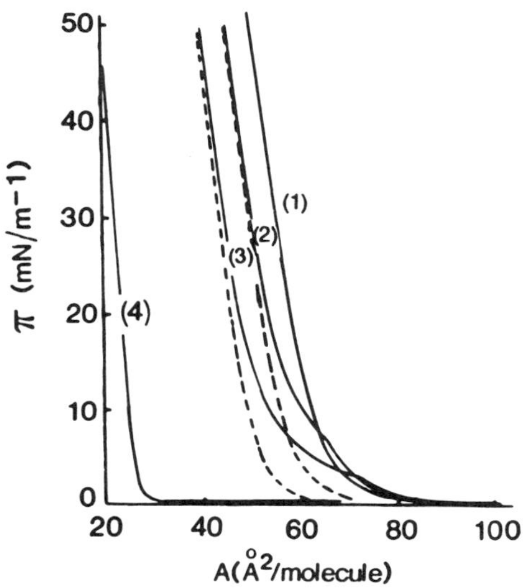

FIG. 4.14 Surface pressure–area per molecule isotherms for 6201-$C_{16}H_{33}SO_4Na$ mixtures (10°C). The mole fractions of 6201 are (1) 1.00; (2) 0.833; (3) 0.501; (4) 0.00. (From Ref. 33. Reproduced by permission of Academic Press, Inc.)

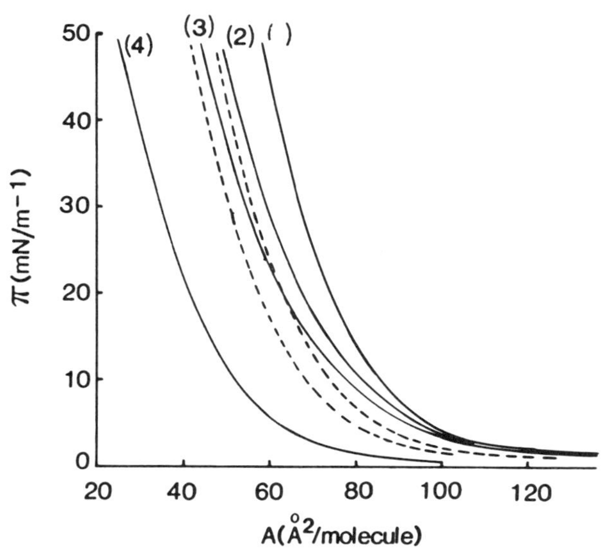

FIG. 4.15 Surface pressure–area per molecule isotherms for FCN4-$C_{18}H_{37}N(CH_3)_3Br$ mixtures (13°C). The mole fractions of FCN4 are (1) 1.00; (2) 0.665; (3) 0.495; (4) 0.00. (From Ref. 33. Reproduced by permission of Academic Press, Inc.)

TABLE 4.4 Surface Tension of Zonyl Fluorinated Surfactants, $R_f = F(CF_2CF_2)_{3-8}$[a]

Zonyl	Structure	Surface tension in water at 25°C (mN/m)		
		0.001%	0.01%	0.1%
FSA	$R_fCH_2CH_2SCH_2CH_2COOLi$	48	22	18
FSP	$(R_fCH_2CH_2O)P(O)(ONH_4)_2$	42	24	21
FSE	$(R_fCH_2CH_2O)_2P(O)(ONH_4)$	45	27	20
UR	$(R_fCH_2CH_2O)P(O)(OH)_2$ $(R_fCH_2CH_2O)_2P(O)(OH)$	46	40	28
FSN	$R_fCH_2CH_2O(CH_2CH_2O)_xH$	29	24	23
FSO	$R_fCH_2CH_2O(CH_2CH_2O)_yH$	22	19	18
FSC	$R_fCH_2CH_2SCH_2CH_2N^+(CH_3)_3CH_3SO_4^-$	58	21	19
FSK	$R_fCH_2CH(OCOCH_3)CH_2N^+(CH_3)_2CH_2CH_2CO_2^-$	40	21	19
TBS	$R_fCH_2CH_2SO_3X$ (X = H and NH_4)	58	38	24

[a] All surfactant concentrations are given in percent solids. Zonyl UR was neutralized with sodium hydroxide to pH 7–8.

The minimum surface tension achievable with a surfactant in solution at equilibrium depends on

1. The number of surfactant molecules accumulated per unit surface area, indicated by the surface excess concentration Γ_m. The number of molecules adsorbed at the surface is inversely proportional to the area occupied by each surfactant molecule (Eq. 13).
2. The surface energy of the monomolecular film of adsorbed surfactant molecules
3. The free energy of adsorption of a surfactant molecule from the bulk solution phase to the surface.

As already discussed in Sec. 4.1, the surface tension of a surfactant has its minimum value at cmc, γ_{cmc}. Variations of surface tension values above cmc are usually small. Hence, cmc represents fairly accurately the surfactant concentration needed to achieve the minimum surface tension.

The effect of surfactant structure on cmc is discussed in Sec. 6.4. The complex relationship between surface tension and cmc depends on the hydrophobe and the hydrophile, including the counterion, of the surfactant. An increase in the chain length of the hydrophobe decreases cmc; branching of the carbon chain increases cmc. Fluorination of the hydrophobe lowers cmc considerably. In addition to the chemical structure of the surfactant, cmc depends on external factors, including electrolyte effects, temperature, and other dissolved or solubilized organic components.

The concentration needed to achieve a given surface tension below cmc depends qualitatively but not solely on the same structural factors as cmc. The concentration needed to lower surface tension decreases with increasing hydrophobicity of the surfactant. The efficiency of a fluorinated surfactant in lowering surface tension is determined mainly by the structure of the hydrophobic group. Hydrophilic groups have a smaller effect.

The Hydrophobe

Fluorination of the hydrophobe decreases markedly the surfactant concentration needed for a substantial surface tension reduction. At 1 g/L concentration, sodium octanesulfonate, $C_8H_{17}SO_3Na$, lowers the surface tension of water at 20°C to 65 nM/m, but sodium perfluorooctanoate, $C_8H_{17}SO_3Na$, at the same concentration lowers surface tension to 32 mN/m [35]. Conventional surfactants with a hydrocarbon-type hydrophobe can lower surface tension to 30–35 mN/m at 0.1% concentration. With fluorinated surfactants surface tensions below 30 mN/m have been achieved at concentrations as low as 10–100 ppm. With some fluorinated surfactants only 100–200 ppm of the surfactant is needed to lower surface tension below 20 mN/m [34,36].

The minimum surface tension achievable is also much lower for fluorinated surfactants than for nonfluorinated surfactants. The minima of the surface tensions of surfactants with a hydrocarbon hydrophobe are in the range 25–35 mN/m [37–39], whereas those of fluorinated surfactants are as low as 15–20 nM/m or even lower [40]. The surface tension of aqueous solutions above cmc varies only slightly with surfactant concentration. Surface tension above cmc decreases with increasing fluorocarbon chain length and depends on the counterion (Table 4.1).

Increasing the carbon chain length increases the efficiency of the surfactant in surface tension reduction [36,37,41] (Fig. 4.16). Surface tension of perfluoroalkanoic acid solutions plotted against their log molar concentration is shown in Fig. 4.17 [42]. The surface tension of perfluoroalkanoic acids obeys Traube's rule

$$\log C_\mathrm{m} = C - m \log K_\mathrm{T} \tag{17}$$

where C_m is the concentration of a homologue surfactant needed for a given surface tension, m is the number of carbon atoms in the hydrophobe C_mF_{2m+1},

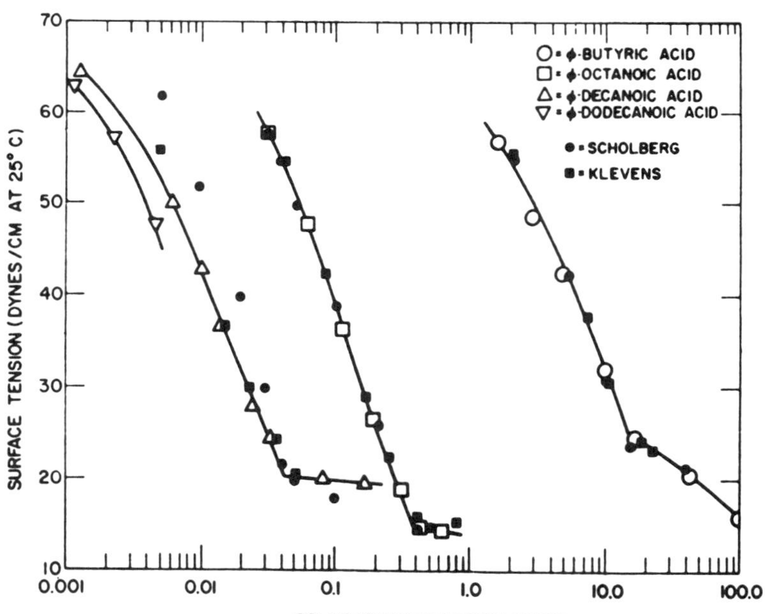

FIG. 4.16 Surface tensions of perfluoroalkanoic acids in aqueous solutions versus log weight percent. (From Ref. 37. Reproduced by permission of American Chemical Society.)

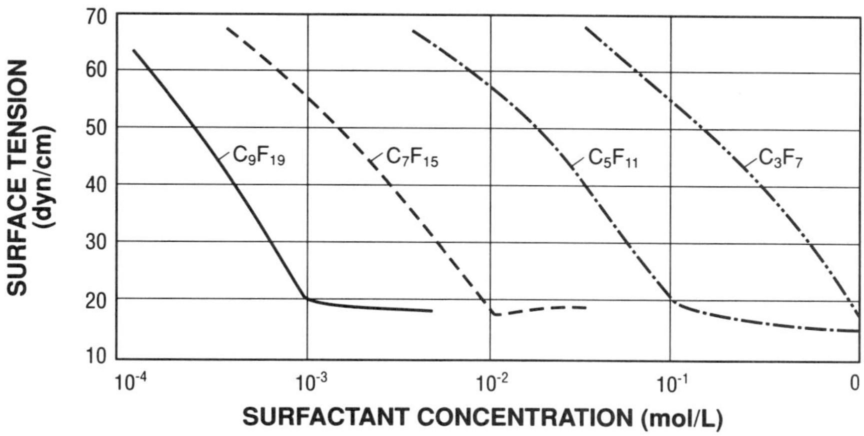

FIG. 4.17 Surface tensions of perfluoroalkanoic acids in aqueous solutions versus log molar concentration. (From Ref. 42. Reproduced by permission of Carl Hanser Verlag.)

K_T is the Traube constant, and C is a constant. The logarithm of the concentration needed to achieve a given surface tension value plotted against the number of carbon atoms in the fluorocarbon hydrophobe yields a straight line (Fig. 4.18).

Traube's rule states that each CH_2 group added to the hydrocarbon chain reduces the concentration required to give a certain surface tension value by approximately a factor of 3. Accordingly, addition of two CH_2 groups lowers the required concentration by a factor of 9. Figure 4.18 shows that two CF_2 groups decrease the required concentration by a factor of about 10, in good agreement with Traube's rule. While the surface tensions of members of a homologous series exhibit certain regularities, Traube's rule seems to hold because of a fortuitous mutual compensation of various factors [12].

The effect of the chain length of some perfluoropolyether surfactants [30] is shown in Figs. 4.7 and 4.8.

Branching of the fluorocarbon chain decreases the efficiency of a fluorinated surfactant in surface tension reduction [43,44]. In analogy, a condensed (spread) monolayer of a perfluorinated *n*-alkanoic acid has a lower critical surface tension than its terminally branched isomer. Bernett and Zisman [45] attributed the effect of branching to different molecular packing and carbon chain adlineation.

The Hydrophile

The effect of the hydrophilic group on surface tension depends on the structure of the hydrophile and, for ionic surfactants, also on the counterion. For a con-

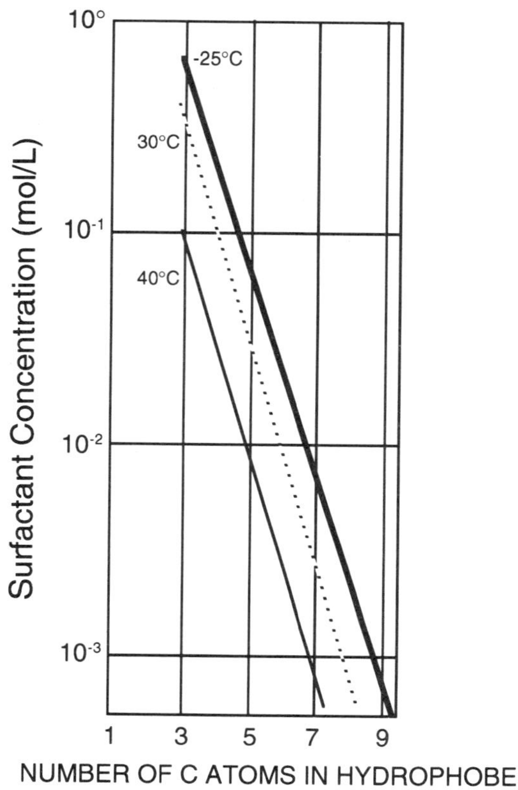

FIG. 4.18 Traube plot for fluorinated surfactants $C_nF_{n+1}COOH$. The log of the concentration required for attaining equal surface tensions is plotted against the number of carbon atoms *n*. (From Ref. 42. Reproduced by permission of Carl Hanser Verlag.)

stant chain length of C_7F_{15} the surface tension of 0.1% solutions of fluorinated surfactants varies between 17 and 47 mN/m, depending on the nature of the hydrophile [42] (Table 4.5). Nonionic fluorinated surfactants usually have lower surface tensions than their ionic counterparts. Nonionic surfactants derived from *N*-methylperfluorosulfonamide lower surface tension to 20 mN/m, whereas the surface tension of perfluorosulfonic acid salt solutions averages 22.8 mN/m [35]. However, the carbamide linkage is chemically not resistant to hydrolysis, which limits the use of these surfactants under severe conditions.

The surface tension of an anionic fluorinated surfactant can be lowered by inserting an oxyethylene group between the fluorinated hydrophobe and the sulfate hydrophile. Greiner and co-workers [46] prepared surfactants having the structure $H[CF_2CF_2]_nCH_2[OCH_2CH_2]_mOSO_3NH_4$. The oxyethylene group has lowered the surface tension by about 5 dyne/cm (Fig. 4.19).

TABLE 4.5 Surface Tensions of Various Fluorinated Surfactants with a C_7F_{15} Perfluoroalkyl Group

Surfactant	Surface tension for 0.1% aqueous solutions (mN/m)
$C_7F_{15}CH_2OSO_3Na$	47
$C_7F_{15}COOH$	40
$C_7F_{15}CH_2O(CH_2CH_2O)_4H$	28
$C_7F_{15}CH_2O(CH_2CH_2O)_{14}H$	25
$C_7F_{15}CO-N(H)(CH_2)_3(CH_2CH_2OH)_2$	17

Source: Ref. 42. Reproduced by permission of Carl Hanser Verlag.

According to Klein and co-workers [47], the minimum surface tension of perfluorooctanesulfonic acid salts depends very little on the nature of the counterion. The minimum surface tension depends mainly on the hydrophobe. However, the concentration needed to achieve the minimum surface tension, presumably γ_{cmc}, varies considerably (Table 4.6). In contrast, Shinoda and co-workers [24] reported large differences between the surface tension minima for *n*-perfluorooctanesulfonic acid salts (Table 4.1). The discrepancy between the results in Tables 4.1 and 4.6 may have been caused by different purities of the surfactants used.

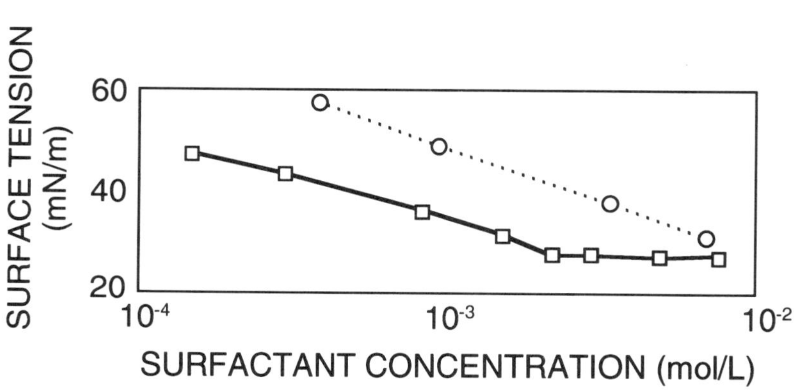

FIG. 4.19 Surface tension of ammonium ω-H-perfluorononyl sulfate (----) and ammonium ω-H-perfluorononyl sulfate with an oxyethylene segment (___). (From Ref. 46. Reproduced by permission of Dr. A. Hüttig Verlag.)

TABLE 4.6 Surface Tension of Perfluorooctanesulfonates

Surfactant	γ_s[a] (mN/m)	C_s (g/L)	t[b] (min) at 1 g/L	Number of phases at 10 g/L
$C_8F_{17}SO_3K$	23.0	4.0	600	2[c]
$C_8F_{17}SO_3NH_4$	22.4	1.2	1050	2[c]
$C_8F_{17}SO_3N(CH_3)_4$	23.1	1.2	200	2
$C_8F_{17}SO_3N(C_2H_5)_4$	22.6	0.25	100	1

[a] γ_s = minimum surface tension at concentration C_s
[b] t = time required to attain γ_s
[c] Solid phase present; limited solubility.
Source: Ref. 47. Reproduced by permission of Carl Hanser Verlag.

Partially Fluorinated Surfactants

Partially fluorinated surfactants are not as surface active as perfluorinated surfactants [37]. Ammonium α-[perfluorooctyl]-ω-sulfato(oxyethylene), $CF_2(CF_2)_6CH_2[OCH_2CH_2)_mOSO_3NH_4$, lowers the surface tension of water to 19.5 dyne/cm, but its ω-hydrogen analogue only to 25 mN/m [46]. The effect of a hydrocarbon segment on the surface activity of a partially fluorinated surfactant depends on the position of the hydrocarbon segment in the molecule. Terminal hydrogen or hydrocarbon segments decrease surface activity in water more than an internal hydrocarbon segment. Very low surface tensions have been achieved with fluorinated surfactants featuring internal hydrocarbon segments. Brace [48] prepared perfluoroalkyl-segmented alkanoic and alkenoic acids and determined their surface tensions as a function of their concentration in water (Fig. 4.20). An increase in the length of the R_f segment in $R_f(CH_2)_mCOONa$ to $R_f = CF_3(CF_2)_6$ decreased the minimum surface tension obtained from 27 to 14.8 mN/m. The longer the hydrocarbon chain (higher value of m), the lower the concentration needed to lower surface tension to a given value.

Kimura and co-workers [49] prepared sulfopropylated *N*-alkylperfluorooctanamides, $C_7F_{15}CON(R)CH_2CH_2CH_2SO_3Na$, where R is H or an alkyl group. The surface activity of the compounds depend on the chain length of the alkyl group. A short-chain (<6 carbons) or a long-chain (12 carbons) alkyl substituent reduces the efficiency of the surfactant, although the effect on effectiveness is small. Alkyl groups of a length in the range of 6 to 10 carbon atoms give the highest surface activity (Table 4.7).

Ishikawa and Sasabe [50] prepared fluorinated surfactants of the structure $(HFPO)_nArSO_3Na$, where $(HFPO)_n$ is an oligo(hexafluoropropene oxide) group and Ar an aryl group. The effect of n on surface tension in water is

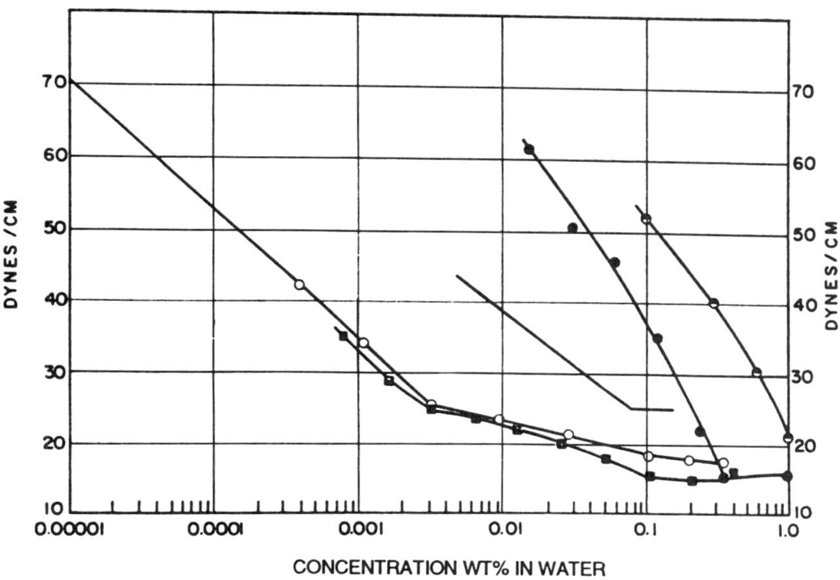

FIG. 4.20 Surface tensions of aqueous solutions of perfluoroalkyl segmented alkanoic acids $CF_3(CF_2)_6CF(CF_3)(CH_2)_{10}COONa$ (○), $CF_3(CF_2)_6CH{=}CH(CH_2)_8COONa$ (■), tetradecanoic acid (—), perfluorooctanoic acid (●), and its ammonium salt (◓). (From Ref. 48. Reproduced by permission of American Chemical Society.)

TABLE 4.7 Surface Tension of Aqueous Solution of $C_7F_{15}CON(R)CH_2CH_2CH_2CH_2SO_3Na$ at 30°C

	Surface tension (mN/m) at each concentration (g/100 mL)							
R	0.01	0.02	0.05	0.1	0.2	0.3	0.4	0.5
H	59.6	54.0	32.8	29.8	23.2	16.4	—	—
CH_3	54.4	50.0	41.0	31.4	22.4	23.8	23.2	22.8
C_2H_5	54.8	44.9	38.2	29.6	23.5	22.7	22.2	21.7
C_3H_7	53.7	47.0	36.8	29.2	21.8	20.9	20.0	19.1
C_4H_9	51.2	44.2	32.0	23.8	21.0	20.0	19.6	19.1
C_6H_{13}	28.0	21.5	18.4	18.0	17.7	17.5	17.4	17.3
C_8H_{17}	22.0	20.3	19.6	19.6	19.4	19.1	18.6	18.6
$C_{10}H_{21}$	28.2	26.8	24.6	24.6	23.0	21.8	20.6	19.6
$C_{12}H_{25}$	53.2	48.2	33.6	33.6	26.6	25.6	23.8	22.2

Source: Ref. 49.

shown in Fig. 4.21 and 4.22. In both series featuring an $(HFPO)_4$ group, surfactants derived from benzene- or toluenesulfonic acid exhibited the highest surface activity, lowering the surface tension of water down to 16 dyne/cm. This observation suggests that oxygen atoms in the perfluorinated oligoether chains are not hydrophilic.

Nonionic fluorinated surfactants which contain both oxyethylene and thioethylene groups, $CF_3(CF_2)_mC_2H_4(SC_2H_4)(OC_2H_4)_nOH$ (series I) and $CF_3(CF_2)_mC_2H_4(SC_2H_4)(OC_2H_4)_p(SC_2H_4)(OC_2H_4)_qOH$ (Series II) [32], have an exceptionally high surface activity (18 mN/m).

Selve at al. [23] prepared fluorinated nonionic surfactants with a two-chain polyoxyethylene hydrophilic head linked to the hydrophobe via an amide bond, $F(CF_2)_l(CH_2)_mC(O)[N(C_2H_4O)_nCH_3]_2$. The surface tension, γ_{cmc}, increased with increasing number of oxyethylene groups, regardless of the hydrophobe structure (16 and 24; 18 and 24; 27 and 30 mN/m for structures $l = 6$, $m = 1$, $n = 2 + 2$ and $4 + 4$; $l = 8$, $m = 1$, $n = 2 + 2$ and $4 + 4$; $l = 0$, $m = 11$, $n = 2 + 2$ and $4 + 4$).

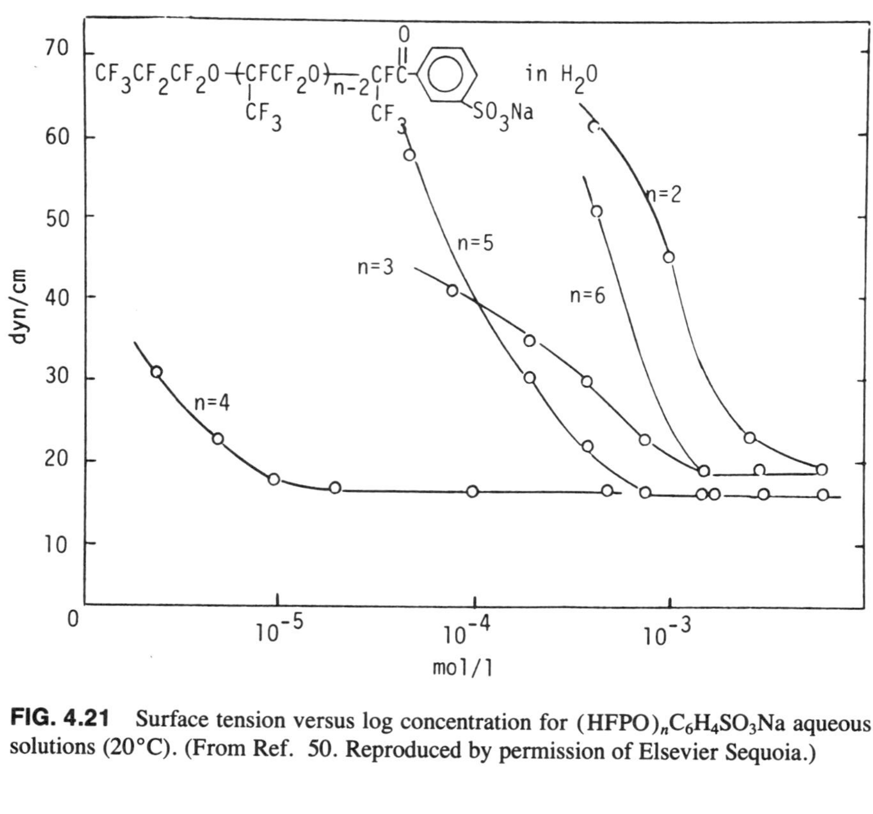

FIG. 4.21 Surface tension versus log concentration for $(HFPO)_nC_6H_4SO_3Na$ aqueous solutions (20°C). (From Ref. 50. Reproduced by permission of Elsevier Sequoia.)

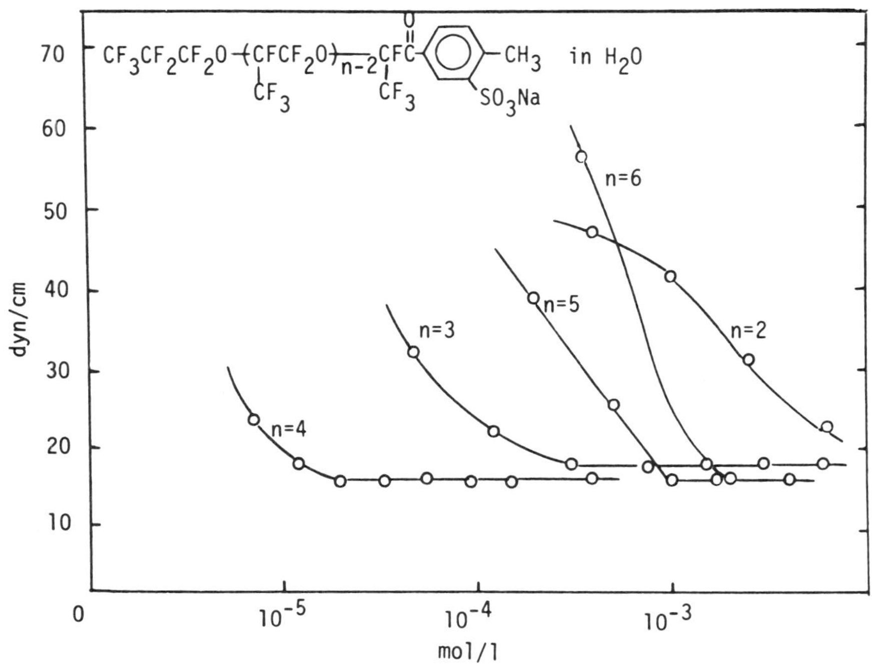

FIG. 4.22 Surface tension versus log concentration for $(HFPO)_nC_6H_3(CH_3)SO_3Na$ aqueous solutions (20°C). (From Ref. 50. Reproduced by permission of Elsevier Sequoia.)

Klevens and Raison [52] studied micellization of sodium dodecyl sulfate and perfluorooctanoic acid. Their data have been reviewed by Mukerjee and Mysels [51] who concluded that partially fluorinated surfactants behave like mixtures of fluorinated and hydrocarbon-based surfactants.

The surface tension data for the mixtures containing sodium dodecyl sulfate and 1.0, 0.75, and 0.20 mole fractions of perfluorooctanoic acid are plotted as a function of perfluorooctanoic acid concentration in Fig. 4.23. All three systems attain the same low and constant surface tension value after a sharp change in the slope corresponding to the formation of perfluorooctanoic acid micelles. The cmc values for the 1.0 and 0.75 molar systems are indistinguishable, indicating that the solubility of sodium dodecyl sulfate is insignificant in those micelles. The curve for the 0.2 mole fraction shows a change of slope and the cmc value is higher than for the other two systems. The differences in the curve shapes suggested that some perfluorooctanoic acid is dissolved in sodium docecyl sulfate micelles, while the rest continues to increase its monomer activity until cmc is reached.

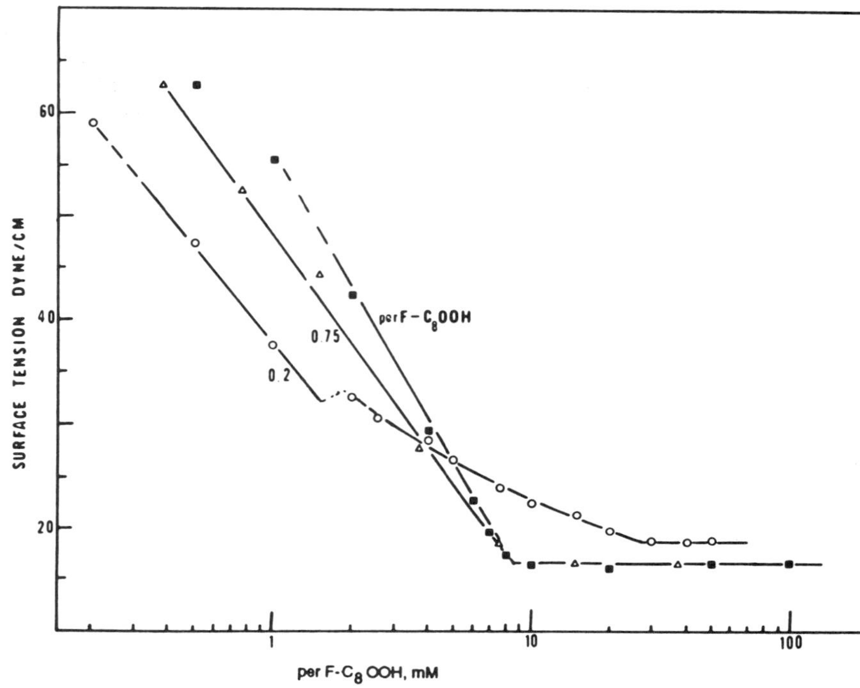

FIG. 4.23 Surface tension of perfluorooctanoic acid and its mixtures with sodium dodecyl sulfate. Mole fractions of perfluorooctanoic acid: 1.00, 0.75, 0.20. (From Ref. 51. Reproduced by permission of American Chemical Society.)

4.4 KINETICS OF ADSORPTION

For practical purposes, it is not only important to know how much of a fluorinated surfactant is needed to reduce surface tension to a desired value. The time required to decrease surface tension is also highly significant. Many industrial processes do not allow sufficient time for a surfactant to attain equilibrium and depend on the kinetics of surfactant adsorption.

The time required for surface tension reduction depends on diffusion processes involved in surfactant adsorption. Kinetic models for surfactant adsorption divide the adsorption process into two steps [53]. The first step is the transport of the surfactant to the subsurface, driven by a concentration gradient or hydrodynamic forces. The second step is the transfer of the surfactant molecule from the subsurface to the adsorbed state. Usually the second step is rapid and the first step determines the adsorption rate, especially when the transport of the surfactant molecule is a diffusion process. Since the bulk aqueous phase contains surfactant micelles, the influence of micelles on adsorption kinetics has to be considered. A kinetic model proposed by Miller [54] takes the formation and dissociation of surfactant micelles into account. Impurities can alter adsorption kinetics markedly and simulate adsorption or desorption barriers [55–57]. Pure surfactants adsorb by a diffusion-controlled adsorption mechanism which excludes barriers to sorption processes [58]. Different mathematical solutions to the problem of diffusion-controlled adsorption kinetics have been reviewed and compared by Miller and Ziller [59].

The kinetics of surfactant adsorption depend on the surfactant structure. Fluorination of the hydrophobe increases the rate ($d\gamma/dt$) of surface tension decrease, but the time needed to attain equilibrium may not be affected considerably. The surface tensions of sodium perfluorooctanesulfonate and its hydrocarbon-type analogue are shown in Fig. 4.24 [60]. Both surfactants reach

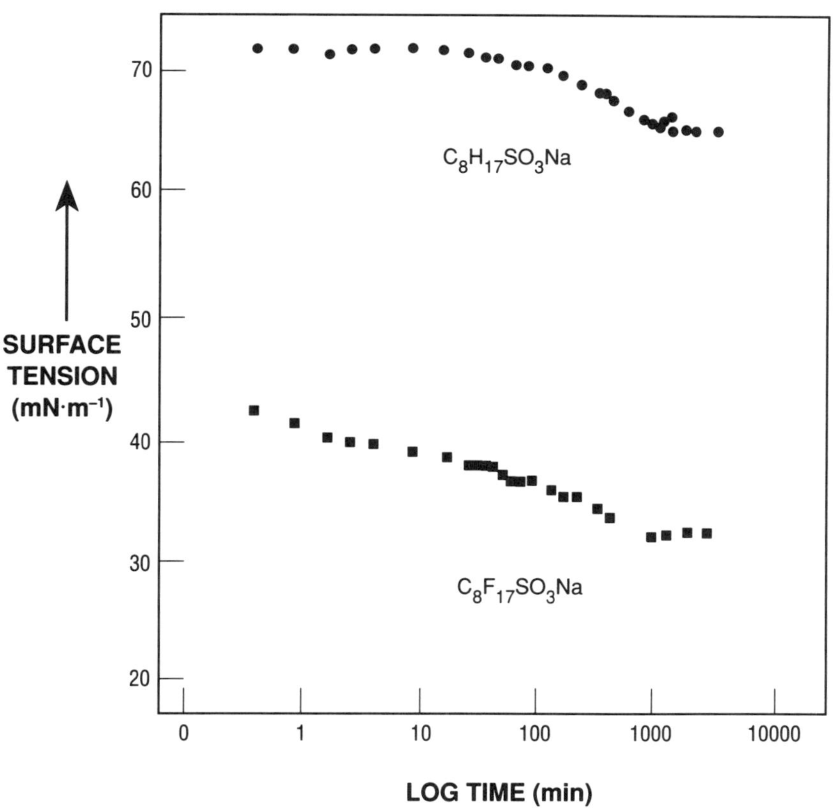

FIG. 4.24 Surface tension of 1 g/L solutions of sodium octanesulfonate and sodium perfluorooctanesulfonate versus log time. (From Ref. 60. Reproduced by permission of Dr. A. Hüttig Verlag.)

their equilibrium surface tension in about the same time. However, the fluorinated surfactant has rapidly decreased the surface tension in less than a half a minute by about 30 dyne/cm.

Dikhtievskaya and Makarevich [61] examined perpendicular and lateral interactions of fluorinated surfactants, $n-C_3F_7O[CF(CF_3)CF_2O]_m-CF=C(CF_3)COONH_4$, at the air-water boundary. The time required to attain equilibrium surface tension increased with increasing fluoroalkyl chain length or decreasing surfactant concentration.

Equilibration is more rapid above cmc than below cmc [47].

The time required for attaining surface tension equilibrium decreases with increasing temperature. Selve et al. [23] have suggested that extremely hydrophobic surfactants at low temperatures and low bulk concentrations have such a strong tendency to escape the aqueous environment that they form a rather disordered superficial layer. Minutes or even hours may be required for the surface film to reorganize. Increasing temperature accelerates the diffusion processes and the equilibrium surface tension is attained in a shorter time. The temperature and concentration effects of adsorption kinetics are illustrated with Figs. 4.25 and 4.26. The adsorption of the fluorinated surfactants F8.1.33 and F10.1.33 (structures are given in Sec. 4.3 [23]) is slow at 5°C but essentially time independent at 45°C.

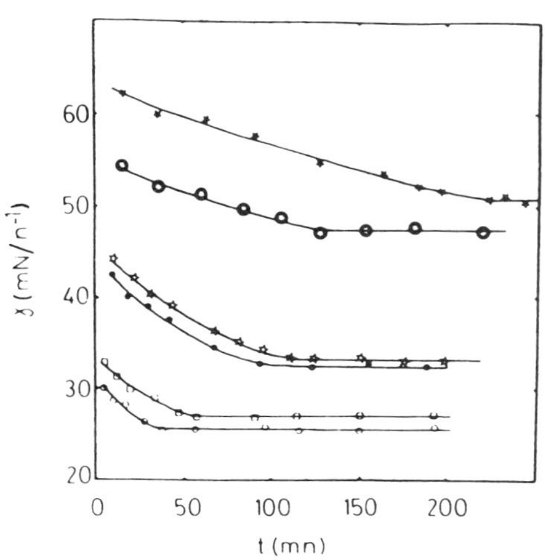

FIG. 4.25 Time dependence of surface tension of two surfactants at various concentrations at 5°C. From top to bottom: F10.1.33 1.0×10^{6}, 1.3×10^{-6}, 3.5×10^{-6} mol/L; F8.1.33 3.9×10^{6}, 8.0×10^{-6}, 1.2×10^{-5} mol/L. (From Ref. 23. Reproduced by permission of Pergamon Press.)

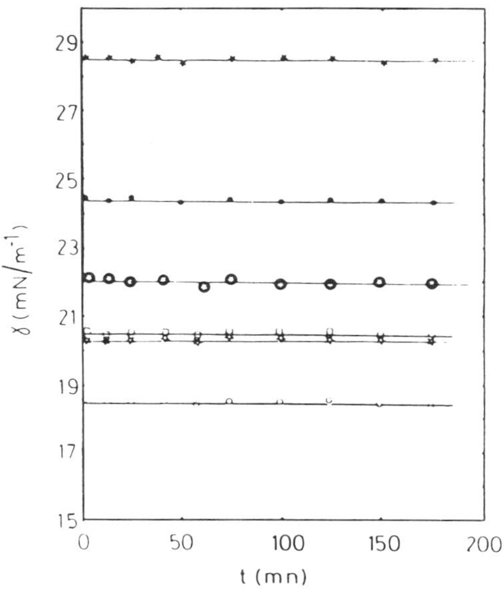

FIG. 4.26 Surface tension of the surfactant solutions shown in Fig. 4.24 but at 45°C. In contrast to Fig. 4.24, adsorption is rapid at 45°C and time dependence of surface tension is not detectable by the method used for measurement. (From Ref. 23. Reproduced by permission of Pergamon Press.)

The rate at which a surfactant reaches equilibrium surface tension also depends on its counterion. Klein et al. found that the tetraethylammonium salt of perfluorooctanesulfonic acid attains surface tension equilibrium more rapidly than the potassium salt (Fig. 4.27) [35], the tetramethylammonium, or the ammonium salt (Table 4.6) [47].

The adsorption rate of a surfactant on the liquid-air interface is affected by the presence of impurities and isomers with different adsorption characteristics. The diffusion rate of the surfactant to the liquid-air interface depends not only on the effective cross section of the isomers but on their solubility as well. A fractionation of the surfactant may occur when the solubility limit is exceeded. If the fluorinated surfactant is a mixture of normal-chain and branched-chain isomers, the normal-chain surfactant components precipitate, while the branched-chain isomers remain in solution [35].

Surfactant adsorption kinetics has been studied by static and dynamic methods. Although static methods can reveal kinetic effects of surface tension, dynamic methods can measure surface tension in a few milliseconds and are eminently suitable for investigating the kinetics of surfactant adsorption (see Chapter 9). The dynamic surface tensions of fluorinated surfactants consisting

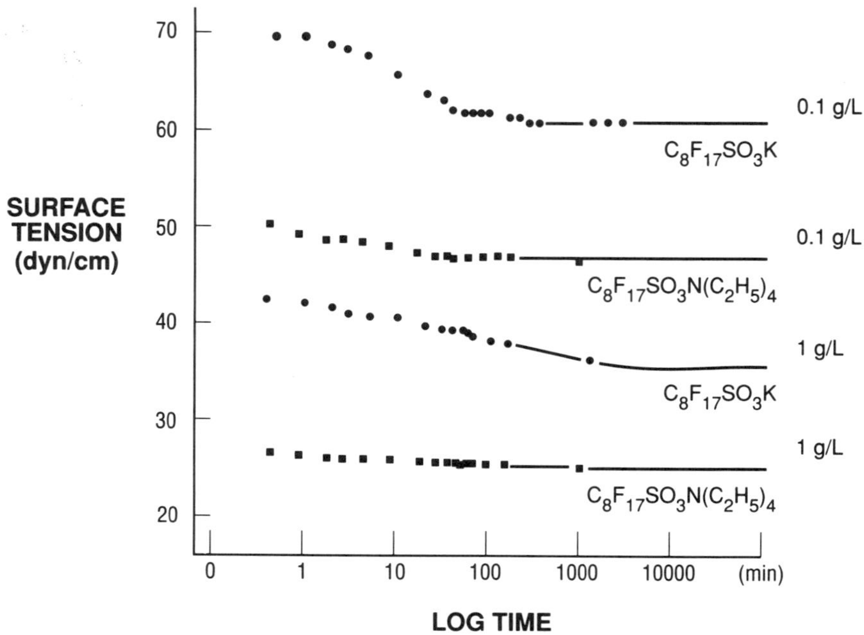

FIG. 4.27 Surface tension versus log time curves for $C_8F_{17}SO_3K$ and $C_8F_{17}SO_{3N}N(C_2H_5)_4$. (From Ref. 35. Reproduced by permission of Carl Hanser Verlag.)

of telomeric mixtures (Zonyl FSA, Zonyl FSN, and Zonyl FSC) were studied at the Textile Research Institute [62] by the maximum bubble pressure method. The surface tension can be measured by varying the flow rate of the gas used to generate a bubble in the liquid. When the bubble grows slowly, surfactant molecules have time to diffuse to the expanding gas-liquid interface. However, when the bubble generation is rapid, the surfactant does not have sufficient time for adsorption and the measured surface tension is higher than the equilibrium surface tension. Although the equilibrium surface tension of 0.30 mM Zonyl FSA, an anionic fluorinated surfactant, was found to be 17 mN/m, the dynamic surface tension at a bubble frequency of 0.1 bubble/s remained above 60 mN/m (Fig. 4.28).

A mixture of Zonyl FSN, a nonionic fluorinated surfactant, and Tergitol NP-10, a nonionic hydrocarbon-type surfactant (nonylphenol with 10 oxyethylene units, Union Carbide Corp.), exhibited surface tension synergism (Fig. 4.29). The dynamic surface tension of the 1:1 mixture is lower than the dynamic surface tensions of the individual components.

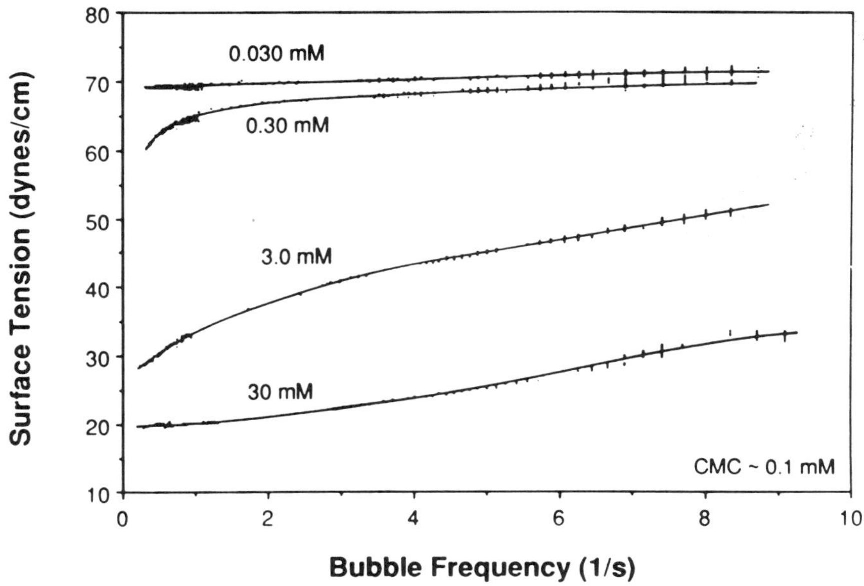

FIG. 4.28 Effect of surfactant concentration on the dynamic surface tension of aqueous solutions of an anionic fluorinated surfactant Zonyl FSA. (From Ref. 62. Reproduced by permission of Elsevier Science Publishers.)

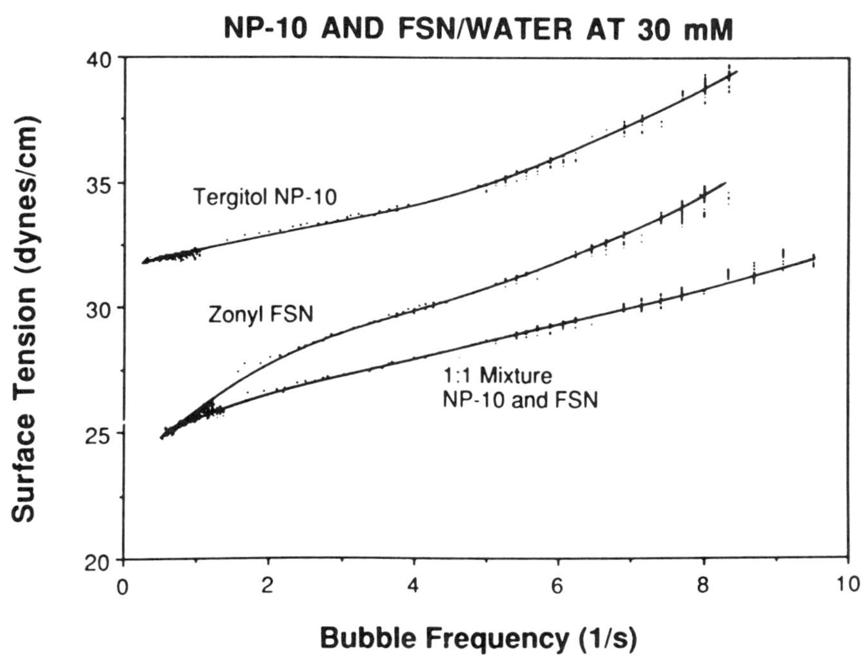

FIG. 4.29 Synergistic effect on the dynamic surface tension of aqueous solutions of Tergitol NP-10 and Zonyl FSN at a total concentration of 30 mM. (From Ref. 62. Reproduced by permission of Elsevier Science Publishers.)

4.5 SURFACE TENSION IN ACIDS AND ALKALI

The outstanding chemical stability of fluorinated surfactants to strong acids and alkali permits their use under conditions in which conventional surfactants would decompose. Certain fluorinated surfactants can be used in the presence of strong oxidants, acids, or alkalies even at elevated temperatures.

The surface tension of Zonyl fluorinated surfactants, featuring the hydrophobe $F(CF_2CF_2)_{3-8}CH_2CH_2-$, in acids and alkali is shown in Table 4.8. Anionic, cationic, nonionic, or amphoteric surfactants listed in Table 4.8 can lower the surface tension of 10% potassium hydroxide, 25% sulfuric acid, 37% hydrochloric acid, or 70% nitric acid below 25 mN/m, unless the surfactant is insoluble in acid.

Anionic Surfactants

Fluorinated surfactants with a carboxylate or phosphate hydrophile precipitate in strongly acid media, whereas surfactants with a sulfonate hydrophile are more soluble and also effective in strongly acid media [63,64]. Talbot [63] measured surface tensions of perfluorinated sulfonic acid potassium salts in acids. Addition of a strong inorganic acid to a 10^{-3} M solution of n-$C_8F_{17}SO_3K$ in water lowered the surface tension to a value of 17 mN/m. The surface tension decreased with decreasing pH of the acid solution. Talbot explained the effect of acids by the formation of free surfactant acid molecules which are more strongly adsorbed on the surface and lower surface tension. Acetic acid had a lesser effect on surface tension of n-$C_8F_{17}SO_3K$ but its effect was proportional to its acidity and ionization.

The surface tensions of some perflourinated anionic surfactants with a C_8F_{17}- hydrophobe in strong acids are given in Table 4.9 [42]. The surfactant $C_8F_{17}SO_3K$ is very effective in concentrated acids, although its solubility in 37% hydrochloric acid is limited to 0.2 g/L.

The surface tension of the tetraethylammonium salt of the perfluorinated surfactant $C_8F_{17}SO_3N(C_2H_5)_4$ in acids is shown in Figs. 4.30 and 4.31 [35].

Glöckner et al. [65] tested the stability of perfluorinated surfactants in acids or alkali under severe conditions. The 0.1% solutions of anionic surfactants in 60% nitric acid or 98% sulfuric acid containing 10 g/L chromic oxide were stored for a time span of 43 days, of which 28 days were storage at 90°C. The surface tensions of the anionic surfactants changed very little during the test (Table 4.10) [65].

The stability of perfluorinated anionic surfactants is also remarkable in alkali. Glöckner et al. [65] found the surface tensions of perfluorinated anionic surfactants in saturated aqueous sodium hydroxide to be constant for at least 100 days (Table 4.11). Salts of perfluoroalkanoic, perfluoroalkanesulfonic, or perfluoroalkenesulfonic acids survived an alkaline fusion [5 g $KNO_3/LiNO_3$ (35–65 mol %), 0.5 g NaOH, 0.05 g surfactant] for 2 h at 200°C, as indicated by the surface tensions of the aqueous solutions of the fusion mass.

TABLE 4.8 Surface Tension (mN/m) of Zonyl Fluorinated Surfactants at 25°C (77°F)

	10% Potassium hydroxide			25% Sulfuric acid			37% Hydrochloric acid			70% Nitric Acid		
Zonyl®	0.001%	0.01%	0.1%	0.001%	0.01%	0.1%	0.001%	0.01%	0.1%	0.001%	0.01%	0.1%
FSA	33	15	16	1[1]	1	1	1	1	1	1	1	1
FSP	30	17	17	1	1	1	1	1	1	1	1	1
FSE	34	19	16	1	1	1	1	1	1	1	1	1
UR**	35	20	21	1	1	1	1	1	1	1	1	1
FSJ	38	24	22	1	1	1	1	1	1	1	1	1
FSN	30	25	23	29	25	23	28	24	23	30	25	23
FSO	25	24	24	25	20	19	32	22	22	41	31[1]	21
FSC	27	18	18	28	19	19	31	20	18	36	25	20
FSK	54	32	23	26	20	18	29	19	17	34	22	18
TBS	51[2]	36[2]	35[2]	30	20	16	29[3]	22[3]	16[3]	50	28	21

1, insoluble; 2, partially insoluble; 3, for 10% HCl.

TABLE 4.9 Surface Tensions of Fluorinated Surfactants (0.01%) in Acids

Surfactant	Surface tension (mN/m)		
	Hydrochloric acid (12.5%)	Sulfuric acid (50%)	Chromic oxide–sulfuric acid
No surfactant	50	71	
$C_8F_{17}SO_3K$	20	17	18
$C_nF_{2n+1}SO_3K$ (n = 6–12)	22	18	19
$(C_nF_{2n+1})_2P(O)OH + C_nF_{2n+1}P(O)(OH)_2$	20	≈40	24

Source: Ref. 42. Reproduced by permission of Carl Hanser Verlag.

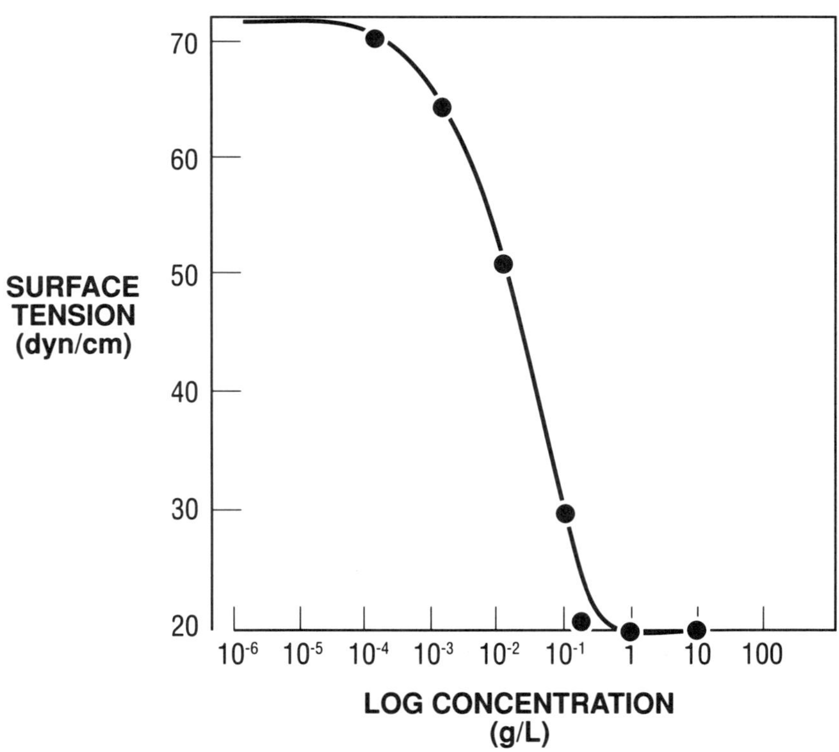

FIG. 4.30 Surface tension curve of $C_8F_{17}SO_{3N}N(C_2H_5)_4$ in 10% nitric acid. (From Ref. 35. Reproduced by permission of Carl Hanser Verlag.)

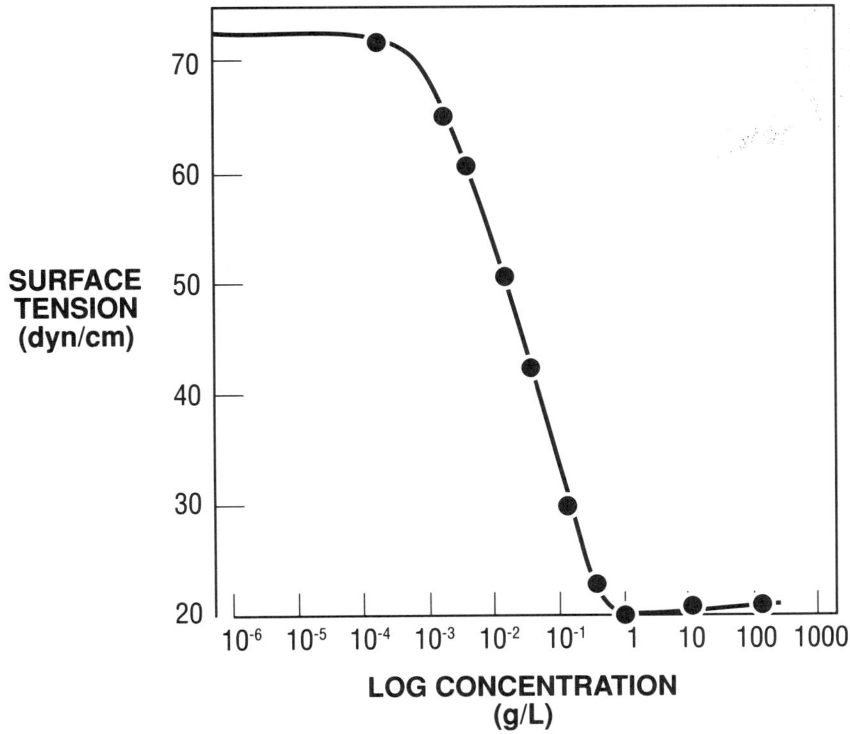

FIG. 4.31 Surface tension curve of $C_8F_{17}SO_{3N}N(C_2H_5)_4$ in 10% phosphoric acid. (From Ref. 35. Reproduced by permission of Carl Hanser Verlag.)

TABLE 4.10 Surface Tension of 0.1% Solutions of Anionic Fluorinated Surfactants in Concentrated Mineral Acids at 22°C

	Surface tension (mN/m)					
	60% HNO_3			90% H_2SO_4 + 10 g/L CrO_3		
	1 d[a]	15 d	43 d[b]	1 d	15 d	43 d[b]
$C_7F_{15}COOH$	33.8	—	33.7	41.4	—	42.6
$H(CF_2)_8COONa$	—	—	—	50.8	(50.5	218 d)
$C_8F_{17}SO_3K$	19.8	18.6	20.9	39.7	38.8	40.1
$C_8F_{17}SO_3N(C_2H_5)_4$	23.2	22.8	22.7	34.6	32.5	32.4
$C_7F_{13}SO_3Na$	42.6	42.6	41.8	43.0	43.3	48.7
$C_nF_{2n-1}SO_3Na$ (n = 10–11)	20.3	19.4	20.1	26.2	27.0	27.8
No surfactant		62.0			63.3	

[a] d = time in days.
[b] Including 28 days at 90°C.
Source: Ref. 65. Reproduced by permission of Carl Hanser Verlag.

TABLE 4.11 Surface Tension of Anionic Fluorinated Surfactants After an NaOH/KOH Fusion at 300°C

Surfactant	Surface tension (mN/m)	
	2 h	5 h
$C_6F_{13}SO_3K$	39.5	46.3
$C_8F_{17}SO_3K$	43.0	57.0
$C_8F_{17}SO_3(C_2H_5)_4$	43.7	55.5
No surfactant	51.3	53.4

Source: Ref. 65. Reproduced by permission of Carl Hanser Verlag.

Nonionic Surfactants

Perfluorinated nonionic surfactants of the structure C_nF_{2n-1}-$(OCH_2CH_2)_mOCH_3$ [65] and nonionics with an ω-H-perfluoro hydrophobe [66] are effective surfactants in acids and in alkali, if stable. Greiner and Herbst [66] prepared nonionic surfactants with the general formula $H(CF_2CF_2)_n$-$CH_2O(CH_2CH_2O)_mH$. The oxyethylated 9-hydroperfluorononanol, at a concentration of 0.3 to 0.5%, lowered the surface tension of 50% sulfuric acid from 76 to 28 mN/m. Heating of the solution for 6 h at reflux did not change its surface tension. Thin-layer chromatography indicated, however, partial degradation of the oxyethylene chain.

Although not significantly soluble in alkali, the oxyethylated 9-hydroperfluorononanol, at a concentration of 0.05%, lowers the surface tension of 20% NaOH to 27 mN/m. Highly oxyethylated (55 EO) 11-hydroperfluoroundecanol has a similar effect on surface tension.

The nonionic fluorinated surfactants are more stable in alkali than in acid. Glöckner et al. [65] found that, unlike perfluorinated anionic surfactants, perfluorinated nonionic surfactants $C_nF_{2n-1}(OCH_2CH_2)_mOCH_3$ are not stable in chromic-sulfuric acid. The solution immediately turns green, indicating decomposition. In spite of the degradation by chromic-sulfuric acid, nonionic surfactants lower the surface tension of the acid substantially. The surface tension does not change significantly even after more than 80 days of storage (Table 4.12). Apparently the degradation of the nonionic surfactant produces surface-active species, perhaps perfluoroalkenyl sulfates or perfluoroalkanoic acids [65].

When selecting a fluorinated surfactant, the stability of the surfactant in the particular working medium has to be considered. Although the C–F bond is stable, the stability of fluorinated surfactants in strong acids and alkali also depends on the stability of other functional groups in the molecule. For example, nonionic surfactants derived from perfluorooctanesulfonamide can lower

TABLE 4.12 Surface Tension of Nonionic Fluorinated Surfactants in Concentrated Chromic-Sulfuric Acid

$C_nF_{2n+1}(OCH_2CH_2)_mOCH_3$		Surface tension (mN/m)		
n	m	γ_o	γ_t	t (days)
10–11	17	42.3	45.5	96
8–11	7.6	36.7	41.2	96
5–8	17	56.4	55.7	86
5–8	8.7	60.7	57.9	90
Monflor 52		42.9	42.2	112
No surfactant				

Source: Ref. 65. Reproduced by permission of Carl Hanser Verlag.

the surface tension of water more than the corresponding sulfonic acid salts. However, their use in alkali is limited by the hydrolytic instability of the sulfonamide linkage [35].

4.6 SURFACE TENSION IN ORGANIC LIQUIDS

Many fluorochemicals are remarkably surface active in organic liquids [67–71]. Scholberg et al. [36] found that perfluorocarboxylic acids lower the surface tensions of organic solvents more than any other chemical ever reported.

Ellison and Zisman screened various compounds for surface activity in organic liquids by using a Langmuir film balance. The compound was spread on an organic liquid and the surface film pressure measured as a function of surface area [71]. The results indicated that the surface activity of fluorinated compounds in organic liquids can be predicted approximately from the Harkins equation for the spreading coefficient

$$S_{b/a} = \gamma_a - (\gamma_b + \gamma_{ab}) \tag{18}$$

where $S_{b/a}$ is the spreading coefficient of liquid b spreading on liquid a and γ_a and γ_b are the surface tensions of liquids a and b, respectively. γ_{ab} is the interfacial tension between liquids a and b. If γ_{ab} is small compared with the surface tension of the organic liquids, γ_{ab} can be neglected. Hence, spreading behavior is approximately indicated by the difference $\gamma_a - \gamma_b$. The greater the surface tension of the solvent and the lower the surface tension of the solute, the greater the surface activity of the solute in the solvent. Since the surface tensions of fluorinated compounds are considerably lower than those of their hydrocarbon analogues, the surface activity of fluorinated compounds in organic solvents is not surprising. However, spreadability is a necessary but not a sufficient condition for surface activity. Even a positive $S_{a/b}$ does not always indicate the extent of surface tension lowering [72].

The effect of surfactants on surface tension is much smaller in solvents than in water. Surfactants can lower the surface tension of water by more than 50 mN/m, to below one quarter of its original value. In aromatic hydrocarbons only a few fluorinated surfactants can decrease the surface tension to half of the initial value. The surface tension depression is smaller, mainly for two reasons. First, the initial surface tension of solvents, 19–30 mN/m for hydrocarbons, is lower to begin with. Second, surfactants are not as effective in organic media as in water because adsorption of the surfactant at the surface of an organic liquid has less effect on the nature of the interface. Fluorinated surfactants are the most effective surfactants in organic media as well as in water. However, even for fluorinated surfactants the lowest surface tension achieved in water is lower than the lowest surface tension in organic liquids.

The structural requirements for a surfactant in an organic liquid are quite different from the structures of typical fluorinated surfactants used in water. In order to function as a surface-active agent, a compound has to be amphiphilic and consist of lyophobic (solvent-insoluble) and lyophilic (solvent-soluble) groups in the molecule [72]. In water the hydrophobic group of the surfactant causes a restructuring of hydrogen-bonded water. The surfactant molecule is transported to the surface, where the molecule orients with the hydrophobic part directed away from water. In organic solvents the situation is different. An alkyl tail, hydrophobic in water, may be lyophilic in a solvent and the hydrophilic head group in water may be lyophobic in a solvent. Conventional surfactants may orient with the hydrophilic lyophobic group away from the solvent surface, resulting in higher surface energy and an increase in surface tension (Fig. 4.32) [73] and (Fig. 4.33) [74]. A fluorinated hydrophobe functions in solvents, as in water, as the lyophobe. A lyophilic group provides solubility in the solvent. As an example, fluorinated carboxylic esters [72] can function as surfactants in organic solvents where the ester group is lyophilic and soluble. The fluorinated chain of the molecule is the lyophobic part.

In aqueous solutions of nonfluorinated surfactants the lowest surface tensions are attained by covering the surface with a close-packed monolayer of vertically oriented hydrocarbon chains forming a continuous layer of $-CH_3$ groups exposed to air [75]. By analogy, the surface tension of a solution of a fluorinated surfactant is at its minimum when the surface is covered with a closely packed monolayer of fluorocarbon groups with their $-CF_3$ groups oriented toward air. The orientation has a considerable effect on surface tension. Vertical orientation exposes the terminal $-CF_3$ groups to air, while tilting of fluorocarbon chains exposes the $-CF_2-$ groups. A surface film consisting of $-CF_2-$ groups has a higher surface tension than a surface film of $-CF_3$ groups [76,77]. Zisman and co-workers found that at the surface of an

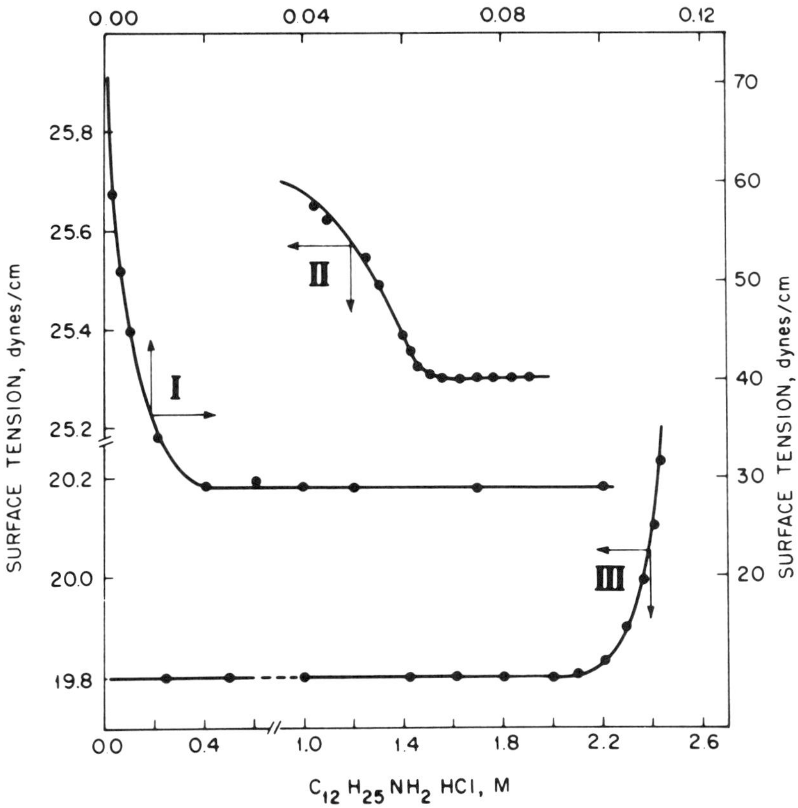

FIG. 4.32 Surface tension of *n*-dodecylammonium chloride in water (curve I), a 50% water-ethanol mixture (curve II), and absolute ethanol (curve III) versus surfactant concentration at 50°C. (From Ref. 73. Reproduced by permission of Wiley Interscience).

organic liquid even the lowest values of area per molecule, calculated for compounds containing only one fluorinated chain, are much larger than the cross-sectional area 24.5 Å determined from the Stuart-Briegleb molecular models [75]. This suggests incomplete surface coverage by a fluorocarbon film, even though the fluorinated chain, at least in some solvents, is oriented away from the surface. Compounds containing two fluorinated chains, for example bis-(φ'octyl)tetrachlorophthalate, have lower values of area per molecule, approaching the cross-sectional value 50 Å indicated by the Stuart-Briegleb models [the symbol φ' is an abbreviation for an $F(CF_2)_nCH_2$- group]. However, the minimum surface tensions found were higher than expected for a surface covered with vertically oriented fluorocarbon chains (Table 4.13). Com-

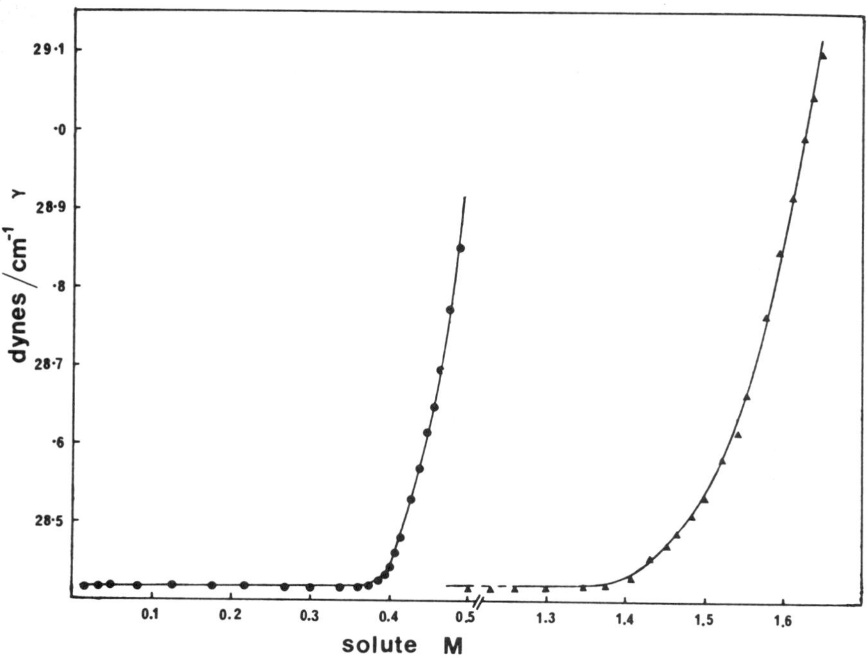

FIG. 4.33 Surface tension of toluene solutions of *n*-dodecylammonium chloride (●) and tri-*n*-dodecylammonium bromide (▲) as a function of surfactant concentration at 50°C. (From Ref. 74. Reproduced by permission of Academic Press.)

pounds with three fluorinated chains behaved similarly. Zisman and coworkers concluded [75] that factors contributing to close packing are (1) structure and solubility of the organophilic group, (2) type and position of the polar group connecting the organophile with the fluorinated organophobe, and (3) fluorination, length, and number of organophobic chains. A terminal hydrogen in an otherwise fluorinated chain increases solubility in polar solvents and decreases adsorption. Longer chains decrease solubility and lead to closer packing at the surface. Too many chains attached to the molecule interfere sterically with each other and increase the area occupied by the molecule.

The solubility of the fluorinated compound has a significant effect on surface activity. Jarvis and Zisman [78] found that the maximum surface tension lowering of some fluorinated compounds is limited by their insufficient solubility, whereas others were too soluble to be adsorbed at the surface of the solution. Fluorinated chemicals must be designed to have a proper balance between soluble and insoluble groups of the molecule. An increase in the organophobic part of the molecule increases the initial slope of the surface tension–concentration curve (Fig. 4.34) but decreases the solubility of the molecule.

TABLE 4.13 Surface Tensions of Saturated Solutions in Various Organic Liquids (at 20°C)

		Surface tension of saturated solution (mN/m)						
Surface-active solute	Surface tension, (mN/m)	Propylene carbonate ($\gamma=41.1$)	Tricresyl phosphate ($\gamma=40.2$)	Alkazene 42 ($\gamma=38.2$)	Nitromethane ($\gamma=36.4$)	Dioxane ($\gamma=32.9$)	Ethylbenzene ($\gamma=28.6$)	Hexadecane ($\gamma=27.6$)
ϕ'-Octyl alcohol	Solid	24.9[a]	27.6[a]	31.2	26.8[a]			25.4
ϕ'-Octyl ethanesulfonate	19.1	23.8[a] (22.3 at 0.16)	22.1[a] (20.9 at 0.14)	23.4	26.7[a]			23.8
Bis-(ϕ'-octoxy)-bis-(*t*-butoxy)silane	18.4	28.3	25.9	24.2	25.7			21.2
Tris-(ϕ'-octoxy)-*t*-butoxysilane	18.5	28.3	38.0	27.4	36.0			24.0
Bis-(ϕ'-octyl)-tetrachlorophthalate	Solid	30.7			31.2	30.4	20.7	
Bis-ϕ'-octyl)-α-*n*-dodecenylsuccinate	19.4	25.3			23.1	21.9[a]	21.9[a]	
Bis-(ϕ'-octyl)-2,4-toluenedicarbamate	Solid	32.5			Too insoluble	24.7	Too insoluble	
Tris-(ϕ'-butyl)-tricarballylate	18.5	24.5[a]			27.1[a] (22.2 at 0.688 M)	27.1[a] (23.6 at 0.559 M)	24.9	
Tris-(ϕ'-octyl)-tricarballylate	Solid	28.2			23.6	22.1	Too insoluble	

[a] At 0.1 mol/L.
Source: Ref. 75.

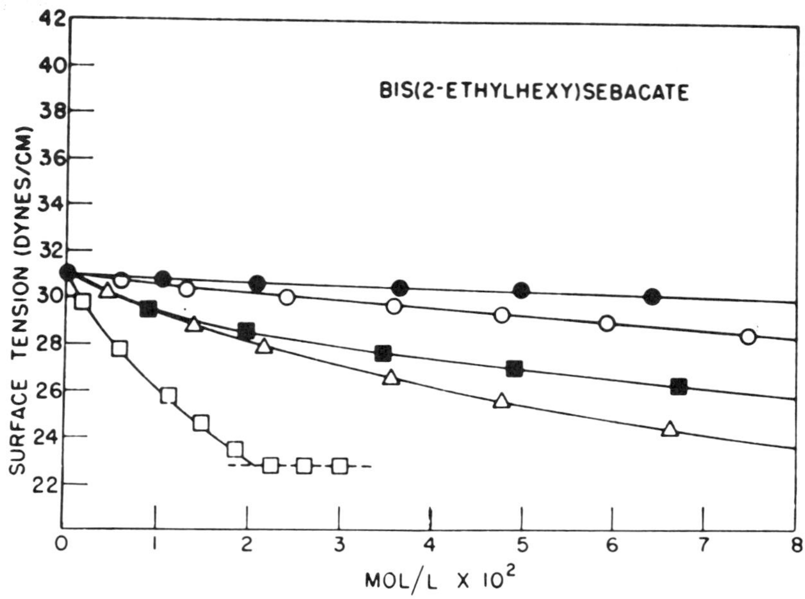

FIG. 4.34 Surface tension–concentration isotherms of partially fluorinated carboxylic esters in organic liquids: bis-(φ'-butyl)3-methylglutarate (○); bis-(φ'-hexyl)3-methylglutarate (△); bis-(φ'-octyl)3-methylglutarate (□); bis-(φ'-heptyl)3-methylglutarate (●); hexylφ-butyrate (▲); 1,2,3,-trimethylolpropane tris-(φ-butyrate) (■). (From Ref. 78. Reproduced by permission of American Chemical Society.)

Eventually, inadequate solubility of the molecule limits the minimum surface tension (maximum surface tension lowering) that can be achieved by increasing the concentration of the surfactant.

Katrizki et al. [79] prepared a series of *N*-(perfluorooctanesulfonyl)-piperazine derivatives (Table 4.14). The most effective member of the series lowered the surface tension of diesel fuel from 27.2 to 21.7 mN/m at a concentration of 1% w/v. Interestingly, the surface tension decrease, 5.5 mN/m, was smaller than the surface tension depression observed for a Zonyl surfactant (Table 4.15) [80]. Zonyl FSJ, a mixture of an amphoteric fluorinated surfactant and a hydrocarbon-based surfactant, was found to be more effective in lowering surface tension than a fluorinated surfactant alone. The synergism between fluorinated and nonfluorinated surfactants is important for lowering interfacial tensions (see Sec. 7.2).

Initially the surface activity in organic solvents was studied with surfactants featuring a fluoroalkyl lyophobe. Recently, fluorinated surfactants with an oligo(hexafluoropropene oxide) chain have been developed. Oil-soluble surfactants with the structure $(HFPO)_n$-Ar, where Ar is an aryl group and $(HFPO)_n$

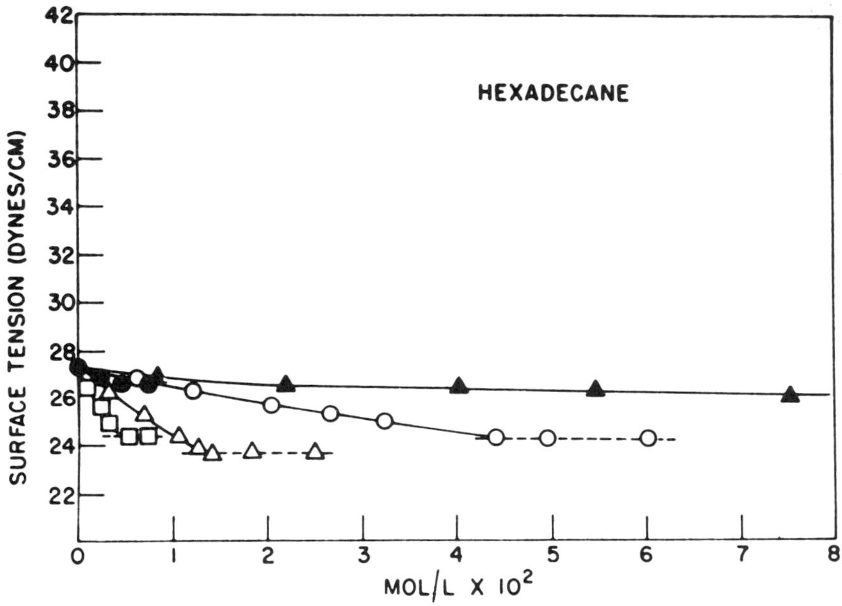

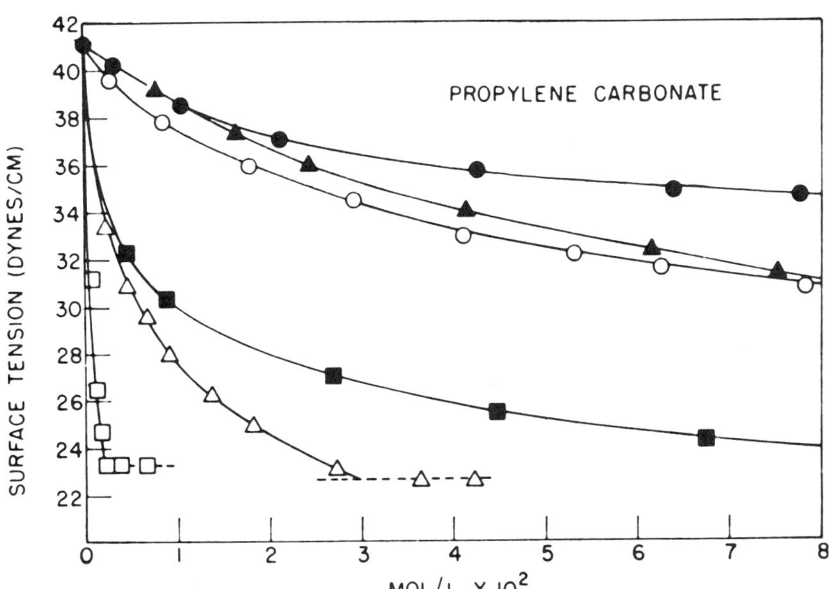

FIG. 4.34 continued

TABLE 4.14 Equilibrium Surface Tensions of 1 wt % Solutions in Diesel Fuel at 24°C

Compound[a]	γ (mN/m)	Compound[a]	γ (mN/m)
Blank	27.2	11	22.8
1	26.2	12	26.7
2	27.2	13	25.8
3	24.6	14	26.0
4	27.2	15	26.2
5	25.1	16	26.2
6	26.5	17	26.9
7	26.6	18	24.3
8	27.0	19	23.9
9	21.7	20	27.0
10	22.8	21	26.8

[a] Compound structures:

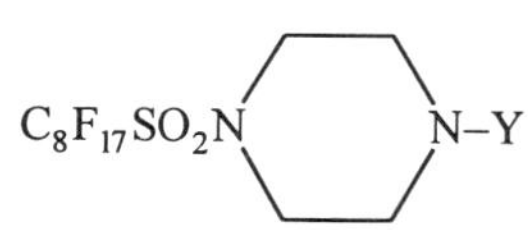

Compound	Y	Compound	Y
1	H	9	$SO_2C_6H_{13}$
2	COC_6H_5	10	$SO_2-4-C_6H_4CH_3$
3	COC_2H_5	11	$SO_2-4-C_6H_4C_8H_{17}$
4	COC_3H_7	12	CH_3
5	COC_5H_{11}	13	C_2H_5
6	COC_8H_{17}	14	C_4H_9
7	$COC_{11}H_{23}$	15	C_6H_{13}
8	SO_2CH_3	16	C_8H_{17}

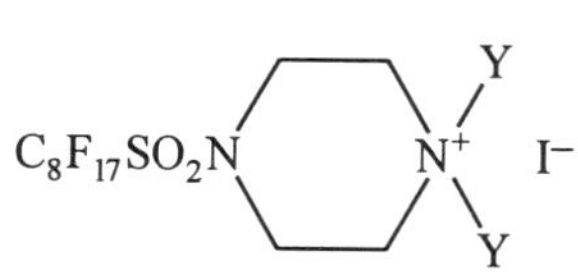

Compound	Y
17	CH_3
18	C_2H_5
19	C_4H_9
20	C_6H_{13}
21	C_8H_{17}

Source: Ref. 79.

TABLE 4.15 Surface Tension Depression by Flourinated Surfactants in Diesel Fuel

Zonyl	Structure ($R_f = C_nF_{2n+1}$; n = 3–8	$\Delta\gamma$ (mN/m)
FSK	$R_fCH_2CH_2CH(OCOCH_3)N^+(CH_3)_2CH_2COO^-$	4.6
FSC	$[R_fCH_2CH_2SCH_2CH_2N^+(CH_3)_3]_2SO_4^{2-}$	6.4
FSN	$R_fCH_2CH_2O(CH_2CH_2O)_xH$	0.4
FSA	$R_fCH_2CH_2SCH_2CH_2COOLi$	0.5
FSE	$(R_fCH_2CH_2O)_2P(O)O^-NH_4^+$	4.9
FSP	$(R_fCH_2CH_2O)P(O)(O^-)_2(NH_4^+)_2$	0.7
FSJ	FSP + nonfluorinated surfactant	9.6

Source: Refs. 79 and 80. Reproduced by permission of the American Chemical Society.

is an oligo(hexafluoropropene oxide) group, n = 2–5, decreased the surface tension of toluene or m-xylene down to 12–14 mN/m at 20°C (Fig. 4.35) [81].

The lowest surface tensions, γ_{min}, the corresponding concentrations for the surfactants $(HFPO)_{2-6}C_6H_5$, and their cmc values are given in Table 4.16. The surface tension of xylene (28.6 mN/m) was lowered considerably even at small surfactant concentrations. The surface tension decrease depends on the length of the perfluorinated chain, exhibiting a maximum decrease at n = 5.

In the $(HFPO)_nC_6H_4CH_3$ series the surfactant $(HFPO)_6C_6H_4CH_3$ was the most effective in lowering the surface tension of toluene and xylene (Table 4.17). Presumably the maximum surface tension lowering corresponds to an optimum balance between the oleophilic p-tolyl group and the oleophobic $(HFPO)_6$ group in the molecule.

The development of fluorinated surfactants for organic solvents is still ongoing. The structural design is limited by (1) the dependence of surface activity on the solubility of the surfactant in the solvent and (2) the initial surface tension of the solvent. Hence, a fluorinated surfactant cannot have the same

TABLE 4.16 Surface Tensions of $(HFPO)_n-C_6H_5$ Solutions in m-Xylene (20°C)

n	γ_{min} (mN/m)	C_{min} (wt %)	CMC (mol/L)
2	17.8	0.58	1.48×10^{-2}
3	12.8	0.21	3.75×10^{-3}
4	12.1	0.23	3.19×10^{-3}
5	9.9	0.23	2.58×10^{-3}
6	11.3	0.29	2.18×10^{-3}

Source: Ref. 81. Reproduced by permission of Elsevier Sequoia.

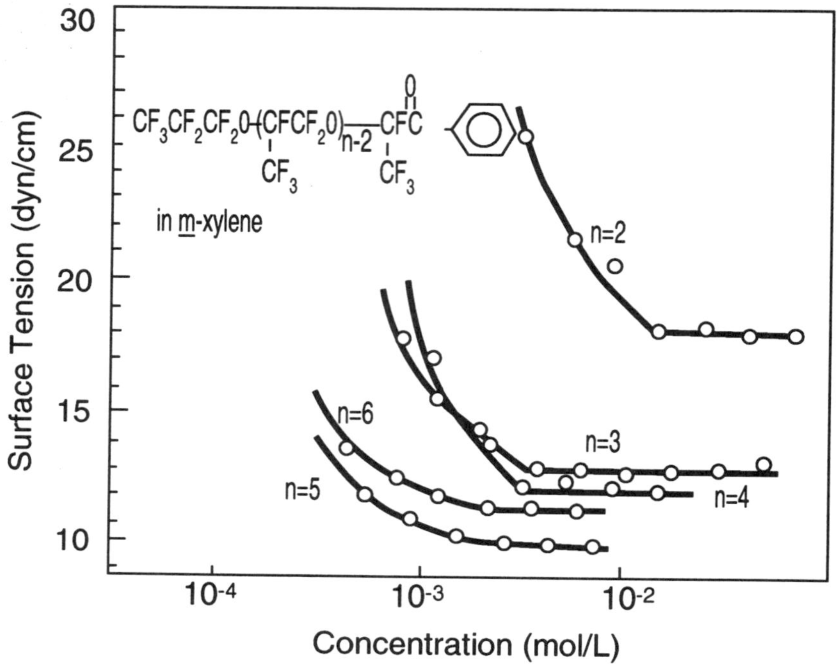

FIG. 4.35 Surface tension–log concentration curves for $(HFPO)_n$-Ph in *m*-xylene. (From Ref. 81. Reproduced by permission of Elsevier Sequoia.)

optimum structure for all organic solvents [78]. For semifluorinated alkanes see Chapter 1.8.

4.7 LIQUID-LIQUID INTERFACE

The interface between two immiscible liquids is ideally a dividing plane between the two phases (Fig. 4.1) In reality, the interface is a region of finite thickness formed by the adsorbed surfactant and the subsurface.

Adsorption of surfactants at a liquid-liquid interface is a more complex process than adsorption at a liquid-gas interface. The simplest case is a system consisting of water, an oil phase, and a surfactant. When the surfactant is soluble only in water, then factors, such as structural effects of the surfactant, which govern the oil-water interfacial tension are similar to those affecting surface tension of water. However, if the surfactant is soluble in both phases, the situation is much more complex. Various interactions and solubilization of oil in water or water in oil have to be considered. The theory of interfacial tension has been reviewed in depth by several authors [2,3,82–84].

Interfacial tension is an important factor in chemical and physical processes involving a large interface, such as formation of emulsions, spreading of one

TABLE 4.17 Surface Tension Mimima (γ_{min}) and Corresponding Concentrations (C_{min}) or Ar-(HFPO)$_n$ in Toluene or Xylene at 20°C

Ar−(HFPO)$_n$ Ar	n	γ_{min} (mN/m)	C_{min} (wt %)
Toluene			
C_6H_5	2	21.0	2.3
	3	12.8	2.3
	4	13.5	2.3
	5	12.2	0.46
	6	12.7	0.46
$p-CH_3C_6H_4$	2	21.4	2.3
	3	13.7	2.3
	4	14.8	0.46
	5	12.7	0.46
	6	12.0	0.46
m-Xylene			
$p-CH_3C_6H_4$	2	21.5	2.3
	3	14.3	2.3
	4	13.8	2.3
	5	12.8	0.46
	6	11.9	0.46

Source: Ref. 81. Reproduced by permission of Elsevier Sequoia.

liquid on another, wetting, foaming, enhanced oil recovery, and other processes of technical interest. Emulsification requires a low, near zero, interfacial tension because a large interface has to be created. A low interfacial tension is a result of positive adsorption of the surfactant at the interface. The effect of a surfactant on interfacial tension depends on the extent of adsorption and the nature of the adsorbed film.

Scholberg et al. [36] observed that fluorinated surfactants exhibit a low interfacial energy between water and a fluorocarbon but, the emulsions of these materials were not stable.

However, Greiner and co-workers [46,66] found fluorinated surfactants of the type $H(CF_2CF_2)_nCH_2O(CH_2CH_2O)_mH$ and their sulfates to lower interfacial tension between hydrocarbons and water and function as effective emulsifiers. The sulfates lowered water-nonane interfacial tension between water and nonane to 5 mN/m.

Motomura et al. [85] measured the interfacial tension of aqueous sodium perfluorooctanoate against hexane as a function of temperature around cmc. The entropy and energy of adsorption were found to be higher for sodium perfluorooctanoate than for dodecylammonium chloride or sodium dodecyl sulfate. The results indicated a weak mutual interaction between the fluorocarbon chain and hexane molecule as well as the strong hydrophobicity of the fluorinated surfactant.

Thoay [86] studied interfacial activity of partially fluorinated surfactants in an oil-water medium. The fluorocarbon group was located either at the end of the hydrocarbon tail, inserted in the middle of the hydrocarbon chain, or attached as a branch to a conventional hydrocarbon type cationic surfactant. Three types of surfactants were used:

1. $C_{18}H_{37}N(CH_2R)CH_2CH_2N(C_2H_5)_2,\ 2C_2H_5Br$

   $R = CF_3,\ C_3F_7,\ C_7F_{15},\ C_7H_{15}$

2. $R_f(CH_2)_{10}X$

   $R_f = C_2F_5,\ C_4F_9,\ C_6F_{13}$

   $X = COONa,\ CH_2OSO_3Na,\ CH_2(OCH_2CH_2)_4OH,\ CH_2N^+(C_2H_5)_3Br^-$

3. $C_nH_{2n+1}OCH_2(CF_2)_2CFHCH_2X$

   $X = N^+(C_2H_5)_3Br^-,\ OSO_3Na$

   $n = 7, 10$

In the first type a fluorocarbon branch was attached to a cationic surfactant. At low concentrations these fluorinated surfactants lower the interfacial tension, γ_i, more than their hydrocarbon analogues, but the limiting value of γ_i is much higher for the fluorinated surfactants than for the hydrocarbon analogues (Fig. 4.36).

The second surfactant type had a terminal fluorocarbon group. Fluorination increased interfacial efficiency but also increased the limiting value of γ_i (Fig. 4.37). Thoay concluded that the main effect of a terminal perfluoro group is to decrease the solubility of the surfactant and increase its interfacial efficiency.

In the third type, a fluorocarbon segment was inserted between a hydrocarbon chain and the hydrophilic group. The interfacial efficiency is similar to that of type II surfactants: greater interfacial efficiency and approximately the same γ_i limit (Fig. 4.38).

None of the fluorinated surfactants was suitable for preparing oil-water microemulsions. Thoay concluded that fluorinated surfactants weaken the CH–CH interchain interaction essential for the formation of stable oil-water microemulsions.

Mukerjee and Handa [87–89] measured surface and interfacial tensions of dilute aqueous solutions of sodium perfluorobutyrate, sodium

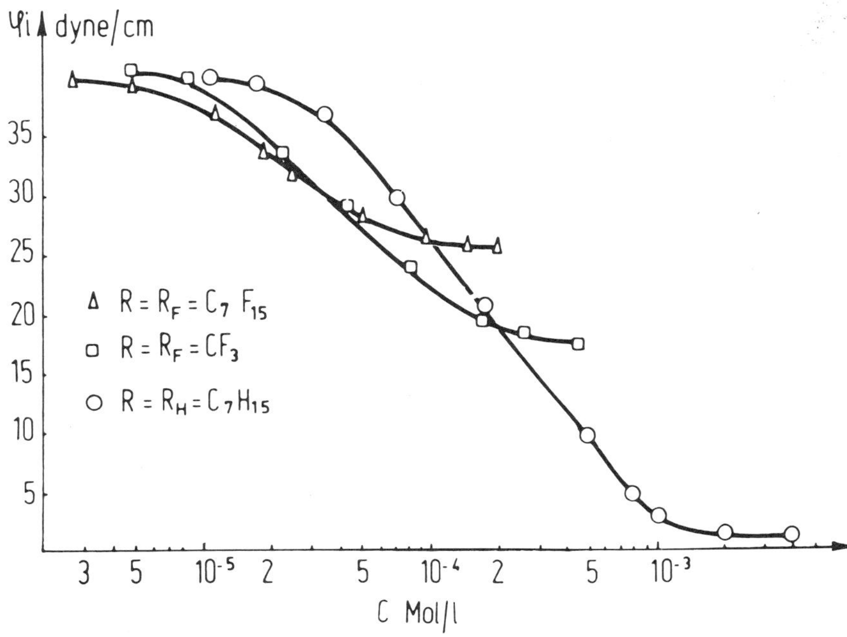

FIG. 4.36 Water-heptane interfacial tension as a function of the concentration of $C_{18}H_{37}N-CH_2CH_2N(C_2H_5)_2$, $2C_2H_5Br$. CH_2R. (From Ref. 86. Reproduced by permission of Academic Press, Inc.)

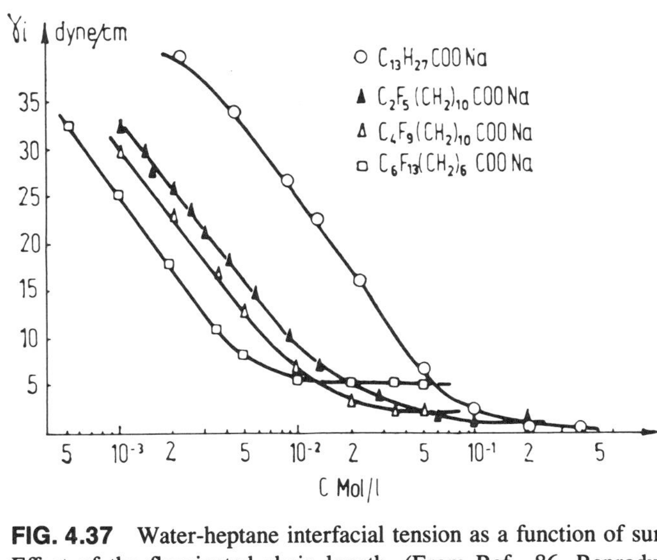

FIG. 4.37 Water-heptane interfacial tension as a function of surfactant concentration. Effect of the fluorinated chain length. (From Ref. 86. Reproduced by permission of Academic Press, Inc.)

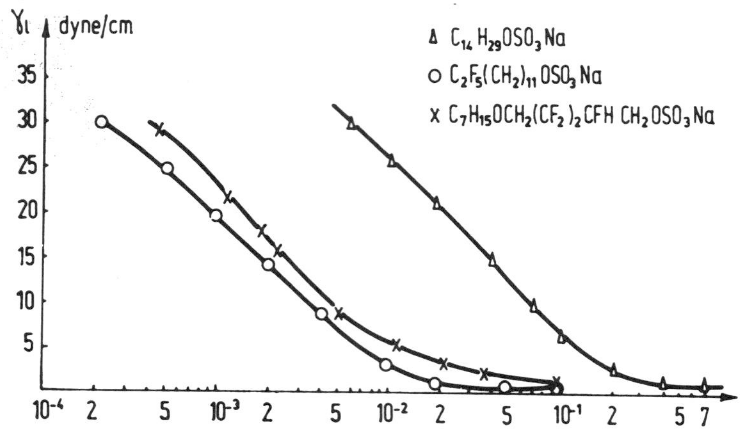

FIG. 4.38 Water-heptane interfacial tension as a function of surfactant concentration. Effect of hydrophobe structure. (From Ref. 86. Reproduced by permission of Academic Press, Inc.)

perfluorooctanoate, sodium perfluorodecanoate, sodium octyl sulfate, and sodium decyl sulfate at the air-water, hexane-water, and perfluorohexane-water interfaces. The surface and interfacial tension values, γ, were converted to interfacial pressures, Π, where $\Pi = \gamma_o - \gamma$. The data obtained for dilute solutions indicated the affinity of the surfactant for the interface.

Interfacial pressures below 2 mN/m increased linearly with surfactant concentration (activity) (Fig. 4.39). The linear relationship is in agreement with a thermodynamic model of adsorption developed by Lucassen-Reynders [90] and others [91–93]. For surfactants which are strong 1:1 electrolytes, the model suggests the limiting form of the equation of state

$$\Pi A = 2kT \tag{19}$$

where A is the surface per surfactant ion, k is the Boltzmann constant, and T is the absolute temperature. For low surfactant concentrations and low Π values, the mutual interactions of adsorbed surfactant molecules at the interface are very weak. The initial $d\Pi/dc$ values therefore indicate the relative affinity of the surfactants for the interface.

A comparison of the sodium perfluorooctanoate and sodium decyl sulfate interfacial pressures at various interfaces showed striking differences between the relative affinities of the surfactant for various interfaces. Although both surfactants have nearly identical cmc values, their adsorption at low surfactant concentrations differ markedly. The perfluorinated surfactant had a higher affinity for the air-water and perfluorohexane-water interfaces than the nonfluorinated surfactant. However, the latter has a higher affinity for the hex-

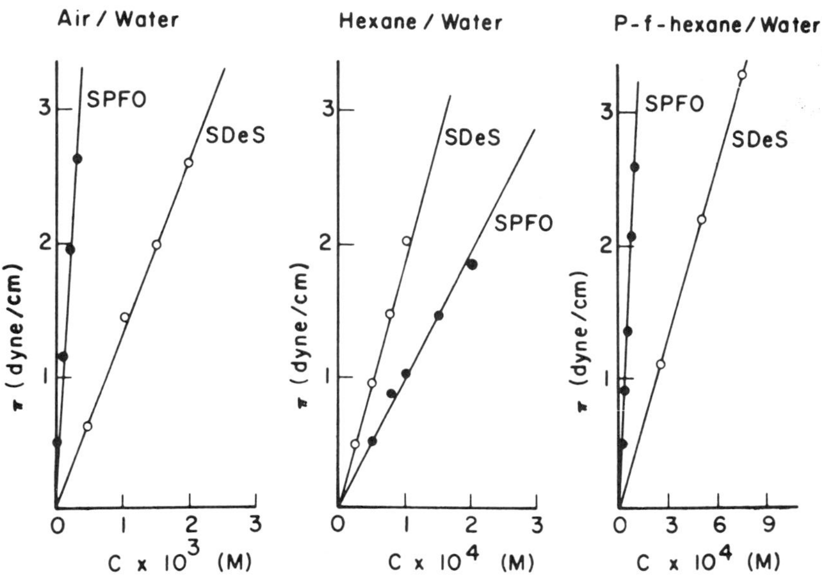

FIG. 4.39 Interfacial pressures, , plotted against the surfactant concentration in the aqueous phase, *C*, at 25°C; (○) SDeS; (●) SPFO. (From Ref. 87. Reproduced by permission of American Chemical Society.)

ane-water interface (Fig. 4.39). The higher affinity of the nonflourinated surfactant for the hydrocarbon-water interface has been attributed to a more favorable interaction of the hydrocarbon chains at the hydrocarbon-water interface.

Mukerjee and Handa [87] estimated the free energies of the surfactants at the air-water, heptane-water, and perfluorooctane-heptane interfaces and calculated the incremental changes in free energy values for the addition of a $-CF_2-$ or a $-CF_2-$ group. The affinity of a surfactant with a hydrocarbon tail for the hexane-water interface was found to be much higher than that for the air-water interface. In contrast, the difference between the affinities of a surfactant with a fluorocarbon tail for the perfluorohexane-water interface and for the air-water interface is not very large. This smaller difference is in accord with the weak interactions among fluorocarbons.

The results obtained by Mukerjee and Handa substantiate the fundamental difference between liquid-air and liquid-liquid interfaces. The surface tension of a surfactant solution depends mainly on interactions with the solvent and adsorption at the liquid-air interface. Interactions with air or vapor are weak. In contrast, a surfactant at a liquid-liquid interface interacts with two liquid phases. The lyophobic part of a surfactant oriented away from one of the liquid phases interacts with the second liquid phase in contact. The interaction between the lyophobic film and the second liquid phase is much more significant than the interaction with air or a vapor.

Zhao Guo-Xi and Zhu Bu-Yao [94] studied the interfacial tension and spreading properties of mixed solutions of anionic (cationic) and cationic (anionic) fluorocarbon and hydrocarbon surfactants. The anionic-cationic system exhibited much higher surface activity than the single components. The aqueous solution of the surfactant mixture had a lower surface tension and a lower oil-water interfacial tension and spread readily on *n*-heptane and light kerosene. The increase in surface activity was attributed to a strong interaction, possibly ion-pair formation, between the surfactant ions with opposite charges.

4.8 EMULSIONS

An emulsion is a reasonably stable mixture of one liquid (the dispersed phase) dispersed in an immiscible liquid (the continuous phase). Aqueous emulsions are oil-in-water (O/W) or water-in-oil (W/O). Here the term *oil* designates a water-immiscible liquid. The basic theory of emulsions has been discussed in several books and articles [12,95–97].

The formation of emulsions increases immensely the interfacial area, A_i, between the two phases and therefore requires a low interfacial tension, γ_i. It can be readily seen that the work, W, required to create the new interface

$$W = \gamma_i A_i \tag{20}$$

is very large unless γ_i is small. However, a low, near-zero, interfacial tension is not a sufficient prerequisite for the formation of a stable emulsion. The stability of an emulsion also depends on nature of the interfacial boundary. The stabilization mechanisms of emulsions are complex and not completely understood. The main mechanisms are steric stabilization, ionic stabilization, and the less effective stabilization by hydrosols. Droplets in the emulsion have to be prevented from approaching each other to a distance where coalescence results. Also, a droplet has to resist deformation during a droplet-droplet collision and a disturbance of the protective film.

For the selection of a suitable surfactant structure the hydrophile-lipophile balance (HLB) has been a useful index [97–101]. The HLB values have played an important part in selecting fluorinated surfactants for emulsification.

A distinction has to be made between kinetically and thermodynamically stable emulsions. Mixtures of water, an oil, a surfactant, and a cosurfactant in proper proportions can spontaneously form transparent, thermodynamically stable emulsions, called microemulsions. The definition is not quite unequivocal. Thermodynamically unstable but kinetically stable systems which form spontaneously have also been called microemulsions [102]. Micellar solutions have been included in the definition of microemulsions and the debate on microemulsion structure is ongoing.

Fluorinated surfactants are unique. The stability of fluorinated surfactants to heat or chemical attack is advantageous when preparing emulsions containing corrosive chemicals. The fluorocarbon part of the molecule is not only hydrophobic but also oleophobic and contributes to a very low free surface energy of the adsorbed surfactant. Fluorinated surfactants are effective emulsifiers for fluorinated monomers, e.g., tetrafluoroethylene, in the polymerization process [103]. Fluorinated surfactants are also used for emulsifying other low-surface-energy liquids, such as silicones and fluorocarbons. However, perfluorinated surfactants are not as effective as hydrocarbon-type surfactants in emulsifying hydrocarbon oils. Synergistic effects of conventional hydrocarbon [104] surfactants have been employed to increase the effectiveness of either surfactant alone and to lower the total surfactant concentration needed. The low efficiency of fluorinated surfactants in emulsifying hydrocarbon oils is beneficial in some applications. Selected fluorinated surfactants can provide wetting without emulsification.

To generate *microemulsions* a surfactant must lower the oil-water interfacial tension to a small value (near zero) [105]. A strong association between the hydrocarbon tails of oil and the hydrophobe of conventional hydrocarbon-based surfactants is essential for forming oil-water microemulsions. Substituting hydrogen by fluorine reduces this hydrocarbon-hydrocarbon interaction and perfluorinated surfactants lack lipophilic groups needed for emulsifying hydrocarbons. Thoay [86] attempted to compensate for the lack of lipophily by preparing partially fluorinated surfactants (see also Sec. 4.6) which have three parts: a lipophilic group, a hydrophilic group, and a fluorocarbon group. However, none of the fluorinated surfactants was found to be effective in generating a microemulsion of hydrocarbons in water. An attempt to prepare a hydrocarbon-water microemulsion produced a gel. Large amounts of the fluorinated surfactants were needed to prevent gel formation.

A keen interest in *microemulsions of fluorocarbons* in water was kindled by the need for synthetic oxygen carriers in blood (see Chapter 10.4). Gerbacia and Rosano [106] prepared a stable fluorocarbon emulsion using a mixture of fluorinated and hydrocarbon-type nonionic surfactants. The droplet size was not determined, however, and the emulsions were not characterized. Oliveros et al. [107] prepared perfluorinated microemulsions consisting of four components: (1) sodium perfluorooctanoate, an anionic fluorinated surfactant; (2) 2,2,3,3,4,4,4-heptafluoro-1-butanol; (3) perfluorohexane; and (4) water. The pseudoternary phase diagram (Fig. 4.40) shows two regions, M_1 and M_2, of respective W/O and O/W microemulsions [107]. These optically transparent microemulsions were characterized by nuclear magnetic resonance spectroscopy.

Anionic perfluorinated and almost completely fluorinated surfactants were used by Ceschin et al. [108] to prepare microemulsions of fluorocarbons. The particle size in monophasic regions is related to surfactant and cosurfactant

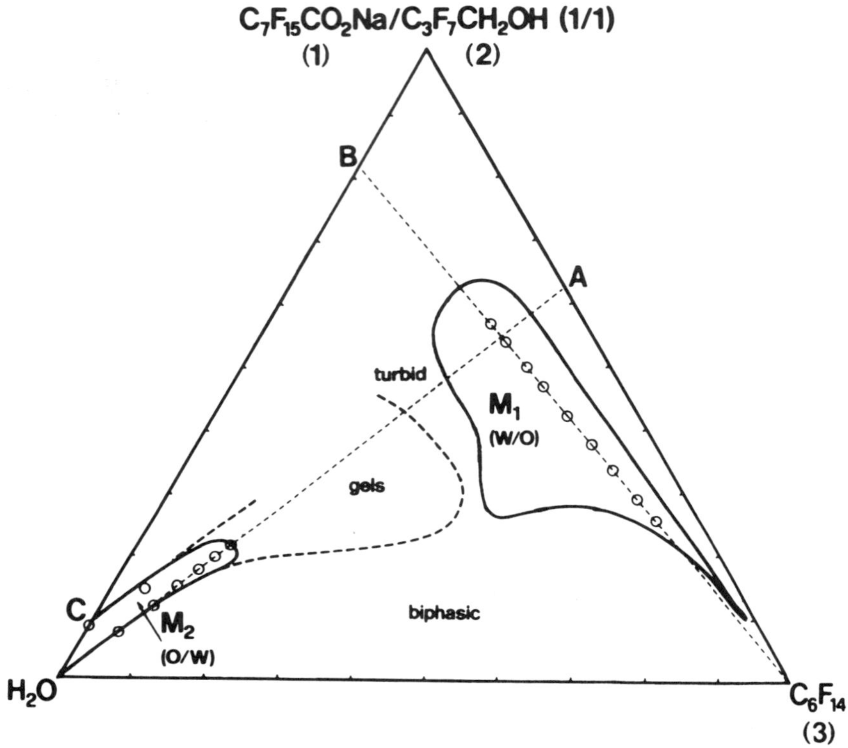

FIG. 4.40 Pseudoternary phase diagram of the system $C_7F_{15}COONa$ (1)–$C_3F_7Ch_2OH$ (2)–C_6F_{14} (3) for a constant surfactant/cosurfactant weight ratio of 1:1 (molar ratio 1:2.2) at 23°C. M_1 and M_2 are the regions of respective W/O and O/W microemulsions. (From Ref. 107. Reproduced by permission of Verlag Helvetica Chimica Acta.)

nature, the weight ratio of surfactant to the oil phase, and the nature of the oil phase.

A French team [109] prepared nonionic microemulsions from a ternary mixture of water, a fluorocarbon (e.g., perfluorodecalin), and a nonionic perfluoroalkylpoly(oxyethylene) surfactant. The microemulsions formed spontaneously when the three components were mixed and shaken by hand. The concentration of the fluorocarbon was high (up to 50% wt/wt), resulting in high solubility of gases in the emulsion. The microemulsions remained stable when diluted with water. Small angle (SANS) measurements gave an average droplet size of about 100 Å and the apparent aggregation number of 2100. The size and shape of droplets remained about the same when the microemulsions were diluted with water. The droplets were visualized, comprising a fluorocarbon core surrounded by the fluorinated surfactant. The surfactant is oriented with its fluorocarbon end toward the fluorocarbon core and its hydrophilic group directed toward water.

According to Shinoda and Kunieda [96], aqueous microemulsions can form from ternary mixtures of a hydrocarbon, a nonionic surfactant, and water when the temperature of the system is kept between the solubilization temperature (lower limit) and the cloud point (higher limit). The area between the solubilization and cloud point curves [109] obtained by the French team represents aqueous microemulsions (Winsor IV solutions [110]). The solubilization temperature and the cloud point increase with increasing fluorocarbon content of the system. Both curves ascend until they join at the phase inversion temperature (Fig. 4.41). To maximize solubilization, it is desirable to have the phase inversion temperature fall into the temperature range where the system is used. Both phase inversion temperature and the cloud point are related to the structure and the hydrophile-lipophile balance (HLB) of the surfactant (see Sec. 6.6).

Mathis et al. [109] prepared surfactants of the general structure $C_mF_{m+1}(CH_2)_p(OC_2H_4)_nOH$. Surfactants with the subscripts $m = 6$, $p = 1$, $n = 5$, and $m = 7$, $p = 1$, $n = 5$, or 6 were selected for their appropriate HLB. The phase diagrams, like the one shown in Fig. 4.42, indicated high sensitivity to small variations of the HLB. This HLB dependence and an increase in solubilization effectiveness [111] mandated the use of monodisperse surfactants.

Robert and Tondre [112] studied reverse microemulsions consisting of water emulsified in a binary mixture of a fluorocarbon and a nonionic fluorinated surfactant. Surfactants of the structure $C_6F_{13}CH_2(OC_2H_4)_nOH$, with $n = 4$, 5, or 6, emulsified large amounts of water (up to 20 wt % water in perfluorodecalin) in fluorocarbons. The solubilization of water by nonionic fluorinated surfactants in fluorocarbons follows the same trends exhibited by hydrocarbon-type systems.

Chittofrati et al. [113] prepared microemulsions of perfluoropolyethers (PFPEs) having the general structure

$$R_fO-(\underset{\displaystyle CF_3}{\underset{|}{C}}F-CF_2-O)_n-(CF_2O)_m-R_f$$

Ammonium salts of PFPE carboxylates are very effective surfactants for the preparation of monophasic, transparent, isotropic PFPE emulsions that behave as true microemulsions. With increasing hydrophobic chain length, the solubility of the surfactant in water decreases and the solubility in oil increases. Short-chain surfactants form O/W microemulsions. In contrast, oil-soluble long-chain surfactants form W/O ternary microemulsions.

Amine(polyfluoroalkoxyacyl)imides [28] (see Sec. 4.2) formed stable perfluorodecalin emulsions when the surface activity of the surfactant was sufficient to lower the surface tension of water below 18 mN/m. Amin-

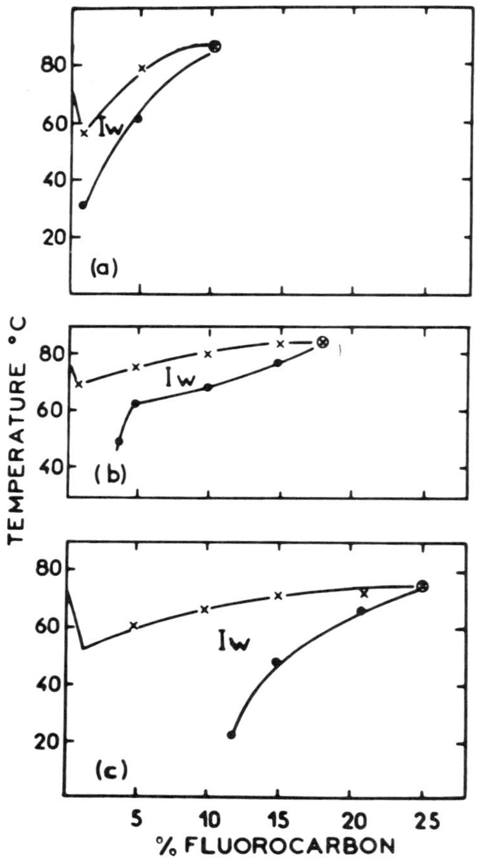

FIG. 4.41 Partial phase diagrams of systems containing 10 wt % $C_6F_{13}C_2H_4(OC_2H_4)_{14}OH$, water, and (a) $C_6F_{13}CH{=}CHC_6F_{13}$, (b) $C_8F_{17}C_2H_5$, (c) $C_6F_{13}CH{=}CH_2$ as a function of temperature. Upper curves are cloud point curves; lower curves are solubilization curves. (From Ref. 109. Reproduced by permission of American Chemical Society.)

acylimides capable of lowering the surface tension of water only to 18 to 25 mN/m formed unstable emulsions that broke after a few hours within a few weeks. Aminacylimides with a higher surface tension in water formed emulsions that were even less stable.

Selve at al. [23] synthesized fluorinated nonionic surfactants with a two-chain polyoxyethylene hydrophilic head linked to the hydrophobe via an amide bond, $F(CF_2)_l(CH_2)_mC(O)N[(C_2H_4O)_nCH_3]_2$. In this surfactant series, the surfactant with the subscripts $l = 10$, $m = 1$, and $n = 33$ has an HLB value

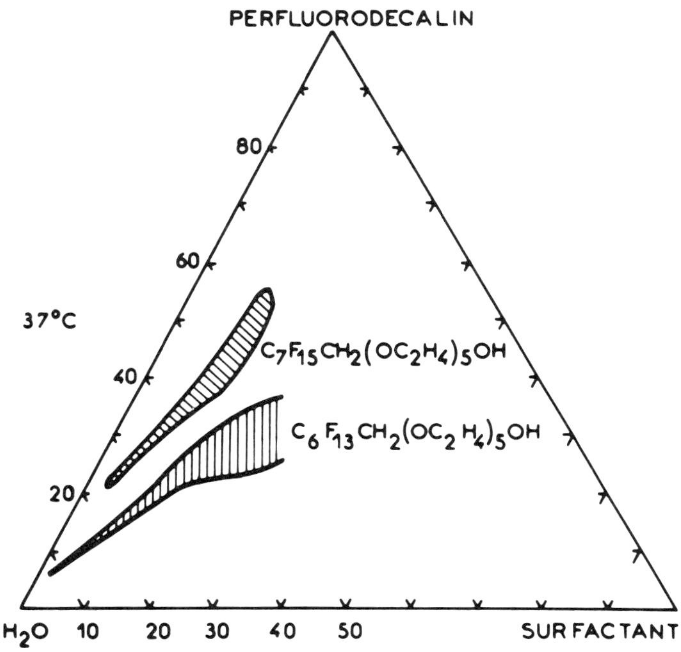

FIG. 4.42 Phase diagrams of the systems water-perfluorodecalin-$C_7F_{15}CH_2(OC_2H_4)_5OH$ and water-perfluorodecalin-$C_6F_{13}CH_2(OC_2H_4)_5OH$ at 37°C (only the microemulsion region is shown, hatch area). (From Ref. 109. Reproduced by permission of American Chemical Society.)

phase suitable for forming microemulsions with inversion temperatures near the ambient temperature. Hence the surfactant emulsified larger amounts (18 wt %) of a perfluorocarbon than other surfactants prepared in this series.

The application of fluorinated surfactants in emulsions and dispersions is covered in Chapter 8. Fluorocarbon emulsions used as oxygen carriers in blood are discussed in Chapter 10.

4.9 FOAMS

Foams are coarse dispersions of gas in a relatively small amount of liquid. In solid foams the continuous phase is a solid which, at one point in time, has been a liquid. Pure liquids cannot form foams and nearly all liquid foams are thermodynamically unstable. Only liquids containing a surface-active component can form foams. The surface-active component may be a solute dissolved in the liquid or insoluble matter at the interface, such as a solid in the form of insoluble particles, a liquid crystal phase, or an insoluble monomolecular film. Surfactants can stabilize but also destabilize foams and cause their collapse.

Foams are complex systems and many factors affect foaming and foam stability [3,11,114–118]. Surface tension trying to pull the gas bubble together is opposed by pressure inside the bubble and the forces within the film between the bubbles, perpendicular to the interface (the disjoining pressure). Foams are subject to mechanical disturbances and environmental stresses that affect the persistence of foam. A foam collapses when (1) the gas from the interior of the bubble diffuses into the gas phase outside the foam or into a larger bubble because of the pressure differences caused by the larger curvature of the smaller bubble, (2) the lamellae are ruptured, or (3) the liquid drains from the lamellae and the lamellae are thinned to below a minimum thickness (about 50–100 Å). Drainage is caused by gravity or by pressure differences across the interface. According to Laplace's equation the pressure difference across a curved interface, ΔP, depends on the curvature of the interface R

$$\Delta P = \gamma \left(\frac{1}{R_1} + \frac{1}{R_2} \right) \tag{21}$$

In the foam bubble the curvature of the interface is largest (smallest radius) where bubbles join (referred to as the plateau border region or the Gibbs triangle) . The pressure in the plateau region is therefore smaller than in the central areas of the lamellae. The pressure difference causes liquid to flow into the plateau border regions and depletes the central parts of the lamellae. An important factor determining foam stability is the ability of foam to resist excessive localized thinning of the lamellae and to repair thinning of the lamellar film caused by drainage or by expansion of the bubble. Gibbs defined this property as foam elasticity, E_s,

$$E_s = \frac{2d\gamma}{d \ln A} \tag{22}$$

An incremental increase of the surface area, A, is counterbalanced by an increase in surface tension. A surface tension gradient formed when the surface area is expanded causes a flow of the liquid toward the thinner areas and hinders stretching of the surface. Ewers and Sutherland [119] proposed a surface transport mechanism in which the surface tension gradient causes the surface monolayer and the subsurface layer to flow from a low- to a high-surface-tension region. The surface elasticity effect arising from a surface tension gradient as proposed by Gibbs is complemented by the Marangoni effect, which results from the time dependence of surface tension. The Gibbs effect deals with foam in the state of equilibrium, whereas the Marangoni effect results from a kinetic time lag of surfactant diffusion which causes a surfactant concentration gradient and consequently a surface tension gradient. Since the diffusion of surfactant from the bulk solution to the interface requires time, the concentration of surfactant in the interface decreases when the surface is expanded. The surface tension increase counteracts the forces stretching the surface. The Marangoni effect is operative only in dilute solutions. Other important factors governing foam persistence are surface viscosity, bulk viscosity, and the electrostatic repulsion between the two sides of the film.

Foams can be desirable or undesirable. It is therefore useful that fluorinated surfactants differ widely in their foaming propensity, as shown for Zonyl sur-

TABLE 4.18 Foaming Action in Water: 0.1% Solids in Water, Ross-Miles Foam Test [ASTM-D1173-53]

		Foam height (mm)					
		Deionized water			300 ppm hard water		
Zonyl	Temperature °C/°F	Initial	3 min	10 min	Initial	3 min	10 min
FSA	41/105	100	95	95			
	68/155	125	120	50			
FSP	41/105	35	30	25		[Surfactant precipitates]	
	68/155	35	30	10			
FSE	41/105	120	110	90			
	68/155	50	40	10			
UR + sodium hydroxide[a]	41/105	15	15	10			
	68/155	15	15	10			
FSJ	41/105	130	120	100	65	30	20
	68/155	140	130	60	60	20	10
FSN	41/105	145	140	135	135	135	130
	68/155	130	130	40	130	120	20
FSN-100	41/105	135	130	130	130	130	130
	68/115	130	100	40	120	95	30
FSO	41/105	74	65	63	35	30	24
	68/155	134	95	62	95	58	35
FSO-100	41/105	90	85	80	78	75	73
	68/155	105	47	33	115	55	35
FSC	41/105	220	220	220	225	225	225
	68/155	220	220	70	200	200	80
FSK	41/105	220	220	220	220	220	220
	68/155	220	220	80	220	200	80
TBS	41/105	228	235	235		Surfactant precipitates	
	68/155	246	264	208			

[a] Aqueous solution prepared by heating 10.0 g Zonyl UR to 50–55°C (122–131°F) in water containing 0.53 g sodium hydroxide and stirring.
Source: Ref. 34.

factants in Table 4.18. (For structure see Table 4.4). Some fluorinated surfactants are effective foaming agents, whereas others prevent foaming. Most effective foaming agents can be found among amphoteric, cationic, and anionic fluorinated surfactants. Their foaming characteristics depend on the structures of the hydrophobe and the hydrophilic group. In general, nonionic fluorinated surfactants, like hydrocarbon-type nonionic surfactants, produce less foam and less persistent foam than ionic fluorinated surfactants.

Surface tension lowering is a prerequisite for foam formation but not a sufficient condition for foam stability. For example, the fluorinated surfactants Zonyl FSA, Zonyl FSP, and Zonyl FSK lower the surface tension of water to about 20 dyne/cm (Fig. 4.43). However, Zonyl FSK is an outstanding foaming agent, whereas Zonyl FSP can function under some conditions as an antifoam agent. The third fluorinated surfactant represented in Fig 4.43, Zonyl FSA, is neither an effective foaming agent nor an antifoam agent. The antifoam activity of Zonyl FSP, a mixture of $(R_fCH_2CH_2O)P(O)(ONH_4)_2$ and $(R_fCH_2CH_2O)_2P(O)(ONH_4)$, is related to its phosphate ester hydrophile.

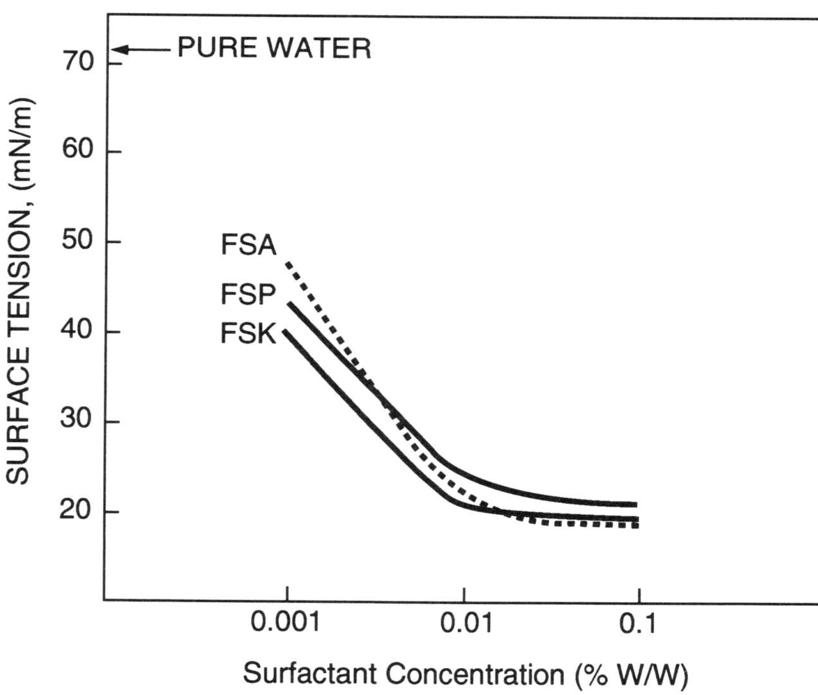

FIG. 4.43 Surface tensions of three Zonyl fluorinated surfactants in water at 25°C. (From Ref. 34.)

Other fluorinated surfactants featuring a phosphate ester–type hydrophile function as antifoam agents. For example, Surflon S-112, a fluorinated surfactant with a phosphate ester hydrophile, depressed foaming of a solution containing a nonionic surfactant from 80 to 38 mm of foam height measured by the Ross-Miles test. After 5 min the foam height was 58 mm in the absence of Surflon S-112 but only 6 mm in the presence of 0.004% Surflon S-112 [120]. Similar results were obtained with Atsurf S-112, a phosphate ester–type fluorinated surfactant [121]. Alkylphosphate esters are known to function as foam inhibitors. Their foam inhibition mechanism is probably related to a large spreading coefficient and drainage promotion in the foam lamellae.

A mixture of an anionic (Fluorad F-95) and a cationic fluorinated surfactant (Fluorad FC-134) has been recommended for foam suppression in organic liquids [122].

Kozhanov et al. [123] studied stability of foams as a function of the concentration of a fluorinated surfactant in the aqueous phase. Foams were made by shaking equal volumes of toluene and water. The stability of foams depended on the final volume of the spreading coefficient of the surfactant solution with respect to toluene. Foams were formed when the spreading coefficient values were positive.

Fluorinated surfactants are used in fire-extinguishing foams which can spread on hydrocarbons and organic solvents (see Chapter 8).

REFERENCES

1. R. C. Brown, Proc. Phys. Soc. London 59, 429 (1947).
2. D. Myers, "Surfactant Science and Technology," VCH Publishers, Weinheim, (1988).
3. S. Ross and I. D. Morrison, "Colloidal Systems and Interfaces," Wiley, New York (1988).
4. D. Attwood and A. T. Florence, "Surfactant Systems," Chapman and Hall, London (1983).
5. G. Némethy and H. S. Sheraga, J. Chem. Phys. 36, 3382 (1962).
6. F. Franks, J. Ravenhill, P. A. Egelstaff, and D. I. Page, Proc. Roy. Soc. (London) A319, 189 (1970).
7. I. Danielssson, B. Lindman, and L. Odberg, Suomen Kemistilehti 42, 209 (1969).
8. B. Tamamushi, in "Colloidal Surfactants," K. Shinoda, T. Nakagawa, B. Tamamushi, and T. Isemura, eds., p. 179, Academic Press, New York (1963).
9. H. Schott, J. Pharm. Sci. 69, 852 (1980).
10. F. Van Voorst Vader, Trans. Faraday Soc. 56, 1067 (1960).
11. M. J. Rosen, "Surfactants and Interfacial Phenomena," 2nd ed., Wiley, New York (1989).
12. A. W. Adamson, "Physical Chemistry of Surfaces," 5th ed., Wiley, New York (1990).

13. J. W. McBain and C. W. Humphreys, J. Phys.Chem. 36, 300 (1932).
14. J. W. McBain and R. C. Swain, Proc. Roy. Soc. (London), A154, 608 (1936).
15. D. J. Salley, A. J. Weith, Jr., A. A. Argyle, and J. K. Dixon, Proc. Roy. Soc.(London) A203, 42 (1950).
16. J K. Dixon, C. M. Judson, and D. J. Salley, "Monomolecular Layers," p. 63, AAAS, Washington, DC (1954).
17. K. Tajima, M. Muramatsu, and T. Sasaki, Bull. Chem. Soc. Japan 43, 1991 (1970).
18. K. Tajima, Bull. Chem. Soc. Japan 43, 3063 (1970).
19. N. H. Steiger and G. Aniansson, J. Phys. Chem. 58, 228 (1954).
20. K. Shinoda and K. Ito, J. Phys. Chem. 65, 1499 (1961).
21. S. J. Rehfeld, J. Colloid Interface Sci. 31, 46 (1969).
22. F. M. Wrenger and S. Wrenn, J. Phys. Chem. 78, 1387 (1974).
23. C. Selve, J. C. Ravey, M. J. Stebe, C. El Moudjahid, E. M. Moumni, and J.J. Delpuech, Tetrahedron 47, 411 (1991).
24. K. Shinoda, M. Hatō, and T. Hayasdhi, J. Phys. Chem. 76, 909 (1972).
25. K. Shinoda and H. Nakayama, J. Colloid Sci. 18, 705 (1963).
26. K. Shinoda and K. Katsura, J. Phys. Chem. 68, 1568 (1964).
27. Th. F. Tadros, J. Colloid Interface Sci. 74, 196 (1980).
28. L. Haywood, S. McKee, and W. J. Middleton, J. Fluorine Chem. 51, 419 (1991)
29. C. H. Arrington and G. D. Patterson, J. Phys. Chem. 57, 247 (1953).
30. G. Caporiccio, F. Burzio, G. Carniselli, and V. Biancardi, J. Colloid Interface Sci. 98, 202 (1984).
31. I. H. Smith and R. H. Ottewill, "Symposium on Surface Active Agents," p. 77, Society of Chemical Industry, London (1979).
32. L. Matos, J. C. Ravey, and G. Serratrice, J. Colloid Interface Sci. 128, 341 (1989).
33. (a) C. H. Zhang, B. Y. Zhu, and G. X. Zhao. J. Colloid Interface Sci. 144, 483 (1991). (b) C. H. Zhang, G. X. Zhao, B. Y. Zhu, ibid. 144, 491 (1991).
34. Du Pont Performance Products, Zonyl Fluorosurfactants (1987).
35. H. G. Klein, J. N. Meussdoerffer, and H. Niederprüm, Metalloberfläche 29, 559 (1975).
36. H. M. Scholberg, R. A. Guenthner, and R. I. Coon, J. Phys. Chem. 57, 923 (1953).
37. M. K. Bernett and W. A. Zisman, J. Phys. Chem. 63, 1911 (1959).
38. A. Lottermoser and B. Baumguertel, Trans. Faraday Soc. 31, 200 (1935).
39. J. Powney, Trans. Faraday Soc. 31, 151 (1935).
40. H. Kunieda and K. Shinoda J. Phys. Chem. 80, 2468 (1976).
41. J. O. Hendrichs, Ind. Eng. Chem. 45, 99 (1953).
42. E. Schuierer, Tenside 13, 1 (1976).
43. H. C. Fielding, in "Organofluorine Chemicals and Their Industrial Applications," R. E. Banks, ed., p. 214, Society of Chemical Industry, London/Ellis Horwood, Chichester (1979).
44. E. G. Shafrin and W. A. Zisman, J. Phys. Chem. 66, 740 (1962).
45. M. K. Bernett and W. A. Zisman, J. Phys.Chem. 71, 2075 (1967).
46. A. Greiner, B. Krüger, and M. Herbst, Z. Chem. 18, 383 (1978).

47. H. G. Klein, J. N. Meussdoerffer, H. Niederprm, and M. Wechsberg, Tenside 15, 2 (1978).
48. N. O. Brace, J. Org. Chem. 27, 4491 (1962).
49. C. Kimura, K. Kashiwaya, M. Kobayashi, and T. Nishiyama, J. Am. Oil Chem. Soc. 61, 105 (1984).
50. N. Ishikawa and M. Sasabe, J. Fluorine Chem. 25, 241 (1984).
51. P. Mukerjee and K. J. Mysels, in "Colloidal Dispersions and Micellar Behavior," Paper Symposium 1974, ACS Symp. Ser. 9, 239 (1975).
52. H. B. Klevens and M. Raison, J. Chim. 51, 1 (1954).
53. A. F. H. Ward and L. Torday, J. Chem. Phys. 14, 453 (1946).
54. R. Miller, Colloid Polym. Sci. 259,1124 (1981).
55. K. J. Mysels and A. T. Florence, Colloid Interface Sci. 43, 577 (1973).
56. K. Lunkenheimer and R. Miller, Tenside 16, 312 (1979).
57. R. Miller and K. Lunkenheimer, Tenside 17, 288 (1980).
58. R. Miller and K. Lunkenheimer, Colloid Polym. Sci. 264, 367 (1986).
59. R. Miller and M. Ziller, Colloid Polym. Sci. 266, 532 (1988).
60. J. N. Meussdoerffer and H. Niederprüm, Chem. Z. 104, 45 (1980).
61. L. V. Dikhtievskaya and N. A. Makarevich, Vestsi Akad. Navuk BSSR, Ser. Khim. Navuk 1991, (1), 24; CA 114, 254357.
62. D. E. Hirt, R. K. Prud'homme, B. Miller, and L. Rebenfeld, Colloids Surfaces 44, 101 (1990).
63. E. L. Talbot, J. Phys. Chem. 63, 1666 (1959).
64. H. G. Bryce, in "Fluorine Chemistry," Vol. V, p. 370, J. H. Simons, ed., Academic Press, New York (1964).
65. V. Glöckner, K. Lunkwitz, and D. Prescher, Tenside 26, 376 (1989).
66. A. Greiner and M. Herbst, Z. Chem. 16, 229 (1976).
67. G. B. Blake, A. H. Albrecht, and H. G. Bryce, Am. Chem. Soc. Div. Petrol. Chem. Gen. Pap., No. 32, 131 (1954).
68. J. G. O'Rear and W. A. Zisman, U.S. Patent 2,824,141 (1958).
69. P. D. Faurote, C. M. Murphy, C. M. Henderson, J. G. O'Rear, and H. Ravner, Ind. Eng. Chem. 48, 445 (1954).
70. A. H. Ellison and W. A. Zisman, J. Phys. Chem. 59, 1233 (1955).
71. A. H. Ellison and W. A. Zisman, J. Phys. Chem. 60, 416 (1956).
72. N. L. Jarvis and W. A. Zisman, J. Phys. Chem. 63, 727 (1959).
73. A. S. Kertes and H. Gutmann, in "Surface and Colloid Science," E. Matijevic, ed., Vol. 8, p. 193, Wiley Interscience, New York (1975).
74. H. Gutmann and A. S. Kertes, J. Colloid Interface Sci. 51, 406 (1975).
75. M. K. Bernett, N. L. Jarvis, and W. A. Zisman, J. Phys. Chem. 66, 328 (1962).
76. H. W. Fox and W. A. Zisman, J. Colloid Sci. 5, 514 (1950).
77. E. F. Hare, E. G. Shafrin, and W. A. Zisman, J. Phys. Chem. 58, 236 (1954).
78. N. L. Jarvis and W. A. Zisman, J. Phys. Chem. 64, 150 (1960).
79. A. R. Katrizki, T. L. Davis, and G. W. Rewcastle, Langmuir 4, 732 (1988).
80. R. Holzleiter, Report to First Diesel Fuel Conference, Aberdeen Proving Grounds, MD, January 1986.
81. N. Ishikawa and M. Sasabe J. Fluorine Chem. 25, 241 (1984).
82. C. A. Miller and P. Neogi, eds., "Interfacial Phenomena," Surfactant Science Series, Vol. 17, Marcel Dekker, New York (1985).

83. F. Tadros, ed., "Surfactants," Academic Press, London (1984).
84. M. Bourrel and R. S. Schechter, Microemulsions and Related Systems: Formulation, Solvency and Physical Properties. Surfactant Science Series, Vol. 30, Marcel Dekker, New York (1988).
85. K. Motomura, I. Kajiwara, N. Ikeda, and M. Aratono, Colloids Surfaces 38, 61 (1989).
86. N. Thoay, J. Colloid Sci. 62, 222 (1977).
87. P. Mukerjee and T. Handa, J. Phys. Chem. 85, 2298 (1981).
88. P. Mukerjee, J. Am. Oil Chem. Soc. 59, 573 (1982).
89. T. Handa and P. Mukerjee, J. Phys. Chem. 85, 3916 (1981).
90. E. H. Lucassen-Reynders, J. Phys. Chem. 70, 1977 (1966).
91. G. L. Gaines, J. Chem. Phys. 69, 2627 (1978).
92. I. D. Robb and A. E. Alexander, J. Colloid Interface Sci. 28, 1 (1966).
93. L. Ter-Minassian-Saraga, Proc. Int. Cong. Surface Activity 2nd, London, I, 36 (1957).
94. Zhao Guo-Xi and Zhu Bu-Yao, Colloid Polym. Sci. 261, 89 (1983).
95. K. J. Lissant, ed., "Emulsions and Emulsion Technology," Marcel Dekker, New York (1974).
96. K. Shinoda and H. Kunieda, in "Microemulsions. Theory and Practice," p. 57, L. M. Prince, ed., Academic Press, New York (1977).
97. P. Becher, ed., "Encyclopedia of Emulsion Technology," Vol. I, Marcel Dekker, New York (1983).
98. W. C. Griffin, J. Soc. Cosmet. Chem. 1, 311 (1949).
99. W. C. Griffin, J. Soc. Cosmet. Chem. 5, 249 (1949).
100. J. T. Davies, Proc. Int. Cong. Surface Activity, 2nd, London, I, 426 (1957).
101. P. Becher, J. Soc. Cosmet. Chem. 11, 325 (1960).
102. S. Friberg, Colloids Surfaces 4, 201 (1982).
103. H. G. Bryce, in "Fluorine Chemistry," J. H. Simmons, ed., Vol. 5, p. 380, Academic Press, New York (1964).
104. M. K. Bernett and W. A. Zisman, J. Phys. Chem. 65, 448 (1961).
105. L. M. Prince, J. Colloid Interface Sci. 23, 165 (1967).
106. W. Gerbacia and H. L. Rosano, U.S. Patent 3,778,381 (1973).
107. E. Oliveros, M.T. Maurette, and A. M. Braun, Helv. Chim. Acta 66, 1183 (1983).
108. C. Ceschin, J. Roques, M. C. Malet-Martino, and A. Lattes, J. Chem. Tech. Biotechnol. 35A, 73 (1985).
109. G. Mathis, P. Leempoel, J.C. Ravey, C. Selve, and J.J. Delpuech, J. Am. Chem. Soc. 106, 6162 (1984).
110. P. A. Winsor, Trans. Faraday Soc. 44, 376 (1948).
111. K. Shinoda and H. Kuneida, J. Colloid Interface Sci 42, 381 (1973).
112. A. Robert and C. Tondre, J. Colloid Interface Sci. 98, 515 (1984).
113. A. Chittofrati, D. Lenti, A. Sanguineti, M. Visca, C, Gambi, D. Senatra, and Zhou Zhen, Colloids Surfaces 41, 45 (1989).
114. J. J. Bikerman, "Foams," Springer-Verlag, New York (1973).

115. E. H. Lucassen-Reynders, in "Anionic Surfactants: Physical Chemistry of Surfactant Action," E. H. Lucassen-Reynders, ed., Surfactant Science Series Vol 11, p. 1. Marcel Dekker, New York (1981).
116. M. J. Schick and I. R. Schmolka, in "Nonionic Surfactants: Physical Chemistry," M. J. Schick, ed., Surfactant Science Series Vol. 23, p. 835. Marcel Dekker, New York (1987).
117. P. Walstra, in "Foams: Physics, Chemistry and Structure," A. J. Wilson, ed., Springer-Verlag Berlin (1989).
118. S. Ross, in "Kirk-Othmer Encyclopedia of Chemical Technology," Vol. 11, p. 127, Wiley, New York (1980).
119. W. E. Ewers and K. L. Sutherland, Aust. J. Sci. Res. A5, 697 (1952).
120. Asahi Glass, Surflon Fluorosurfactants (1986).
121. Imperial Chemical Industries, Ltd., "Monflor" Fluorochemical Surface Active Agents (1987).
122. 3M Company, Technical Information , 3M Brand Fluorochemical Surfactants (1967).
123. B. P. Kozhanov, M. Yu. Pletnev, and I. G. Vlasenko, Kolloidn. Zh. 45(3), 563, 1983; CA 99, 59291

5
Solid-Liquid Boundary

5.1 WETTING

Since the fluorocarbon tail of fluorinated surfactants is not only hydrophobic but also oleophobic, the interactions of fluorinated surfactants with solid surfaces differ from those of hydrocarbon surfactants. Fluorinated surfactants, depending on their structure and the system in which they are used, either facilitate wetting or function as repellents and hinder wetting. The effect of fluorinated surfactants on wetting depends on lowering of interfacial tensions and adsorption of the surfactant on the liquid-vapor and solid-liquid interfaces. A brief overview of wetting fundamentals may elucidate the complexities involved.

Wetting of solid surfaces and fibrous assemblies, such as textiles, involves several primary processes: immersion, capillary sorption, adhesion, and spreading [1–4]. During immersion (Fig. 5.1a) or capillary sorption (Fig. 5.1b) a solid-vapor interface disappears and a solid-liquid interface appears. By convention, the work of immersion, W_i, or the work of penetration, W_p, performed during capillary sorption is defined as the free energy change [5] when the solid and liquid in contact are separated (reversal of wetting).

For spontaneous penetration, e.g., a positive capillary rise, the work of penetration has to be positive. This is the case when the interfacial energy of the solid in contact with vapor exceeds the interfacial tension between the solid and the liquid for interfaces of unit area.

$$W_i = W_p = \gamma_{SV} - \gamma_{SL} \quad (1)$$

Adhesion is attraction between two surfaces in contact (Fig. 5.2). When the surfaces in contact are those of a solid and a liquid, the work of adhesion, W_A,

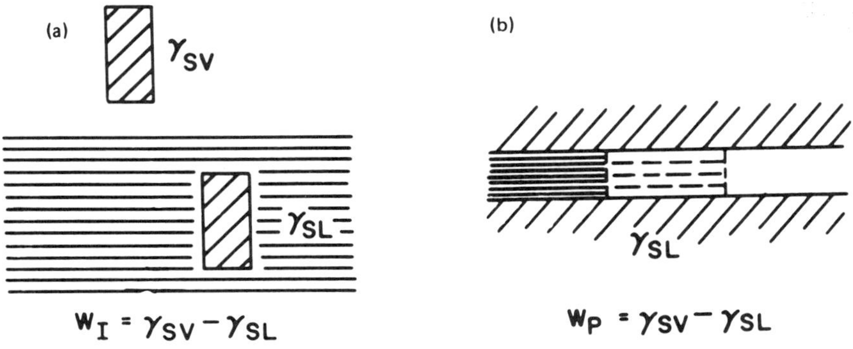

FIG. 5.1 (a) Immersion of a solid in a liquid. (b) Capillary sorption.

is equal to the change of surface free energy of the system when the liquid and the solid are separated. The separation results in loss of their interface with interfacial tension, γ_{SL}, and formation of two new surfaces with surface tensions γ_{SV} and γ_{LV}. The work of adhesion is given by the Dupré equation [6] per unit area of interfaces.

$$W_A = \gamma_{SV} + \gamma_{LV} - \gamma_{SL} \tag{2}$$

Spreading is the flow of liquid at least two molecular layers thick over a solid. During spreading (Fig. 5.3), the solid-liquid and liquid-vapor interfaces increase, whereas the solid-vapor interface decreases. Again by definition, the work of spreading, W_S, is the reversible work equal to the free energy change that occurs when the solid and liquid are separated (reversal of spreading) per unit area.

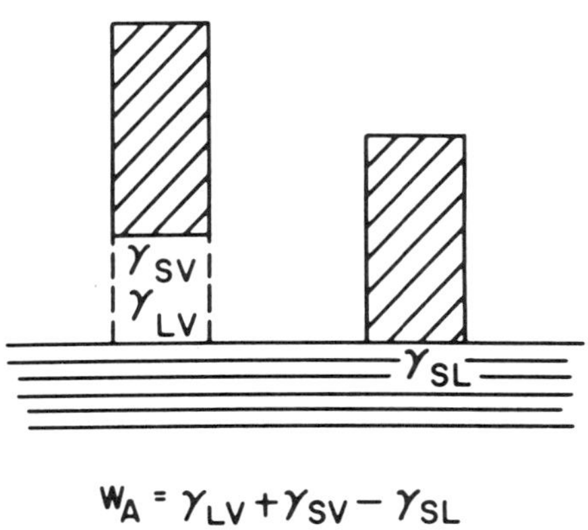

FIG. 5.2 Adhesion between a liquid and a solid.

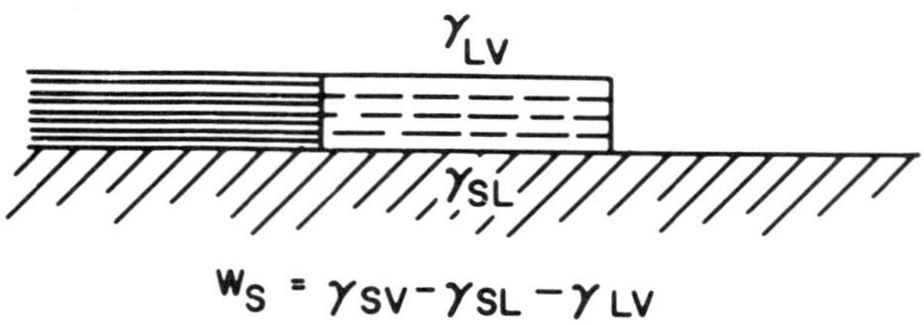

FIG. 5.3 Spreading of liquid on a solid.

$$W_S = \gamma_{SV} - \gamma_{LV} - \gamma_{SL} \tag{3}$$

For spreading to be spontaneous, the work of spreading, W_S, has to be positive. The work of spreading has also been called the spreading coefficient.

A drop placed on a solid flattens when it spreads. As a result, the liquid-vapor interface per unit area of the solid beneath the liquid decreases with decreasing curvature of the drop. The decrease in the liquid-vapor interface is relatively small and is usually neglected in the Eq. 3. Figure 5.3 shows a simplified model with a liquid film of constant thickness and shape.

Equations 1 to 3 are valid only for ideal, smooth, homogeneous, impermeable, and nondeformable surfaces. Since real substrates do not have such ideal surfaces, their wetting phenomena are more complicated. In addition, the prediction of wetting phenomena, e.g., spreading, from wetting energetics is difficult because a direct method for determining γ_{SV}, a term found in Eqs. 1, 2, and 3, is not available. It is more convenient to use the forces in balance at a three-phase (solid, liquid, vapor) boundary as an indication of wettability.

A drop of liquid placed on a solid surface spreads on or covers only a limited area of the surface and assumes a shape that appears to be constant (Fig. 5.4). The angle θ exhibited by the drop is called the contact angle and is con-

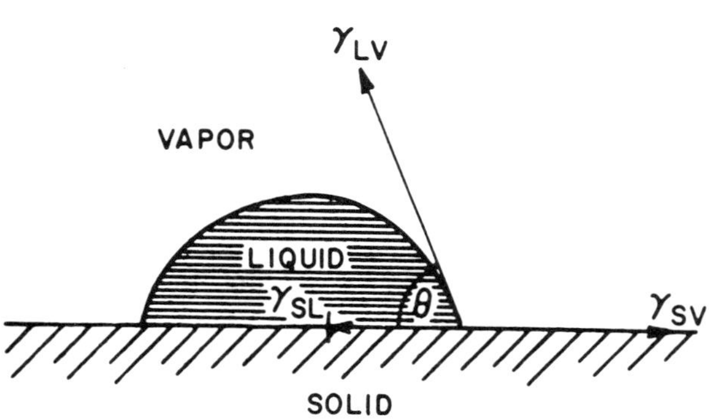

FIG. 5.4 Equilibrium contact angle.

sidered to be characteristic of the particular liquid-solid interaction. Therefore, the equilibrium contact angle serves as an indication of the wettability of the solid by the liquid. The emphasis here is on equilibrium, because valid conclusions can be drawn from the value of the contact angle only when equilibrium is assured.

Young [7] proposed that a liquid drop on a plane solid surface (Fig. 5.4) is subject to the following equilibrium forces:

$$\gamma_{SV} = \gamma_{SL} + \gamma_{LV} \cos\theta \tag{4}$$

where θ is the contact angle in the liquid at the solid-liquid-vapor boundary. The validity of Young's equation has been questioned [8,9], but thermodynamic derivations [10–12] have shown it to be correct for ideal systems in equilibrium. The surfaces in such systems have to be smooth, homogeneous, impermeable, and nondeformable. Equations 4 and 2 can be combined and if the adsorption of vapor on the solid surface is included,

$$W_A = \gamma_{LV}(1 + \cos\theta) \tag{5}$$

Equation 5 is more useful than Eq. 2, because it includes measurable quantities. Equation 5 relates adhesion to cohesion of the liquid, since $\gamma_{LV} = 1/2\ W_C$, the work of cohesion. Equation 5 implies that when the contact angle θ is 0, adhesion is equal to cohesion of the liquid, $2\gamma_{LV}$, and the spreading coefficient, W_S, is equal to zero. Such a conclusion would not be correct, however, because Eq. 5 applies to an equilibrium condition, which spreading is not. It is more appropriate to visualize that the wettability of a solid is higher for liquids that exhibit a smaller contact angle when placed on the solid; when the contact angle approaches zero, the wettability has its maximum limit and the liquid spreads spontaneously.

The contact angle concept has been very useful, but nevertheless somewhat controversial, mainly because the equilibrium contact angle in an ideal system has been confused with a measured contact angle in real systems. Whereas Eqs. 4 and 5 deal with ideal systems, in real systems the contact angle is not single valued. The contact angle displayed after the liquid front has advanced is usually larger than the contact angle observed after the liquid has receded from a previously wet surface. The difference between the advancing and the receding contact angles is contact angle hysteresis, which has been attributed to adsorption of the liquid on the solid with a consequent change of the surface energy of the solid [13], surface heterogeneity [14], or roughness of the solid surface [15]. Real solid surfaces are not absolutely smooth but appear rough on microscopic examination.

Johnson and Dettre [16,17] analyzed the theories of Wenzel [15], Cassie and Baxter [18], Shuttleworth and Bailey [11], and Good [19] in terms of an idealized sinusoidal rough surface. Contact angle hysteresis was attributed to energy barriers between metastable states of a drop, as first suggested by Shuttleworth and Bailey [11]. Because of the existence of these metastable states, a contact angle measured on a rough surface will usually not obey Eq. 6 but will depend on the direction of movement of the three-phase boundary prior to measurement of the angle. Inhomogeneity of the surface, adsorption on the solid surface, or contamination of the liquid can also give rise to contact angle hysteresis [14,20]. Johnson and Dettre [21,22] postulated, in analogy with their roughness theory, that the contact angle that the drop assumes on a heterogeneous surface also depends on energy barriers between the metastable states.

Miller [23] has cautioned that the contact angle is not an cause of wetting but its consequence, determined by three independent attractive forces. An observed change of the contact angle is not always simple to interpret.

Most of the wetting studies are concerned with the equilibrium contact angle. Practical wetting phenomena are dynamic, however, and involve movement of a liquid relative to the solid surface. The shape of the contact angle is not governed only by interfacial forces but may depend on gravitational, viscous, and inertial forces as well [24–31]. The observed contact angle is affected by the relative speed of the liquid in motion.

Since wetting is a dynamic process, photographic techniques have been employed to record the movement of a liquid. Initially a still camera was used to measure capillary sorption by sequential photography [4]. Now, video cameras are being utilized [32] to study wetting kinetics and the interface. Although the number of publications on wetting phenomena is already large, the undiminished research activity indicates that the understanding of wetting kinetics and the dynamic contact angle is far from complete.

5.2 CRITICAL SURFACE TENSION AND SURFACE ENERGY

Eqs. 1, 2, 3, and 4 indicate that wetting depends on the surface free energies of the solid and the liquid at the interface. This suggests that the wettability of a solid by a liquid could be predicted if the surface free energy of the solid were known.

Zisman and his co-workers [33] have contributed extensively to the fundamental knowledge of wettability of low-energy solids (fluorocarbons, hydrocarbons). Fox and Zisman [34] measured advancing contact angles of homologous series of liquids on low-energy surfaces and plotted the $\cos\theta$ values against the surface tensions of the liquids (Fig. 5.5). The lines representing homologous series of liquids extrapolated approximately to the same γ_{LV} value at $\cos\theta = 1$. Zisman named this surface tension value the *critical surface tension* of the solid, γ_c, and proposed that only liquids having surface tensions below this value spread on the solid. Liquids with a surface tension above γ_c form a finite

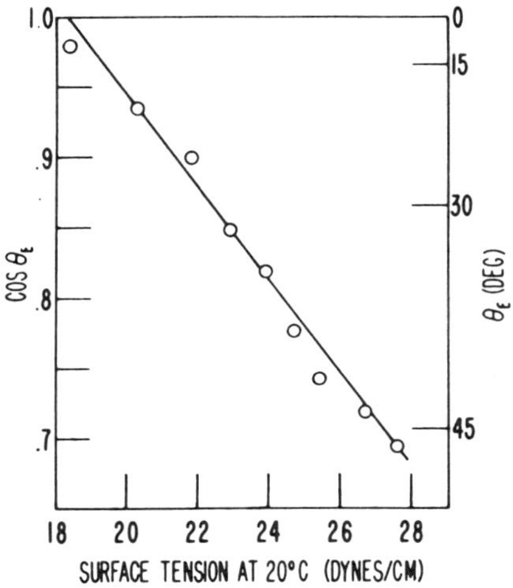

FIG. 5.5 Wettability of poly(tetrafluoroethylene) by *n*-alkanes. (From Ref. 33. Reproduced by permission of American Chemical Society.)

contact angle on the solid. When the plot of $\cos\theta$ against γ_{LV} resulted in a rectilinear band (Fig. 5.6), the intercept of the lower limb of the band at $\cos\theta = 1$ was chosen as the critical surface tension of the solid. Zisman [33] used the critical surface tension principle to correlate wettability with the constitution of low- energy surfaces. Zisman concluded that the critical surface tension and the wettability are determined by the nature and packing of the exposed surface atoms of the solid and are otherwise independent of the nature and arrangement of the underlying atoms [33]. This concept has predicted correctly the wettability of repellent surfaces in the absence of external forces.

Although the critical surface tension concept has been criticized for its empirical nature [23], critical surface tension data have been very useful in developing water- and oil-repellent finishes for textiles and paper. One reason for this practical impact may be the similarity between the systems studied by Zisman and the surfaces of water- and oil-repellent finishes.

Zisman was one of the first to recognize that the critical surface tension concept is strictly empirical and to suggest that γ_c needs to be replaced by parameters having a thermodynamic or statistical mechanical basis [19]. Fox and Zisman [34] have cautioned that γ_c varies between liquid types and that it is not a measure of the surface energy of the solid γ_{So}. For some solids the γ_c values determined with pure liquids differ significantly from the γ_c values obtained using surfactant solutions. These discrepancies have stimulated several investi-

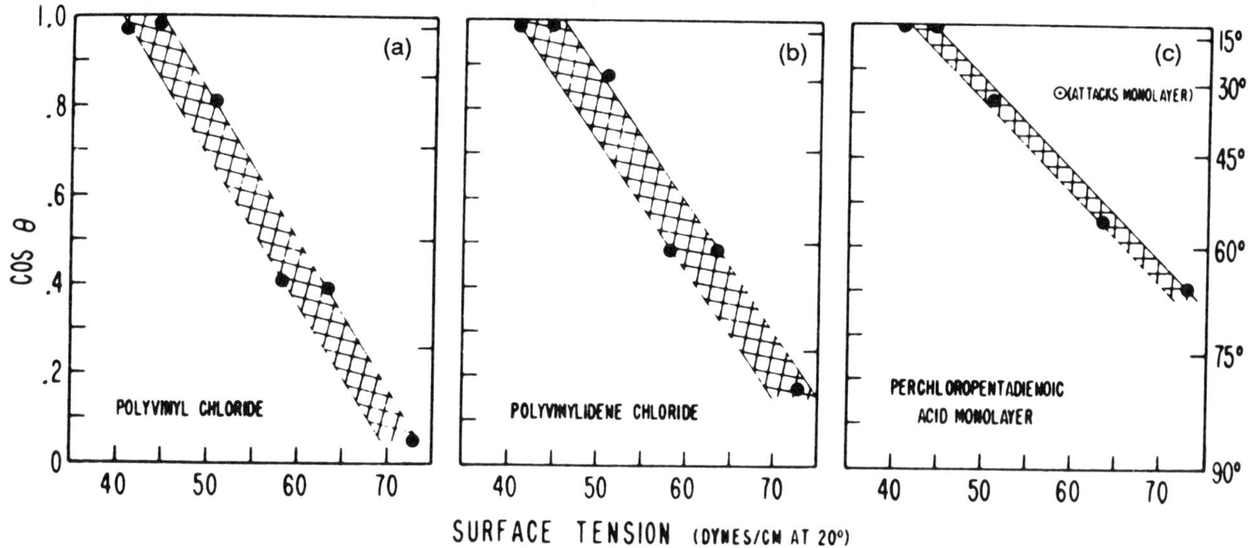

FIG. 5.6 Wettability by various liquids on surface of (a) poly(vinyl chloride); (b) poly(vinylidene chloride); (c) close-packed monolayer of perchloropentadienoic acid. (From Ref. 33. Reproduced by permission of American Chemical Society.)

gations of surfactant adsorption on low-energy surfaces and the anomalies of fluorinated surfactant adsorption (see Sec. 5.3).

The critical surface tension method for determining wettability also has some practical limitations. Several contact angle measurements involving a series of liquids are necessary to determine γ_c of a single solid. Another limitation is the curvature of the $\cos\theta$ versus γ_{LV} plot. Although the plot should yield a straight line, the relation is curvilinear even for an homologous series of liquids. This can cause a considerable error when liquids with very low contact angles are not available and extensive extrapolation is necessary.

Because of the deficiencies and limitations of the critical surface tension concept, the need to determine the surface free energy of solids has remained. Since the surface free energy of solids is difficult to measure, attempts have been made to estimate the surface free energy from interaction with liquids [2,35–43].

5.3 ADSORPTION FROM SOLUTION

The general theory of adsorption at a solid-liquid interface has been reviewed in several monographs and journal articles [43–61].

Adsorption at the solid-liquid interface is associated with a decrease in free surface energy. The adsorbed surfactant is held at the solid surface by physical, ionic, or chemical forces.

Physical adsorption is caused primarily by van der Waals forces. The adsorption mechanisms may involve dispersion forces, hydrophobic bonding, charge transfer, or hydrogen bonding.

Ionic adsorption results from an ionic bond formed between ionic sites at the surface and oppositely charged surfactant ions.

Chemisorption involves the formation of a covalent chemical bond. Chemisorption provides the strongest bond, van der Waals forces the weakest [58].

Adsorption at the solid-liquid interface depends on several factors:

1. The structure of the surfactant. Strength of the adsorptive bond formed by the adsorbed surfactant depends on the structure of the hydrophobic and hydrophilic groups of the surfactant and decreases usually in the order: cationic > anionic > nonionic surfactant.
2. The nature of the substrate. Substrates are nonpolar, polar, or metallic (high surface energy). The surface characteristics of the substrate surface, such as the polarity, chemical composition, and surface geometry (porosity), are major factors affecting adsorption.
3. The composition of the liquid phase. The nature of the solvent and the dissolved electrolytes or additives influences adsorption.
4. The physical conditions: temperature, pressure, and agitation.

Adsorption can be characterized by the amount of the surfactant adsorbed as a function of equilibrium concentration. The quantitative amount of the adsorbed surfactant is usually determined by analyzing the liquid phase by spectroscopic, calorimetric, and electrophoretic methods [47]. The physical and chemical state of the adsorbed layer can be examined by spectroscopy, electron diffraction, x-ray diffraction, electron spectroscopy for chemical analysis (ESCA), interferometry, and electron microscopy. The orientation of the adsorbed surfactant molecules depends on the nature of the adsorbent and the surfactant, the interaction between the adsorbent and the surfactant, the lateral interaction between the adsorbed surfactant molecules, and the concentration of the surfactant in solution. The adsorbed layer may be homogeneous or consist of coadsorbed surfactants of different composition. Solvent molecules may also be adsorbed at the solid interface.

The amount of a surfactant adsorbed at a constant temperature and pressure increases with the surfactant concentration to an adsorption limit or exhibits maxima and minima. Adsorption isotherms have been classified by Giles et al. [62,63] using letters and numbers to indicate subgroups (Fig. 5.7). However, two shapes of isotherms, 2L and 3L, are most common [55].

Zisman and his co-workers [33,34,64–72] characterized the wettability of low-energy solids, adsorbed monolayers, and Langmuir-Blodgett films of fluorinated surfactants by measuring the contact angle θ and plotting $\cos\theta$ against the surface tension (see Sec. 5.2). Condensed monolayers were prepared on smooth surfaces of polished platinum, brass, copper, or glass by adsorption from solution or by the retraction method [64–68]. Data obtained by these methods [19] are tabulated in Table 5.1.

The surface comprised of closely packed -CF_3 groups has the lowest γ_c and, if γ_c correlates with the surface free energy, the lowest surface energy. Substitution of one of the fluorine atoms by a hydrogen atom more than doubles γ_c. The effect of fluorine atoms on wettability of the surface is also evident on progressive replacement of hydrogen atoms in polyethylene by fluorine. The critical surface tension decreases linearly with the increasing fluorine substitution. Replacement of hydrogen atoms by chlorine increases γ_c.

The wetting characteristics of adsorbed films indicate that the adsorption of perfluorinated alkanoic acids and the wettability of their monolayers depend on the length of the pefluorinated carbon chain. Hare et al. [65] prepared monomolecular films of perfluorinated butyric, valeric, caproic, caprylic, capric, and lauric acids by adsorption on platinum from their solutions in *n*-decane. The short-chain perfluoroalkanoic acids were adsorbed rapidly but the two longest-chain acids required a longer time to form a complete monolayer. The wetting properties of the monomolecular films were characterized by contact angle measurements for a number of liquids. The cosine of the contact angle θ_E for *n*-alkanes plotted against the number of carbon atoms of the per-

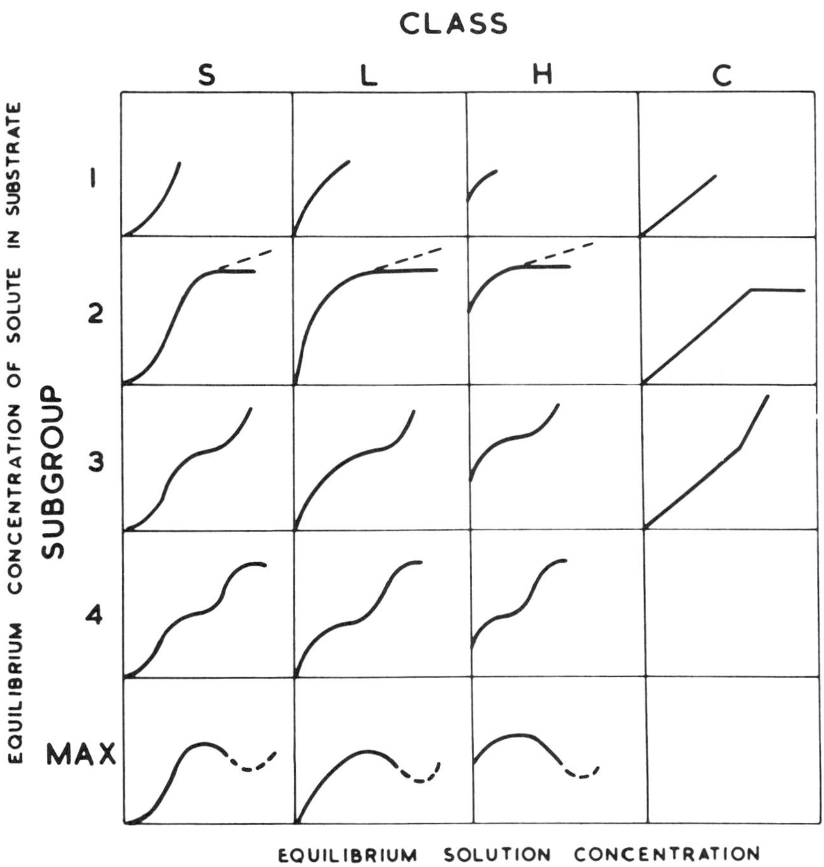

FIG. 5.7 Classification of isotherm shapes. (From Ref. 53. Reproduced by permission of Marcel Dekker, Inc.)

fluorinated acids (Fig. 5.8) fell along parallel straight lines. The remarkable repellency of the perfluoroalkanoic acid monolayers was attributed to an exposed plane rich in perfluoromethyl groups, shown for perfluorododecanoic acid by electron diffraction.

The wettability of the monolayer-coated surfaces is determined mainly by the nature and packing of outermost atoms in the exposed surface. The dominant effects of fluorinated terminal groups on the wettability of the adsorbed monolayer has been demonstrated with partially fluorinated fatty acids. Shafrin and Zisman [69] studied wetting characteristics of heptadecanoic acids which featured perfluoroethyl, perfluoropropyl, perfluoropentyl, and perfluoroheptyl groups in the terminal 17- position. Monomolecular films were prepared by

TABLE 5.1 Critical Surface Tensions of Low-Energy Surfaces

Surface constitution	γ_c (mN/m at 20°C)
Fluorocarbon surfaces	
$-CF_3$	6
$-CF_2H$	15
$-CF_3$ and $-CF_2-$	17
$-CF_2-$	18
$-CH_2-CF_3$	20
$-CF_2-CFH-$	22
$-CF_2-CH_2-$	25
$-CFH-CH_2-$	28
Hydrocarbon surfaces	
$-CH_3$ (crystal)	22
$-CH_3$ (monolayer)	24
$-CH_2-$	31
$-CH_2-$ and $-CH=$ (conjug.)	33
$-CH-$ (aromatic)	35
Chlorocarbon surfaces	
$-CClH-CH_2-$	39
$-CCl_2-CH_2-$	40
$=CCl_2$	43

Source: Ref. 33. Reproduced by permission of the American Chemical Society.

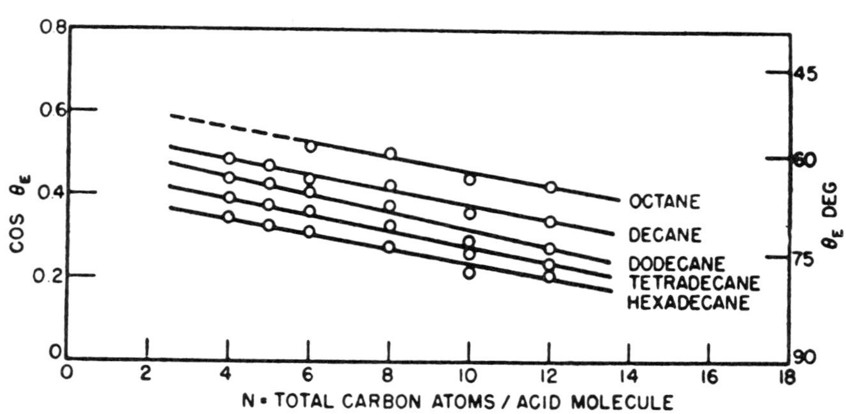

FIG. 5.8 Contact angles observed for *n*-alkanes on monolayers of perfluorinated alkanoic acid homologues. (From Ref. 65. Reproduced by permission of American Chemical Society.)

adsorption from the melt on chromium. 17-(Perfluoroheptyl)heptadecanoic acid was also adsorbed onto platinum, nickel, quartz, and soda-lime glass. The films adsorbed on siliceous surfaces were more wettable than films on metals. A comparison of partially fluorinated acids featuring a terminal perfluoroheptyl chain with perfluorinated acids revealed that the wettability of the monolayer is determined by the fluorinated outermost segment in the exposed surface plane. The adjacent alkyl segment, if adequately shielded by the fluorocarbon segment, does not affect the wetting characteristics of partially fluorinated surfactants with a terminal fluorinated segment. The effect of the fluorinated segment on wettability is illustrated in Fig. 5.9 with a plot of γ_c versus the number of fluorine-substituted carbon atoms in the hydrophobe. Straight lines represent the γ_c values for surfactants featuring more than three fluorine- substituted carbon atoms. The line for partially fluorinated surfactants intercepts the line for perfluorinated surfactants at seven carbon atoms. The data suggest that the perfluoroheptyl segment is sufficiently long to shield the hydrocarbon segment from the external boundary and the wettability of the monolayer is determined mainly by the packing of fluorocarbon chains.

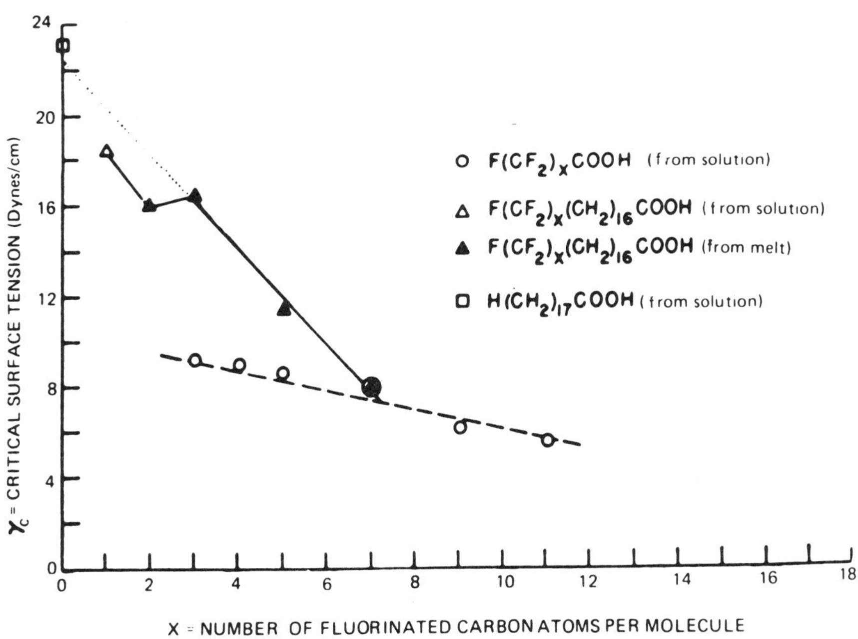

FIG. 5.9 Effect of fluorination of the adsorbed alkanoic acid monolayer on the critical surface tension. (From Ref. 69. Reproduced by permission of American Chemical Society.)

In accord with the dominant effect of the terminal group on wettability, terminal branching and terminal chlorine substitution increase the wettability of the adsorbed monolayers formed by perfluoroalkanoic acids [68]. Condensed monolayers of two homologous series of acids, $(CF_3)_2CF(CF2)_nCOOH$ and $CF_2Cl(CF_3)(CF_2)_nCOOH$, were adsorbed on chromium or platinum by the melt retraction method [70]. This technique is based on the observation that a drop of nonspreading acid rolled over the metal surface leaves an adsorbed monomolecular film behind. The increased wettability resulting from terminal branching or chlorine substitution was explained by changes in adsorption site spacing, molecular packing, and chain adlineation of the monolayers.

The packing and configuration of the adsorbed monolayer depend mainly on the relative sizes of the fluorocarbon and hydrocarbon segments and the cohesive forces between adjacent segments. Shafrin and Zisman [69] considered three models for the acids with a perfluorinated segment of three carbon atoms or longer (Fig. 5.10). Models A and B assume that the perfluoro segments are in a closely packed extended position. The larger diameter of the fluorinated segments separates the hydrocarbon segments and reduces the otherwise strong cohesive forces between then. Model A postulates a short-range order for the hydrocarbon segments, a condition similar to the liquid state. Model B assumes a somewhat greater order for the hydrocarbon chains. Model C describes a high degree of order and close packing of both the hydrocarbon and fluorocarbon segments. To reduce steric hindrance, hydrocarbon segments have to be inclined. The main objection to model C is the unlikely parallel alignment required for two sets of planes representing the hydrocarbon and fluorocarbon segments.

The wetting of a solid surface by a surfactant solution depends on the adsorption of the surfactant on the liquid-vapor and liquid-solid interfaces. If a surfactant is equally adsorbed on liquid-vapor and liquid-solid interfaces and the solid-vapor surface tension, γ_{SV}, is not affected by the liquid, the equilibrium contact angle θ should correspond to the surface tension of the liquid, γ_{LV}, in accordance with Young's equation:

$$\cos\theta = \frac{\gamma_{SV} - \gamma_{SL}}{\gamma_{LV}} \tag{6}$$

If this assumption is correct, the wettability of a solid should depend solely on the surface tension, γ_{LV}, of a pure liquid or a surfactant solution, regardless of the structure of the surfactant used to lower the surface tension of water. However, this assumption was found to be incorrect. When Bernett and Zisman [71] used surfactant solutions instead of pure liquids in their wetting studies, fluorinated surfactants behaved differently than conventional surfactants with a hydrocarbon chain. Furthermore, the wetting characteristics of a fluorinated surfactant solution depend on the nature of the solid surface.

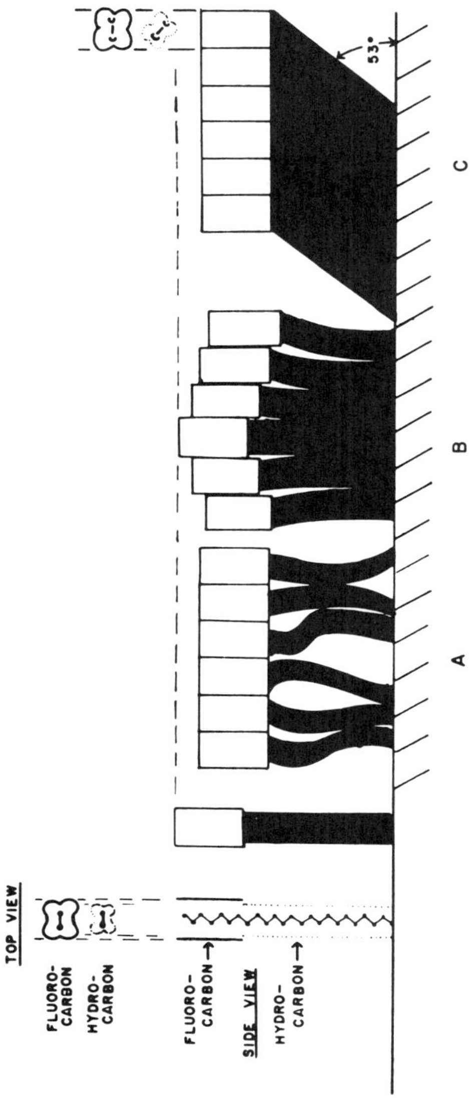

FIG. 5.10 Schematic configurations of adsorbed molecules of 17-(perfluoroheptyl)-heptadecanoic acid consistent with close packing of perfluoroheptyl groups. (From Ref. 69. Reproduced by permission of American Chemical Society.)

According to the critical surface tension concept, the equilibrium contact angle θ and the ability of the solution to spread on polyethylene and polytetrafluoroethylene (Teflon) are related to the critical surface tension γ_c of the solid and the surface tension γ_{LV} of the solution. Surfactants with a hydrocarbon-based hydrophobe cannot lower the surface tension of water below 18 mN/m, the critical surface tension γ_c of Teflon. Hence aqueous solutions of hydrocarbon surfactants cannot wet Teflon. Fluorinated surfactants can decrease the surface tension of water below the critical surface tension of Teflon (18 mN/m) and wet it completely [72] (Fig. 5.11). The critical surface tension value (19 mN/m) obtained for Teflon by plotting $\cos\theta$ versus γ_{LV} is in good agreement with the anticipated value of 18 mN/m. However, a similar plot for fluorosurfactant solutions on polyethylene gave a critical surface tension (γ_c) value for polyethylene of 20 mN/m, instead of the anticipated value

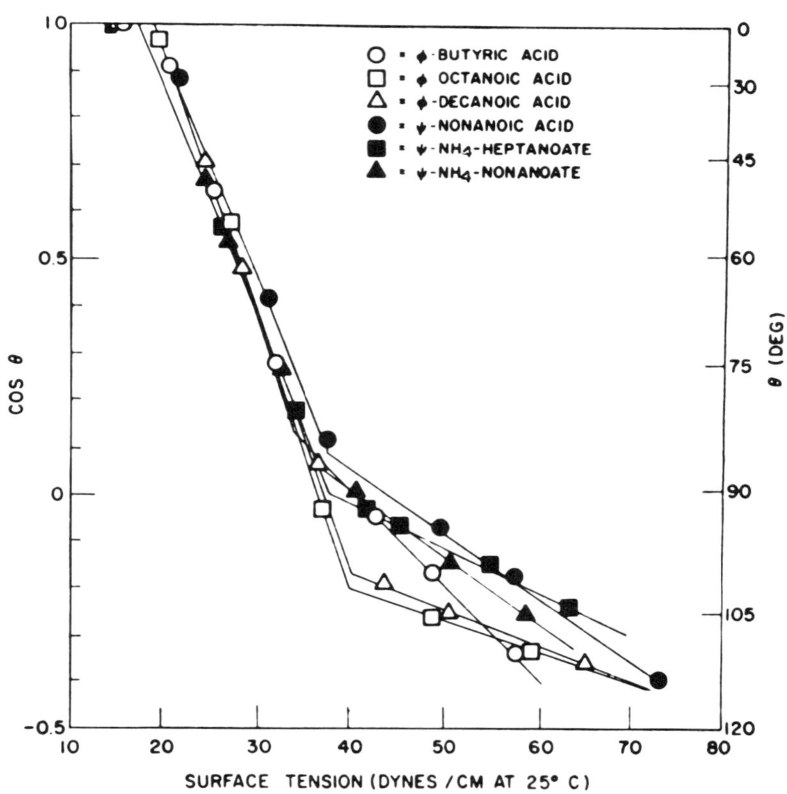

FIG. 5.11 Wettability of Teflon by aqueous solutions of highly fluorinated compounds. (From Ref. 72. Reproduced by permission of American Chemical Society.)

of 31 mN/m obtained with pure liquids (Fig. 5.12). Bernett and Zisman explained the apparent shift of γ_c by adsorption of fluorinated surfactants on polyethylene and a decrease of the solid-vapor interfacial tension γ_{SV}. Contact angle data for hexadecane were used to propose a more tenacious adsorption of fluorinated surfactants on polyethylene than on Teflon (Fig. 5.13). According to the hypothesis, fluorinated surfactants adsorbed on polyethylene are oriented with their fluorinated hydrophobe toward water and render polyethylene more hydrophobic. The rationale of this explanation, disputed later by others [73], was based on the assumption that the change in free energy is larger when a surface consisting of $-CF_2-$ or $-CF_3-$ groups replaces a surface consisting of $-CH_2-$ groups, instead of being deposited on a already fluorinated $-CF_2-$ surface. Since fluorinated surfactants adsorbed on the water-vapor interface are oriented with the hydrophile toward water, the hypothesis invokes

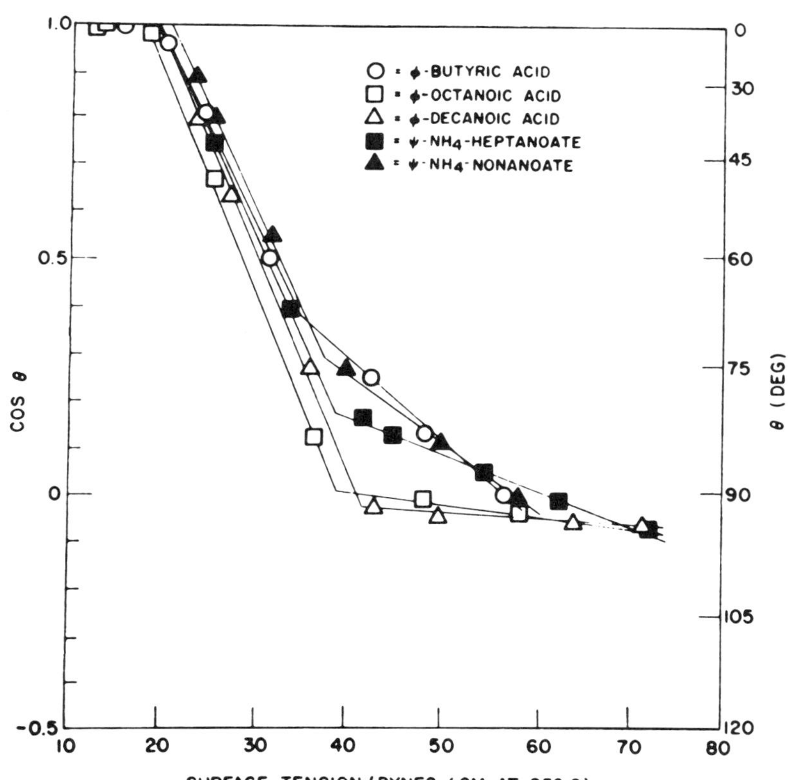

FIG. 5.12 Wettability of polyethylene by aqueous solutions of highly fluorinated compounds. (From Ref. 72. Reproduced by permission of American Chemical Society.)

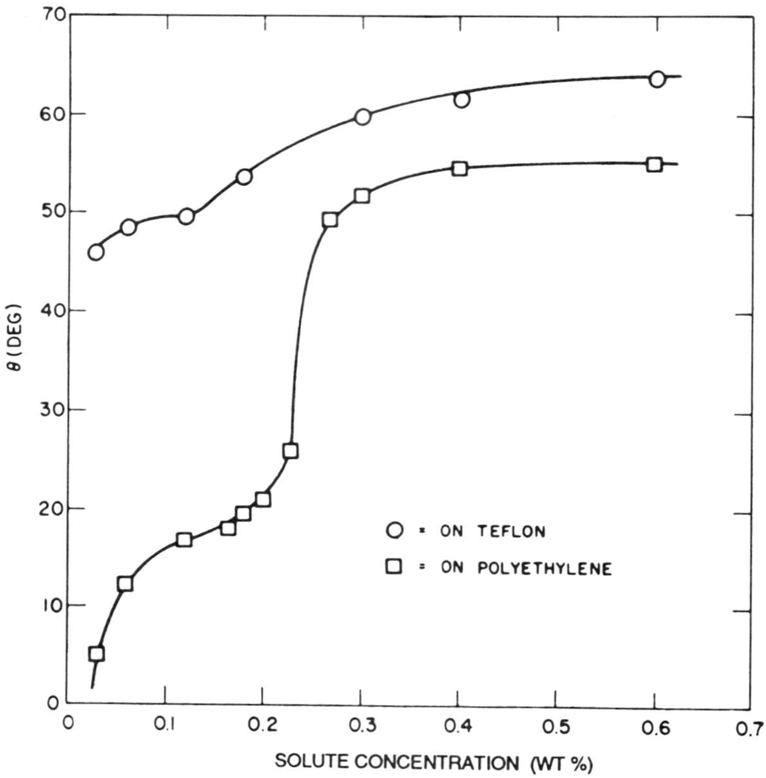

FIG. 5.13 Hexadecane contact angles on polymer surfaces with perfluorooctanoic acid adsorbed from aqueous solutions. (From Ref. 72. Reproduced by permission of American Chemical Society.)

unequal adsorption of fluorinated surfactants at water-vapor and solid-liquid interfaces.

Using the Young and Gibbs equation (Eq. 6) and Fowkes and Harkins' [74] assumption that

$$\frac{d\gamma_{SV}}{d\gamma_{LV}} = 0 \tag{7}$$

Johnson and Dettre [48] derived the equation

$$\cos\theta = \frac{2\gamma_C - \gamma_{LV}}{\gamma_{LV}} \tag{8}$$

The equation states that the adsorption at both the solid-liquid and liquid-vapor interfaces is similar if it is governed by interactions between water molecules and the surfactant hydrophobic tails and not by attractive forces at the interface. If this assumption is valid and Eq. 8 holds, then the γ_c values obtained from solvent contact angle data should be equal to the γ_c values obtained using surfactant solutions. A comparison of γ_c values (Table 5.2) shows that the agreement is reasonably good for hydrocarbon surfactants on nonpolar surfaces and for a fluorinated surfactant on Teflon. In other cases, large discrepancies of γ_c values indicate that adsorption to the solid-liquid and liquid-vapor interfaces is not equal.

The unequal adsorption to the liquid-vapor and solid-liquid interfaces has been the subject of several studies [73,75,77]. In contrast to Bernett and Zisman [72], Pyter et al. [73] explained the different wetting characteristics of hydrocarbon surfactants and fluorinated surfactants by low adsorption of fluorinated surfactants on nonpolar solids. The higher contact angles exhibited by solutions of fluorinated surfactants on polyethylene were explained by poor adsorption of fluorinated surfactants instead of the strong adsorption proposed by Bernett and Zisman [72].

Pure liquids and aqueous surfactant solutions having the same surface tension, γ_{LV}, do not always exhibit the same contact angle. Pure liquids are often better wetting agents than surfactant solutions. The differences in wetting have been attributed to differences in γ_{SL}. The value of γ_{SL} is higher for the surfac-

TABLE 5.2 Critical Surface Tension Values (mN/m)

Liquid system	Solid				
	Paraffin	Polyethylene	Teflon	PMMA[a]	Nylon
Pure liquids	25.5 [48][b]	31 [71]	18 [71]	40 [77]	43 [76]
Hydrocarbon surfactant, in water	23	30 [71]	22, 18 [71]	26	26
Fluorinated surfactant, in water	15	20 [72]	13, 18 [72]	11	
Predicted by Eq. 8 and solvent θ data	28	33	17	47	48

[a] PMMA, poly(methyl methacrylate).
[b] Numbers in brackets are reference numbers.
Source: Values without reference numbers are from Ref. 73. Reproduced by permission of Academic Press, Inc.

tant solution than for the pure liquid. This suggests lower adsorption of surfactants to the solid-liquid interface than to the liquid-vapor interface. For hydrocarbon surfactants these differences are larger for semipolar solids such as poly(methyl methacrylate) than for nonpolar solids, such as polyethylene. For a perfluorinated surfactant, such as perfluorooctanoic acid, the effect is significant for semipolar and nonpolar solids as well, for example, paraffin and polyethylene. Fluorinated surfactant solutions exhibit higher contact angles on nonpolar solids, such as polyethylene and paraffin, than solutions of hydrocarbon- derived surfactants having the same surface tension (Fig. 5.14). The higher contact angles of fluorinated surfactants are in accord with the lower adsorption of fluorinated hydrocarbon surfactants on Graphon [73], used as a hydrocarbon-type model substrate for comparing the adsorption of fluorinated and hydrocarbon surfactants.

The orientation, thickness, and surface coverage of adsorbed fluorinated surfactants have been studied by x-ray photoelectron spectroscopy (XPS, also known as ESCA). Gerenser et al. [78] determined the orientation of a cationic fluorosurfactant, Zonyl FSC, on SiO_2 and poly(ethylene terephthalate) surfaces. XPS indicated that at low surfactant concentrations SiO_2 is covered by a uniform monolayer. The surfactant orientation changed as the packing density was increased. On poly(ethylene terephthalate) the surface coverage was incomplete and the surfactant orientation did not change when the surfactant

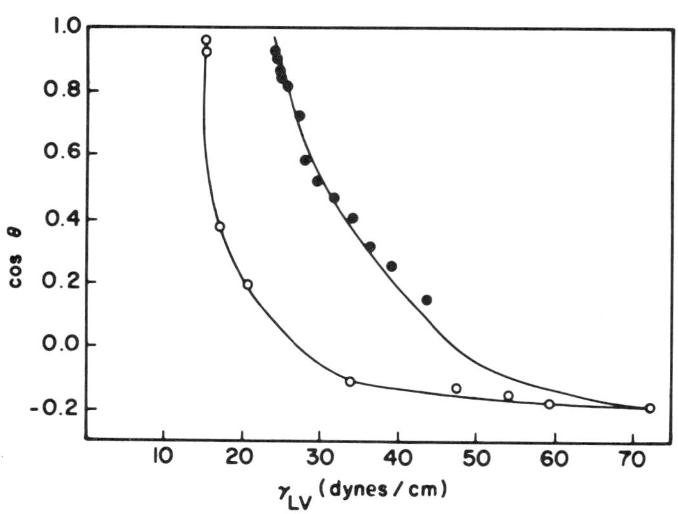

FIG. 5.14 Cos θ versus surface tension of aqueous surfactant solutions on paraffin at 25°C. (●) Aerosol OT in 0.1 M NaCl; (○) perfluorooctanoic acid. (From Ref. 73. Reproduced by permission of Academic Press.)

coverage was increased. Oxidation of the surface via corona discharge resulted in a uniform and continuous distribution of the surfactant over the surface [79]. These results suggested chemisorption of the fluorinated surfactant on the oxidized surface by ion exchange.

Batts [80] studied the adsorption of the fluorosurfactant FC134 at a gelatin-air interface by XPS. Their data indicated that less fluorinated surfactant is needed for decreasing the critical surface tension, γ_c, of dried gelatin layers to its minimum value than for lowering the surface tension of water, γ_{LV}, to the minimum values of the aqueous surfactant solution.

Clayfield et al. [81] attempted to prepare monolayers of fluorinated surfactants FC134 and FX161 on glass by adsorption from acetone solutions. Stable and reproducible contact angles were obtained with *n*-alkanes, but polar liquids such as water caused desorption of the surfactant from the solid-liquid interface.

Virtually uncharged fluorocarbon surfaces on smooth sheets of muscovite mica have been prepared using the double-chain cationic surfactant *N*-(α-trimethylammoniumacetyl)*O,O'*-bis(1*H*, 1*H*, 2*H*, 2*H*-perfluorodecyl)*L*-glutamate chloride [82]. Surfaces were coated with the fluorinated surfactant either by Langmuir-Blodgett deposition or by adsorption from solution [83]. Slow adsorption from solution gave an initially very thin surface layer of surfactant molecules lying flat on the mica surface. The surface neutralization was essentially complete, and it took 2–5 days for a reasonably well-packed uncharged monolayer to form. The initially uncharged monolayer is a transient state. With time a surface charge builds up, caused by additional adsorption or loose association of surfactant molecules. This adsorption mechanism casts some doubt on the stepwise mechanism of surfactant adsorption on solid surfaces. In contrast, Langmuir-Blodgett films gave deposited monolayers which did not change within the time period of the study. Because the surfactant did not adsorb to Langmuir-Blodgett monolayers their charge did not vary with time during the adsorption process.

The adsorption of fluorinated surfactants on coal was investigated by Lai and Gray [84] by measuring the contact angles, the spreading coefficient, and the effect of the fluorinated surfactant on flotation. The fluorinated surfactants used were Zonyl FSN (nonionic), Zonyl FSC (cationic), Zonyl FSK (amphoteric), and Zonyl FSA (anionic). The contact angle of water on coal is about 38–42 degrees. The anionic fluorinated surfactant increased the contact angle to 80–90 degrees. The other fluorinated surfactants were less effective. The adsorption of the fluorinated surfactants on coal increased in the order nonionic < cationic < amphoteric < anionic. Lai and Gray proposed an adsorption mechanism involving acid-to-acid dimerization. According to the mechanism, the carboxylic group of the anionic fluorinated surfactant Zonyl FSA combines with carboxylic acid groups on the surface of coal. Lai and Gray considered the acid-to-acid dimerization adsorption to be stronger than the ionic or hydrogen bonding type of adsorption. As a result of strong adsorption, the anionic fluorinated surfactant is most effective in transferring coal from the aqueous phase into the air phase for froth flotation.

The adsorption of fluorinated surfactants from aqueous solutions of dye particles depends on the structure of the fluorinated surfactant and the polarity of the dye surface. Hartmann [85] investigated the adsorption of the anionic surfactant $C_8F_{17}SO_3N(C_2H_5)_4$ and nonionic surfactants $C_nF_{2n+1}O(CH_2CH_2O)_mH$ ($m = 6, 9, 13$; $n = 7, 10$) as single solutes by contact angle and surface tension measurements. The anionic fluorinated surfactant was found to be a more effective dispersant than the nonionic ones. Hartmann observed that, in contrast to hydrocarbon surfactants, fluorinated surfactants are adsorbed more strongly on polar surfaces than on nonpolar surfaces and are therefore better dispersants than hydrocarbon- type surfactants for certain dyes. The interesting question, how fluorinated surfactants stabilize the dye dispersion, remains to be answered. The adsorbed fluorinated surfactants increase the contact angle of water on dye particles. This suggests that the surfactant was adsorbed with its hydrophile oriented toward the dye surface and the fluorocarbon tail toward water. Hence the adsorption of the fluorinated surfactant reduced the surface energy and wettability of the dye particles by water, in analogy to autophobic wetting.

The phenomena observed by Hartmann et al. [85,86] differ from stabilization mechanisms of dye dispersions with hydrocarbon-type surfactants. Kissa [87, 88] used a water-immiscible solvent, in which the dye but not the surfactant is soluble, as a surface probe for dye dispersions. Dispersed dye particles favor the water phase in a solvent-water two-phase system because the adsorbed dispersant renders the dye particles more hydrophilic. Fluorinated surfactants function as dispersants for fluorocarbon polymers by adsorbing on the particles with the hydrophile oriented toward water. Clearly, the adsorption phenomena observed by Hartmann et al. for dye dispersions need to be elucidated and the mechanisms of stabilizing dispersions with fluorinated surfactants investigated.

Although much progess has been made, more fundamental information on the adsorption of fluorinated surfactants on solid surfaces is needed. Since most commercial applications involve surfactant interactions at the solid-liquid boundary, the adsorption phenomena are of great practical interest. Fluorinated surfactants are usually mixtures of homologues and are frequently employed as mixtures with hydrocarbon surfactants. Therefore a thorough understanding of adsorption mechanisms for surfactant mixtures on solid substrates is also very important (see Sec. 5.4).

5.4 ADSORPTION OF MIXED SURFACTANTS

Adsorption of binary surfactant mixtures at a solid-solution interface has been exploited commercially in many applications. In spite of the widespread use of hydrocarbon surfactant–fluorinated surfactant mixtures, very few articles have been published on their adsorption at a solid-solution interface.

The interaction of hydrocarbon and fluorocarbon surfactants on the surface of dispersed particles has been studied through a flocculation and redispersion process [89–91]. Dispersions of positively charged particles can be flocculated with an anionic surfactant. An excess of the anionic surfactant forms a bilayer on the particle surface and causes redispersion of the flocculated sol. This flocculation reversal was used to study the interaction between mixed surfactants on a solid surface. A dispersion of iron(III) oxide hydrate particles was flocculated with an anionic hydrocarbon or fluorocarbon surfactant at pH 3.5, where the sols had a positive zeta potential. Subsequently, a second fluorocarbon or hydrocarbon surfactant was added to the flocculated sol. The extent of redispersion depended on the interaction between the two surfactants on the solid particle surface.

Changes in zeta potential and turbidity of iron(III) oxide hydrate sols flocculated by sodium dodecyl sulfate (SDS) are shown in Fig. 5.15. When SDS was added to the sols, the positive zeta potential of the sol became negative. The turbidity increased sharply when more than 8×10^{-2} mM SDS was added. The turbidity increase indicated that the sol flocculated by SDS was redispersed by SDS in excess of the optimum flocculation concentration of 8×10^{-2} mM. When the iron(III) oxide hydrate sols were flocculated with 8×10^{-2} mM SDS and sodium 4-[2-heptafluoroisopropyl-1,3-bis(trifluoromethyl)butenoxy]benzenesulfonate (NF100) or 3-heptafluoroisopropyl-2,4-bis(trifluoromethyl)-5,8,11,14,17,20,23,26-octoxa-heptaicos-3-ene (NF7) was added, the sols flocculated by SDS were redispersed. The turbidity of the sols increased. The zeta potential of the sols decreased when NF100 was added. Addition of NF7, a nonionic surfactant, had very little effect on the zeta potential.

Figure 5.16 shows changes in zeta potential and turbidity of iron(III) oxide hydrate sols flocculated with NF100. The optimum flocculation concentration was about 3×10^{-3} mM NF100. The sols were redispersed by NF7 or NP7.5, a hydrocarbon-type nonionic surfactant (polyoxyethylene nonylphenyl ether with a polyoxyethylene chain of average 7.5 EO). The turbidity increased sharply. The zeta potential changed only a little, as expected for a nonionic surfactant. Sols flocculated by NF100 were not redispersed by SDS. The inability of SDS, an anionic hydrocarbon surfactant, to redisperse the sols was attributed to very low adsorption of SDS on the sols flocculated by NF100, a fluorinated surfactant.

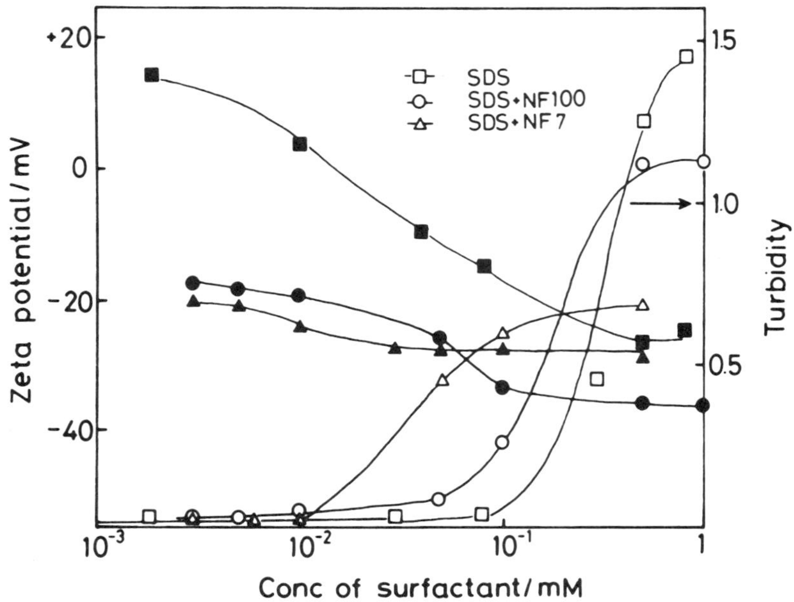

FIG. 5.15 The zeta potential and turbidity of α-Fe_2O_3 sols flocculated by SDS as a function of surfactant concentration; closed symbols, zeta potential; open symbols, turbidity; arrow, turbidity of sols in the absence of surfactant. (From Ref. 89. Reproduced by permission of Elsevier Science Publishers.)

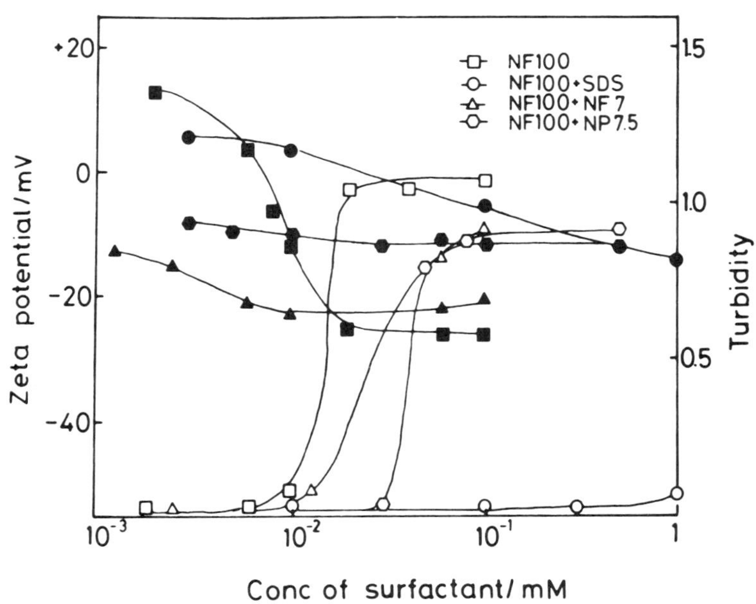

FIG. 5.16 The zeta potential and turbidity of α-Fe_2O_3 sols flocculated by NF100 as a function of surfactant concentration; closed symbols, zeta potential; open symbols, turbidity. (From Ref. 89. Reproduced by permission of Elsevier Science Publishers.)

Changes in zeta potential and turbidity of iron(III) oxide hydrate sols flocculated with lithium perfluorooctanesulfonate (LiFOS) are shown in Fig. 5.17. The nonionic surfactants NF7 and NP7.5 redispersed the sols. However, the anionic hydrocarbon surfactant LiDS (lithium dodecyl sulfate) had no significant effect. Accordingly, sols flocculated by LiDS were redispersed by a nonionic surfactant, NF7, but not by the anionic hydrocarbon surfactant LiFOS (Fig. 5.18).

Esumi et al. [89–91] explained the flocculation and redispersion mechanisms by adsorption processes. Low concentrations of anionic surfactants neutralize the positive charge of iron(III) oxide hydrate sols and cause flocculation. The adsorbed anionic surfactant is oriented with its hydrophilic groups toward the particle surface and the hydrophobic groups toward water. If the second addition of a surfactant results in adsorption on the flocculated sols, the sols redisperse. The adsorption is caused by hydrophobic interactions and the surfactant is oriented with its hydrophilic groups toward water.

The amounts of the surfactants adsorbed [91] indicate that the formation of mixed bilayers between anionic (hydrocarbon or fluorocarbon) surfactants and

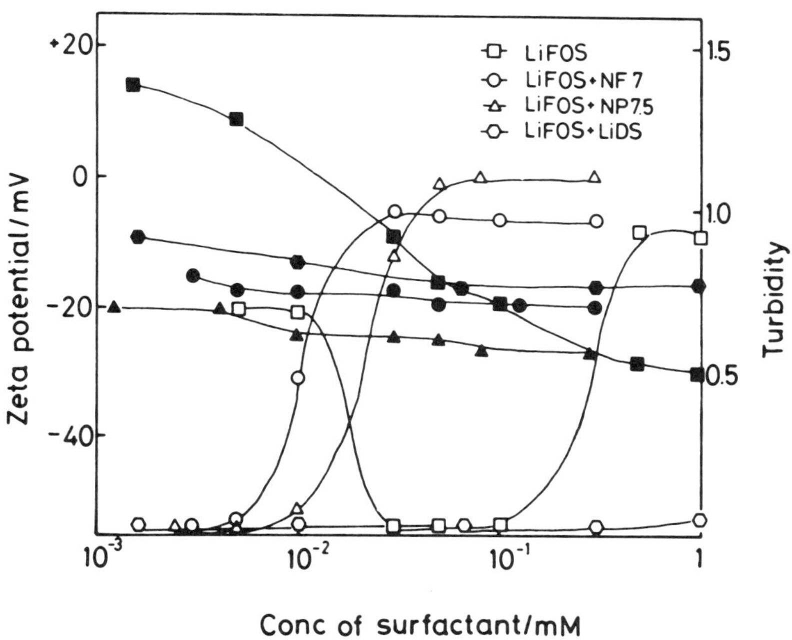

FIG. 5.17 The zeta potential and turbidity of α-Fe_2O_3 sols flocculated by LiFOS as a function of surfactant concentration; closed symbols, zeta potential; open symbols, turbidity. (From Ref. 89. Reproduced by permission of Elsevier Science Publishers.)

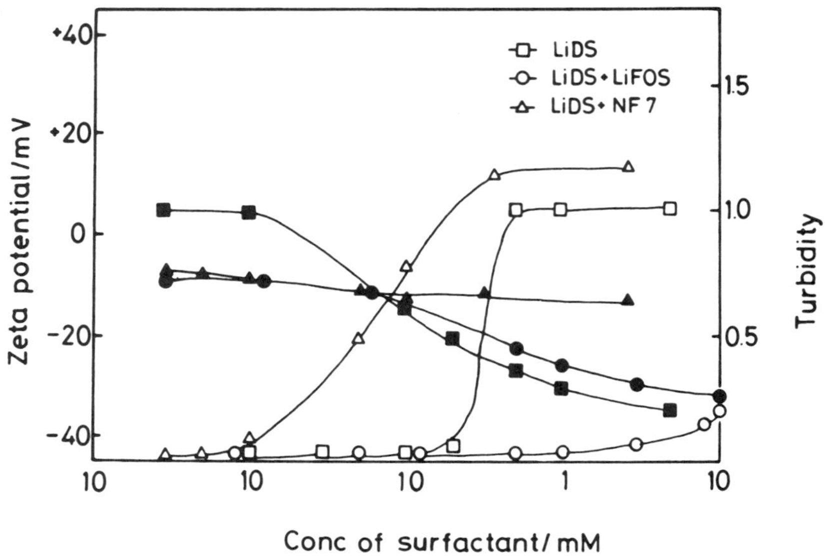

FIG. 5.18 The zeta potential and turbidity of α-Fe_2O_3 sols flocculated by LiDS as a function of surfactant concentration; closed symbols, zeta potential; open symbols, turbidity. (From Ref. 89. Reproduced by permission of Elsevier Science Publishers.)

nonionic (hydrocarbon or fluorocarbon) surfactants is more favorable than the formation of mixed bilayers between anionic fluorocarbon and anionic hydrocarbon surfactants on ferric hydrosols.

The adsorption of LiDS and LiFOS on alumina depends on the feed concentration and the LiFOS/LiDS mole ratio [92]. Below a feed concentration of 5 mmol/L the total amounts of the surfactant adsorbed from the binary solution are larger than those from a single surfactant solution. Above a feed concentration of 6 mmol/L the amounts adsorbed from the binary solution are smaller than the amounts adsorbed from single surfactant solutions. Esumi et al. [92] related the adsorption decrease at high concentrations to the formation of a mixed LiFOS-LiDS micelle in the bulk phase.

In contrast to the LiDS-LiFOS system, the total adsorption of SDS and 4-[2-heptafluoroisopropyl-1,3-bis(trifluoromethyl)- butenoxy]benzenesulfonate on alumina was lower than that of the individual surfactants adsorbed separately [93]. The decrease in adsorption suggests an unfavorable interaction between these surfactants in the adsorbed layer.

It is noteworthy and perhaps puzzling that very few fundamental papers have been published on the adsorption of mixed fluorocarbon and hydrocarbon surfactants at the liquid-solid boundary, in spite of its great practical importance. One reason for this apparent dormancy may be the proprietary nature of knowledge. The results of research conducted by industrial laboratories may be found in the patent literature (Chapter 8).

REFERENCES

1. C. A. Miller and P. Neogi, "Interfacial Phenomena," Vol. 17, Surfactant Science Series, Marcel Dekker, New York (1985).
2. E. Kissa, in "Handbook of Fiber Science and Technology: Chemical Processing of Fibers and Fabrics," Vol. II, Part B, p. 143, Marcel Dekker, New York (1984).
3. E. Kissa, Pure Appl. Chem. 53, 2255 (1981).
4. E. Kissa, J. Colloid Interface Sci. 83, 265 (1981).
5. J. W. Gibbs, Trans. Connecticut Acad. 3 (1876–1878); "Collected Works," Vol. I, Longmans, Green, New York (1928).
6. A. Dupré, "Théorie Méchanique de la Chaleur," p. 369, Gauthier-Villars, Paris (1869).
7. T. Young, Philos. Trans. Roy. Soc. (London) 95, 255 (1858).
8. J. J. Bikerman, Kolloid Z. 218, 52 (1967).
9. A. Marmur, Adv. Colloid Interface Sci. 19, 75 (1983).
10. C. G. Sumner, "Symposium on Detergency," p. 15, Chemical Publishing Co., New York (1937).
11. H. Shuttleworth and G. L. Bailey, Disc. Faraday Soc. 3, 16 (1948).
12. R. E. Johnson, J. Phys. Chem. 63, 1655 (1959).
13. N. K. Adam, The Chemical Structure of Solid Surfaces as Deduced from Contact Angles, in "Contact Angle, Wettability and Adhesion," R. F. Gould, ed., Adv. Chem. Ser., Vol. 43, p. 52, American Chemical Society, Washington, DC (1964).
14. F. M. Fowkes and W. D. Harkins, J. Am. Chem. Soc. 62, 3377 (1940).
15. R. N. Wenzel, Ind. Eng. Chem. 28, 988 (1936).
16. R. E. Johnson and R. H. Dettre, Contact Angle Hysteresis, in "Contact Angle, Wettability and Adhesion," R. F. Gould, ed., Adv. Chem. Ser., Vol. 43, p. 112, American Chemical Society, Washington, DC (1964).
17. R. H. Dettre and R. E. Johnson, Contact Angle Hysteresis, Part II, in "Contact Angle, Wettability and Adhesion," R. F. Gould, ed., Adv. Chem. Ser., Vol. 43, p. 136, American Chemical Society, Washington, DC (1964).
18. A. B. D. Cassie and S. Baxter, Trans. Faraday Soc. 40, 546 (1944).
19. R. J. Good, J. Am. Chem. Soc. 74, 504 (1952).
20. D. C. Pease, J. Phys. Chem. 49, 107 (1945).
21. R. E. Johnson and R. H. Dettre, J. Phys. Chem. 68, 1744 (1964).
22. R. H. Dettre and R. E. Johnson, J. Phys. Chem. 69, 1507 (1965).
23. B. Miller, The Wetting of Fibers, in "Surface Characteristics of Fibers and Textiles," M. J. Schick, ed., Marcel Dekker, New York (1977).
24. A. M. Schwartz and S. B. Tejada, J. Colloid Interface Sci. 38, 359 (1972).
25. R. L. Hoffman, J. Colloid Interface Sci. 50, 228 (1975).
26. R. Burley and B. S. Kennedy, in "Wetting, Spreading, and Adhesion," J. F. Padday, ed., p. 327, Academic Press, New York (1978).
27. E. V. Gribanova and L. I. Mulchanova, Kolloidn. Zh. 40, 30, 217 (1978)
28. J. G. Petrov and B. P. Radoev, Colloid Polym. Sci. 259, 753 (1981).

29. S. Wu, "Polymer Interface and Adhesion," Marcel Dekker, New York (1982).
30. T. D. Blake, Wetting, in "Surfactants," Th. F. Tadros, ed., p. 221, Academic Press, London (1984).
31a. G. Ström, M. Fredriksson, P. Stenius, and B. Radoev, J. Colloid Interface Sci. 134, 107 (1990).
31b. G. Newcombe and J. Ralston, Langmuir 8, 190 (1992).
32. T. Kawase, S. Sekoguchi, T. Fujii, and M. Minagawa, Textile Res. J. 56, 409 (1986).
33. W. A. Zisman, Relation of the Equilibrium Contact Angle to Liquid and Solid Constitution, in "Contact Angle, Wettability and Adhesion," R. F. Gould, ed., Adv. Chem. Ser., Vol. 43, p. 1, American Chemical Society, Washington, DC (1964).
34. H. W. Fox and W. A. Zisman, J. Colloid Sci. 7, 109 (1956).
35. L. A. Girifalko and F. J. Good, J. Phys. Chem. 61, 904 (1956).
36. R. J. Good, L. A. Girifalko, and G. Kraus, J. Phys. Chem. 62, 1418 (1958).
37. R. J. Good and L. A. Girifalko, J. Phys. Chem. 64, 561 (1960).
38. S. Wu, J. Colloid Interface Sci. 71, 605 (1979).
39. F. M. Fowkes, J. Phys. Chem. 66, 382 (1962).
40. F. M. Fowkes, J. Phys. Chem. 67, 2538 (1963).
41. F. M. Fowkes, Ind. Eng. Chem. 56, 40 (1964).
42. D. K. Owens and R. C. Wendt, J. Appl. Polym. Sci. 13, 1741 (1969).
43. J. A. Finch and G. W. Smith, Contact Angles and Wetting, in "Anionic Surfactants. Physical Chemistry of Surfactant Action," E. H. Lucassen-Reynders, ed., Surfactant Science Series, Vol. 11, p. 317, Marcel Dekker, New York (1981).
44. A. Cahn and J. L. Lynn, Jr., in Kirk-Othmer, Encyclopedia of Chemical Technology, Third ed., Vol. 22, p. 332 (1983).
45. B. Tamamushi, in "Colloidal Surfactants," by K. Shinoda, T. Nakagawa, B. Tamamushi, and T. Isemura, Physical Chemistry, Vol. XII, p. 179, Academic Press, New York, 1963.
46. F. M. Fowkes, ed., "Contact Angle, Wettability, and Adhesion," Adv. Chem. Ser. 43, American Chemical Society, Washington, DC (1964).
47. J. J. Kipling, "Adsorption from Solution of Nonelectrolytes," Academic Press, London, 1965.
48. R. E. Johnson, Jr. and R. H. Dettre, in "Surface and Colloid Science," Vol. 2. p. 81, E. Matijević, ed., Wiley-Interscience, New York (1969).
49. D. H. Everett, in "Colloid Science," D. H. Everett, ed., Vol. 1, p. 49, The Chemical Society, London, 1973.
50. C. E. Brown and D. H. Everett, in "Colloid Science," Vol. 2, p. 52, The Chemical Society, London, 1975.
51. A. W. Adamson, "Physical Chemistry of Surfaces," 5th ed., Wiley, New York (1990).
52. K. l. Mittal, "Adsorption at Interfaces," ACS Symposium Series, No. 8, American Chemical Society, Washington, DC 1975.
53. C. H. Giles, Adsorption at Solid/Liquid Interfaces, in "Anionic Surfactants. Physical Chemistry of Surfactant Action," E. H. Lucassen-Reynders, ed., Surfactant Science Series, Vol. 11, p. 143, Marcel Dekker, New York (1981).

54. D. Attwood and A. T. Florence, "Surfactant Systems," p. 20, Chapman & Hall, London, 1983.
55. G. D. Parfitt and C. H. Rochester, in "Adsorption from Solution at the Solid/Liquid Interface," G. D. Parfitt, ed., Academic Press, New York (1983).
56. R. Aveyard, in "Surfactants," Th. F. Tadros, ed., p. 153, Academic Press, London, 1984.
57. D. K. Chattoray and K. S. Birdi, "Adsorption and the Gibbs Surface Excess," Plenum, New York 1984.
58. W. von Rybinski and M. J. Schwuger, in "Nonionic Surfactants. Physical Chemistry," M. Schick, ed., Surfactant Science Series, Vol. 23, p. 45, Marcel Dekker, New York, 1987.
59. D. Myers, "Surfactant Science and Technology," VCH Publishers, New York, 1988.
60. M. J. Rosen, "Surfactants and Interfacial Phenomena," 2nd ed., Wiley, New York 1989.
61. B. T. Ingram and R. H. Ottewill, Adsorption of Cationic Surfactants at Interfaces, in "Cationic Surfactants. Physical Chemistry," D. N. Rubingh and P. M. Holland, eds., Surfactant Science Series Vol. 37, p. 87, Marcel Dekker, New York (1991).
62. G. H. Giles, T. H. Mac Ewan, S. N. Nakhwa, and D. Smith, J. Chem. Soc. 3973 (1960)
63. G. H. Giles, D . Smith, and A. Huitson, J. Colloid Interface Sci. 47, 755 (1974)]
64. F. Schulman and W. A. Zisman, J. Colloid Sci. 7, 465 (1952).
65. E. F. Hare, E. G Shafrin, and W. A. Zisman, J. Phys. Chem. 58, 236 (1954).
66. E. G. Shafrin and W. A. Zisman, J. Phys. Chem. 64, 519 (1960).
67. C. O. Timmons and W. A. Zisman, J. Phys. Chem. 59, 984 (1965).
68. M. K. Bernett and W. A. Zisman, J. Phys. Chem. 71, 2075 (1967).
69. E. G. Shafrin and W. A. Zisman, J. Phys. Chem. 66, 740 (1962).
70. R. L. Cottington, E. G. Shafrin, and W. A. Zisman, J. Phys. Chem. 62, 513 (1958).
71. M. K. Bernett and W. A. Zisman, J. Phys. Chem. 63, 1241 (1959).
72. M. K. Bernett and W. A. Zisman, J. Phys. Chem. 63, 1911 (1959).
73. R. A. Pyter, G. Zografi, and P. Mukerjee, J. Colloid Interface Sci. 89, 144 (1982).
74. F. M. Fowkes and W. D. Harkins, J. Am. Chem. Soc. 62, 3377 (1940).
75. E. H. Lucassen-Reynders, J. Phys. Chem. 67, 969 (1963).
76. A. El-Shimi and E. D. Goddard, J. Colloid Interface Sci. 48, 242 (1974).
77. W. J. Murphy, M. W. Roberts, and J. R. H. Ross, J. Chem. Soc. Faraday Trans. I 68, 1190 (1972).
78. L. J. Gerenser, J. M. Pochan, M. G. Mason, and J. J. Elman, Langmuir 1, 305 (1985).
79. L. J. Gerenser, J. M. Pochan, J. F. Elman, and M. G. Mason, Langmuir 2, 765 (1986).
80. G. N. Batts, Colloids Surf. 22, 133 (1987).
81. E. J. Clayfield, A. G. Dixon, A. W. Foulds, and R. J. L. Miller, J. Colloid Interface Sci. 104, 500 (1985).
82. M. Claesson and H. K. Christenson, J. Phys. Chem. 92, 1650 (1988).

83. H. K. Christenson, P. M. Claesson, J. Berg, and P. C. Herder, J. Phys. Chem. 93, 1472 (1989).
84. R. W. Lai and H. L. Gray, in "Interfacial Phenomena in Biotechnology and Materials Processing," p. 293, Y. A. Attia, B. M. Moudgil, and S. Chander, eds., Elsevier Science Publishers, Amsterdam (1988).
85. G. Hartmann, Colloids Surf. 57, 295 (1991).
86. G. Hartmann, H.J. König, and D. Prescher, Textiltechnik 36, 541 (1986)]
87. E. Kissa, Langmuir 6, 478 (1990).
88. E. Kissa, Langmuir 6, 1217 (1990).
89. K. Esumi, Y. Ono, M. Ishizuka, and K. Meguro, Colloids Surf. 32, 139 (1988).
90. K. Esumi, Y. Sakamoto, K. Yoshikawa, and K. Meguro, Bull. Chem. Soc. Jpn. 61, 1475 (1988).
91. K. Esumi, Y. Sakamoto, K. Yoshikawa, and K. Meguro, Colloids Surf. 36, 1 (1989).
92. K. Esumi, H. Otsuka, and K. Meguro, Langmuir 7, 2313 (1991).
93. K. Esumi, H. Otsuka, and K. Meguro, J. Colloid Interface Sci. 142, 582 (1991).

6
Fluorinated Surfactants in Solution

6.1 SOLUBILITY

The solubilities of fluorinated surfactants are related to the unusual properties of the fluorine atom and the C−F bond. Fluorine is the most electronegative element and is very difficult to polarize. Fluorine can form a very stable bond with hydrogen or carbon (see Sec. 3.1). The rigidity of the C−F bond causes stiffening of the perfluoroalkane chain and limits interactions with other molecules. Because of their small size, fluorine atoms can shield the perfluorinated carbon atom without steric stresses. Perfluoromethyl or perfluoromethylene groups therefore form compounds with very weak intermolecular forces. As a consequence of weak interactions, perfluoroalkanes are insoluble in common organic solvents. Perfluoroalkanes are more hydrophobic than hydrocarbons, evidenced by solubility data: CF_4 is seven times less soluble in water than CH_4 [1,2]. Water is almost 7 times less soluble in perfluoroheptane than in heptane on a equal weight basis [3] and 25 times more on a molar basis.

The solubility of fluorinated surfactants depends on the hydrophile of the surfactant, in addition to the structure of the fluorinated group. The solubility of *perfluoroalkanoic acids* in water, like the solubility of nonfluorinated alkanoic acids, decreases with increasing chain length. At 25°C perfluorohexanoic acid and shorter-chain perfluoroalkanoic acids are miscible with water in all proportions. However, perfluorooctanoic acid and perfluorodecanoic acid are only slightly soluble in water [4].

The solubility of alkali metal *salts of perfluorinated alkanoic acids* in water decreases with increasing chain length. Some metal salts of perfluoroacetic and perfluorobutyric acids are soluble in water and in organic solvents as well

(Table 6.1) [5]. Alkali metal salts of perfluorocarboxylic acids having a chain length of four or more carbon atoms form micellar solutions. Their solubility in water can be characterized by their Krafft point (see Sec. 6.3) and critical micelle concentration (cmc) (see Secs. 6.2 and 6.6).

The *perfluoroalkanesulfonic acids* having one to six carbon atoms are colorless oils which fume in moist air and form monohydrates [6]. The C_2 and C_3 acids are miscible with water in all proportions. The solubility of perfluoroalkanesulfonic acids in water decreases with increasing chain length and the C_7 and C_8 acids are only moderately soluble in water. Their surface activity increases with decreasing solubility in water.

The solubilities of the *salts of perfluoroalkanesulfonic acids* depend on the nature of the counterion. The sodium and potassium salts of C_1, C_2, and C_3 perfluoroalkanesulfonic acids are soluble in water but their solubilities decrease with increasing chain length [6]. The tetraethylammonium salt of perfluorooctanesulfonic acid is more soluble in water than the potassium salt and is soluble in organic solvents such as alcohols, dimethyl formamide (DMF) or dimethyl sulfoxide (DMSO) as well [7].

The solution behavior of pentaerythritol tetraperfluorobutyrate reveals an extremely weak attraction between fluorocarbon and hydrocarbon groups [8,9]. Critical compositions are unsymmetrical and in agreement with values calculated for regular solutions using ideal entropy instead of Flory-Huggins entropy.

TABLE 6.1 Solubilities of Perfluoroalkanoates

	Solubility (g/100 mL)					
Salt	Water	Methanol	Ether	Benzene	Petroleum ether	Acetone
Perfluoroacetates						
Sodium	>10	4–10	<2	<2	<2	4–10
Mercury	>10	>10	>10	>10	4–10	4–10
Aluminum	>10	>10	<2	<2	<2	4–10
Barium	>10	>10	<2	<2	<2	>10
Nickel	4–10	>10	4–10	<2	<2	4–10
Perfluorobutyrates						
Sodium	>10	>10	<2	<2	<2	>10
Lithium	>10	>10	4–10	<2	<2	>10
Calcium	>10	>10	<2	<2	<2	>10
Lead	>10	>10	4–10	<2	<2	>10
Silver	>10	>10	>10	>10	<2	>10

Source: Ref. 5.

The solubilities of some commercial fluorinated surfactants are shown in Table 6.2 [10]. Nonionic fluorinated surfactants of the structures shown in Table 6.2 are soluble in several organic solvents, including acetone, ethyl acetate, toluene, tetrahydrofuran, and methyl chloroform. The amphoteric surfactant Zonyl FSK is soluble in acetone and ethyl acetate but only very slightly soluble in tetrahydrofuran. The surfactant Zonyl TBS, featuring a sulfonate hydrophile, is insoluble in organic solvents other than alcohols, in which it is slightly soluble. All other surfactants listed in Table 6.2 are soluble in isopropyl alcohol and in methanol. None of the fluorinated surfactants listed in Table 6.2 are soluble in aliphatic hydrocarbons.

6.2 MICELLE FORMATION

The solution behavior of surfactants is a result of their amphiphilic nature. Surfactants have lyophobic and lyophilic groups (the term lyophilic denotes a solvent compatible species, the term lyophobic indicates the opposite). Surfactants used in water have one or several hydrophilic head groups compatible with water and one or several hydrophobic tails which repel water. The surfactant therefore orients itself at the water-air interface with its hydrophilic part directed toward water and the hydrophobic tail pointing toward air. In water, surfactant molecules associate to form micelles or aggregates. The hydrophobic tails of the surfactant molecules form the interior of the micelle and the hydrophilic head groups are exposed to water (Fig. 6.1). In organic solvents the orientation of surfactant molecules is reversed. The hydrophilic groups are lyophobic and form the interior of the aggregate, whereas the lyophilic segments are oriented toward the solvent.

The term micelle describes an equilibrium aggregate of colloidal dimensions [11–13]. Franses et al. [14] have suggested that the term *micelle* stands for stable, disjoint, cooperative, closed, equilibrium colloidal aggregates. The term *stable* means constancy of the overall macroscopic properties of the aggregate population, although individual aggregates are not permanent. By *disjoint* is meant that the extent of aggregates is limited in all three dimensions and the aggregates remain clearly identifiable even when closely packed together. The term *cooperative* refers to the association pattern, the dependence of the free energy of aggregation on the aggregate size. The term *equilibrium* states that the aggregates form spontaneously and reversibly and are limited in size. The definition requires that a closed surface can be visualized to divide the hydrophilic and hydrophobic moieties. This condition is not always met and should be dropped from the otherwise precise definition.

Micelle formation or micellization occurs at a narrow surfactant concentration range, called the critical micelle concentration. Below the cmc the surfactant dissolves in molecular state. At cmc monomeric surfactant molecules asso-

TABLE 6.2 Solubility of Zonyl Fluorinated Surfactants

	Zonyl											
	FSA	FSP	FSE	UR[a]	FSJ	FSN	FSN-100	FSO	FSO-100	FSC	FSK	TBS
Stability, 25°C(77°F) ("S" = stable, "I" = insoluble)												
10% Potassium hydroxide	S	S	S	S	S	S	S	S	S	S	S	PS[b]
25% Sulfuric acid	I	I	I	I	I	S	S	S	S	S	S	S
37% Hydrochloric acid	I	I	I	I	I	S	S	S	S	S	S	I[c]
70% Nitric acid	I	I	I	I	I	S	S	S	S	S	S	S
Solubility, 25°C(77°F) (g solids/100 g solution)												
Water	>2	>2	>2	2	>2	>2	>2	0.1	0.1	>2	>2	0.1
Isopropyl alcohol	0.1	0.7	0	2	0.4	>2	>2	>2	>2	>2	>2	0.1
1:1 Water isopropyl alcohol	>2	>2	>2	2	>2	>2	>2	>2	>2	>2	>2	0.1
Methyl alcohol	>2	>2	0.5	>2	>2	>2	>2	>2	>2	>2	>2	0.1
Acetone	0	0.1	0	2	~0.1	>2	>2	>2	>2	1	1	0
Ethyl acetate	0	0	0	2	~0.1	>2	>2	>2	>2	0	>2	0
Toluene	0	0	0	0	0	0	>2	0	>2	0	0	0
n-Heptane	0	0	0	0	0	0	0	0	0	0	0	0
Tetrahydrofuran	0	0	0	0	0	>2	>2	>2	>2	0	0.1	0
Methyl chloroform	0	0	0	0	0	0	>2	0.1	>2	0	0	0

[a] Neutralized with sodium hydroxide to pH 7–8. Unneutralized, Zonyl UR is soluble in isopropyl alcohol, methyl ethyl ketone, methyl isobutyl ketone, and mixtures of acetone and isopropyl alcohol.

[b] Partially soluble.

[c] Soluble in 10% HCl.

Chemical Structures

$R_f = F(CF_2CF_2)_{3-8}$

Zonyl

FSA	$R_fCH_2CH_2SCH_2CH_2CO_2Li$
FSP,FSE	$(R_fCH_2CH_2O)P(O)(OHN_4)_2$ $(R_fCH_2CH_2O)_2P(O)(OHN_4)$
UR	$(R_fCH_2CH_2O)P(O)(OH)_2$ $(R_fCH_2CH_2O)_2P(O)(OH)$
FSJ	Zonyl FSP + nonfluorinated surfactant
FSN	$R_fCH_2CH_2O(CH_2CH_2O)_xH$
FSN-100	$R_fCH_2CH_2O(CH_2CH_2O)_xH$
FSO	$R_fCH_2CH_2O(CH_2CH_2O)_yH$
FSO-100	$R_fCH_2CH_2O(CH_2CH_2O)_yH$
FSC	$R_fCH_2CH_2SCH_2CH_2N^+(CH_3)_3$ $CH_3SO_4^-$
FSK	$R_fCH_2CH(OCOCH_3)CH_2N^+(CH_3)_2CH_2CO_2^-$
TBS	$R_fCH_2CH_2SO_3X$ (X = H and NH_4)

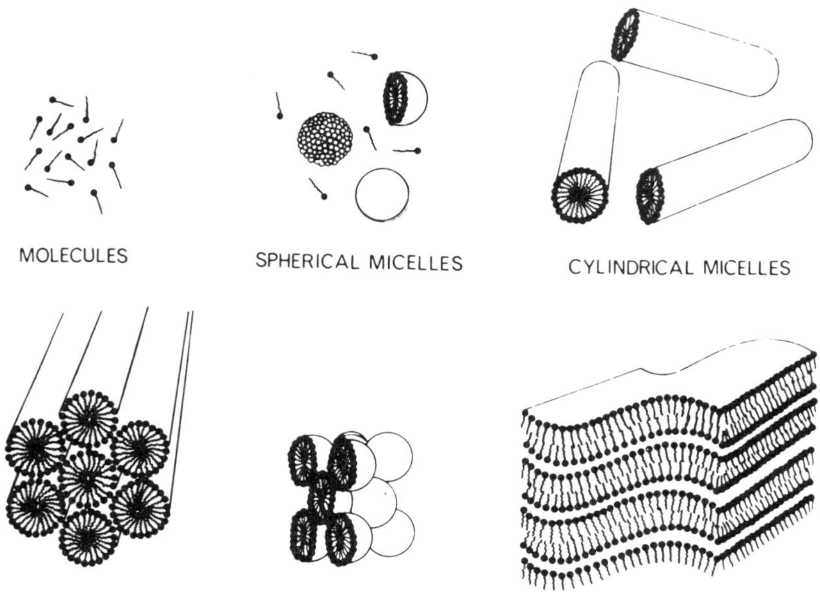

FIG. 6.1 Idealized structures that may exist in solutions of surfactants. (From Ref. 194. Reproduced by permission of Academic Press.)

ciate to form micelles and the solubility of the surfactant increases abruptly. The physical properties of the surfactant solution, such as surface tension, electric conductivity, osmotic pressure, and light scattering, also exhibit a sharp discontinuity at the cmc (Fig. 6.2).

Micelles are in a dynamic state [15,16]. Surfactants in a micelle are mobile. Above the cmc the molecularly dissolved surfactant molecules are in a dynamic equilibrium with the associated surfactant molecules in a micelle. A surfactant molecule may leave one micelle and adjoin another micelle.

Micelle formation has been explained by several theories which regard the micelle as either a chemical species or a separate phase. The simplest to understand and probably the most adequate is the mass action model [17–29], which regards the micelle as a chemical species. The mass action model is based on association of monomeric surfactant molecules in dynamic equilibrium with the micelle

$$nA \overset{K_m}{\rightleftarrows} A_n \tag{1}$$

where n is the number of surfactant molecules in the micelle and K_m is the association constant.

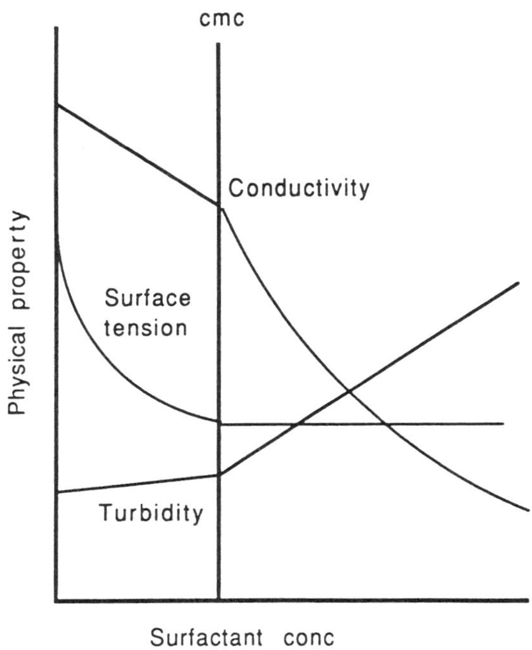

FIG. 6.2 Changes of physical properties occurring at the critical micelle concentration.

The stepwise association theory, developed by Aniansson and co-workers [30–36], describes micellization as a stepwise process involving the association and dissociation of a surfactant monomer with a micelle. In a stepwise process, one monomer at a time enters or leaves the micelles,

$$A + A_{n-1} \rightleftarrows A_n \qquad n = 2, 3, 4, \ldots \tag{2}$$

where A_n is a micelle with the aggregation number n.

Since the association of the monomeric surfactant molecules occurs stepwise, the mass action model requires an association constant, K_m, for every association step. Because of experimental limitations, numerical values for each association constant cannot be determined and have to be assumed [26,37,38]. Usually, as an approximation, a micellar solution is described with one K_m value as if the solution were monodisperse.

Burchfield and Woolley [39,40] have described a one-step mass action model that includes only one micellar species, with the assumption that the surfactant is a strong 1:1 electrolyte at infinite dilution. A surfactant solution having a concentration above cmc is considered a mixed electrolyte. The mass action equilibrium is given by

$$(n\beta)M^+ + nA \rightleftarrows M_{n\beta}A_n^{-n(1-\beta)} \tag{3}$$

where n is the surfactant aggregation number and β is the fraction of surfactant counterions associated with the micelle. The surfactant is represented as $[M^+, A^-]$ and the micellar aggregate as $[M_{n\beta}A_n^{-n(1-\beta)}]$. Equation 3 is applicable to cationic surfactants by reversing the charge signs of the surfactant and its counterion. Other parameters of the model are the screening factor δ for the micellar charge and the Guggenheim ion interaction parameters for counterion-monomer and counterion-micelle interactions, B_1 and B_2.

The two-phase (phase separation) model [24,41–43] regards the micelle as a separate phase, albeit a small entity of microscopic dimensions. The cmc is considered to correspond to the maximum solubility of the monomeric surfactant. If the saturation concentration is exceeded, a new phase, the micelle, appears. The micelle is thermodynamically stable and reversible. The phase separation model assumes that the activity of the surfactant molecule [44–51] and/or the surface tension [50,52] of the surfactant solution remains constant above its cmc. This assumption is not correct, however [29]. Furthermore, the phase separation model is not consistent with the number of degrees of freedom given by the Gibbs phase rule [29]. In spite of these difficulties, the two-phase model has explained solubilization and mixed micelle formation reasonably well and is therefore widely used.

Other theories have emerged, such as the charged phase separation model for ionic surfactants [50] and the theory based on thermodynamics of small systems [53–57]. The debate on micellization theories is still going on.

Solutions of fluorinated surfactants have been investigated and their micellar nature has been confirmed [58–59]. The substitution of the larger and highly electronegative fluorine atom for the smaller hydrogen increases the amphiphilic nature of the surfactant and lowers surface tension and cmc. The alkali and ammonium salts of perfluoroalkanoic acids exhibit surfactant properties and form micelles for a chain length of four carbon atoms, whereas eight carbon atoms are needed for the nonfluorinated alkanoates.

The size and the structure of micelles formed by fluorinated surfactants are discussed in Chapter 7. The interior of typical micelles appears to have a liquid-like structure. Micellar solutions therefore have solvent characteristics not exhibited by molecular solutions. Aqueous micellar solutions can dissolve water-insoluble substances by incorporating their molecules in or on the micelle (see Sec. 6.6).

The formation and dissociation of micelles is a very rapid process, usually occurring in a fraction (10^{-2} to 10^{-9}) of a second. Since the rate constants of such extremely rapid processes cannot be measured by conventional techniques, kinetic studies of micellization began when relaxation methods had been developed [36]. Kinetics of fluorinated surfactant micellization has been investigated by NMR relaxation, chemical relaxation, electric birefringence, and luminescence probing methods (see Chapter 9).

Sams et al. [60] proposed a two-state kinetic model which assumed a monomeric state and an associated state consisting of aggregates in various sizes larger than the monomer. The model describes only the fast process and assumes that the rate constants for association and dissociation are independent of the micelle size. A revised version of the two-state model [61–62] assumed micelle formation to be an adsorption phenomenon with micelles at equilibrium with monomers adsorbing and escaping from the surface of micelles.

A more detailed kinetic model utilized in studies of micellization kinetics is based on the theory by Aniansson and Wall [30–32] and modified by Kahlweit and co-workers [34,63–65]. The theoretical model developed by Aniansson and Wall [30,31] assumes a stepwise aggregation of surfactant monomers to form micelles [36]. When a micellar solution at equilibrium is perturbed by an instant change of temperature or pressure, the size distribution of micellar aggregates is shifted. The reestablishment of the equilibrium is characterized by two relaxation times, representing a fast process and a slow process.

The fast process involves an exchange of monomeric surfactants between the micelles and the intermicellar solution. A monomer or several monomers dissociate from or associate to existing micelles. As a result, the micellar distribution curve shifts without a change in the number of micelles [60,61,66,67].

In the slow process the number of micelles changes as a result of micelle formation from monomers or a complete breakdown of micelles to monomers. The concentration of micelles relaxes to a new equilibrium value. The aggregation number does not change, however.

The theory formulated by Aniansson and Wall and its modifications by Kahlweit and co-workers [34,63–65] have been the basis for most interpretations of chemical relaxation times and provided valuable information on kinetics of micellization.

The relaxation time constant, τ_1 for the rapid relaxation process is given by the expression [30,31]

$$\tau_1^{-1} = \frac{k^-}{\sigma^2} + \frac{k^- a(1 + c_o)}{n} \tag{4}$$

where k^- is the rate constant for the dissociation of the surfactant monomer from the micelle, σ^2 is the variance of the Gaussian distribution curve of micellar sizes, and c_o is the average deviation from equilibrium (usually less than 1%). The factor a is given by

$$a = \frac{A_{exc}}{A_1} \quad \text{and} \quad A_{exc} = A_{tot} - A_1 \tag{5}$$

where A_{tot} is total surfactant concentration and A_1 is the monomer concentration.

As an approximation, A_1 can be equated with cmc and c_o can be neglected

$$\tau_1^{-1} = \frac{k^-}{\sigma^2} + \frac{k^-}{n[(A_{tot} - \mathrm{cmc})/\mathrm{cmc}]} \tag{6}$$

Alternatively, the Aniansson and Wall equation can be written as

$$\tau_1^{-1} = \frac{k^+ A_{tot}}{n} + \frac{k^-}{n}\{(\mathrm{n}/\sigma^2) - 1\} \tag{7}$$

where k^+ is the association constant.

The relaxation time of the slow process τ_2 is given by

$$\tau_2^{-1} = \frac{k^- c_i n^2}{A_1 + (\sigma^2/n)(A_{tot} - \mathrm{cmc})} \tag{8}$$

where c_i is the micelle concentration at the micellar distribution curve minimum.

Rassing et al. [61] studied kinetics of sodium perfluorooctanoate micellar systems using the ultrasonic relaxation method. They observed a fast relaxation process attributed to a micelle formation. The ultrasonic relaxation times revealed that periodic fluctuations in temperature and pressure caused by the acoustic wave are several magnitudes less than the temperature or pressure perturbations of jump techniques. Rassing et al. [61] suggested that the ultrasonic and jump methods measure different modes of micelle formation whose relaxation times differ by several orders of magnitude. Ultrasonic absorption techniques [68] have also been used to measure relaxation spectra of cesium perfluorooctanoate in water and in deuterium [69].

The kinetics of micellization of perfluorinated surfactants has been investigated by Hoffmann and co-workers [70–76] by pressure jump and a shock wave method with conductivity detection [70–76]. Hoffmann and Ulbricht [71] also used a temperature jump relaxation technique [77] with optical detection, utilizing a pH indicator (thymol blue) to observe relaxation processes of a 1:1 mixture of perfluorooctanoic acid and its sodium salt. For micellar systems in which fast relaxation times could be measured, the parameters k^-/n, k^-/σ^2, σ^2/n, and k^+/n were calculated.

The exchange of surfactant monomers between the micellar and the monomeric state is diffusion controlled for monomers with hydrophilic counterions. The exchange rate is slower for monomers with hydrophobic counterions. For lithium and substituted ammonium salts of perfluorooctanesulfonic acid [72] the values of the micellar distribution curve, σ^2/n, increased with increasing hydrophobicity of the counterion or with a decrease of temperature. Both the association rate constant, k^+/n, and dissociation rate constant, k^-/n, decreased with increasing counterion hydrophobicity. The increase in k^-/n values was larger than the decrease in cmc values, suggesting that hydrophobic counterions form a barrier to the monomer exchange process.

A log-log plot of k^-/n versus cmc is linear, with the exception of the point representing the surfactant with a $C_3H_7NH_3^+$ counterion (Fig. 6.3) [72]. The system with a $C_3H_7NH_3^+$ counterion also has a different temperature dependence, shown in Fig. 6.4 [72] with an Arrhenius plot. The temperature dependence of the rate constant is given by

$$\frac{k^-}{n} = \left(\frac{D}{l^2}\right) \exp\left(\frac{-\epsilon}{RT}\right) \tag{9}$$

where D is the diffusion coefficient of the monomer, l the length of a CH_2- group, and ϵ the energy of monomer transfer from the micelle interior to the outside of the micelle.

The energy of momomer transfer calculated from Eq. 9 includes the heat of activation of the diffusion process plus the heat of transfer of the monomer

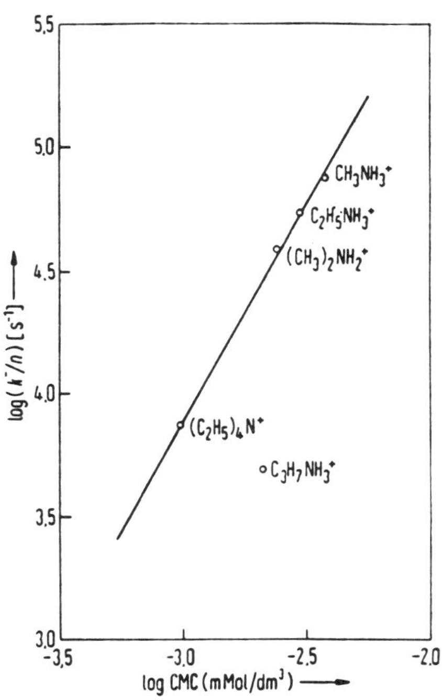

FIG. 6.3 Plot of k^-/n versus log cmc for perfluorooctanesulfonates at 35°C. (From Ref. 72. Reproduced by permission of Akademische Verlagsgesellshaft.)

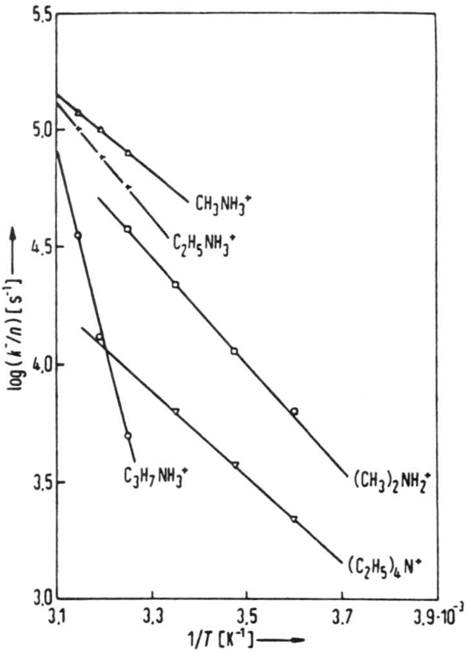

FIG. 6.4 Plot of k^{-}/n versus $1/T$ for perfluorooctanesulfonates. (From Ref. 72. Reproduced by permission of Akademische Verlagsgesellshaft.)

from the micellar interior to the transition state. Both values appear to be lower than the corresponding values for hydrocarbon surfactants.

The reciprocal relaxation time of the fast process increases linearly with increasing surfactant concentration (Fig. 6.5) [71,73] in accord with Eq. 6.

Relaxation measurements yield information on the size and shape of micelles. Hoffmann et al. [73] observed that below a certain temperature the amplitudes of the two relaxation processes decreased rapidly and amplitudes of new relaxation processes appeared. The new processes were attributed to the relaxation effects of another type of micelles, which appeared to be emulsion droplet–like giant molecules. The residence time of the surfactant molecules in the new micelle was unusually long, explained by incorporation of ion pairs, formed by the surfactant and its counterion, in the micelle. The existence of giant micelles has been disputed by Fontell and Lindman [78]. Subsequent studies by Hoffmann et al. have indicated that the giant aggregates are probably dispersions of liquid crystalline mesophases (see Sec. 7.1).

Hydrocarbon-derived surfactants with a fluorine-containing counterion do not belong strictly to the fluorinated surfactant class but nevertheless have

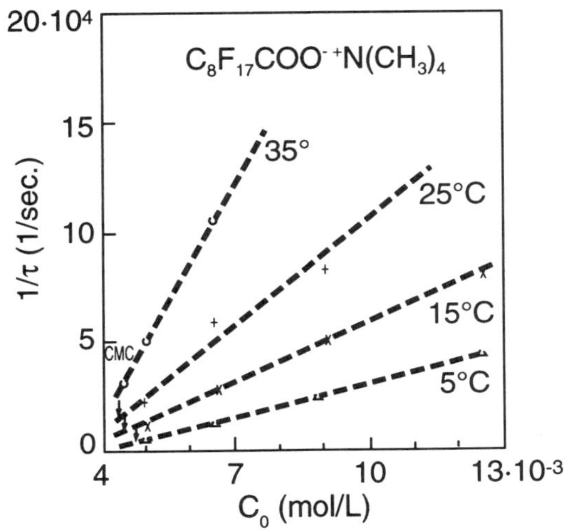

FIG. 6.5 Reciprocal values for the short relaxation time τ_1 plotted against the total concentration c_o for $C_8F_{17}COON(CH_3)_4$ at different temperatures. (From Ref. 73. Reproduced by permission of Akademische Verlagsgesellshaft.)

some unusual properties. Hoffmann et al. [74,76] investigated micelle formation kinetics in solutions of cationic surfactants having short-chain perfluoroalkanoate counterions. The surfactants consisted of a docedylammonium, tetradecylammonium, dodecylpyridinium, or tetradecylpyridinium cation and a perfluoroacetate, perfluoropropionate, or perfluorobutyrate counterion. Pressure jump and shock wave techniques revealed two relaxation processes in accord with the Aniansson-Wall theory. The residence time of a monomer in the micelle, the association and dissociation rate constants, the mean aggregation number of the micelles, the width of the micelle size distribution curve, and the concentration of the micellar nuclei were determined. The kinetic data suggested that the short-chain perfluorinated anions, CF_3COO^- and $C_2F_5COO^-$, are incorporated into the micelle containing dodecylpyridinium or tetradecylpyridinium cations but the longer-chain perfluorinated anions, $C_3F_7COO^-$, are located at the micelle surface [74]. The perfluoroacetate and perfluoropropionate ions facilitate the exchange of surfactant monomers between the micelle and the intermicellar solution. In contrast, the perfluorobutyrate counterions retard the surfactant monomer exchange.

When the chain length of the fluorinated counterion is increased further, the counterion assumes surfactant characteristics in its own right. The cationic surfactant then becomes a mixed surfactant consisting of a fluorinated surfactant anion and a hydrocarbon surfactant cation (see Sec. 7.2).

6.3 KRAFFT POINT

The solubility of surfactants in water increases gradually with increasing temperature until, at the Krafft point, the solubility increases abruptly [79]. The Krafft point is the temperature at which the solubility of monomeric surfactant molecules is equal to the cmc at that temperature. The Krafft point can also be defined as the temperature at which the solubility versus temperature curve intersects the cmc versus temperature curve [29,80]. Below the Krafft point the surfactant solution is molecular and the solubility of the surfactant is low. In a saturated solution the monomeric surfactant molecules are in equilibrium with a crystalline solid phase. Above the Krafft point the surfactant forms micelles and the crystalline solid phase, if present, is in equilibrium with the liquid phase containing micelles and monomeric molecules of the dissolved surfactant (Fig. 6.6) [81].

Shinoda and Hutchinson treated the micelle as a separate phase [41,45,82] and proposed that the Krafft point is the temperature above which the solid hydrated surfactant melts and dissolves as micelles in water [83,84]. In a phase diagram of an aqueous surfactant, the Krafft point is the "triple point" at which monomolecular surfactant coexists with micelles and the solid surfactant [42]. At the Krafft point micelles are in equilibrium with monomeric surfactant molecules [83,85–87]. If micelles are treated as a separate phase, the equilibrium is univariant because the phase rule allows only one degree of freedom and the equilibrium

$$\text{Hydrated solid} \rightleftarrows \text{monomers} \rightleftarrows \text{micelles} \tag{10}$$

is fixed at a given pressure. Moroi at al. [29] and La Mesa and Coppola [88] have argued that this assumption is incorrect because three phases coexist all along the Krafft line (Fig. 6.7) [88]. Experimental evidence that the cmc changes with pressure [89,90] also contradicts the assumption that the Krafft point is invariable.

The mass action model (Eqs. 1 and 2) gives two degrees of freedom at the Krafft point, since a solution phase is in equilibrium with a solid [88]. The number of components is three: water, molecularly dissolved surfactant, and micelles:

$$f = c - p + 2 - r \tag{11}$$

where f is the degrees of freedom, c is the number of components (water, monomeric surfactant, and micelle), p is the number of phases (surfactant solution phase, and surfactant solid phase), and r is the number of equilibrium reactions (one) [29]. The two degrees of freedom are in accord with the evidence that solubility of a surfactant in water is defined by temperature and pressure. However, this conclusion based on the mass action model cannot be

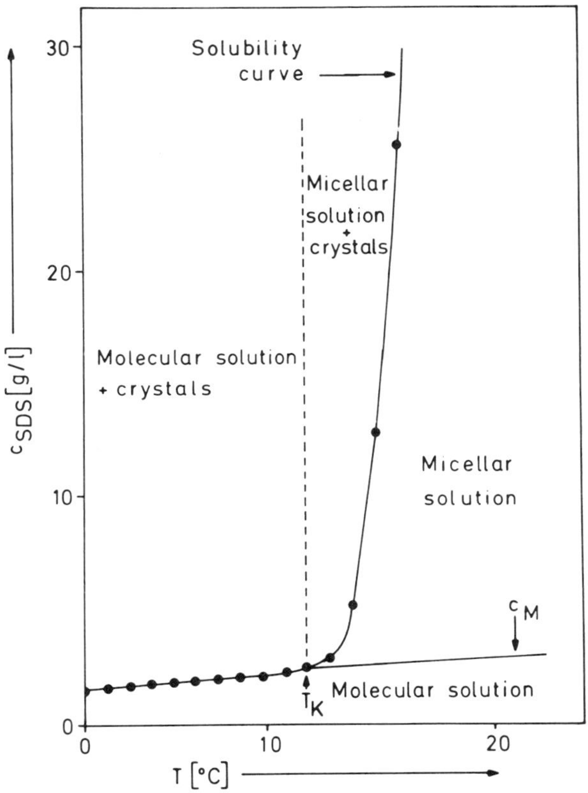

FIG. 6.6 Phase diagram of the sodium dodecyl sulfate–water system near the Krafft point T_K. (From Ref. 81. Reproduced by permission of Pergamon Press.)

strictly correct either because an ionic micelle is surrounded by counterions. The condition of electroneutrality would reduce the degrees of freedom to one. Moroi has suggested that the electroneutrality problem can be overcome if the solubility trend around the Krafft point is known. The overall solubility above the Krafft point is then given by

$$[S_T] = [S_k] \times \left(\frac{[S]}{[S]_k} \right)^m \tag{12}$$

where $[S_T]$ is the analytical concentration of micelles at temperature T, [S] is the solubility of the monomer, $[S_k]$ and $[S]_k$ are the analytical concentrations of micelles and monomeric concentrations at the Krafft point, respectively, and m is the aggregation number.

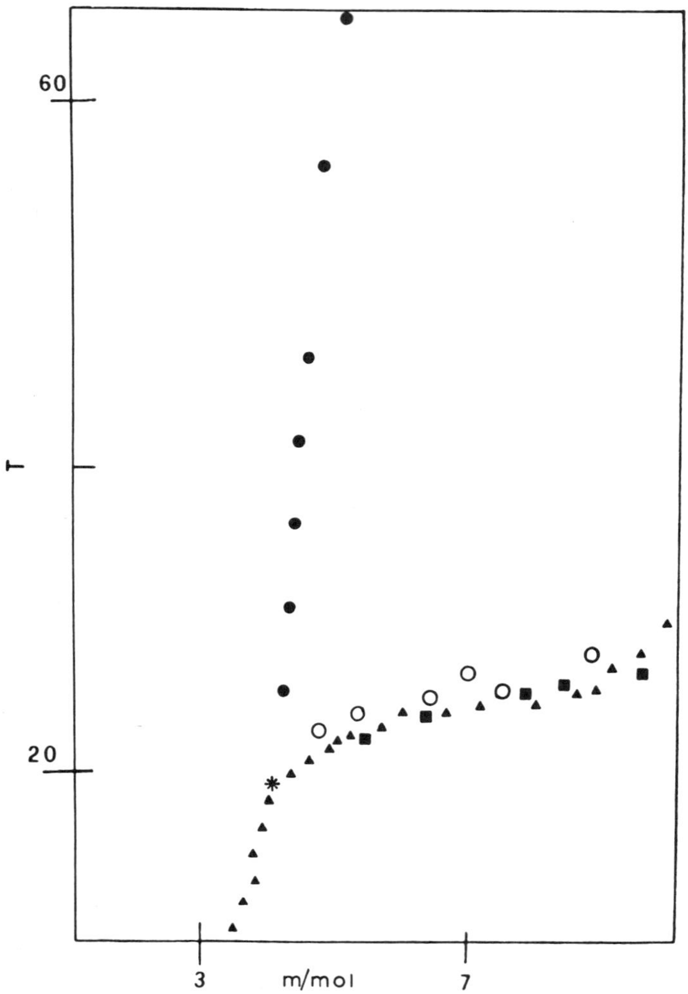

FIG. 6.7 Partial phase diagram of the water-NaPFN system, obtained by using electrical conductivity (▲), turbidity (■), and density data (○). Critical micelle concentrations at different temperatures (●) were obtained from electrical conductance or surface tension data. (*) Krafft point. (From Ref. 88. Reproduced by permission of Elsevier Science Publishers.)

Equation 12 states that the overall solubility above the Krafft point is related to the formation of micelles and depends on micelle aggregation numbers.

In spite of some complications, the mass action model serves as a useful approximation for describing micellization processes.

The Krafft points of perfluoroalkanoic and perfluoroalkanesulfonic acids are given in Table 6.3 [59,91]. The Krafft points of carboxylates are generally lower than those of sulfonates. In a homologous series Krafft points increase approximately with increasing order of the melting points. Increasing the carbon chain length elevates the Krafft point of a fluorinated surfactant, in analogy to that of a hydrocarbon-derived surfactant. The incremental increase of Krafft points for an increase of the chain length by one carbon atom is much larger than that for the corresponding hydrocarbon-derived surfactant.

Branching of the carbon chain lowers the Krafft point [59]. A moderately branched surfactant has a lower Krafft point and melting point than a straight-chain surfactant (Table 6.4).

Krafft points depend on the counterion of perfluorosulfonates (Fig. 6.8) and perfluoroalkanoates (Fig. 6.9) [59]. Krafft points are affected by several factors, including the nature of the bond between the surfactant anion and the counterion, the structure of the hydrated solid, and the degree of hydration. Shinoda and co-workers [59] have shown that the incremental increase of the Krafft point corresponding to an increase of the carbon chain by one carbon atom depends on the hydration of the counterion. The more hydrated the counterion is, the steeper the slope. For univalent counterions the slope decreases in the order $H > Li > Na > K \approx NH_4 \approx Rb$.

La Mesa [92] measured the temperature dependence of cmc for ionic surfactants including Li perfluorononanoate. Minima of cmc values indicated that metastable micellization processes function below the Krafft point.

The Krafft point depends on pressure and increases with increasing critical solution temperature (see Sec. 6.6).

The Krafft point is affected by other ingredients present in the surfactant solution. Another surfactant added to the solution of a surfactant depresses the Krafft point [59]. Shinoda at al. determined the Krafft points of binary mixtures of perfluorosulfonates (Fig. 6.10) and perfluoroalkanoates (Fig. 6.11) from solubility data. They observed that depression of the Krafft point increases with decreasing Krafft point of the added surfactant. Shinoda and co-workers found the change in the Krafft point to be analogous to the melting point depression in accord with Raoult's law. The Krafft point is also depressed by alcohols. Small amounts of hexanol, heptanol, and octanol increase the solubility of dodecyl sulfate in water and depress the Krafft point. Nakayama et al. [87] have proposed that the melting point of hydrated solid is depressed by the formation of a mixed micelle consisting of surfactant and alcohol. However, the melting point model of a hydrated solid for explaining the Krafft point is not valid [24].

TABLE 6.3 Krafft Points of Fluorinated Surfactants

Compound	Krafft point (°C)
$n\text{-}C_6F_{13}COOLi$	<0
$n\text{-}C_6F_{13}COONa$	<0
$n\text{-}C_6F_{13}COOK$	16.2
$n\text{-}C_6F_{13}COORb$	49
$n\text{-}C_7F_{15}COOH$	20
$n\text{-}C_7F_{15}COOLi$	<0
$n\text{-}C_7F_{15}COONa$	8.6
$n\text{-}C_7F_{15}COOK$	25.6
$n\text{-}C_7F_{15}COORb$	20.2
$n\text{-}C_7F_{15}COOCs$	<0
$n\text{-}C_7F_{15}COONH_4$	2.5
$n\text{-}C_8F_{17}COOH$	48.3
$n\text{-}C_8F_{17}COOLi$	<0
$n\text{-}C_8F_{17}COONa$	24.6
$n\text{-}C_8F_{17}COOK$	35.3
$n\text{-}C_8F_{17}COORb$	69
$n\text{-}C_8F_{17}COONH_4$	10.6
$n\text{-}C_8F_{17}COONH_3C_2H_4OH$	<0
$n\text{-}C_8F_{17}COONH(C_2H_4OH)_3$	<0
$n\text{-}C_{10}F_{21}COOLi$	<0
$n\text{-}C_{10}F_{21}COONa$	58.3
$n\text{-}C_{10}F_{21}COOK$	56.0
$n\text{-}C_{10}F_{21}COORb$	93
$n\text{-}C_{10}F_{21}COONH_4$	33
$n\text{-}C_{10}F_{21}COONH_3C_2H_4OH$	18
$n\text{-}C_{10}F_{21}COONH(C_2H_4OH)_3$	20
$n\text{-}C_{12}H_{25}COOLi$	42
$n\text{-}C_{12}H_{25}COONa$	89
$n\text{-}C_{12}H_{25}COOK$	75.5
$(CH_3)_2CF(CF_2)_4COOH$	<0
$(CH_3)_2CF(CF_2)_4COONa$	<0
$(CH_3)_2CF(CF_2)_4COOK$	<0
$(CH_3)_2CF(CF_2)_4CH{=}CHCH_2COOK$	<0
$n\text{-}C_7F_{15}SO_3Na$	56.5
$n\text{-}C_8F_{17}SO_3H$	<0
$n\text{-}C_8F_{17}SO_3Li$	<0
$n\text{-}C_8F_{17}SO_3Na$	75
$n\text{-}C_8F_{17}SO_3K$	80
$n\text{-}C_8F_{17}SO_3NH_4$	41
$n\text{-}C_8F_{17}SO_3NH_3C_2H_4OH$	<0
$n\text{-}C_9F_{19}SO_3\tfrac{1}{2}Mg \cdot 2H_2O$	41
$n\text{-}C_{11}F_{23}SO_3\tfrac{1}{2}Mg \cdot 2H_2O$	90 ± 5

Source: Refs. 59 and 91.

TABLE 6.4 Melting Points and Krafft Points of Normal and Branched Chain Perfluorooctanoic Acids

Surfactant	Melting point (°C)	Krafft point (°C)
n-$C_7F_{15}COOH$	56.4–57.9	20
$(CF_3)_2CF(CF_2)_4COOH$	13–14	<0

Source: Ref. 59. Reproduced by permission of American Chemical Society.

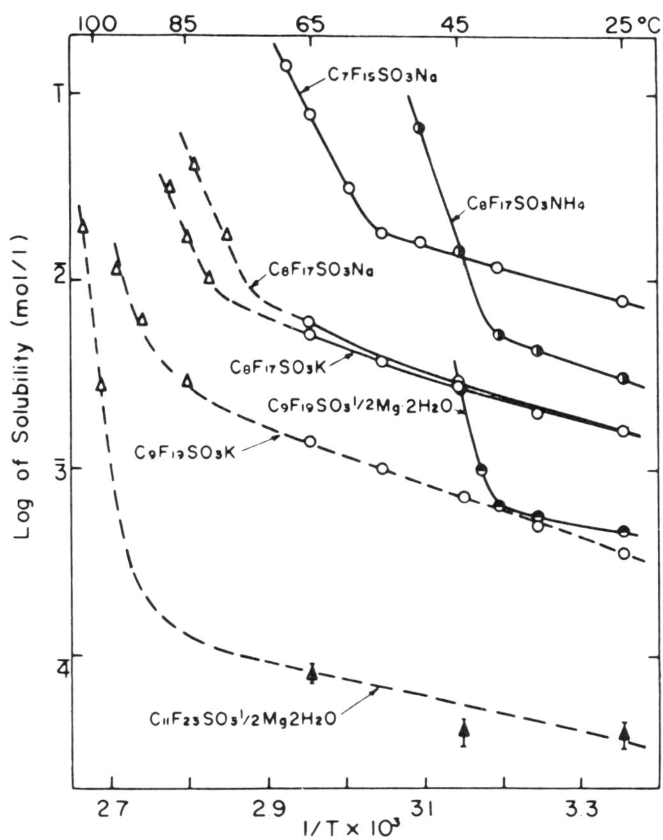

FIG. 6.8 Solubilities of perfluoroalkanesulfonates, $C_nF_{2n+1}SO_3M$, in water as a function of temperature. $C_9F_{19}SO_3K$ is less pure than the other surfactants. (From Ref. 59. Reproduced by permission of American Chemical Society.)

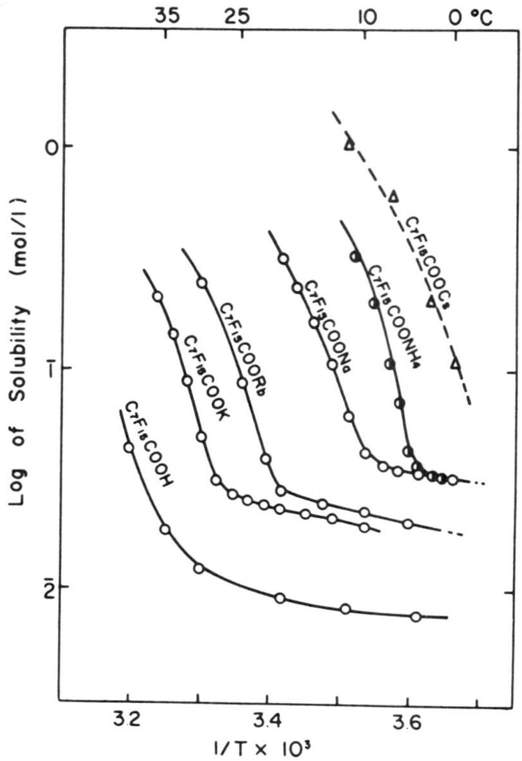

FIG. 6.9 Solubilities of $C_7F_{15}COOM$ in water as a function of temperature. (From Ref. 59. Reproduced by permission of American Chemical Society.)

The Krafft point can be lowered by (1) increasing the solubility of the monomer or (2) decreasing the cmc value of the surfactant.

Krafft points for fluorinated nonionic surfactants have not been reported. However, the widespread view that nonionic surfactants do not have a Krafft point is incorrect. Although only a few nonionic surfactants exhibit a Krafft point, an implication that the Krafft point is related to the ionic charge of a surfactant is misleading. Nonionic surfactants have low cmc values and the solubility of monomeric nonionic surfactants does not increase considerably with increasing temperature [93]. Krafft points are not observed when the (fictitious) Krafft point is below the freezing point of water. Krafft points have been reported for $C_{16}E_7$ [94], Brij 56, polyoxyethylated (10) cetyl alcohol [95,96], and Brij 76, polyoxyethylated (10) stearyl alcohol [96]. Kuwamura [97] found that Krafft points of nonionic surfactants having a glycine residue steadily decrease with increasing number of ethylene oxide units. L-Alanine raises the

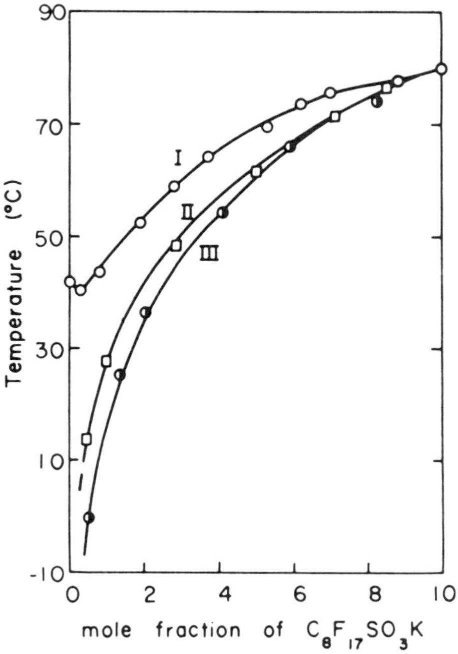

FIG. 6.10 Krafft point of binary surfactant mixtures: (I) $C_8F_{17}SO_3NH_4$ + $C_8F_{17}SO_3K$; (II) $C_8F_{17}SO_3H$ + $C_8F_{17}SO_3K$; (III) $C_8F_{17}SO_3NH_3C_2H_4OH$ + $C_8F_{17}SO_3K$. (From Ref. 59. Reproduced by permission of American Chemical Society.)

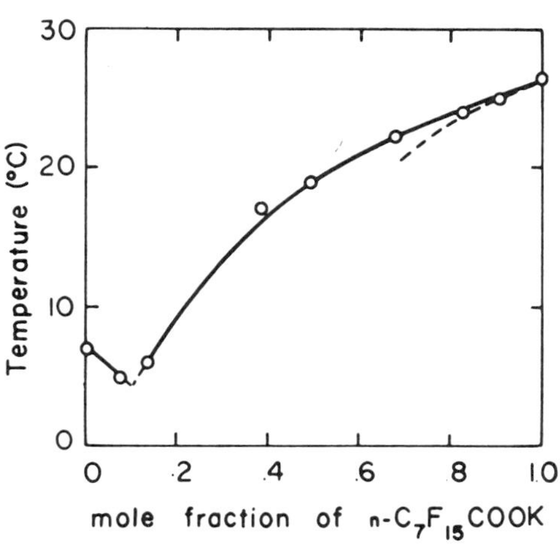

FIG. 6.11 Krafft point of binary surfactant mixture n-$C_7F_{15}COONa$ + n-$C_7F_{15}COOK$. (From Ref. 59. Reproduced by permission of American Chemical Society.)

Krafft point somewhat, whereas sarcosine has only a slight effect. This suggests that a hydrogen bond–forming secondary amino linkage in a nonionic surfactant molecule increases crystallinity and hydrophilicity of a nonionic surfactant.

6.4 CLOUD POINT

Aqueous solutions of nonionic surfactants become suddenly turbid when the temperature is raised. The temperature at which the sudden onset of turbidity occurs is called the cloud point. The cmc decreases with increasing temperature and the micellar weight increases [98–101]. At the cloud point the aggregates have become so large that turbidity becomes perceptible even to the naked eye [93,101]. The surfactant solution separates into two coexistent phases: a surfactant-rich phase and a phase in which the surfactant is depleted. The cloud point depends on the surfactant concentration but the concentration effect is usually weak.

The cloud point and the surfactant-water systems have been extensively studied for nonfluorinated nonionic surfactants. The phase separation at the cloud point has been explained by dehydration of the polyoxyethylene chain of the nonionic surfactant with increasing temperature [102]. The dehydration theory has lost some of its appeal in view of light-scattering, viscosity, and vapor pressure data [101]. Although it is true that hydration of oxyethylene ether oxygens decreases with increasing temperature, the amount of water entrapped by the polyoxyethylene chains may increase [101].

Since the cloud point is related to solubility of the nonionic surfactant in water, structural features of the surfactant molecule affect the cloud point. Nonionic surfactants having a longer polyoxyethylene chain have a higher cloud point. At constant oxyethylene content, the cloud point is lowered by (1) decreasing molecular weight of the surfactant (at a constant oxyethylene content, decereasing the size of the hydrophobe), (2) broader distribution of polyoxyethylene chain lengths, (3) branching of the hydrophobic group, (4) a central position of the hydrophilic group in the surfactant molecule, (5) replacement of the terminal hydroxyl by a methoxy group, or (6) replacement of the ether bond between the hydrophilic and hydrophobic groups by an ester bond [93,103].

A cloud point curve determined by Mathis et al. [104] for a fluorinated surfactant is shown in Fig. 6.12. Aliquots (1–2 mL) of surfactant solutions were heated in capped vials and the temperature was gradually raised until the solution became turbid. The curve exhibits a sharp decline and a flat minimum, followed by a gradual increase when the mass fraction of the surfactant is increased from 0 to 1%. The lowest point on the curve is the cloud point of the surfactant in the absence of any additives.

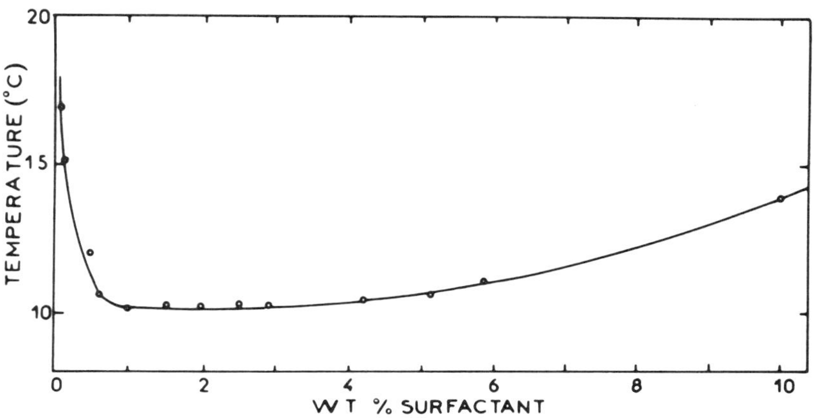

FIG. 6.12 Cloud point curve for the system $C_6C_{13}CH_2(OC_2H_4)_6OH/H_2O$ system. (From Ref. 104. Reproduced by permission of American Chemical Society.)

Critical micelle concentrations of fluorinated surfactants are approximately equal to those of their hydrocarbon analogues with a 1.5 to 1.8 times longer hydrocarbon chain (see Section 6.6). Thus the cmc of the fluorinated surfactant $C_6F_{13}C_2H_4(OC_2H_4)_{14}OH$ (4.5×10^{-4} M) is of the same order of magnitude as that of surfactant $C_{12}H_{25}(OC_2H_4)_{12}OH$ (1.4×10^{-4}). This rule does not appear to hold for cloud points of fluorinated surfactants. Mathis et al. [104] found a cloud point of 75°C for the fluorinated surfactant $C_6F_{13}C_2H_4(OC_2H_4)_{14}OH$. The estimated cloud point is about 120°C for surfactant $C_{12}H_{25}(OC_2H_4)_{12}OH$ and about 140°C for surfactant $C_{12}H_{25}(OC_2H_4)_{14}OH$. Although cloud points vary with the homologue distribution of nonionic surfactants, the observed discrepancy is significant. Hence the cloud points of fluorinated surfactants are lower than anticipated on the basis of cloud points for hydrocarbon-type surfactants with comparable cmc values.

The cloud point is affected by the presence of other solutes. Salts may increase or decrease the cloud point, depending on the nature of the anion and cation. Alcohols, fatty acids, and phenols depress the cloud point [101].

6.5 THERMODYNAMICS OF MICELLIZATION

The thermodynamic description of micelle formation is complicated by the dynamic nature of micellar solutions and ambiguities of models defining the contents of micelles and the micellization process (see Section 6.2). The mass action model, which regards micellization as a reversible aggregation, has problems with the definition of standard states, polydispersity of micellar solutions, and the dependence of the aggregation number on temperature and surfactant concentration. The phase separation model, on the other hand, has difficulties with the degrees of freedom given by the Gibbs phase rule and the definition of the heat of micellization [24,54,105,106].

According to the original mass action model, micellization is a reversible association of micelles [20,107]

$$nA \rightleftarrows A^n \tag{13}$$

where n is the number of surfactant molecules.

The association constant, K_M, is given by

$$K_M = \frac{a_M}{(a_A)^n} \tag{14}$$

where a_M is the activity of micelles and a_A is the activity of surfactant molecules.

At low surfactant concentrations the activities of the surfactant and its micelles can be replaced with their concentrations [27]:

$$K_M = \frac{c_M}{(c_A)^n} \tag{15}$$

where c_M and c_A are the micelle and surfactant concentrations, respectively.

Assuming ideal behavior, the chemical potentials, μ_A and μ_M, of the surfactant monomers and micelles, respectively, are

$$\mu_A = \mu_A^o + RTc_A \tag{16}$$

$$\mu_M = \mu_M^o + RTc_M \tag{17}$$

At equilibrium

$$n\mu_A = \mu_M \tag{18}$$

The free energy of micellization, ΔG^o, is then given by

$$\begin{aligned}\Delta G^o &= RT \ln c_A - (RT/n) \ln c_M \\ &= RT \ln c_A + (RT/n) \ln n - (RT/n) \ln nc_M\end{aligned} \tag{19}$$

When n is large,

$$\Delta G^o \approx RT \ln c_A = RT \ln \text{cmc} \tag{20}$$

If the aggregation number n varies with temperature or, as in some cases, with surfactant concentration, Eq. 20 is not valid. If the variation of the aggregation number n with temperature is assumed to be small and can be neglected

[27,45,47,54,98,106,108–113] an equation of the Clausius-Clapeyron type can be written:

$$\Delta H^{\circ} = -T^2 \left(\frac{\delta \Delta G^{\circ}/T}{\delta T} \right)_p = RT^2 \left(\frac{\delta \ln \mathrm{cmc}}{\delta T} \right)_p \tag{21}$$

Equation 21 is not strictly valid for calculating the heat of micellization because certain assumptions made in its derivation do not hold here. The equation implies that the micelle is at equilibrium near cmc in a standard state [27,54]. However, micelles are not definite stoichiometric entities but aggregates of different sizes that are in dynamic equilibrium with themselves and surfactant monomers. The aggregation number may vary with temperature. An extended mass action model describes micellization as a multiple equilibrium characterized by a series of equilibrium constants (see Section 6.2). Since these equilibrium constants cannot be determined, the micellar equilibrium is usually described by one equilibrium constant and an average aggregation number is used as an approximation.

Equation 21 describes a highly idealized model for the association of nonionic surfactant molecules or ions. The simple mass action model assumes monodispersity of micelles. Counterions of surfactant ions are not included. In reality, large numbers of counterions are associated with micelles of anionic or cationic surfactants. Hence, if the surfactant dissociates into ions, the counterion and the degree of dissociation have to be considered [24]:

$$n\mathrm{A}^- + m\mathrm{X}^+ \rightleftarrows \mathrm{M}^{(n-m)-} \tag{22}$$

or

$$n\mathrm{A}^+ + m\mathrm{X}^- \rightleftarrows \mathrm{A}^{(n-m)+} \tag{23}$$

and the equilibrium constant, K_M, is given for the equilibrium by

$$K_M = \frac{[\mathrm{M}]}{[\mathrm{A}^-]^n[\mathrm{X}^+]^m} \tag{24}$$

The activity coefficients are assumed to be unity and $m < n$, since the micelles have a net charge. For cationic surfactants (Eq. 23) the same equation (Eq. 24) applies after reversing the charge signs.

The free energy of micellization is given for ionic surfactants by

$$\Delta G^{\circ} = RT \ln c_A + \left(\frac{RTm}{n} \right) \ln c_X - \left(\frac{RT}{n} \right) \ln c_M \tag{25}$$

The use of Eq. 25 to calculate ΔG° for ionic systems is complicated by the uncertain value of *m/n*, especially in the presence of electrolytes.

The degree of dissociation of micellar counterions, α, has been calculated from electrical conductivity data:

$$\frac{d\kappa}{dc} = \alpha(\lambda + F\mu) \tag{26}$$

where κ is the specific conductivity, c is the surfactant concentration, λ is the equivalent conductance of the counterion, F is the Faraday constant and μ is the electrophoretic mobility of the micelles [114]. The value of α for sodium perfluorooctanoate was estimated to be 0.45 at 30°C at 1 atm. This is significantly higher than the value of α, 0.28, for sodium dodecyl sulfate.

The degree of binding of counterions, β, is obtained as the absolute slope of the plot of ln cmc versus ln c_x, where c_x is the total counterion concentration [24,25,42,114,115]. The experimental value for sodium perfluorooctanoate, 0.56, is fairly close to the value 0.55, obtained by assuming that $\beta = 1 - \alpha = 1 - 0.45 = 0.55$. Some β values and aggregation numbers for perfluorinated surfactants are given in Table 6.5 [115].

According to the mass action model of micellization of ionic surfactants [24], the equilibrium between surfactant monomers and micelles is given by [25]

$$n[S^-] + \beta n[Na^+] \rightleftarrows [M]^{-(1-\beta)n} \tag{27}$$

where S^- denotes the anionic surfactant monomer, M the micelle, and n the aggregation number of the micelle, and the concentrations are expressed in mole fraction units.

In accord with this model, the standard free energy of micelle formation per monomer unit, ΔG^o_m, is given by

$$\frac{\Delta G^o_m}{RT} = \ln[S^-] + \beta \ln[Na^+] - \left(\frac{1}{n}\right) \ln[M^{-(1-\beta)n}] \tag{28}$$

Mukerjee at al. [116] have emphasized that calculations based on the mass action model or the pseudophase separation model give significantly different enthalphy values. However, the discrepancies between both models diminish if the aggregation number increases to infinity [102,117].

TABLE 6.5 Values of Aggregation Number, n, and Degree of Binding, β, for Salts of Perfluorooctanoic Acid (PFO) and Perfluorononanoic Acid (PFN)

Surfactant	n	β	Reference
NaPFO	20	0.56	114
NaPFO	15	0.54	116
KPFO	—	0.52	118
LiPFN	20 ± 2	0.60 ± 0.002	80

Source: Ref. 115. Reproduced by permission of Elsevier Science Publishers.

Data published on thermodynamic properties of fluorinated surfactants are scarce. Shinoda and Hutchinson [42] treated the micelle as a separate pseudo-phase of very small dimensions. The phase separation model describes the micelle as a separate phase which begins to form at cmc. If micelle formation is analogous to phase separation, the heat and entropy of micellization can be calculated from the temperature dependence of the activities of micelle-forming species.

Shinoda and Katsura [118] calculated the entropy of micellization from the temperature dependence of cmc for potassium perfluorooctanoate under the condition of a definite counterion concentration. If micelle is considered a phase, the entropy, S_M, and enthaply, H_M, of micellization are given by [42,102,118]

$$\Delta S_M = \frac{\Delta H_M}{T} = -RT\left[\left(\frac{\delta \ln X_2}{\delta T}\right)_{\Delta\mu} + K_g\left(\frac{\delta \ln X_3}{\delta T}\right)_{\Delta\mu}\right] \tag{29}$$

where T is the absolute temperature, R is the gas constant, X_2 and X_3 are the mole fractions of the surfactant and counterions, respectively, and K_g is a constant.

If X_3 is kept constant, the second term in Eq. 29 is zero and the activity of micelle-forming species is dependent only on the surfactant ion concentration.

In the absence of added salts, $X_2 = X_3$, and

$$\Delta S_M = -(1 + K_g)RT\left(\frac{\delta \ln X_2}{\delta T}\right)_{\delta\mu} \tag{30}$$

The phase separation model has been criticized by proponents of the mass action model [24,119] (see Section 6.3).

Tadros [120] determined some physical parameters for the Monflor industrial fluorinated surfactants. The free energy, enthalpy, and entropy of micellization for Monflor fluorinated surfactants were obtained from the γ versus log c (surface tension plotted against log concentration) curves at various temperatures. The nonionic, anionic, and cationic surfactants used in the study were Monflor 51, $C_{10}F_{19}O(CH_2CH_2O)_nC_{10}F_{19}$, where $n \approx 23$; Monflor 31, $C_{10}F_{19}OC_6H_4SO_3Na$; and Monflor 71, $C_{10}F_{19}OC_6H_4N^+Me_3MeSO_4^-$.

Tadros calculated the standard free energy of micellization, ΔG^o_M, by the equation

$$\Delta G^o_M = zRT \ln x_{cmc} \tag{31}$$

where x_{cmc} is the mole fraction of the surfactant at cmc and z is the number of ions formed by the surfactant molecule (for a nonionic surfactant $z = 1$).

The standard enthalpy of micellization was calculated using

$$\Delta H^o_M = RT^2\left(\frac{d \ln x_{cmc}}{dT}\right) \tag{32}$$

The standard free entropy of micellization, S^o_M, was calculated from

$$\Delta G^o_M = \Delta H^o_M - T\Delta S^0_M \tag{33}$$

Equations 32 and 33 suggest that ΔH^o_M and S^o_M can be calculated from the variation of log x_{cmc} with $1/T$. However, plots of log x_{cmc} versus $1/T$ were not linear (Fig. 6.13), indicating a decrease in enthalpy of micellization with increasing temperature. ΔH^o_M was therefore calculated at each temperature from the tangent to the curve. The results are given in Tables 6.6 and 6.7 for Monflor 51, a nonionic surfactant, and Monflor 31, an anionic surfactant, respectively.

The free energy of transfer, ΔG^o_M, for a $-CF_2-$ group from an aqueous environment to a micelle is similar to that of corresponding hydrocarbon surfactants [80,121] (Table 6.8). However, the free energy of adsorption for a $-CF_2-$ group at the air–aqueous solution interface is much higher than the

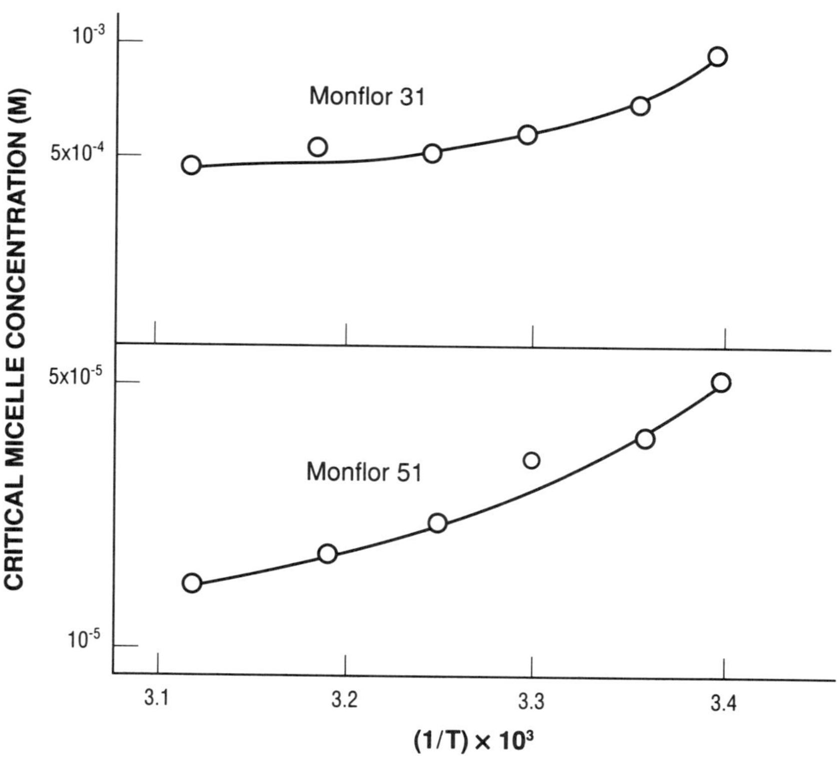

FIG. 6.13 The temperature dependence of cmc for Monflor 51 and 31. (From Ref. 120. Reproduced by permission of Academic Press, Inc.)

TABLE 6.6 Area/Molecule, cmc Values and Thermodynamic Parameters of Micellization for Monflor 51

Temperature (°C)	Γ (mol/m^2 × 10^6)	Area/ molecule (nm^2)	cmc (M × 10^6)	$-\Delta G$ (kcal/mol)	ΔH (kcal/mol)	$T\Delta S$ (kcal/mol)
20	2.97	0.56	4.79	8.1	14.1	22.2
25	3.12	0.53	3.47	8.4	10.8	19.3
30	3.19	0.52	3.02	8.7	9.1	17.8
35	3.3	0.49	2.09	9.0	6.5	17.8
40	(3.17)	(0.52)	1.74	9.3	4.5	13.8
45	(2.87)	(0.58)	1.45	9.6	4.5	13.9

Source: Ref. 120. Reproduced by permission of Academic Press.

TABLE 6.7 Area/Molecule, cmc Values, and Thermodynamic Parameters of Micellization for Monflor 31

Temperature (°C)	Γ (mol/m^2 × 10^6)	Area/ molecule (nm^2)	cmc (M × 10^6)	$-\Delta G$ (kcal/mol)	ΔH (kcal/mol)	$T\Delta S$ (kcal/mol)
20	4.71	0.35	9.12	12.8	17.8	21.3
25	4.56	0.36	6.61	13.4	8.4	21.8
30	4.37	0.38	5.63	13.8	8.7	22.5
35	4.32	0.39	5.01	14.2	8.1	22.3
40	4.17	0.40	5.25	14.3	3.9	18.3
45	4.07	0.41	4.68	14.7	3.1	17.8

Source: Ref. 120. Reproduced by permission of Academic Press.

TABLE 6.8 Free Energy of Transfer (ΔG^o_M) for a $-CF_2-$ Group from Aqueous to Micellar Environment and Free Energy of Adsorption for a $-CF_2-$ Group (ΔG^o_{ads}) at the Air-Aqueous Solution Interface

ΔG^o_M = −762 ± 25 cal/mol, at 25°C
ΔG^o_M = −825 cal/mol, at 30°C, for KPFO[a]
ΔG^o_M = −706 cal/mol, at 30°C, for KDS

ΔG^o_{ads} = −1.3 kcal/mol, at 25°C
ΔG^o_{ads} = −1.2 kcal/mol, at 30°C, for KPFO
ΔG^o_{ads} = −620 cal/mol, at 30°C, for KDS

[a] KPFO, potassium perfluorooctanoate [122]; KDS, Potassium decyl sulfate.
Source: Ref. 80. Reproduced by permission of American Chemical Society.

free energy of adsorption for a corresponding hydrocarbon surfactant, in accord with the greater surface activity of fluorinated surfactants. The behavior of perfluorinated surfactants has been explained [80,122] by chain-chain and chain-solvent interactions. Micellization processes involve weaker chain-chain interactions. Transfer from an aqueous environment to the air-aqueous interface is driven by the chain-solvent interaction, which is stronger for perfluorinated surfactants than for hydrocarbon surfactants. La Mesa and Sesta [80] have suggested that the thermodynamic bulk and surface properties of perfluorinated surfactants are related to the hydrophobic effect of perfluoromethylene chains.

Fisicaro et al. [115] determined the enthalpy of micellization for sodium perfluorooctanoate from heats of dilution using an LKB flow microcalorimeter. The enthalphy value, 8.70 kJ/mol at 25°C, calculated by using the pseudophase transition model is in good agreement with the enthalphy value calculated by using the Woolley and Burchfield one-step mass action model (see Section 6.2). The value reported by Johnson and Olofson [123], $\Delta H_M = 6.8$kJ/mol at 25°C, for lithium perfluorononanoate is considerably lower than expected for the substitution of lithium for sodium and an increase of the chain length by one $-CF_2-$ unit.

The enthalpies and entropies of fluorinated surfactants are positive and higher than those of their hydrocarbon counterparts. The higher positive enthalpies and entropies have been attributed to the higher hydrophobicity of the fluorocarbon chain relative to the hydrocarbon chain. Micellization involves the transfer of the hydrophobic chain of the surfactant from an aqueous environment to the micelle. The positive entropy change indicates an increase in randomness which is unlikely to result from aggregation of surfactant monomers [98]. The increase in entropy is more likely to arouse from the destruction of "iceberg" structured water around the hydrophobic chain [13,24,117,120,124]. Micellization is endothermic [125] and ΔH decreases with increasing temperature because at higher temperatures there is less water structure to break [120]. Tadros attributed the difference between enthalpies and entropies of fluorinated surfactants and those of their hydrocarbon counterparts to hydrophobic interactions. The fluorocarbon chain has a higher water structure–promoting ability than a hydrocarbon chain. Hence the hydrogen-bonded water structure around fluorocarbon chains is more extensive and micellization requires breaking more water structure around a fluorinated surfactant chain than around the corresponding hydrocarbon chain. This condition leads to higher positive enthalpies and entropies of micellization.

The involvement of water structure in the micellization process is also indicated by molar volume changes. The volume change associated with micellization is substantially greater for a perfluorinated surfactant than for a corresponding hydrocarbon surfactant [126,127]. A higher value of ΔV_m has been attributed to a more pronounced volume change caused by structural perturbations of water.

The volume change upon micellization reported for lithium perfluorononanoate by La Mesa and Sesta [80], $\Delta V_M = 18 \pm 2$ mL/mol, corresponds to the ΔV_M value of 20 mL/mol estimated by Sugihara and Mukerjee for sodium perfluorooctanoate [126]. However, the V_M value of 14.2 mL/mol reported by Johnson and Olofsson [123] for lithium perfluorononanoate is considerably lower than the value obtained by La Mesa and Sesta. The $V_M = 21.5 \pm 1$ mL/mol value estimated for perfluorooctanoic acid [83] is higher than that of its sodium salt and $\approx$ 5 mL/mol higher than expected for purely Coulombic interactions of H^+ at the micelle surface. Shinoda and Soda [83] explained the apparent discrepancy by substantial covalent bonding between the perfluorooctanoate anions and H^+ ions at the micelle surface.

6.6 CRITICAL MICELLE CONCENTRATION

The critical micelle concentration (cmc) denotes the surfactant concentration range at which surfactant molecules associate to form micelles. The term cmc has been defined as the surfactant concentration in a solution in which just half of the total surfactant present is in the monomeric form [128]. Since several physical properties of micellar solutions change abruptly at cmc, the value of cmc and its dependence on surfactant structure, composition of the solution, and physical conditions are of great practical importance. Wetting, detergency, solubilization, and other important processes depend on the cmc of the micellar solution.

The critical micelle concentration of a surfactant solution can be determined by measuring surface tension, electric conductivity, osmotic pressure, light scattering, viscosity, dye solubilization, and other physical properties (see Chapter 9). A critical discussion of the methods for the determination of cmc values has been published by Mukerjee and Mysels [129].

Hydrophobe Structure

The cmc of fluorinated or hydrocarbon surfactants decreases exponentially with increasing number of carbon atoms, according to the empirical equation [130–132]

$$\log \text{cmc} = A - Bn \tag{34}$$

where concentrations are expressed in mol/L, n is the number of atoms in the hydrophobe carbon chain of the surfactant, and A and B are constants for a particular homologous series and temperature. The constant A varies with the nature and number of the hydrophilic group or substituents in the hydrophobic chain. B is reasonably constant (about 0.29 for alkane hydrocarbon chains having one ionic hydrophile) but varies with the number of hydrophiles [102]. B has a value of about 0.50 for nonionic hydrocarbon-type surfactants.

In analogy to surfactants with a hydrocarbon chain, the critical micelle concentrations for fluorinated surfactants in an aqueous solution depend on the fluorocarbon chain length and the counterion (Table 6.9) [58,59, 71,73,91,128,133–141] and decrease with increasing hydrophobe chain length.

A plot of log cmc versus the number of carbon atoms in the fluorocarbon chain of anionic fluorinated surfactants is linear (Fig. 6. 14) [135] in accord with Eq. 34. The same relationship also seems to hold for nonionic fluorinated surfactants [142]. Examples of cmc values for some nonionic fluorinated surfactants are given in Table 6.10.

TABLE 6.9 Critical Micelle Concentrations for Anionic Fluorinated Surfactants[a]

Compound	cmc (mmol/L)	References
CF_3COOH	2600	58
C_2F_5COOH	2060	139
n-C_3F_7COOH	710, 740, 750	58, 134, 137
n-C_4F_9COOH	530	139
n-$C_5F_{11}COOH$	82, 51	58, 138
n-$C_6F_{13}COOH$	50	139
n-$C_7F_{15}COOH$	8.7, 9.0, 10.5, 9.1, 8.7	58, 59, 71, 134,137
n-$C_8F_{17}COOH$	2.8, 5.6	59, 139
n-$C_9F_{19}COOH$	0.89, 0.78, 0.85	59, 134, 137
n-$C_{10}F_{21}COOH$	0.48	139
n-$C_6F_{13}COOLi$	98	91
n-$C_8F_{17}COOLi$	10.6, 10.8	91, 73
n-$C_{10}F_{21}COOLi$	0.39	91
n-$C_6F_{13}COONa$	171	91
n-$C_7F_{15}COONa$	36, 32	36, 71
n-$C_8F_{17}COONa$	9.1	91
n-$C_{10}F_{21}COONa$	0.43	91
n-C_4F_9COOK	700	140
n-$C_5F_{11}COOK$	500	58
n-$C_6F_{13}COOK$	129, 62	91, 140
n-$C_7F_{15}COOK$	26.3, 27	58, 59
n-$C_8F_{17}COOK$	9.1	140
n-$C_9F_{19}COOK$	0.9	58
n-$C_{10}F_{21}COOK$	0.34	91
n-$C_7F_{15}COORb$	28	91

TABLE 6.9 (*Continued*)

Compound	cmc (mmol/L)	References
n-$C_5F_{11}COONH_4$	110	134
n-$C_7F_{15}COONH_4$	33	59
n-$C_8F_{17}COONH_4$	6.7	91
n-$C_{10}F_{21}COONH_4$	0.48	91
n-$C_8F_{17}COONH_3C_2H_4OH$	6.5	91
n-$C_{10}F_{21}COONH_3C_2H_4OH$	6.1	91
n-$C_{10}F_{21}COONH(C_2H_4OH)_3$	0.54	91
n-$C_8F_{17}COON(CH_3)_4$	4.5	73
n-$C_8F_{17}COO\frac{1}{2}Mg$	2.7	73
$(CF_3)_2CF(CF_2)_4COOH$	8.5	59
$(CF_3)_2CF(CF_2)_4COONa$	32	59
$(CF_3)_2CF(CF_2)_4COOK$	30	59
$(CF_3)_2CF(CF_2)_4CH{=}CHCH_2COOK$	≈1.5	59
$H(CF_2)_6COOH$	150	133
$H(CF_2)_8COOH$	30, 90	133, 134
$H(CF_2)_6COONH_4$	250, 110	133, 134
$H(CF_2)_8COONH_4$	38, 28	133, 134
$H(CF_2)_{10}COONH_4$	9	133
n-$CF_3(CH_2)_8COONa$	187	128
n-$CF_3(CH_2)_8COOK$	164	128
n-$CF_3(CH_2)_{10}COONa$	51	128
n-$CF_3(CH_2)_{11}COONa$	24	128
$(C_3F_7)_2P(O)ONa$	220	141
$(C_4F_9)_2P(O)ONa$	88	141
$(C_5F_{11})_2P(O)ONa$	27	141
n-$C_7F_{15}SO_3Na$	17.5	59
n-$C_8F_{17}SO_3Li$	6.3, 7.5	59, 71
n-$C_8F_{17}SO_3Na$	8.5	59
n-$C_8F_{17}SO_3K$	8.0	59
n-$C_8F_{17}SO_3NH_4$	5.5	59
n-$C_8F_{17}SO_3NH_3C_2H_4OH$	4.6	59
n-$C_8F_{17}SO_3N(C_2H_5)_4$	7.5	71
n-$C_9F_{17}SO_3\frac{1}{2}Mg \cdot 2H_2O$	0.64	59
n-$C_{11}F_{23}SO_3\frac{1}{2}Mg \cdot 2H_2O$	≈0.14	59

[a] Methods used for the determination of cmc: conductivity [58,71,73,91]; surface tension [58,59,134,137]; solubility vs. temperature curve [59]; dye titration [133]; fluorine NMR [128]. Temperature: the temperatures at which cmc values were determined are given in the original references. The temperatures vary from 2.5 to 100°C.

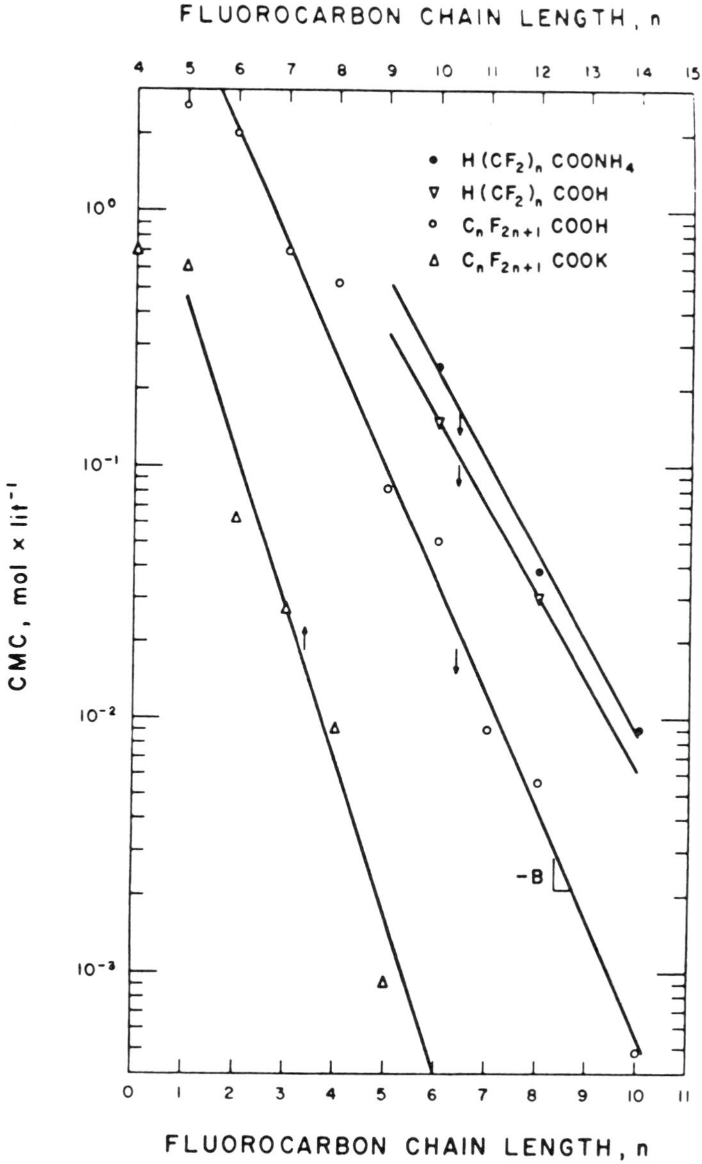

FIG. 6.14 Log cmc of fluorinated surfactants plotted against their number of CF_2 groups in the hydrophobic chain. (From Ref. 135. Reproduced by permission of American Chemical Society.)

TABLE 6.10 Critical Micelle Concentrations of Nonionic Fluorinated Surfactants in Water

Compound	cmc (mmol/L)	Temperature (°C)	Reference
$C_6F_{13}CH_2(OCH_2CH_2)_5OH$	0.35		104
$C_7F_{15}CH_2(OCH_2CH_2)_5OH$	0.048		104
$C_6F_{13}CH_2CH_2(OCH_2CH_2)_{11.5}OH0.45$	20	143	
$C_6F_{13}CH_2CH_2(OCH_2CH_2)_{14}OH$	0.61	20	143
$C_8F_{17}C_2H_4N(C_2H_4OH)_2$	0.16	20	143
$F(CF_2)_6CH_2CON[(C_2H_4O)_3CH_3)_2$	0.55	25	144
$F(CF_2)_8CH_2CON[(C_2H_4O)_3CH_3)_2$	0.012	25	144
$F(CF_2)_{10}CH_2CON[(C_2H_4O)_3CH_3)_2$	≈0.0003	25	144

Anionic surfactants with a oligo(oxyhexafluoropropylene) hydrophobe obey the relationship expressed by Eq. 34. [145]. Caporiccio at al. [145] prepared two series of surfactants. Monocarboxylic acids (series A) were prepared by photooxidation of hexafluoropropylene. Dicarboxylic acids (series B) were prepared by reductive cleavage of perfluoropolyperoxide resulting from tetrafluoroethylene photooxidation (see also Chapter 2). The log cmc values of the monocarboxylic acids, the dicarboxylic acids, and their ammonium salts correlated linearly with the number of oxyhexafluoropropylene units in the hydrophobic segment of the molecule (Figs. 6.15 and 6.16)

Since Eq. 34 is empirical, attempts have been made to develop a theoretical basis for this relationship. Shinoda et al. [59,102] expressed the change in cmc as a function of hydrophobic chain length by the equation

$$\ln \mathrm{cmc} = -\frac{m\omega}{(1 + K_g)kT} + K \tag{35}$$

where K_g and K are empirical constants, k is the Boltzmann constant, T the absolute temperature, m the number of carbon atoms in the fluorocarbon chain, and ω the free energy difference per $-CF_2-$ group (or $-CH_2-$ group) between micellar state and monomeric state.

B is given by Eqs. 34 and 35:

$$B = \frac{\omega}{(1 + K_g)2.303kT} \tag{36}$$

The ω value can be determined from the slope of the straight line obtained by plotting log cmc values versus the number of carbon atoms in a perfluoroalkane carbon chain (Fig. 6.17) [91]. The slope, B, of the straight lines is the same for the Na and K salts. Substituting $K_g = 0.52$, obtained for

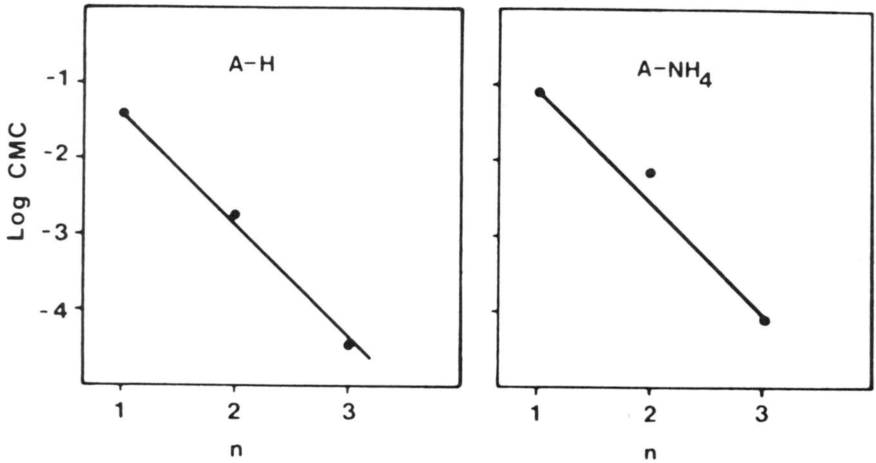

FIG. 6.15 Log cmc versus number of oxyhexafluoropropylene units in the hydrophobic chain of monocarboxylic surfactants. (A-H) $CF_3(OC_3F_6)_nOCF_2COOH$, $n = 1, 2, 3$; ($A\text{-}NH_4$) $CF_3(OC_3F_6)_nOCF_2COONH_4$, $n = 1, 2, 3$. (From Ref. 145. Reproduced by permission of Academic Press, Inc.)

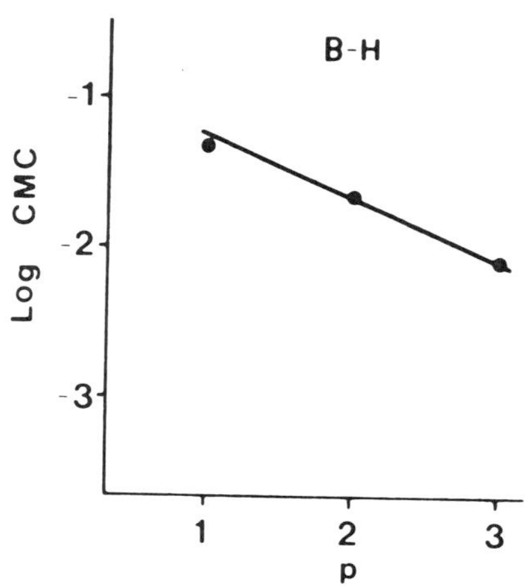

FIG. 6.16 Log cmc vs. number of oxyhexafluoropropylene units in the hydrophobic chain of dicarboxylic surfactants, $HOOCCF_2[(OC_2F_4)-(OCF_2)]_pOCF_2COOH$. (B1-H), $p = 1$; (B2-H) $p = 2$; (B3-H) $p = 3$. (From Ref. 145. Reproduced by permission of Academic Press, Inc.)

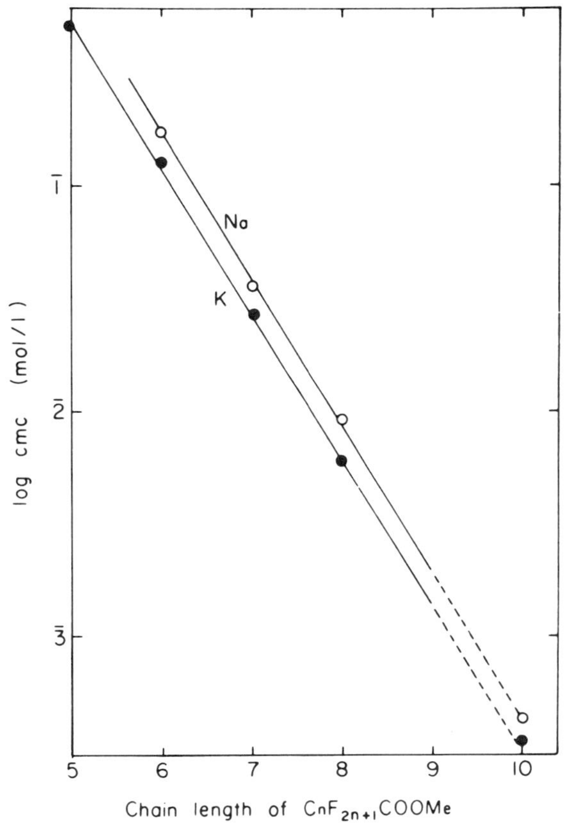

FIG. 6.17 Log cmc plotted against the number of carbon atoms in the hydrophobic chain of perfluoroalkanoates. (From Ref. 91. Reproduced by permission of American Chemical Society.)

n-$C_7F_{15}COOK$ [118], into Eq. 36, Kunieda and Shinoda [91] obtained $\omega = 2.21kT$. This value is about twice the ω value ($1.08kT$) [41] found for the corresponding hydrocarbon surfactant.

The critical micelle concentrations of fluorinated surfactants are compared with those of hydrocarbon surfactants in Table 6.11 [59,144]. The cmc values for C_7 and C_8 fluorinated surfactants are approximately equal to those for C_{11} and C_{12} hydrocarbon surfactants, respectively. The greater hydrophobicity of the fluorocarbon chain enhances the amphiphilic character of the surfactant and increases surface activity, reflected in surface tensions below 20 dyne/cm and lower cmc values [1,146]. Perfluorination reduces cmc values about four times per $-CF_2-$ group. The A value (or K value in Eq. 35) does not change con-

TABLE 6.11 Critical Micelle Concentrations of Fluorinated Surfactants and Those of Ordinary Surfactants

Compound	cmc (mmol/L)	Temperature (°C)	Compound	cmc (mmol/L)	Temperature (°C)
$C_8F_{17}SO_3Na$	8.5	75	$C_{12}F_{25}SO_3Na$	8.1	25
$C_8F_{17}SO_3K$	8.0	80	$C_{12}F_{25}SO_3K$	8.1	25
$C_7F_{15}COONa$	36	8	$C_{11}H_{23}COONa$	26	25
$C_7F_{15}COOK$	27	25.6	$C_{11}H_{23}COOK$	25.5	25
$F(CF_2)_6CH_2CON[(C_2H_4O)_3CH_3)]_2$	0.55	25	$H(CH_2)_{11}CON[(C_2H_4O)_3CH_3)]_2$	0.51	25

Source: Refs. 59 and 144.

siderably with the type of the surfactant. Since the relation expressed by Eq. 34 is exponential, the cmc value of a fluorinated surfactant is approximately equal to that of a hydrocarbon surfactant with a hydrocarbon chain 1.5 to 1.7 times longer than the fluorocarbon chain [59,144,147].

At equal chain length the critical micelle concentrations of perfluorinated surfactants are lower than those of their hydrocarbon analogues [1,71,129,148]. Potassium perfluorooctanoate has a 13-fold lower cmc than potassium octanoate. The difference between cmc values decreases with decreasing chain length and is only threefold for the, albeit poorly documented, hexanoates.

Hydrophile Structure

The cmc of a fluorinated surfactant also depends on the nature of the hydrophile but to a lesser effect than on the hydrophobe structure (Table 6.9). The carboxylates have higher cmc values than sulfonates (Table 6.11). This is in accord with the order of decreasing cmc values—carboxylates > sulfonates > sulfates—observed by Klevens [132] for hydrocarbon-type surfactants.

The critical micelle concentrations of nonionic fluorinated surfactants are much lower than those of their ionic analogues, The cmc values of nonionic fluorinated surfactants increase with increasing length of the oxyethylene chain

$$\log \text{cmc} = A + Bn \tag{37}$$

where n is the number of oxyethylene units in the hydrophile and A and B are constants [149,150]. The value of B calculated from the cmc values of fluorinated surfactants $C_6F_{13}CH_2CH_2(OC_2H_4)_nOH$, with n = 12 and 11.5, is 0.054 (or 0.125 when natural logarithms are used). The B value found is in good agreement with that of hydrocarbon surfactants (0.056), suggesting that the number of oxyethylene groups has the same effect on critical micelle concentrations whether the hydrophobe is fluorinated or not [104,143].

Matos et al. [151] investigated the surface activity of nonionic fluorinated surfactants which contained both oxyethylene and thioethylene groups:

$CF_3(CF_2)_mC_2H_4(SC_2H_4)(OC_2H_4)_nOH$ [series I]

$CF_3(CF_2)_mC_2H_4(SC_2H_4)(OC_2H_4)_p(SC_2H_4)(OC_2H_4)_qOH$ [series II]

The critical concentrations C_1, (formation of micelles or mesophase particles) indicated that one inserted thioethylene group cancels out the effect of 1.6 oxyethylene groups. All the data fit the same line (Fig. 6.18) when plotting log C_1 versus the effective number of oxyethylene groups

$$n = (p + q - 1.6) \tag{38}$$

The slope of the line is the constant *B* in Eq. 37. The value of *B*, 0.056, is in good agreement with the *B* value reported by Mathis et al. [104].

The foregoing examples indicate that oxyethylene groups increase the cmc of a fluorinated surfactant. However, oxyethylene groups inserted between a hydrophobe and a sulfate group have the opposite effect—the cmc value is reduced. Greiner et al. [152] determined cmc values of ether sulfate-type fluorinated surfactants derived from telomer alcohols $H[CF_2CF_2]_nCH_2[OCH_2CH_2]_mOSO_3NH_4$, where $n = 2, 3, 4$ and the average

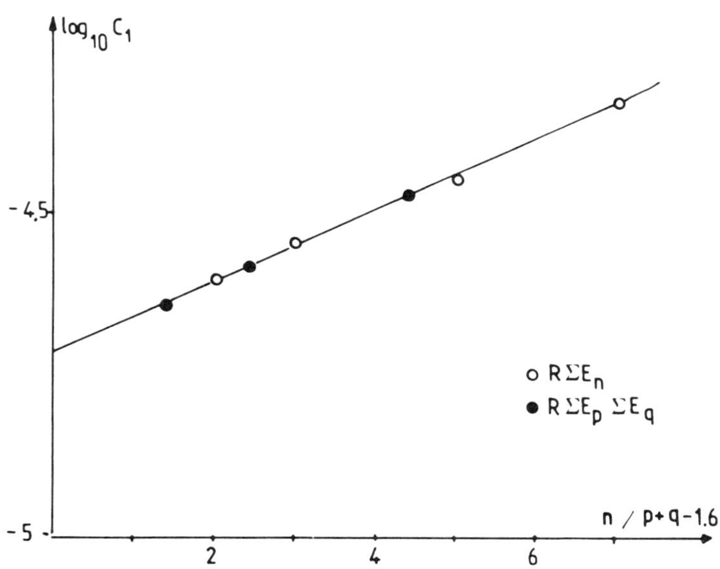

FIG. 6.18 Plot of log C_1 vs. the effective number of oxyethylene units *n* (○) or $n = p + q$ (●). (From Ref. 151. Reproduced by permission of Academic Press, Inc.)

of m is 3. The cmc values for surfactants with the oxyethylene segment are lower than those for ammonium perfluoroalkyl sulfate without the oxyethylene segment (Fig. 6.19). The anomalous effect, attributed to a slight increase of hydrophobicity of the molecule, is similar to that observed with alkyl sulfates [153,154].

Counterion

The cmc values of ionic fluorinated surfactants are affected by the nature of their counterion. The lines of the log cmc versus chain length plot (Fig. 6.17) have the same slope, B, but the cmc values for Na salts are higher than those for the corresponding K salts. Although the cmc values reported in the literature (Table 6.9) vary somewhat with the conditions and the method of their determination, it is evident that cmc values decrease in the order of the counterions Li > Na > K. The cmc values decrease with increasing atomic radii, increasing ion binding, and decreasing hydration of the counterion. The cmc value of tetraethylammonium perfluorooctaonoate is 4.2 times smaller than that of the sodium salt [155], in accord with the hydrophobicity of the $(C_2H_5)_4N^+$ ion.

The effect of hydrophobicity of the counterion is also evident when an organic counterion of a hydrocarbon-type surfactant is fluorinated. Hoffmann et al. [74,76,156] studied cationic surfactants with perfluorinated counterions and observed that perfluorination of an organic counterion lowers cmc, in analogy to perfluorination of the hydrophobe. Perfluorination of the anion lowers the cmc as much as doubling the length of the hydrocarbon chain of the anion, as illustrated with the cmc values of dodecylpyridinium with $C_3F_7COO^-$ (cmc = 0.70 mM) or $C_7H_{15}SO_3^-$ (cmc = 0.52) counterions [156]. The surface tension and cmc values decreased with increasing chain length of the perfluorinated anion in the order $CF_3COO^- > C_2F_5COO^- > C_3F_7COO^-$. The relation between the log of cmc and the number of CF_2- groups is linear. Kinetic measurements have indicated that the shorter-chain anions, CF_3COO^- and $C_2F_5COO^-$, are incorporated into micelles but the anions with a longer chain, $C_3F_7COO^-$, form a cohesive film on the micelle surface and hinder a micellar interchange [74].

The valence of the counterion has a considerable effect on cmc values. Surfactants with divalent counterions have a lower cmc than surfactants with monovalent counterions.

In solutions of ionic surfactants the activity of the micelle-forming species depends on the concentrations of both the surfactant ions and the counterions. The cmc values decrease with increasing counterion concentration (Fig. 6.20) [118]. If the change in the aggregation number is neglected or is very large

$$\ln \text{cmc} = -K_g \ln C_g + \text{const} \tag{39}$$

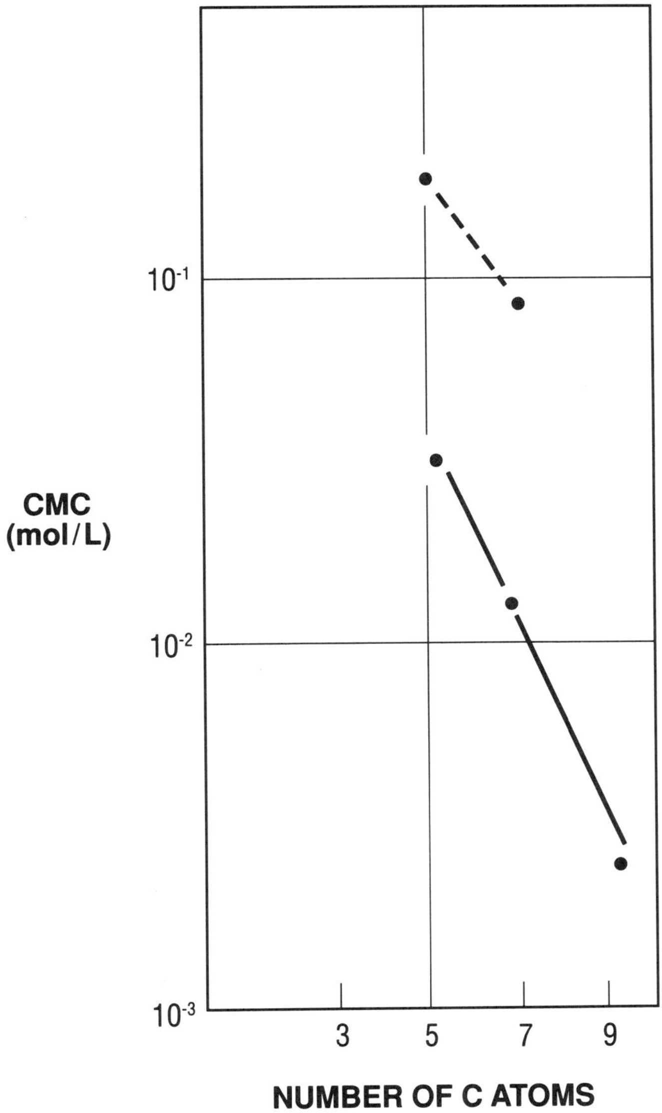

FIG. 6.19 Log cmc vs. the number of carbon atoms in the hydrophobic chain of (___) $H(CF_2CF_2)_nCH_2(OCH_2CH_2)_mOSO_3NH_4$, $n = 2,3,4$; m (average) $= 3$; (-----) $H(CF_2CF_2)_nCH_2OSO_3NH_4$, $n = 2$ and 3; $m = 0$. (From Ref. 152. Reproduced by permission of Dr. Alfred Hüttig Verlag.)

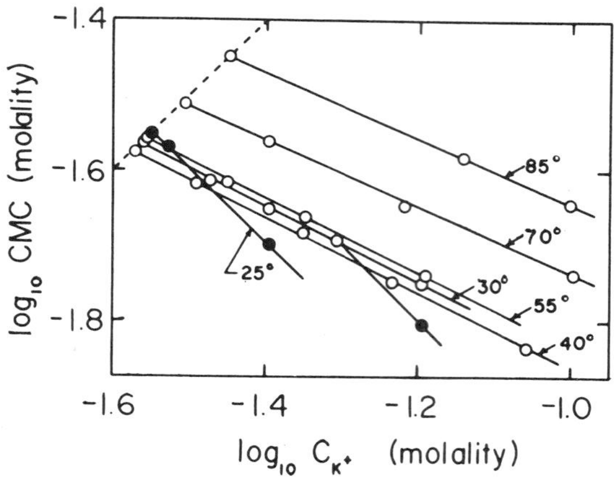

FIG. 6.20 The effect of counterion concentration and temperature on cmc in aqueous solutions of potassium perfluorooctanoate. (From Ref. 118. Reproduced by permission of American Chemical Society.)

where K_g is an experimental constant and C_g the counterion concentration [117,157,158].

Hydrophile-Hydrophobe Balance

Micelle formation is apparently controlled by two opposing factors, the cohesive forces between hydrophobic surfactant tails and the affinity of the hydrophilic groups for water. At a given temperature, the balance between these factors determines the size, shape, and charge of the micelle. Lin and co-workers have shown that the effective chain length is more relevant than the actual chain length [135,136,159–162]. Branching of the hydrophobe chain increases cmc.

Lin and Marszall [136] defined the hydrophobicity index, HI, as

$$\mathrm{HI} = n_{\mathrm{eff}}/n \tag{40}$$

where n is the nominal chain length expressed by the number of $-CF_2-$ groups in the hydrophobe chain.

The slope of the log cmc versus n plot (Fig. 6.14) yields n_{eff} values [135]:

$$\log \mathrm{cmc} = A - Bn_{\mathrm{eff}} = A - \frac{n\varphi'}{2.303kT(1 + K_g)} \tag{41}$$

where A and B are empirical constants, k is Boltzmann's constant, and $n\varphi'$ is the free energy change associated with the transfer of the hydrophobic chain from the aqueous environment into the interior of the micelle.

The use of the effective chain concept lead to a correlation between the hydrophilic-hydrophobic (lipophilic) balance (HLB) values and cmc values. In a homologous series of surfactants, chain length determines the HLB and cmc values. It is not surprising, therefore, that HLB values of fluorinated surfactants correlate with cmc values (Figs. 6.21 and 6.22), like those of hydrocarbon surfactants.

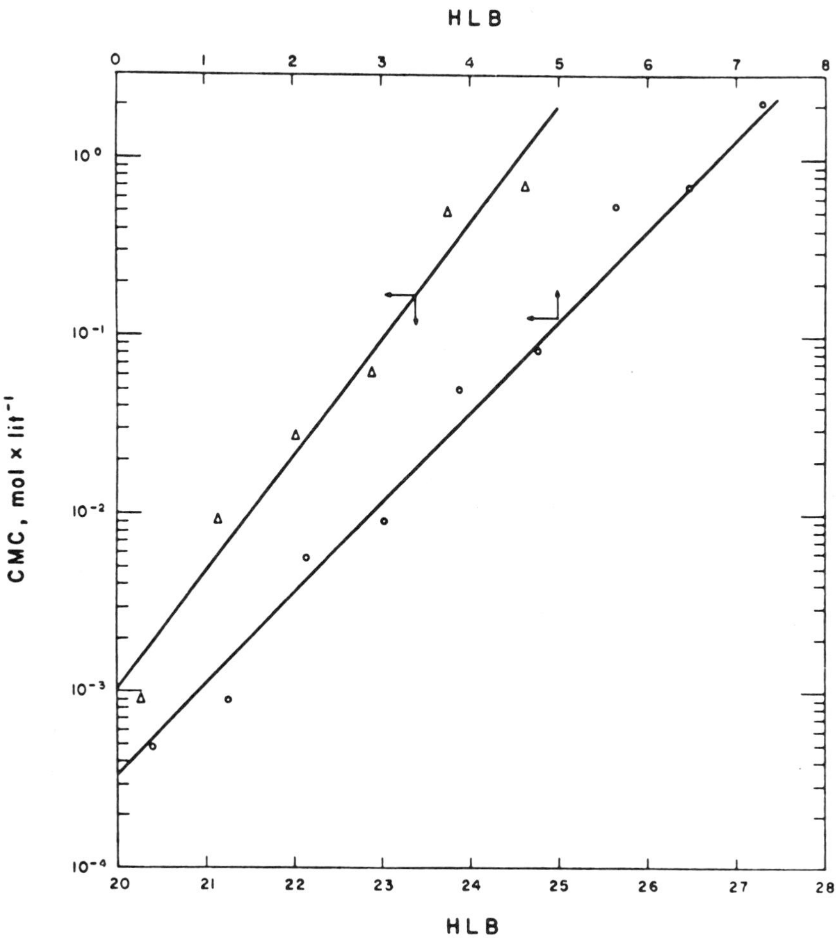

FIG. 6.21 Log cmc of fluorinated surfactants plotted against the corresponding HLB values (●) $C_nF_{2n+1}COOH$; (△) $C_nF_{2n+1}COOK$. (From Ref. 135. Reproduced by permission of American Chemical Society.)

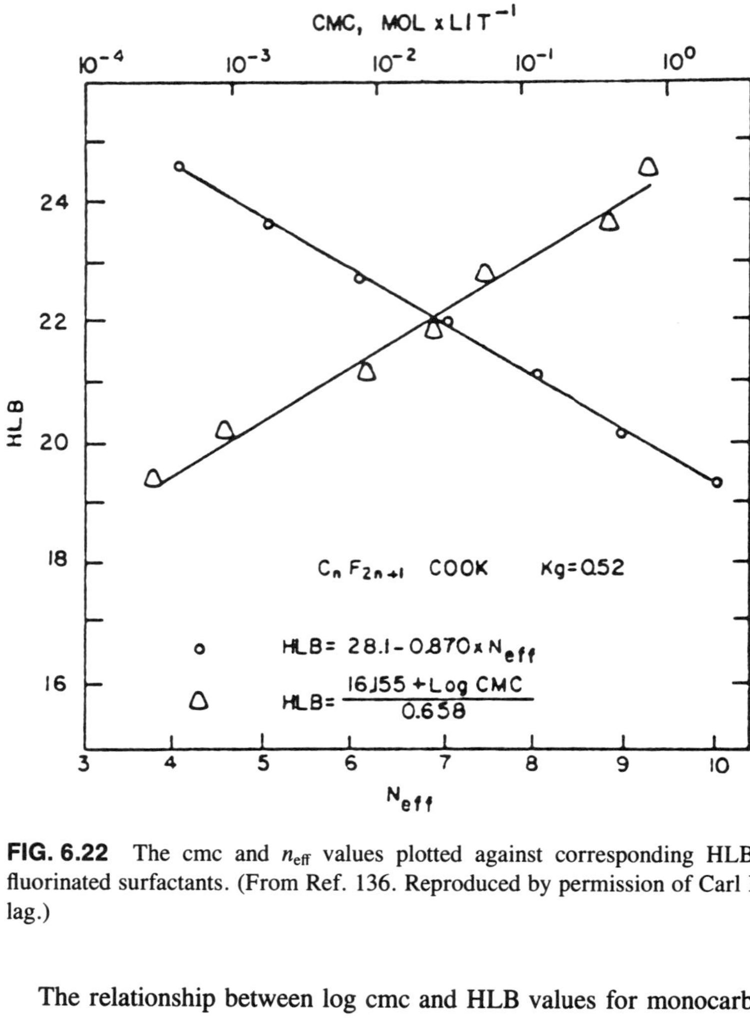

FIG. 6.22 The cmc and n_{eff} values plotted against corresponding HLB values for fluorinated surfactants. (From Ref. 136. Reproduced by permission of Carl Hanser Verlag.)

The relationship between log cmc and HLB values for monocarboxylic and dicarboxylic perfluoropolyether surfactants with a poly(oxyhexafluoropropylene) hydrophobe is also linear [145].

The general relationship between HLB and cmc values is given by [163]

$$\log \text{cmc} = a + b(\text{HLB}) = a + \frac{(\text{HLB})}{1 + K_g} \tag{42}$$

where a and b are experimental constants and K_g is the ratio of counterions to surfactant ions in the micelle.

Lin and Somasundaran [163] related HLB values to the free energy change related to the transfer of hydrophobic chains from an aqueous environment into the micelle

$$\text{HLB} - 7 = \Sigma(\text{hydrophilic group numbers}) - \frac{n\varphi'}{2.303kT} \tag{43}$$

The HLB group number for a CF_2 group was found to be 0.87 for long-chain, unbranched polar molecules with terminal CF_2 or CF_3 groups [135]. The value for the $-CF_2-$ group is higher than the group number of 0.475 for the $-CH_2-$ group (Table 6.12) [136,164]. The difference between the two group numbers is in accord with the higher hydrophobicity of the fluorinated alkane chain.

Introducing the group number for the $-CF_2-$ group, HLB becomes a linear function of the perfluorinated alkane chain length:

$$\text{HLB} - 7 = \Sigma(\text{hydrophilic group numbers}) - 0.87n_{\text{eff}} \tag{44}$$

Lin [135] concluded that the relationship between cmc and HLB for fluorinated surfactants is analogous to that of hydrocarbon surfactants. The empirical HLB values relate to the free energy change associated with the transfer of the hydrophobe from the aqueous environment into the micelle.

Partially Fluorinated Surfactants

Partially fluorinated surfactants exhibit at least two apparent anomalies [1]:

1. Their cmc values are higher than expected from the effect perfluorination usually has on cmc. Muller at al. [128,166–170] found that perfluorination of the terminal CH_3 group of a surfactant's alkyl chain approximately doubles the cmc instead of lowering it.
2. The ω-trifluoromethyl group appears to be in a more aqueous environment than expected for its location in the micellar core.

TABLE 6.12 HLB Group Numbers for Hydrophilic and Hydrophobic Groups

Hydrophilic group	Group number	Lipophilic group	Group number
$-SO_4Na$	+38.7	$=CH-$	−0.475
$-COOK$	+21.1	$-CH_2-$	−0.475
$-COONa$	+19.1	$-CH-$	−0.475
$-SO_3Na$	+11.0	$-CH_3$	−0.475
NR_3 (tertiary amine)	+9.4	$-CF_2$	−0.870
Ester	+2.4	$-CF_3$	−0.870
$-COOH$	+2.1	Phenyl	−1.662
$-OH$	+1.9		
$-O-$	+0.5		

Source: Refs. 136 and 164.

Both facts are in accord with the mutual phobicity of hydrocarbons and fluorocarbons, evidenced by the limited mutual solubility and the nonideality of the solutions [1]. This mutual phobicity reduces micellization of partially fluorinated surfactants and explains the aqueous environment of the terminal ω-trifluoromethyl group.

If partially fluorinated surfactants were to behave ideally, fluorination of the hydrocarbon chain would decrease the cmc value of the surfactant linearly with increasing number of hydrogens replaced by fluorine. Mukerjee and Mysels [1] found instead that substitution of a fluorine in the ω-trifluoro group by a single hydrogen increases the cmc value by a factor of 3 or 5, bringing it closer to the cmc of the hydrocarbon surfactant. Replacement of three hydrogens in the terminal $-CH_3$ group with fluorine increases cmc, instead of lowering it [1]. The anomalous cmc values suggest that the ω-trifluoromethyl group has a "much more aqueous environment" than expected for the core average. The mutual phobicity of the fluorocarbon and hydrocarbon sections in the surfactant molecule counteract aggregation driven by the hydrophobicity of the hydrocarbon chain and the still stronger hydrophobicity of the fluorocarbon chain [1].

Anomalies have also been observed with branched-chain anionic fluorinated surfactants. Kimura et al. [171] investigated micellization of sulfopropylated *N*-alkylperfluorooctanamides, $C_7F_{15}CON(R)CH_2CH_2CH_2SO_3Na$, where R is H or an alkyl group. The cmc values of the compounds depend on the chain length of the alkyl group. A short (<6 carbons) or a long (12 carbons) alkyl substituent increases the cmc of the surfactant. Alkyl groups of a length in the range of 6 to 10 carbon atoms give the lowest cmc values (Table 6.13). The cmc minimum results from phobic interactions. The mutual phobicity between hydrocarbon and fluorocarbon groups in the surfactant molecule reduces micellization of the partially fluorinated surfactant.

TABLE 6.13 Critical Micelle Concentrations (cmc) and Krafft Points for Fluorinated Surfactants $C_7F_{15}CON(R)CH_2CH_2CH_2SO_3Na$

R	cmc (mmol/L)	Krafft point (°C) 1%
H	—	45
CH_3	4.03	10
C_2H_5	3.59	<0
C_3H_7	2.84	<0
C_4H_9	2.28	<0
C_6H_{13}	1.75	>100
C_8H_{17}	1.15	>100
$C_{10}H_{21}$	2.44	>100
$C_{12}H_{25}$	3.27	>100

Source: Ref. 171. Reproduced by permission of American Oil Chemists Society.

Surfactants consisting of an oligo(oxyhexafluoropropylene) hydrophobe and an aryl group exhibited an irregular cmc dependence on the hydrophobe chain length. Ishikawa and Sasabe [172] prepared oil-soluble surfactants $(HFPO)_n$-Ar, where Ar is an aryl group and $(HFPO)_n$ is an oligo(hexafluoropropylene oxide) group, $n = 2$–5. The logarithms of cmc values estimated for surfactants, $(HFPO)_n$-phenyl, from the bending point of surface tension curves in *m*-xylene decreased linearly with n increasing from 3 to 6. However, the cmc value for the surfactant with $n = 2$ was much higher than predicted by the linear relationship (Fig. 6.23). The cmc values of the sulfonated surfactants, $(HFPO)_n$-aryl′-SO_3Na, in water exhibited an irregular behavior with deep minima.

In contrast, fluorinated surfactants with a predominantly fluorinated alkyl chain obey the cmc–chain length relationships expressed by Eqs. 34 and 37. Selve at al. [144] studied the surface activity of fluorinated nonionic surfactants with a two-chain polyoxyethylene hydrophilic head linked to the hydrophobe via an amide bond, $F(CF_2)_l(CH_2)_mC(O)N[(C_2H_4O)_nCH_3]_2$. The cmc values of surfactants with a structure $m = 1, n = 3$, and $l = 6$, 8, and 10, are 0.55, 0.012, and 0.0003 mmol/L, respectively. The ratio of the cmc values for the surfactants with $l = 6$ and 8 is 46, which corresponds to a 1.7-fold cmc decrease per C_2F_4 unit. As expected, the cmc values of the fluorinated surfactant with six CF_2 groups are very similar to the cmc values of a hydrocarbon surfactant with 10 CH_2 groups (Table 6.10).

Temperature

The variation of the cmc with temperature is shown in Fig. 6.24 [116], and Fig. 6.20 [118]. The cmc of most ionic surfactants exhibits a minimum as the temperature is varied from about 0 to 80°C. The cmc-temperature curves of nonionics do not usually exhibit minima, probably because their (fictitious) minima are at higher temperatures where the determination of cmc is limited by phase separation and the cloud point.

The observed temperature dependence of cmc is a result of two opposing factors. An increase in temperature reduces hydration of the hydrophile and consequently favors micelle formation. The decrease of cmc with increasing temperature (the left side of the curve in Fig. 6.24) is probably caused by a decrease of hydration of the monomer. However, an increase in temperature also lessens interactions between water molecules and reduces the hydrophobic effect-related free energy difference between a $-CF_2-$ in a monomeric and a micellar state. The cmc increase with temperature on the right side of the curve in Fig. 6.24 suggests that at higher temperatures the disruption of water structures around the hydrophobe of the surfactant becomes the dominant factor.

As an added complication, the aggregation number and the shape of micelles may vary with temperature. Finally, the experimental method

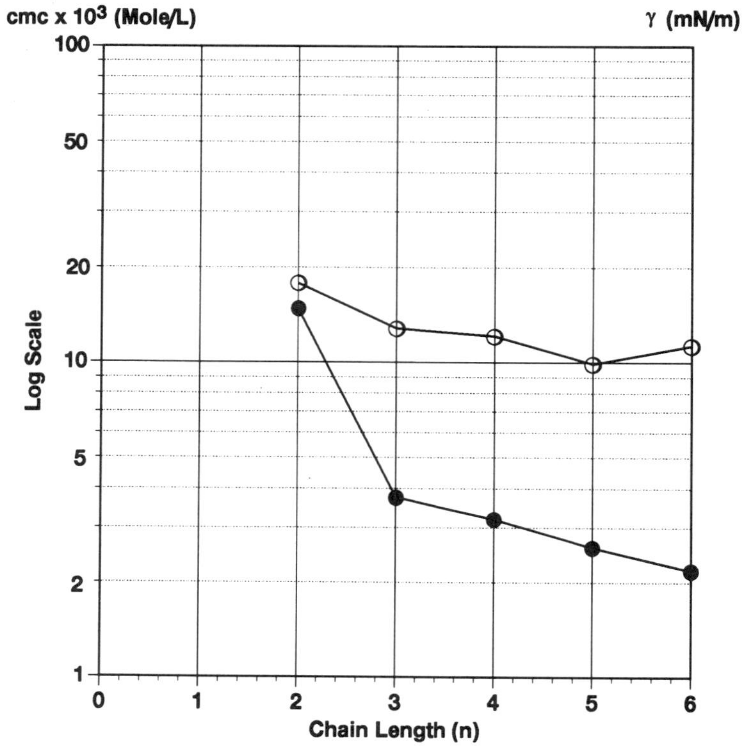

FIG. 6.23 Log cmc values plotted against the number, n, of hexafluoropropylene oxide groups ○ = surface tension, γ; ● = cmc. (Data from Ref. 172.)

employed to determine cmc values as a function of temperature may itself be affected by the changing temperature. For example, when cmc is determined by conductance measurements, a change in conductivity may be reflecting a change in fluidity of water [116].

Equation 20 implies a linear relationship between log cmc and $1/T$. This assumption has been used to calculate the enthalpy of micellization, ΔH_m, by a Clausius-Clapeyron type of equation (Eq. 21). Tadros [120] attempted to calculate enthalpy values for Monflor fluorinated surfactants but found the dependence of log cmc on $1/T$ to be nonlinear. The Clausius-Clapeyron equation holds only when the surfactant concentration is constant [127].

Undoubtedly, the temperature dependence of cmc is a complex relationship affected by several factors, some having an opposite effect.

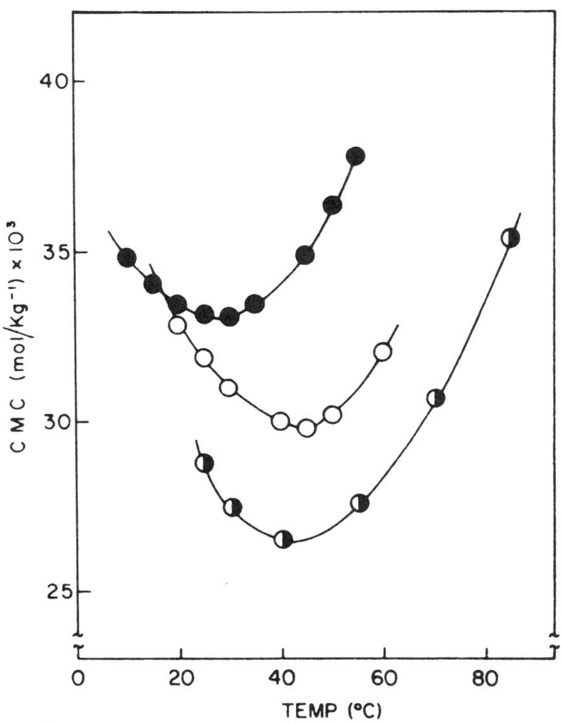

FIG. 6.24 Effect of temperature on cmc: sodium decyl sulfate (●), sodium perfluorooctanoate (O), potassium perfluorooctanoate (O). (From Ref. 116. Reproduced by permission of American Chemical Society.)

Pressure

The cmc values are pressure dependent. Sugihara and Mukerjee [126] measured electrical conductivities of sodium perfluorooctanoate solutions as a function of pressure. The cmc values were found to increase with increasing pressure from 1 atm to 2000 kg/cm^2 and then decrease slightly at higher pressures (Fig. 6.25) [126]. Below cmc, the differential conductivity, $d\kappa/dc$, where c is the molar concentration, decreases linearly with pressure, in qualitative accord with increasing viscosity and decreasing mobility of single ions (Fig. 6.26) [126]. Above cmc, the differential conductivity increases linearly with pressure. The increase of $d\kappa/dc$ with increasing pressure above cmc indicates that the degree of dissociation of micelles increases with increasing pressure.

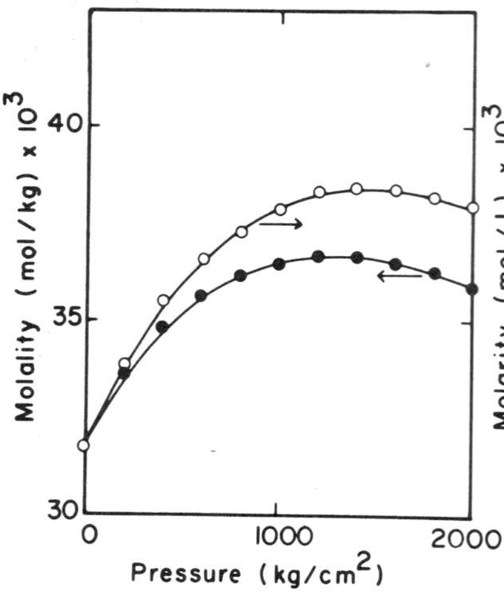

FIG. 6.25 Effect of pressure on the cmc of sodium perfluorooctanoate at 30°C, using molality and molarity scales.(From Ref. 126. Reproduced by permission of American Chemical Society.)

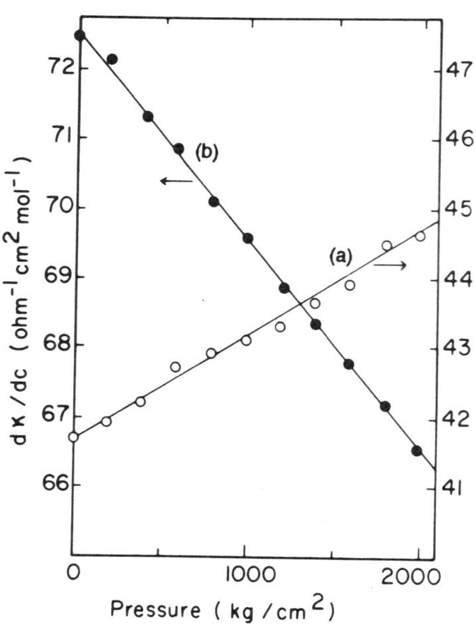

FIG. 6.26 Effect of pressure on $d\kappa/dc$ (a) above and (b) below cmc. (From Ref. 126. Reproduced by permission of American Chemical Society.)

The volume change related to micelle formation has been calculated using the charged phase separation model [173]:

$$\Delta V_{\mathrm{m}} = RT(1+\beta)\left[\frac{\delta \ln(cmc)}{\delta P}\right]T \tag{45}$$

where R is the molar gas constant, T is the absolute temperature, P is pressure, and β is the degree of counterion binding.

For the mass action model, the volume change $\Delta V_{\mathrm{m}}^{\mathrm{o}}$ is given by Eq. 46, assuming that 98% of the surfactant molecules are present as monomeric species at the experimentally determined cmc value [126].

$$\Delta V_{\mathrm{m}}^{0} = RT\left(1+\beta\frac{-1}{n}\right)\left[\frac{\delta \ln(\mathrm{cmc})}{\delta P}\right]_{T} + RT\left(\frac{\delta\beta}{\delta P}\right)_{T} \ln[0.98(\mathrm{cmc})] \tag{46}$$

When the aggregation number n is large, $1/n$ is small with respect to $1 + \beta$ and the first term of Eq. 46 is then approximately equal to the right-hand side of Eq. 45 derived for the charged phase separation model.

The estimated volume changes resulting from micellization are shown as a function of pressure in Fig. 6.27. At a low pressure the volume changes calculated using the charged phase separation model or the mass action model are similar, but at high pressures the effect of $(\delta\beta/\delta P)_T$ is more significant and the

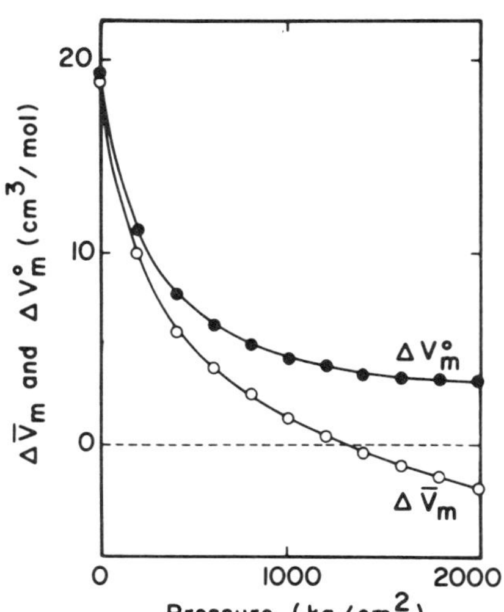

FIG. 6.27 Estimated volume changes on micelle formation: $V_{\mathrm{m}}^{\mathrm{o}}$, using the charged phase separation model, Eq. 45; V_{m}, using the mass action model, Eq. 46. (From Ref. 126. Reproduced by permission of American Chemical Society.)

curves diverge. Since the determination of β at high pressures is experimentally difficult, absolute values of $(\delta\beta/\delta P)_T$ may not be exact but the trend is probably real.

A number of factors affect the net volume change on micellization. It is therefore difficult to relate ΔV_m^0 to molecular interactions.

Ikawa et al. [127] determined the critical solution temperature (Krafft point) and the critical solution pressure (Tanaka pressure) of sodium perfluorodecanoate in water. A phase diagram of sodium perfluorodecanoate versus pressure at 55°C is shown in Fig. 6.28. The curves of solubility versus pressure (aQb) and of cmc versus pressure (dQe) intersect at point Q, representing the Tanaka pressure. The phase diagram is divided into three regions: solution of monomolecular species (S), the micellar solution (M), and the hydrated solid

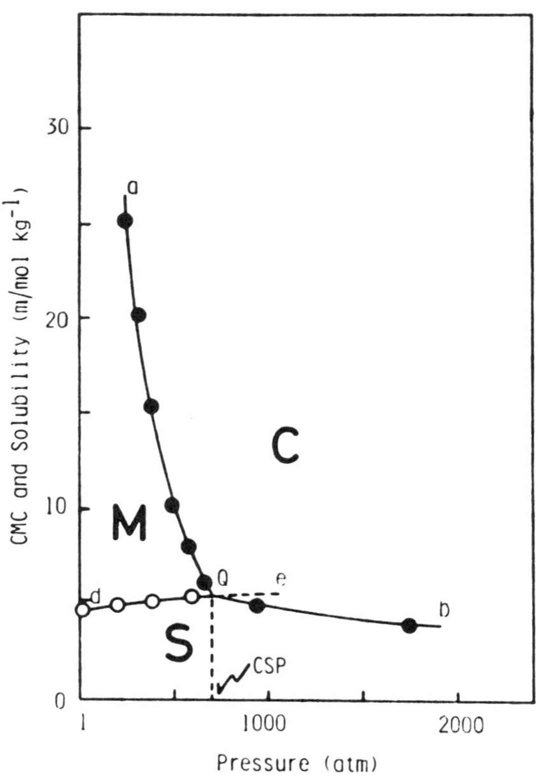

FIG. 6.28 The phase diagram of sodium perfluorodecanoate concentration vs. pressure at 55°C. M, S, and C denote the micellar, singly dispersed, and hydrated solid states, respectively; Q, a triple point; CSP, critical solution pressure. (From Ref. 127. Reproduced by permission of Plenum Publishing.)

(C). The rapid decrease of solubility with increasing pressure (curve aQ) was attributed to the transfer of surfactant from micelles to the hydrated solid phase, which is accompanied by a large decrease in partial molar volume.

The Krafft point increases with increasing Tanaka pressure (Fig. 6.29). Sodium perfluorodecanoate (SPFDe) has the highest Krafft point of the surfactants included in Fig. 6.28 at any pressure applied at a constant temperature such as 50°C. Hence the range where micelles can exist is narrower for sodium perfluorodecanoate than for the other surfactants shown.

Electrolytes and Additives

The critical micelle concentration is affected by the presence of electrolytes and other components of the surfactant solution [129,154,174]. Electrolytes depress the cmc of ionic surfactants and increase the micellar size by decreasing the thickness of the ionic cloud around the ionic groups and reducing the electrostatic repulsion between them. For example, the cmc of lithium perfluoro-

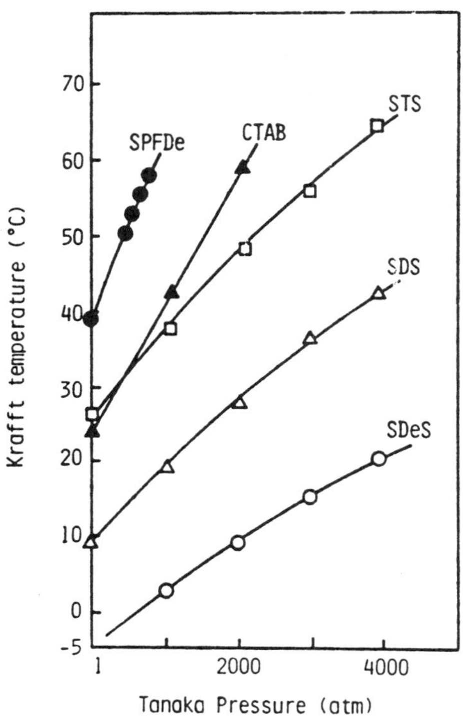

FIG. 6.29 Krafft temperature as a function of critical solution pressure for various surfactants. (From Ref. 127. Reproduced by permission of Plenum Publishing.)

octanoate in water at 25°C is 0.0341 M, but in 1.012 M LiCl the cmc value is 0.00277 M [155]. Partial fluorination of the hydrocarbon tail of a surfactant does not alter the effect of the electrolyte concentration on cmc and micelle size of a cationic surfactant (Table 6.14) [175].

The effect of NaCl on the cmc of $CF_3(CH_2)_8COONa$ is given by the equation [166]

$$\log \text{cmc} = -1.24 - 0.58 \log(\text{Na}^+) \tag{47}$$

The linear relationship on logarithmic scales is based on the law of mass action, applied to the formation of a single micelle which includes bound counterions [166]. The points for NaOH deviate from the straight line representing the points for NaCl (Fig. 6.30) [166], indicating that at higher electrolyte concentrations the nature of the ions has a significant effect on cmc.

The cmc of a fluorinated surfactant depends on the solvent composition in a manner similar to that of its hydrocarbon counterpart [167]. The effect of an organic additive is complex, including changes in the dielectric constant and consequently in electrostatic forces, stabilization of the monomer by modifying the water structure, and hydrophobic interactions between the monomer and the additive. Additives may affect the composition and the aggregation number of micelles. Fluorine NMR has shown that additives, such as urea, glycine, glycerol, acetamide, methanol, ethanol, acetone, dioxane, and tetrahydrofuran, vary in their ability to penetrate into the micelle [167]. Even for the least penetrating additives, such as urea or glycine, the effect on cmc is not solely related to the variation of the dielectric constant.

TABLE 6.14 Effect of Electrolyte on cmc Values and Aggregation Numbers of Cationic Surfactants DTAB and TDTAB

Surfactant	NaBr molality	Aggregation number	cmc × 10^2 (mol/kg) LS	DS
DTAB	0.10	83	0.46	
	0.50	94	0.16	0.23
	1.00	106	0.14	
	1.50	112	0.09	
TDTAB	0.10	74	0.90	
	0.50	90	0.40	0.51
	1.00	89	0.22	
	1.50	104	0.17	

[a] Abbreviations: DTAB, dodecyltrimethylammonium bromide; TDTAB, 12,12,12-trifluorododecyltrimethylammonium bromide; LS, light scattering; DS, dye solubilization.
Source: Ref. 175. Reproduced by permission of Academic Press.

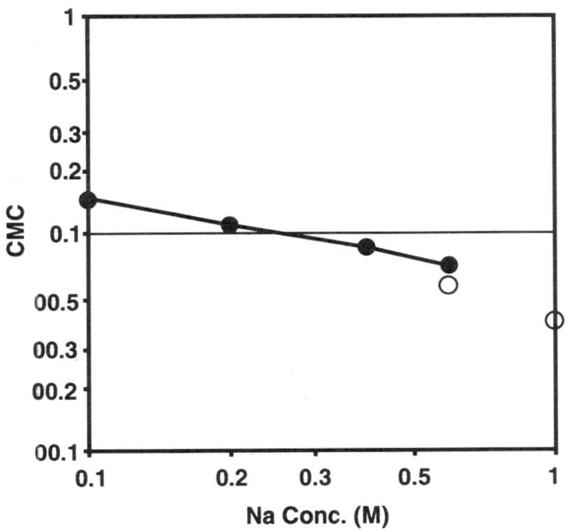

FIG. 6.30 The cmc values of $CF_3(CH_2)_8COONa$ vs. the total concentration of sodium ions at the cmc, on logarithmic scales. Filled circles obtained using NaCl; empty circles obtained using NaOH. (From Ref. 166. Reproduced by permission of American Chemical Society.)

6.7 SOLUBILIZATION

Micelles in surfactant solutions can incorporate in or upon themselves molecules of substances which are otherwise insoluble in the solvent, usually water. The solutions are optically isotropic, clear, and thermodynamically stable. McBain [79,176] termed this process *solubilization*. Solubilization mechanisms are complex and, in addition to sorption in or on micelles, other mechanisms may be operative. It is sometimes difficult to differentiate between solubilization as viewed by McBain and other processes causing a solubility increase, such as hydrotropy and comicellization. Attwood and Florence [174] have defined solubilization more broadly as "the preparation of a thermodynamically stable isotropic solution of a substance normally insoluble or very slightly soluble in a given solvent by the introduction of an additional amphiphilic component or components." The definition includes dilute or concentrated solutions, hydrotropy, and comicellization. The definition does not assume a similar solubilization mechanism, as long as the resulting solution is isotropic and thermodynamically stable. A definition given by Myers [153] is even broader: the preparation of a thermodynamically stable, isotropic solution of a substance normally insoluble or very slightly soluble in a given solvent by the addition of one or more amphiphilic compounds at or above their critical micelle concentration. This definition includes systems such as micellar emulsions and microemulsions.

A molecule solubilized by a nonionic surfactant may be located in the core of the micelle, at the core–polyoxyethylene chain interface, in the polyoxyethylene layer, or on the surface of the micelle [93,153] (Fig. 6.31).

A molecule solubilized by an ionic surfactant may be in the core of the micelle, between the hydrophobic chains, in the layer formed by the counterions, or on the surface of the micelle. Since the micelle is in a dynamic state, the loci of the solubilizate in a micelle cannot be sharply defined. Like surfactant molecules or ions in the micelle, molecules of the solubilizate are mobile and a rapid interchange between different sites may be possible. It is believed, however, that the preferred location of the solubilizate in a given micelle depends on the structure of the solubilizate as well.

The amount of a solubilizate which can be solubilized depends on several factors. The dominant variables are the structures of the surfactant and the

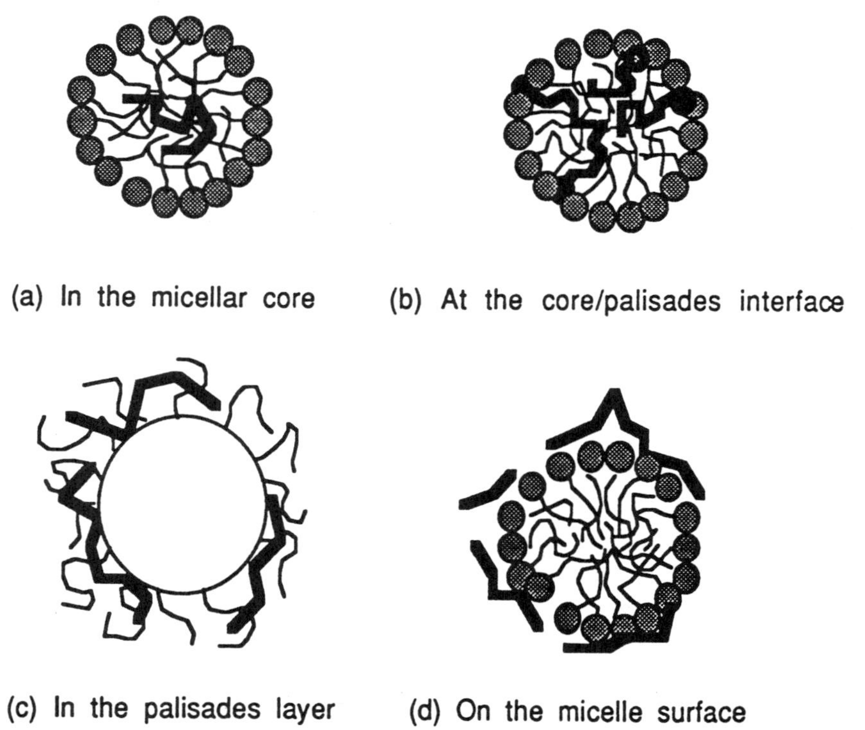

FIG. 6.31 Solubilization: (a) in the micellar core; (b) at the core-palisades interface; (c) in the palisades layer; (d) on the micelle surface. (From Ref. 125. Reproduced by permission of VCH Publishers.)

solubilizate. Both the structure of the hydrophobic chain and the nature of the counterion can affect solubilization. Although the relation between solubilization and surfactant structure is complex, it is clear that the interactions between a solubilizate molecule and the lipophobic hydrophobe of a fluorinated surfactant must be different from interactions between the solubilizate and the lipophilic hydrophobe of a hydrocarbon surfactant. Solubilization by fluorinated surfactants is therefore of great theoretical as well as practical interest. The published information on solubilization by fluorinated surfactants is, however, sparse.

Considering the nature of the fluorocarbon chain, the most likely compounds solubilizable by fluorinated surfactants must be nonpolar and have mainly dispersion forces as intermolecular forces. The mutual phobicity between hydrocarbons and fluorocarbons suggests fluorocarbons or chlorofluorocarbons as preferred solubilizates for fluorinated surfactants. Kunieda and Shinoda [91] solubilized CCl_2FCClF_2 in aqueous solutions of perfluoroalkanoates having different counterions and fluorocarbon chain lengths. CCl_2CClF_2, a nonpolar liquid, was chosen for the solubilization study because its relatively small molal volume was expected to increase the amount solubilized. Various amounts of CCl_2FClF_2 were added to a known volume of surfactant solution in an ampoule and the boundary between the two phases was measured. The authors assumed a linear dependence of solubilization on surfactant concentration and considered solubilization curves (Fig.6.32) [91] based on only a few experimental points valid. Solubilization depended on the nature of the counterion. Since the chain length and the nature of the counterion affect the hydrophile-lipophile balance (HLB) of the surfactant, the authors concluded that a well-balanced HLB may be essential for effective solubilization. Li perfluorodecanoate ($\gamma = 20.5$ mN/m) appeared to be too hydrophilic for solubilization, whereas the HLB of the ethanolammonium salt ($\gamma = 13.8$) seemed to be favorable.

Gerry and co-workers [175] found that terminal perfluorination of dodecyltrimethylammonium bromide reduces solubilization of Orange OT by the surfactant (Fig. 6.33). Since terminal fluorination did not reduce the aggregation number of the micelles substantially, a decrease in micelle size did not appear to be a cause of reduced solubilization. The mutual phobicity between terminal CF_3 groups and the hydrocarbon groups of the solubilizate, the dye molecule, was the more likely explanation.

Teddy and Wheeler [177] have described the three-component phase diagram for the ammonium perfluorooctanoate/octanol/water system. The phase structure was determined by optical microscopy. Octanol was found to be four times less soluble in ammonium perfluorooctanoate solutions than in sodium octanoate solutions. The low solubility of octanol was attributed to a low mutual solubility of fluorocarbon and hydrocarbon chains and not to

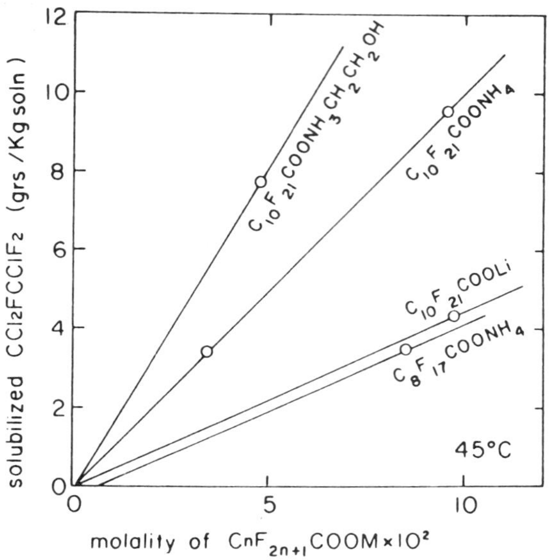

FIG. 6.32 The effect of the counterion of perfluoroalkanoates on the solubilization of CCl_2CClF_2 at 45°C. (From Ref. 91. Reproduced by permission of American Chemical Society.)

different counterions, ammonium versus sodium. However, the ammonium counterion may have contributed to the high solubility of ammonium perfluorooctanoate in octanol.

The pseudophase model (see Section 6.2) for micellar solutions makes it possible to establish a partitioning coefficient for the partitioning of the solubilizate between the aqueous solution and the micellar pseudophase. Treiner at al. [178–180] studied partitioning of alcohols and phenol in an aqueous solution of a fluorinated surfactant or of mixed anionic hydrocarbon and fluorocarbon surfactants. The fluorinated surfactants used in their studies were potassium or sodium perfluorooctanoate and lithium perfluorooctane sulfonate.

Nakayama et al. [87] suggested that partitioning coefficients between water and a micellar pseudophase can be calculated from the depression of Krafft points by nonelectrolyte additives. Kaneshina et al. [181] showed that partitioning coefficients obtained by the Krafft point method were in excellent agreement with those determined directly by gas chromatography. Kaneshina used a Clausius-Clapeyron type equation:

$$P = -[55.5 \times \Delta H_F \times (T - T_0)]/m_n R T_0^2 \tag{48}$$

where ΔH_F is the heat of fusion of the hydrated solid surfactant, T and T_0 are, respectively, the Krafft point in the presence of a nonelectrolyte additive and in

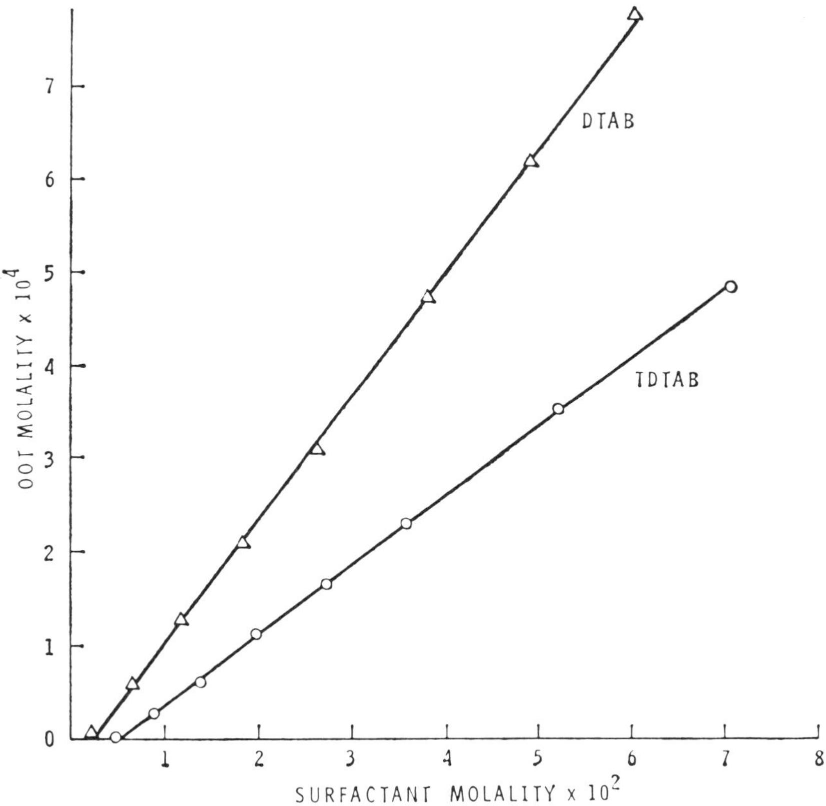

FIG. 6.33 Solubilization of Orange OT by solutions of 12,12,12-trifluorododecyltrimethylammonium bromide (TDTAB) and dodecyltrimethylammonium bromide (DTAB) in 0.5 M NaBr. (From Ref. 175. Reproduced by permission of Academic Press, Inc.)

the absence of any additive, and m_n is the molality of the additive. Treiner and Chattopadhyay [178] used the method of Kaneshina et al. [181] but determined Krafft points and cmc values from conductance data. The Krafft point (T_0 = 299.2 K) and the heat of fusion of the hydrated solid (ΔH_F = 9.6 kcal/mol) for potassium perfluorooctanoate are known. Since the heat of fusion of sodium perfluorooctanoate was not known, an indirect method [182] was used to obtain the partition coefficient P. In a dilute micellar solution

$$\log(\mathrm{cmc}_0/\mathrm{cmc}) = K_M m_N \qquad (49)$$

where cmc_0 and cmc are the critical micelle concentrations in the absence and presence of a solute at molality m_N. The micellization coefficient K_M is given by

$$K_M = \frac{1}{2}\left(k_S^N + \frac{55.5P}{2.303F}\right) \tag{50}$$

where k_S^N is the Setchenov constant [182] of the solute in the premicellar region and P is the partition coefficient. Both k_S^N and P are expressed on the mole fraction scale. The coefficient F expresses nonideal interactions between the solute and the micelle, in analogy to an activity coefficient.

Figure 6.34 [178] shows a plot of log (cmc_0/cmc) versus solute molality for sodium perfluorooctanoate. The semilogarithmic plot is linear, in accord with Eq. 48.

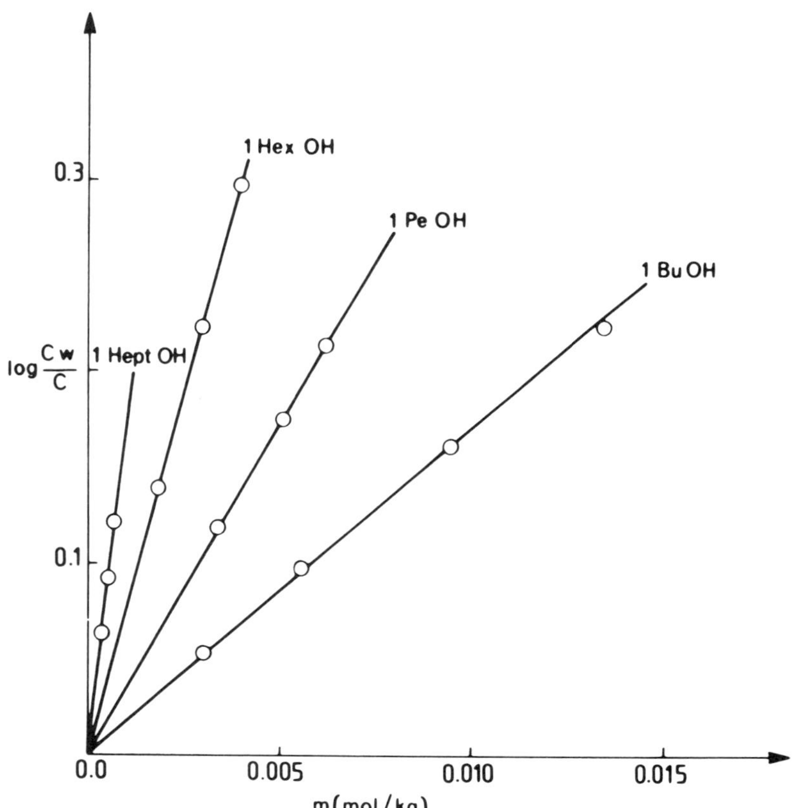

FIG. 6.34 The effect of nonelectrolyte concentration on the cmc of sodium perfluorooctanoate at 298.15 K. c_w is cmc_0, c is cmc in the presence of solute. (From Ref. 178. Reproduced by permission of Academic Press, Inc.)

The partition coefficients and cmc values, represented as the K_M coefficient, are given in Table 6.15.

The partition coefficients yield the standard free energy of solubilization ΔG_s^o:

$$\Delta G_s^o = -RT \ln P \tag{51}$$

For n-alkanols solubilized by potassium perfluorooctanoate

$$\Delta G_s^o = -1.848 - 0.438n \tag{52}$$

and for n-alkanols solubilized by sodium perfluorooctanoate

$$\Delta G_s^o = -2.081 - 0.434n \tag{53}$$

where n is the number of methylene groups. The slope, which is the same for both perfluorooctanoic acid salts, gives the contribution of methylene group to ΔG_s^o. The intercept for $n = 0$ represents the contribution of a methyl plus a hydroxyl group. The standard free energy of transfer of a methylene group to perfluorooctanoate micelles is about −440 cal/mol. The standard free energy of transfer of a methylene group to hydrocarbon-type surfactants is more negative: −600 cal/mol for sodium dodecyl sulfate micelles and −660 cal/mol for trimethyldodecylammonium bromide micelles. The difference, related to CH_2/CF_2 phobicity, is, however, counterbalanced by the strong hydroxyl-carboxylate interaction. The electron-withdrawing CF_2 groups increase the aci-

TABLE 6.15 Partition Coefficients P (Expressed as Mole Fractions) and Micellization Constants K_M for n-Alkanols in Sodium or Potassium Perfluorooctanoate Solutions at 25.0°C

Alkanol	P	K_M	F^a
In potassium perfluorooctanoate solutions			
1-Butanol	195 ± 10	1.02	0.72
1-Pentanol	480 ± 30	2.24	0.72
1-Hexanol	890 ± 50	5.02	0.70
1-Heptanol	1860 ± 50	10.6	0.69
In sodium perfluorooctanoate solutions			
1-Butanol	320	1.72 ± 0.06	
1-Pentanol	620	3.43 ± 0.10	
1-Hexanol	1390	7.63 ± 0.20	
1-Heptanol	2900	16.1 ± 0.3	

[a] F is equivalent to an activity coefficient. It expresses nonideal interactions between the solute and the micelle [178].
Source: Ref. 178.

dity of the carboxylic acid and consequently the ion-dipole interaction of the carboxylate head group with the hydroxyl group of the alkanol.

Carlsfors and Stilbs [183] investigated solubilization of a series of alkanols, benzyl alcohol, and benzene in deuterium oxide containing sodium perfluorooctanoate or a mixture of sodium perfluorooctanoate and sodium decanoate. The partition coefficients were obtained from multicomponent self-diffusion coefficients [184], determined by the Fourier transform NMR pulsed-gradient spin-echo (FT-PGSE) method [184,185]. The self-diffusion of fluorinated components was monitored by fluorine NMR, the self-diffusion of hydrocarbon components by proton NMR. The observed diffusion coefficient, D^{obs}, is a time average between free and solubilized molecules:

$$D^{obs} = pD^{bound} + (1 - p)D^{free} \tag{54}$$

The quantity p indicates the amount of a compound solubilized.

The partition coefficient

$$P = X_A/X_B \tag{55}$$

where X_A is the mole fraction of the solubilizate in the micellar pseudophase and X_B is the mole fraction of the solubilizate in the aqueous pseudophase, was calculated from the p values and known solubilizate and surfactant concentrations.

The fraction solubilized, by sodium perfluorooctanoate and the partition coefficient, increase with increasing chain length of the alcohol (Table 6.16) [183]. The increase in solubilization was attributed to the increase in hydrophobicity of the alcohols with increasing chain length. The partition ratio of benzene was found to approximate that of butanol. This unexpected result suggested that the solubilization mechanism, and possibly the site of benzene in the micelle, is different from that of alcohols [183].

The partition coefficients in the sodium perfluorooctanoate–sodium decanoate system (Table 6.16) are similar to those of the pure perfluorooctanoate system. For benzene, however, the partition coefficient was higher in the mixed system, suggesting again a different solubilization mechanism.

The differences in thermodynamic parameters of solubilization between fluorinated surfactants and their hydrocarbon counterparts have been reviewed by Fisicaro et al. [115] and the large discrepancies between partition coefficient values obtained by different techniques pointed out (compare the P values for alkanols in sodium perfluorooctanoate given in Tables 6.15 and 6.16). The limited reproducibility appeared to be related to the nature of the micellar systems studied: the equilibrium between different kinds of mixed micelles can easily be perturbed by the procedure used for the determination of partition coefficients and the equilibrium is slowly restored.

Bongiovanni et al. [186] solubilized perfluorocarbons in aqueous solutions of commercial grade perfluorooctanoic acid (about 80% C_8) and its sodium

TABLE 6.16 Partitioning in Micellar Solutions of Sodium Perfluorooctanoate (0.101 M) and a Mixed Micellar Solution of Sodium Perfluorooctanoate (0.0997 M) and Sodium Decanoate (0.0404 M)

	Partitioning coefficient (ratio of mole fractions)	
Solubilizate (mmol/kg)	In sodium perfluorooctanoate	In sodium perfluorooctanoate + sodium decanoate
Propanol (21)	80 ± 20	81 ± 6
Butanol (24)	280 ± 30	280 ± 10
Pentanol (32)	480 ± 20	430 ± 20
Hexanol (27, 36)	1000 ± 100	810 ± 40
Heptanol (25)	4300 ± 500	3200 ± 300
Benzene (13)	230 ± 10	340 ± 10
Benzyl alcohol (13)	280 ± 20	190 ± 50

Source: Ref. 183.

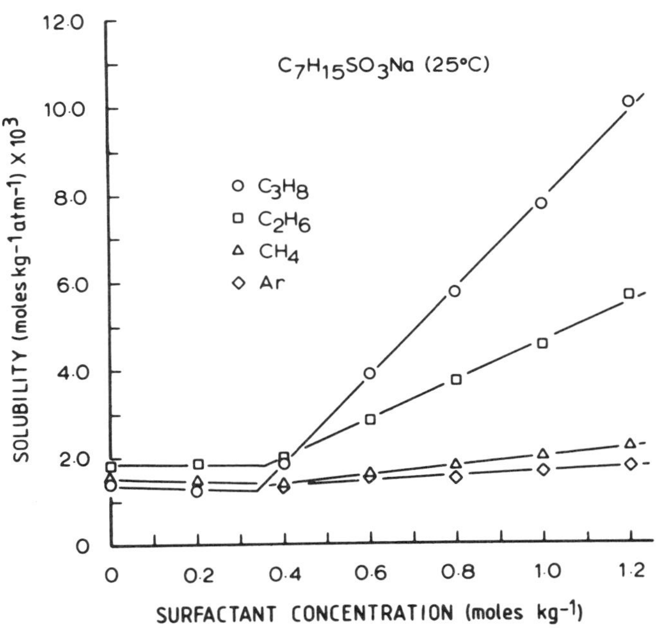

FIG. 6.35 Solubility of gas (moles of gas absorbed in 1000 g of water containing sodium 1-heptanesulfonate) as a function of surfactant concentration at 25°C. (From Ref. 190. Reproduced by permission of Academic Press, Inc.)

salt. The observed increase in solubility of perfluorooctanoic acid when neutralized with sodium hydroxide was related to lowering of the Krafft point. W/O and O/W isotropic monophasic liquid systems were obtained by changing the pH of the system.

Micelles in aqueous surfactant solutions provide nonpolar submicroscopic regions for sorption of gases [187–189]. Solubilization of a gas in aqueous surfactant solutions depends on several variables, including the nature of gas, the presence of amphiphilic substances, size of the hydrophobic group, nature of the counterion of ionic surfactants, perfluorination of the hydrocarbon chain, salinity, temperature, pressure, and pH of an ionic surfactant solution.

Prapaitrakul and King [190,191] studied solubilization of gases in aqueous solutions of sodium 1-heptanesulfonate, lithium perfluorooctanoate, sodium

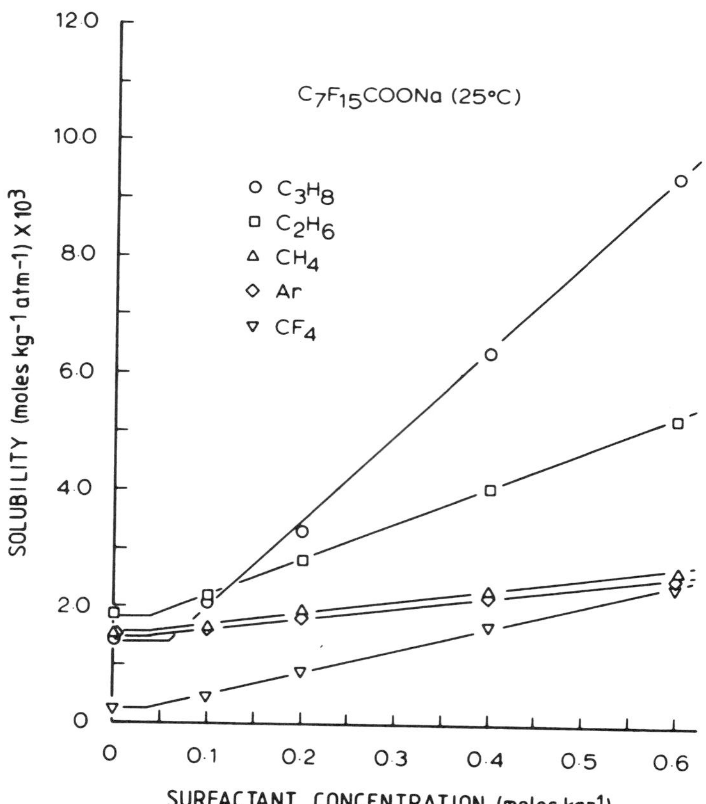

FIG. 6.36 Solubility of gas (moles of gas absorbed in 1000 g of water containing sodium perfluorooctanoate) as a function of surfactant concentration at 25°C. (From Ref. 190. Reproduced by permission of Academic Press, Inc.)

perfluorooctanoate, and lithium perfluorodecanoate. The solubility of each gas (oxygen, argon, methane, ethane, propane, and carbon tetrachloride) in lithium perfluorooctanoate and lithium perfluorodecanoate obeys Henry's law at all surfactant concentrations. The solubility of each gas in aqueous solutions of sodium 1-heptanesulfonate remained constant until the surfactant concentration reached cmc (Fig 6.35) [190]. Above the cmc, the solubility of gas increases linearly with increasing surfactant concentration, evidencing micellar solubilization (Fig. 6.36 [190] and Fig. 6.37 [191]). Substituting sodium for lithium did not affect solubilization (Fig 6.38) [191]. Lithium perfluorodecanoate is a more effective solubilizing agent than lithium or sodium perfluorooctanoate, in accord with evidence that a longer fluorocarbon chain increases solubilization.

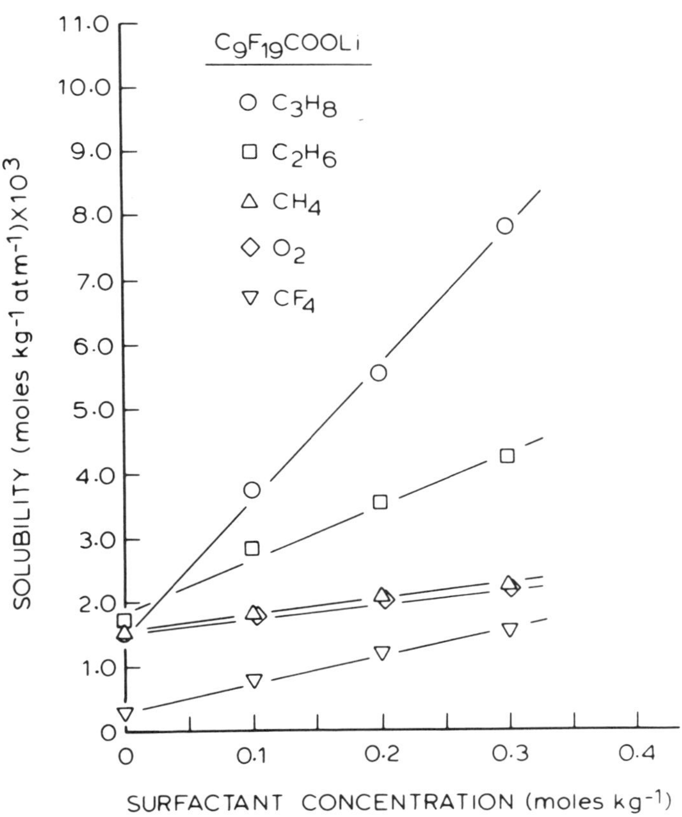

FIG. 6.37 Solubility of gas (moles of gas absorbed in 1000 g of water containing lithium perfluorodecanoate) as a function of surfactant concentration at 25°C. (From Ref. 191. Reproduced by permission of Academic Press, Inc.)

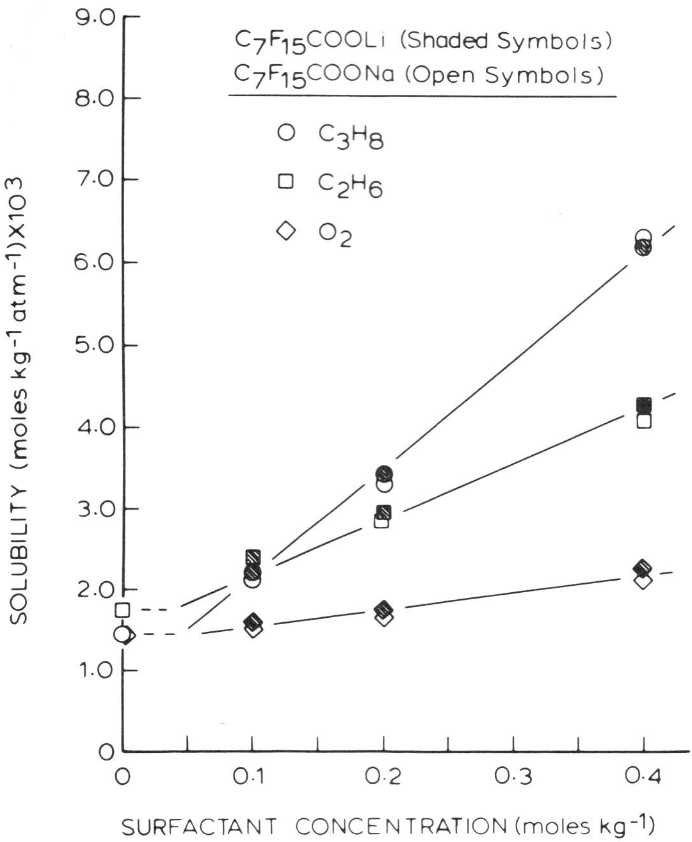

FIG. 6.38 Solubility of gas, expressed as moles of gas absorbed in 1000 g of water containing lithium perfluorooctanoate (shaded symbols) or sodium perfluorooctanoate (open symbols) and plotted as a function of surfactant concentration at 25°C. (From Ref. 191. Reproduced by permission of Academic Press, Inc.)

The constant-pressure (one-atmosphere) solubilities of gases increase linearly when plotted logarithmically as a function of gas critical temperature. The solubilities of CO_2 and NO_2 in sodium perfluorooctanoate solutions are higher than predicted by this linear relationship obtained for other gases [192]. The increase in solubility of CO_2 and NO_2, also observed for hydrocarbon-type surfactants, suggests highly specific adsorption phenomena at the surface regions of micelles.

Parallel trends in solubility data for fluorocarbons and perfluorinated surfactants suggest that the micellar interior, where solubilization of the gas occurs, has solvent properties similar to those of a bulk perfluorocarbon solvent, such as perfluoroheptane [190]. The solubilities of gases in surfactant solutions are lower than the corresponding solubilities in perfluoroheptane but the diminution in micellar solubility is a function of micelle size alone. The discrepancy increases with decreasing length of the fluorocarbon chain, measured in terms of carbon number. The Laplace pressure model used to explain solubilization of gases in hydrocarbon surfactants [122,193] also applies to solubilization of gases in fluorinated surfactants [190,191].

Solubilization of dyes and luminescence probes (see Chapters 7 and 9) by surfactants has an important role in the determination of micellar structure. Solubilization of a dye in mixed fluorinated and nonfluorinated surfactant systems is discussed in Sec. 7.2.

REFERENCES

1. P. Mukerjee and K. J. Mysels, Pap. Symp. 1974 ACS Symp. Ser. 9, 239 (1975).
2. J. H. Hildebrand, J. M. Prausnitz, and R. L. Scott , "Regular and Related Solutions," p. 204, Van Nostrand Reinhold, New York (1970).
3. F. Linke, "Solubilities," pp. 1135–1137, Van Nostrand, Princeton, NJ (1958).
4. E. A. Kauck and A. R. Diesslin, Ind. Eng. Chem. 43, 2332 (1951).
5. J. O. Hendrichs, Ind. Eng. Chem. 45, 99 (1953).
6. T. Gramstad and R. N. Haszeldine, J. Chem. Soc. 2640 (1957).
7. H. G. Klein, J. N. Meussdoerffer, H. Niederprün, and M. Wechsberg, Tenside 15, 2 (1978).
8. K. Shinoda and J. H. Hildebrand, J. Phys. Chem. 62, 481 (1958).
9. K. Shinoda and J. H. Hildebrand, J. Phys. Chem. 65, 1885 (1961).
10. Zonyl Fluorosurfactants, Du Pont Performance Products, Du Pont, Wilmington, DE (1987).
11. IUPAC Manual of Symbols and Terminology, Pure Appl. Chem. 31, 577 (1972).
12. G. S. Hartley, in "Micellization, Solubilization, and Microemulsions," Vol. 1, p. 23, K. L. Mittal, ed., Plenum, New York (1977).
13. G. Tanford, "The Hydrophobic Effect: Formation of Micelles and Biological Membranes," p. 36, Wiley & Sons, New York (1973).
14. E. I. Franses, H. T. Davis, W. G. Miller, and L. E. Scriven, J. Phys. Chem. 84, 2413 (1980).
15. B. C. Bennion and E. M. Eyring, J. Colloid Interface Sci. 32, 286 (1970).
16. T. Yasunaga, K. Takeda, S. Harada, J. Colloid Interface Sci. 42, 457 (1973).
17. E. R. Jones and C. R. Bury, Philos. Mag. 4, 841 (1927).
18. J. Grindley and C. R. Bury, J. Chem. Soc. 679 (1929).
19. D. G. Davies and C. R. Bury, J. Chem. Soc. 2263 (1930).
20. R. C. Murray and G. S. Hartley, Trans. Faraday Soc. 31, 183 (1935).
21. G. S. Hartley, "Aqueous Solutions of Paraffin Chain Salts," Hermann et Cie, Paris (1936).
22. J. N. Phillips, Trans. Faraday Soc. 51, 561 (1955).
23. H. Elworthy, and K. J. Mysels, J. Colloid Interface Sci. 21, 331 (1966).
24. P. Mukerjee, Adv. Colloid Interface Sci. 1, 241 (1967).

25. P. Mukerjee, K. J. Mysels, and P. Kapauan, J. Phys. Chem. 71, 4166 (1967).
26. P. Mukerjee, J. Phys. Chem. 76, 565 (1972).
27. K. S. Birdi, in "Colloidal Dispersions and Micellar Behavior," K. Mittal, ed., Plenum, New York (1975).
28. J. B. Rosenholm, T. E. Burchfield, and L. J. Hepler, J. Colloid Interface Sci. 78, 198 (1980).
29. Y. Moroi, R. Sugii, and R. Matuura, J. Colloid Interface Sci. 98, 184 (1984).
30. E. A. G. Aniansson and S. N. Wall, J. Phys. Chem. 78, 1024 (1974).
31. E. A. G. Aniansson and S. N. Wall, J. Phys. Chem. 79, 857 (1975).
32. E. A. G. Aniansson, S. N. Wall, J. Lang, C. Tondre, R. Zana, H. Hoffmann, I. Kielmann, and W. Ulbricht, J. Phys. Chem. 80, 905 (1976).
33. E. A. G. Aniansson, Ber. Bunsenges. Phys. Chem. 82, 981 (1978).
34. M. Kahlweit and M. Teubner, Adv. Colloid Interface Sci. 13, 1 (1980).
35. S. N. Wall and E.A.G. Aniansson, J. Phys. Chem. 84, 727 (1980).
36. E. A. G. Aniansson, in "Aggregation Processes in Solution," E. Wyn-Jones and J. Gormally, eds., Studies in Physical and Theoretical Chemistry, Vol. 26, Elsevier, Amsterdam (1983).
37. P. Mukerjee, in "Micellization, Solubilization, and Microemulsions," K. L. Mittal, ed., Vol. I, p. 171, Plenum, New York (1977).
38. E. Ruckenstein and R. Nagarajan, J. Phys. Chem. 85, 3010 (1981).
39. T. E. Burchfield and E. M. Woolley, J. Phys. Chem. 88, 2149 (1984).
40. E. M. Woolley and T. E. Burchfield, J. Phys. Chem. 88, 2155 (1984).
41. K. Shinoda, Bull. Chem. Soc. Jpn. 26, 101 (1953).
42. K. Shinoda and E. Hutchinson, J. Phys. Chem. 66, 577 (1962).
43. Y. Moroi, N. Nishikido, H. Uehara, and R. Matuura, J. Colloid Interface Sci. 50, 254 (1975).
44. G. Stainsby and A. E. Alexander, Trans. Faraday Soc. 46, 587 (1950).
45. E. Hutchinson, A. Inaba, and L. G. Bailey, Z. Phys. Chem. (Frankfurt) 5(NF), 344 (1955).
46. E. D. Goddard and G. C. Benson, Can. J. Chem. 35, 986 (1957).
47. E. Matijevic and B. A. Pethica, Trans. Faraday Soc. 54, 587 (1958).
48. E. Hutchinson, Z. Phys. Chem. (Frankfurt) 21?, 38 (1959).
49. K. J. Mysels, P. Mukerjee, and M. A.-H. Hamdiyyah, J. Phys. Chem. 67, 1943 (1963).
50. T. Sasaki, M. Hattori, J. Sasaki, and K. Nukina, Bull. Chem. Soc. Jpn. 48, 1397 (1975).
51. M. Koshinuma and T. Sasaki, Bull. Chem. Soc. Jpn. 48, 2755 (1975).
52. H. Elworthy and K.J. Mysels, J. Colloid Interface Sci. 21, 331 (1966).
53. T. L. Hill, "Thermodynamics of Small Systems," Vol. 1, W. A. Benjamin, New York (1964).
54. D. G. Hall and B. A. Pethica, in "Nonionic Surfactants," M. Schick, ed., Surfactant Science Series Vol. 1, p. 516, Marcel Dekker, New York (1967).
55. D. G. Hall, Trans. Faraday Soc. 3, 66, 1351 (1970).
56. D. G. Hall, Trans. Faraday Soc. 66, 1359 (1970).
57. J. N. Israelachvili, D. J. Mitchell, and B. W. Ninham, J. Chem. Soc. Faraday Trans. 2, 72, 1925 (1976).

58. H. B. Klevens and M. Raison, J. Chim. Phys. 51, 1 (1954).
59. K. Shinoda, M. Hato, and T. Hayashi, J. Phys. Chem. 76, 909 (1972)
60. P. J. Sams, E. Wyn-Jones, and J. Rassing, Chem. Phys. Lett. 13, 233 (1972)
61. J. E. Rassing, P. J. Sams, and E. Wyn-Jones, J. Chem. Soc. Faraday Trans. 2, 70(7), 1247 (1974).
62. P. J. Sams, J. E. Rassing, and E. Wyn-Jones, Adv. Mol. Relaxation Processes 6, 255 (1974).
63. E. Lessner, M. Teubner, and M. Kahlweit, J. Phys. Chem. 85, 1529 (1981).
64. M. Kahlweit, J. Colloid Interface Sci. 90, 92 (1982).
65. M. Kahlweit, Pure Appl. Chem. 53, 2069 (1981).
66. E. Graber, J. Lang, and R. Zana, Kolloid Z. 238, 470 (1970).
67. E. Graber, J. Lang, and R. Zana, Kolloid Z. 238, 479 (1970).
68. G. T. Tiddy, M. F. Walsh, and E. Wyn-Jones, J. Chem. Soc. Commun. 252 (1979).
69. J. Gettins, P. L. Jobling, M. F. Walsh, and E. Wyn-Jones, J. Chem. Soc. Faraday Trans. 2 76, 794 (1980).
70. G. Platz and H. Hoffmann, Ber. Bunsenges. Phys. Chem. 76, 491 (1972).
71. H. Hoffmann and W. Ulbricht, Z. Phys. Chem. Neue Folge 106, 167 (1977).
72. H. Hoffmann and B. Tagesson, Z. Phys. Chem. Neue Folge 110, 113 (1978).
73. H. Hoffmann, W. Ulbricht, and B. Tagesson, Z. Phys. Chem. Neue Folge 113, 17 (1978).
74. H. Hoffmann, B. Tagesson, and W. Ulbricht, Ber. Bunsenges. Phys. Chem. 83, 148 (1979).
75. H. Hoffmann, G. Platz, H. Rehage, K. Reizlein, and W. Ulbricht, Makromol. Chem. 182, 451 (1981).
76. H. Hoffmann, G. Platz, and W. Ulbricht, J. Phys. Chem. 85, 1418 (1981).
77. M. Eigen and L. de Maeyer, in "Technique of Organic Chemistry," Vol. VIII, p. 895, S. L. Friess, E. S. Lewis, and A. Weissberger, eds., Wiley-Interscience, New York (1963).
78. K. Fontell and B. Lindman, J. Phys. Chem. 87, 3289 (1983).
79. M. E. L. McBain and E. Hutchinson, "Solubilization and Related Phenomena," Academic Press, New York (1955).
80. C. La Mesa and B. Sesta, J. Phys. Chem. 91,1450 (1987).
81. K. Shinoda, Pure Appl. Chem. 52, 1195 (1980).
82. A. E. Alexander, Trans. Faraday Soc. 38, 54 (1942).
83. K. Shinoda and T. Soda, J. Phys. Chem. 67, 2072 (1963).
84. M. Hato and K. Shinoda, Bull. Chem. Soc. Jpn. 46, 3886 (1973).
85. M. Hato and K. Shinoda J. Phys. Chem. 77, 378 (1973).
86. K. Tsujii and J. Mino, J. Phys. Chem. 82, 1610 (1978).
87. H. Nakayama, K. Shinoda, and E. Hutchinson, J. Phys. Chem. 70, 3502 (1966).
88. C. La Mesa and L. Coppola, Colloids Surf. 35, 325 (1989).
89. R. F. Tuddenham and A. E. Alexander, J. Phys. Chem. 66, 1839 (1962).
90. S. Kaneshina, M. Tanaka, T. Tomida, and R. Matuura, J. Colloid Interface Sci. 44, 525 (1974).
91. H. Kunieda and K. Shinoda, J. Phys. Chem. 80, 2468 (1976).
92. C. La Mesa, Colloids Surf. 35, 329 (1989).

93. R. A. Mackay, in "Nonionic Surfactants. Physical Chemistry," M. J. Schick, ed., Marcel Dekker, New York (1987).
94. D. Attwood, J. Phys. Chem. 72, 339 (1968).
95. S. K. Han, S. M. Lee, M. Kim, and H. Schott, J. Colloid Interface Sci. 132, 444 (1989).
96. H. Schott and S. K. Han, J. Pharm. Sci. 65, 979 (1976).
97. T. Kuwamura, in "Structure/Performance Relationships in Surfactants," M. J. Rosen, eds., ACS Symp. Ser. No. 253, p. 27 (1984).
98. M. J. Schick, J. Phys. Chem. 67, 1796 (1963).
99. M. E. Ginn, F. B. Kinney, and J. C. Harris, J. Am. Oil Chem. Soc. 37, 183 (1960).
100. J. M. Corkill, J. F. Goodman, and S. P. Harrold, Trans. Faraday Soc. 60, 202 (1964).
101. T. Nakagawa, in "Nonionic Surfactants," M. J. Schick, ed, Surfactant Science Series, Vol. 1, Marcel Dekker, New York (1967).
102. K. Shinoda, T. Nakagawa, B. Tamamushi, and T. Isemura, "Colloidal Surfactants," Academic Press, New York (1963).
103. H. Schott, J. Pharm. Soc. 58, 1443 (1969); R. A. Mackay, in "Nonionic Surfactants. Physical Chemistry," M. J. Schick, ed. Surfactant Science Series, Vol. 23, Marcel Dekker, New York (1987).
104. G. Mathis, P. Leempoel, J. C. Ravey, C. Selve, and J. J. Delpuech, J. Am. Chem. Soc. 106, 6162 (1984).
105. A. Couper, in "Surfactants," Th. F. Tadros, ed., Academic Press, London (1983).
106. D. G. Hall, in "Nonionic Surfactants. Physical Chemistry," M. J. Schick, ed., p. 233–296, Surfactant Science Series, Vol. 23, Marcel Dekker, New York (1987).
107. M. J. Vold, J. Colloid Sci. 51, 561 (1955).
108. G. Stainsby and A. E. Alexander, Trans. Faraday Soc. 46, 587 (1950).
109. E. D. Goddard and G. C. Benson, Can. J. Chem. 62, 1083 (1958).
110. J. M. Corkill, J. F. Goodman, and R. H. Ottewill, Trans. Faraday Soc. 57, 1627 (1961).
111. B. D. Flockhart, J. Colloid Sci. 16, 484 (1961).
112. K. W. Herrmann, J. Phys. Chem. 66, 295 (1962).
113. J. M. Corkill, J. F. Goodman, S. P. Harrold, and J. R. Tate, Trans. Faraday Soc. 63, 240 (1967).
114. G. Sugihara and P. Mukerjee, J Phys. Chem. 85 1612 (1981).
115. E. Fisicaro, E. Pelizzetti, R. Bongiovanni, and E. Borgarello, Colloids Surf. 48, 259 (1990).
116. P. Mukerjee, K. Korematsu, M. Okawauchi, and G. Sugihara, J. Phys. Chem. 89, 5306 (1985).
117. K. Shinoda, M. Kobayashi, and N. Yamaguchi, J. Phys. Chem. 91, 5292 (1987).
118. K. Shinoda and K. Katsura, J. Phys. Chem. 68, 1568 (1964).
119. P. Mukerjee, J. Phys. Chem. 66, 1375 (1962).
120. Th. F. Tadros , J. Colloid Interface Sci. 74, 196 (1980).
121. I. Danielsson, J. B. Rosenholm, P. Stenius, S. Backlund, S. Prog. Colloid Polym. Sci 61, 1 (1976).

122. P. Mukerjee. Kolloid. Z. Z. Polym. 236, 76 (1970).
123. I. Johnson and G. Olofsson, J. Chem. Soc. Faraday Trans. 1 84, 551 (1988).
124. J. M. Corkill, J. F Goodman, and J. R. Tate, Trans. Faraday Soc. 69, 986 (1964).
125. C. Nemathy and H. A. Sheraga, J. Chem. Phys. 36, 3401 (1962).
126. G. Sugihara and P. Mukerjee, J. Phys. Chem. 85, 1612 (1981).
127. Y. Ikawa, S. Tsuru, Y. Murata, M. ˜kawauchi, M. Shigematsu, and G. Sugihara, J. Solution Chem. 17, 125 (1988).
128. N. Muller and R. H. Birkhahn, J. Phys. Chem. 71, 957 (1967).
129. P. Mukerjee and K. J. Mysels, "Critical Micelle Concentrations of Aqueous Surfactant Systems," National Standard Reference Data Service, Vol. 36, U.S. National Bureau of Standards, Washington, DC (1971).
130. H. B. Klevens, J. Phys. Colloid Chem. 52, 130 (1948).
131. M. L. Corrin, J. Colloid Sci. 3,333 (1948).
132. H. B. Klevens, J. Am. Oil Chem. Soc. 30, 74 (1953).
133. C. H. Arrington, Jr., and G. D. Patterson, J. Phys. Chem. 57, 247 (1953).
134. M. K. Bernett and W. A. Zisman, J. Phys. Chem. 63, 1911 (1959).
135. I. J. Lin, J. Phys. Chem. 76, 2019 (1972).
136. J. J. Lin and L Marszall, Tenside 15, 243 (1978).
137. H. M. Scholberg, R. A. Guenther, and R. I. Coon, J. Phys. Chem. 57, 923 (1953).
138. H. B. Klevens and J. Vergnoble, Proc. Int. Congr. Surface Activ., 2nd, 1, 395 (1957)
139. H. B. Klevens, Kolloid Z. Z. Polym. 53,158 (1959).
140. W. A. Zisman, in "Kirk-Othmer Encyclopedia of Chemical Technology," 2nd ed., pp. 707–738 (1963).
141. N. N. Kalibabchuck, L. K. Dyachek, and V. A. Semenh, Colloid J. USSR, 38, 138 (1976). CA 84, 17100.
142. A. Greiner and M. Herbst, Z. Chem. 16, 229 1976).
143. G. Mathis, J. C. Ravey, and M. Buzier, Microemulsions (Proc. Conf. Phys. Chem. Microemulsions, 1982) (1983).
144. C. Selve, J. C. Ravey, C. El Moudjahid, E. M. Moumni, and J. J. Delpuech, Tetrahedron 47, 411 (1991).
145. G. Caporiccio, F. Burzio, G. Carniselli, and V. Biancardi, J. Colloid Interface Sci. 98, 202 (1984).
146. P. Mukerjee and K. J. Mysels, in "Colloidal Dispersions and Micellar Behavior," K. L. Mittal, ed., p. 239, ACS Symp. Ser. 9 (1975).
147. J. C. Ravey, A. Gherbi, and M. J. Stebe, Prog. Colloid Polym. Sci. 259, 761 (1981).
148. J. N. Meussdoerffer and H. Niederprüm, Chem. Z. 104, 45 (1980).
149. H. Lange, in "Proceedings, 3rd International Congress of Surface Activity," Vol. I, p.279 (1960).
150. P. Becher, in "Nonionic Surfactants," M. J. Schick, ed., Marcel Dekker, New York (1966)
151. L. Matos, J.-C. Ravey, and G. Serratrice, J. Colloid Interface Sci. 128, 341 (1989).

152. A. Greiner, B. Krüger, and M. Herbst, Z. Chem. 18, 383 (1978).
153. D. Myers, "Surfactant Science and Technology," VCH Publishers, New York (1988).
154. M. J. Rosen, "Surfactants and Interfacial Phenomena," 2nd ed., John Wiley & Sons, New York (1989).
155. P. Mukerjee, M. J. Gumkowski, C. C. Chan, and R. Sharma, J. Phys. Chem. 94, 8832 (1990).
156. M. Angel and H. Hoffmann, Z. Phys. Chem. Neue Folge 139, 153 (1984).
157. K. Shinoda, J. Phys. Chem. 59, 432 (1955).
158. D. F. Evans and P. J. Whightman, J. Colloid Interface Sci. 86, 515 (1982).
159. I. J. Lin, J. P. Friend, and Y. Zimmels, J. Colloid Interface Sci. 45, 164 (1973).
160. I. J. Lin, B. M. Moudgil, and P. Somasundaran, Colloid Polym. Sci. 252, 407 (1974).
161. I. J. Lin, Israel J. Technol. 9, 621 (1971).
162. I. J. Lin, Trans. AIME 250, 225 (1971).
163. I. J. Lin and P. Somasundaran, J. Colloid Interface Sci. 37, 731 (1971).
164. J. T. Davies, "Proceedings, 2nd International Congress of Surface Activity," Vol. 1, p. 426, Butterworths, London (1957).
165. J. T. Davies and E. K. Rideal, "Interfacial Phenomena," 2nd ed. , Academic Press, New York (1963).
166. N. Muller and R. H. Birkhahn, J. Phys. Chem. 72, 583 (1968).
167. N. Muller and T. W. Johnson J. Phys. Chem. 73, 2042 (1969).
168. N. Muller and F. E. Platko, J. Phys. Chem. 75, 547 (1971).
169. N. Muller and H. Simsohn, J. Phys. Chem. 75, 942 (1971).
170. N. Muller, J. H. Pellerin, and W. W. Chen, J. Phys. Chem. 76, 3012 (1972).
171. C. Kimura, K. Kashiwaya, M. Kobayashi, and T. Nishiyama, J. Am. Oil Chem. Soc. 61, 105 (1984).
172. N. Ishikawa and M. Sasabe, J. Fluorine Chem. 25, 241 (1984).
173. S. Kaneshina, M. Tanaka, T. Tomida, and R. Matuura, J. Colloid Interface Sci. 48, 450 (1974).
174. D. Attwood and A. T. Florence, "Surfactant Systems: Their Chemistry, Pharmacy and Biology," Chapman & Hall, London (1983).
175. H. E. Gerry, P. T. Jacobs, and E. W. Anacker, J. Colloid Interface Sci. 62, 556 (1977).
176. J. W. McBain, "Colloid Science," D. C. Heath, Boston (1950).
177. G. J. T. Teddy and B.A. Wheeler, J. Colloid Interface Sci. 47, 59 (1974).
178. C. Treiner and A. K. Chattopadhyay, J. Colloid Interface Sci. 98, 447 (1984).
179. C. Treiner, J. F. Bocquet, and C. Pommier, J. Phys. Chem. 90, 3052 (1986).
180. C. Treiner, A. A. Khodja, and M. Fromon, Langmuir 3, 729 (1987).
181. S. Kaneshina, H. Kamaya, and I. Ueda, J. Colloid Interface Sci. 83, 589 (1981).
182. C. Treiner, J. Colloid Interface Sci. 90, 444 (1982).
183. J. Carlfors and P. Stilbs, J. Colloid Interface Sci. 103, 332 (1985).
184. P. Stilbs, J. Colloid Interface Sci. 87, 385 (1982).
185. E. L. Hahn, Phys. Rev. 80, 580 (1950).
186. R. Bongiovanni, E. Borgarello, F. M. Carlini, E. Fisicaro, and E. Pellizetti, Colloids Surf. 48, 277 (1990).

187. J. W. McBain and J. J. O'Conner, J. Am. Chem. Soc. 62, 2855 (1940).
188. J. W. McBain and J. J. O'Conner, J. Am. Chem. Soc. 63, 875 (1941).
189. J. W. McBain and A. M. Soldate, J. Am. Chem. Soc. 64, 1556 (1942).
190. W. Prapaitrakul and A. D. King, J. Colloid Interface Sci. 112, 387 (1986).
191. W. Prapaitrakul and A. D. King, J. Colloid Interface Sci. 118, 224 (1987).
192. D. W. Ownby, W. Prapaitrakul and A. D. King, J. Colloid Interface Sci. 125, 526 (1988).
193. I. B. C. Matheson and A. D. King, J. Colloid Interface Sci. 66, 464 (1978).
194. J. M. Corkill and J. F. Goodman, Adv. Colloid Interface Sci. 2, 297 (1969).

7
Structure of Micelles and Mesophases

7.1 STRUCTURE OF MICELLES

Micelles are in a dynamic state. Their characteristics, size and shape vary with the structure of the surfactant and the solution conditions, such as the concentration, ionic strength, temperature, pressure, and the nature of additives. The aggregation number of the micelles is not independent of the method by which it is measured. The experimentally determined size and shape of micelles are therefore not unequivocally defined parameters of a surfactant solution but descriptive characteristics of micelles.

The characteristics of micelles formed by hydrocarbon-type surfactants have been extensively studied. In dilute solutions, micelles are considered to be spherical or nearly spherical. The minimum space of the micelle interior has been calculated with the assumption that the radius of the interior region is approximately equal to the length of the fully extended hydrophobic chain and empty space in the micelle interior is absent. The latter assumption is in accord with the liquid like character of the micelle interior, supported by substantial evidence. However, micelles formed by hydrocarbon-type surfactants are larger than the space required for the hydrocarbon chain. Oblate or prolate micelle models have been proposed. At higher surfactant concentrations a surfactant may form rod-like, cylindrical, disk-shaped, or lamellar micelles [1] or vesicles, which are lamellar micelles arranged in a spherical shape with water in between the lamellas (see Fig. 6.1). (Mesophases and liquid crystals, see Chapter 7.4.)

Micellar solutions of hydrocarbon-type surfactants have been characterized by several techniques, including light scattering, tensiometry, electric conductivity, solubilization, nuclear magnetic resonance (NMR), fluorescence quenching, small-angle neutron scattering (SANS), electric birefringence, ultrafiltration, sedimentation rates in the ultracentrifuge, and x-ray scattering [2–6] (see Chapter 9). The size and shape of many hydrocarbon-type surfactants are known as a function of a variety of conditions. The aggregation numbers of hydrocarbon type micelles in water range from 25 to over 10,000 [6]. The size of micelles increases with increasing length of the hydrophobic chain but decreases with increasing size of the hydrophilic group. A neutral electrolyte decreases the aggregation number of anionic surfactants. The effect of electrolytes on nonionic surfactant solutions is more complex. The aggregation numbers of ionic surfactants decrease with increasing temperature but polyoxyethylated nonionic surfactants aggregate extensively at higher temperatures (cloud point, Sec. 6.3).

In contrast to the extensive information on hydrocarbon type surfactants, the knowledge of fluorinated surfactant micelles is sparse, probably because the fluorinated surfactant solutions are more difficult to investigate. Light scattering has been the most important tool for the study of hydrocarbon-type micelles. Unfortunately, the refractive indexes of fluorocarbons are close to that of water (see Chapter 3) and the determination of micellar shape and size of perfluorinated surfactants by conventional light-scattering techniques is difficult, if not impossible. Less sensitive or more complicated techniques have to be used for the study of perfluorinated surfactants. Micellar aggregation numbers of partially fluorinated surfactants have been determined by light scattering, however [7].

X-ray scattering techniques have been used extensively in the investigation of micellar solutions [8–14]. In solutions of hydrocarbon surfactants three diffraction bands have been attributed to the surfactant:

1. The "M" or micelle thickness band
2. The "S" or short spacing band
3. The "I" or long spacing band, dependent on surfactant concentration

The x-ray scattering pattern of fluorinated surfactants differs from that of hydrocarbon-type surfactants. M band spacings have been observed for potassium tetradecanoate but not for ammonium salts of C_7 and C_9 perfluoroalkanoic acids (Fig. 7.1) [15]. Arrington and Patterson [15] explained the lack of M spacing by a higher electron density of the fluorocarbon chain.

Luminescence and electronic energy transfer have been used frequently to determine aggregation numbers of hydrocarbon type surfactants [3]. In contrast, only a few luminescence studies have been made for fluorinated surfactants [16,17].

Turro and Lee [16] employed the luminescence method to determine the aggregation number for sodium perfluorooctanoate micelles. The aggregation

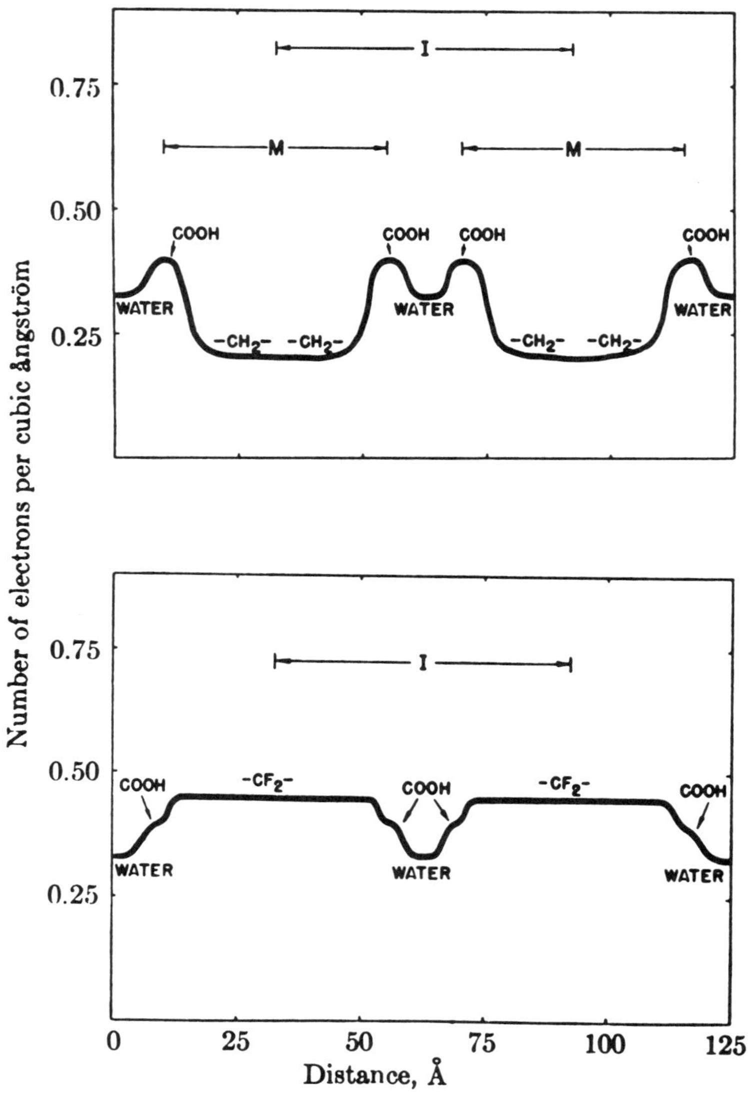

FIG. 7.1 Electron density: potassium tetradecanoate (upper figure);ammonium perfluoroalkanoate (lower figure). (From Ref. 15. Reproduced by permission of American Chemical Society.)

number calculated from quenching of $Ru(bpy)_3^{2+}$ phosphorescence was found to be seven. Although this numerical value may be inexact, the aggregation number of perfluorooctanoate micelles is undoubtedly much smaller than that of micelles of corresponding hydrocarbon surfactants. In spite of the small size, sodium perfluorooctanoate micelles exhibit characteristics of a conventional micelle, such as solubilization of hydrophobic substances and enhanced quenching by quenchers solubilized in micelles containing energy donors.

Muto et al. [17] determined the aggregation number, N, for Li perfluorooctanesulfonate by fluorescence decay of pyrene solubilized in micelles. The aggregation number was found to have a very low value at low concentrations, $N = 6$ at 2×10^{-2} M. The aggregation number increased rapidly with increasing concentration up to 0.1 M and slowly beyond this concentration (Fig. 7.2). The small micelle size is in accord with conductivity and kinetic data of Hoffmann and Tagesson [18]. The degree of ionization was found to be large at cmc, indicating small micelles. Muto et al. [17] explained the small micellar size by two antagonistic effects. First, the cmc values of fluorinated surfactants are lower than those of hydrocarbon surfactants with an equal chain length, because of the strong hydrophobicity of a fluorocarbon chain. Second, the rigidity of the fluorocarbon chain may prevent the formation of spherical micelles right at the cmc, in contrast to hydrocarbon surfactants.

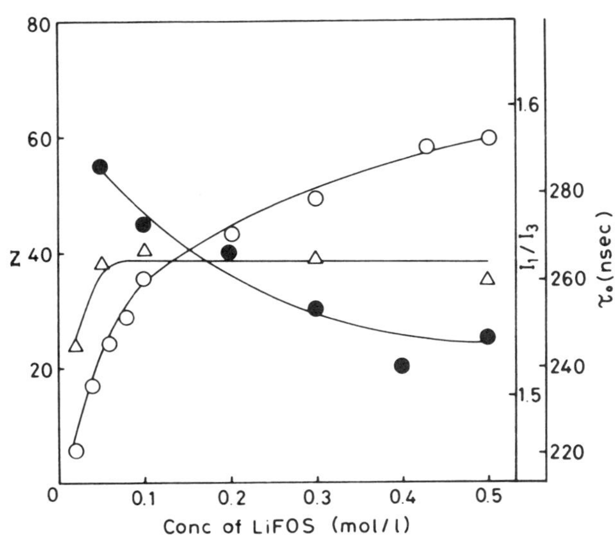

FIG. 7.2 Micelle aggregation number N (○), pyrene fluorescence lifetime τ_0 (△), and fluoresecnece intensity ratio I_1/I_3 (●) as a function of lithium perfluorooctanesulfonate concentration at 25°C. (From Ref. 17. Reproduced by permission of Academic Press, Inc.)

Chemical relaxation data have been utilized to calculate aggregation numbers for salts of perfluorooctanesulfonic acid [18,19]. Aggregation numbers have been calculated from the kinetic data, based on the assumption that the micellar distribution curve obeys a Gaussian distribution function. This procedure was validated with aggregation numbers for hydrocarbon-type surfactants, which were in agreement with those obtained from light-scattering data.

The aggregation numbers for the tetraethylammonium salt were calculated from kinetic data with the assumptions that the incorporation of surfactant molecules in a micelle is diffusion controlled, the rate constant for the association of a surfactant molecule to the micelle $k^+ = 10^9$ L/mol · sec at 15°C, and the energy of activation is 8 kcal (Table 7.1). The aggregation numbers increased with decreasing temperature and increasing hydrophobicity of the counterion. The aggregation numbers were considerably larger than those expected for spherical micelles. The length of the carbon chain would have limited the aggregation number to $\leqslant 20$. Hoffmann and Tagesson [18] concluded therefore that the shape of the micelle must be nonspherical. The unusual relaxation behavior of magnesium and dimethylammonium salts of perfluorononanoic acid was attributed by Hoffmann et al. [21] to emulsion-droplet-like giant micelles containing about 1000 monomers. More recent studies by Hoffmann and co-workers have indicated, however, that the giant micelles are probably dispersions of liquid crystalline mesophases [22].

TABLE 7.1 Association and Dissociation of Some Perfluorinated Surfactants[a]

Surfactant	Temperature (°C)	k^+ (L/mol · s)	k^- (s^{-1})	n	σ
$C_8F_{17}SO_3N(C_2H_5)_4$	5	4.8×10^8	$4.6 \times 10_5$	210	26
	10	6.0×10^8	5.7×10^5	204	28
	15	7.3×10^8	6.8×10^5	183	29
	20	8.2×10^8	7.6×10^5	155	30
	25	1.0×10^9	9.2×10^5	140	32
$C_7H_{15}COOH$	30	1.2×10^9	1.3×10^7	$11{,}000^m$	48
$C_7F_{15}COO(H/Na = 1{:}1)$	15	7.3×10^8	1.0×10^7	$10{,}000^m$	43
	25	1.0×10^9	1.4×10^7	$10{,}000^m$	50
$C_7F_{15}COONa$ [20]	25	1.0×10^9	3.2×10^7	12	<5

[a] Symbols: k^+, rate constant for a micelle entering the micelle; k^-, rate constant for a monomer exiting the micelle; n, average number of monomers in the micelle; σ, variance of the number of monomers in micelle; ^{m}, mesophases.

Source: Ref. 19. Reproduced by permission of Z. Phys. Chem. Neue Folge.

Nuclear magnetic resonance is a very powerful tool for investigating micellar structures of fluorinated surfactants. NMR spectroscopy yields values of the free energy of micellization, ΔG^{o}_{m}, and the corresponding enthalpy and entropy changes, ΔH^{o}_{m} and ΔS^{o}_{m} [23–29].

The effect of the environment on fluorine atoms in the micelle has been studied by NMR. Muller et al. [23,25,28] determined the chemical shifts of five structurally nonequivalent types of fluorine atom in salts of perfluorooctanoic acid to indicate the "degree of hydrocarbon-like character" of an atom in the micelle. They described the effect of the environment of CF_3 numerically by the parameter Z, defined as

$$Z = \frac{\delta(S_m) - \delta(S)}{\delta(F) - \delta(S)} \tag{1}$$

where S represents the surfactant ion and m the aggregation number, $\delta(S_m)$ the chemical shift of the fluorine nuclei in the micellized surfactant, $\delta(S)$ the chemical shift of the monomeric surfactant, and $\delta(F)$ the chemical shift of the surfactant in a fluorocarbon medium. In the absence of water in the environment of the fluorocarbon groups $\delta(S_m) = \delta(F)$ and consequently $Z = 1$.

If the surfactant is not soluble in a fluorocarbon, a correct value for $\delta(F)$ is difficult to obtain. Empirically,

$$Z = \frac{\delta(S_m) - \delta(S)}{3.8 - \delta(S)} \tag{2}$$

The Z values obtained by Muller et al. (Table 7.2 [30]) suggested that the polar hydrophilic groups of the surfactant do not cover the micelle surface completely but leave some of the hydrophobic surface exposed. The environment is more organic for the ω-carbon than for the β-carbon and the α-carbon, which is the least organic environment. The α-carbon is exposed to water because of its location vicinal to the hydrophilic group. The neighboring β-

TABLE 7.2 Degree of Organic-Like Character

	Z values		
	α-CF_2	β-CF_2	ω-CF_3
C_3F_7COOH [26]	0.71	0.77	0.82
$C_7F_{15}COONa$ [28]	0.37	0.66	0.84
$CF_3(CH_2)_nCOONa$[a] [23]			0.53

[a] $n = 9, 11, 12$

Source: Ref. 30. Reproduced by permission of American Chemical Society.

carbon is exposed to water to some extent. The Z values suggested that even the ω-carbon of an ω-trifluoro group is slightly but significantly hydrated.

^{19}F spin-lattice relaxation times of heptafluorobutyric acid in water and D_2O also have indicated that the fluorocarbon chains in the micelle are to some extent exposed to water [31]. However, the penetration of water into the interior of the micelle has been disputed by others [30,32]. Berr and Jones [32] found the model in which water does not penetrate into the fluorocarbon core to agree best with their SANS data.

Partially fluorinated surfactants also have been investigated by NMR. Muller and Birkhahn [24] found the fluorine chemical shift for *anionic terminally fluorinated surfactants*, $CF_3(CH_2)_8COONa$ and $CF_3(CH_2)_{10}COONa$, to be essentially independent of ionic strength. An added electrolyte depressed cmc but did not affect the chemical shifts for the surfactant anions in the micelle. They explained their results by the formation of prolate-shaped spheroid micelles in the presence of sufficient electrolyte. An alternative explanation that electrolytes do not increase the micellar size was considered to be unlikely.

Muller and Johnson [25] studied the effect of organic additives on sodium 12,12,12-trifluorododecyl sulfate micelles. Mixed solvents used in the study included aqueous urea, glycine, glycerol, acetamide, methanol, ethanol, acetone, dioxane, and tetrahydrofuran. Changes in the shapes of the dilution-shift curves indicated a penetration of additive into micelles and a strong effect of the additives on the micellar size. In some cases the aggregation numbers were decreased to a value below 5. The demicellizing effectiveness of the additives was not related solely to a change in the dielectric constant. The effects of cosolvents, as well as temperature effects, were qualitatively similar to those of the corresponding nonfluorinated surfactants.

In water the tetramethylammonium salt of trifluorododecyl sulfate forms smaller micelles than the sodium or potassium salts. In 2.0 M dioxane, the sodium or potassium salts form smaller micelles than in water, with aggregation numbers about 10–20. The tetramethylammonium salt forms larger micelles in water containing dioxane than in water alone. In 2.0 m tetrahydrofuran the three salts gave similar results, but at the same concentration tetrahydrofuran reduces the micellar size more than dioxane, to a very low aggregation number, possibly 3 to 4 [28].

The NMR spectra of *terminally fluorinated nonionic surfactants* resemble those of anionic surfactants with the same hydrophobe [27]. The surfactants 8,8,8-trifluorooctylhexaoxyethylene glycol monoether and 8,8,8-trifluorooctyl methyl sulfoxide yielded the same $\delta(S)$ value, which is identical to that of corresponding anionic surfactants. Hence the nearby environment of the CF_3 group appears to be unaffected by the chemical constitution, size, or electric charge of the hydrophile. The aggregation number for 8,8,8-trifluorooctylhexaoxyethylene glycol monoether was found to be 16 at 27°C. At temperatures above 42°C the micelles were larger. Addition of urea also caused an increase in the micelle size.

The fluorine NMR chemical shifts of *terminally fluorinated cationic surfactants* 12,12,12-trifluorododecyltrimethylammonium bromide and 10,10,10-trifluorodecyltrimethylammonium bromide are similar to those of the corresponding anionic and nonionic surfactants with a terminal trifluoromethyl group [29]. When the counterion is fluoride, instead of bromide, the chemical shift of the fluoride counterion is concentration dependent. The trifluoromethyl chemical shifts were interpreted utilizing a double equilibrium model. The aggregation number was estimated to be 25 in dilute solutions from the curvature of the chemical shift concentration plots near the first cmc. Above the "second cmc" the aggregation number was assumed to be 60, considered the spatial limit for 12-carbon surfactant chains in a spherical micelle.

The investigation of perfluorinated surfactants by *light scattering* is hampered by the low refractive index of fluorocarbons. Partially fluorinated surfactants are amenable, however, to the light-scattering technique. Gerry et al. [7] measured aggregation numbers of 12,12,12-trifluorododecyltrimethylammonium bromide and dodecylammonium bromide in NaBr solutions by light scattering (Table 6.14). The aggregation numbers for the surfactant with the terminal CF_3-group were only slightly smaller than those for the corresponding unfluorinated surfactant, although fluorination had nearly doubled the cmc value. This result appeared to be puzzling at first, because fluorocarbons are more hydrophobic than hydrocarbons and fluorination of a surfactant should favor micellization. Surfactants with a completely fluorinated hydrophobic chain behave as expected. However, partially fluorinated surfactants are different. An ω-trifluoromethyl group resists mixing with hydrocarbon chains and the resulting mutual phobicity causes the CF_3- group to seek the water-micellar core interface, instead of the hydrocarbon-like core [30,33]. The micellar size constancy in the 0.5 to 1.0 M NaBr concentration range suggests a geometrically defined upper size limit of the spherical micelle. The increase of the aggregation number at NaBr concentrations above 1.0 M is related to a micellar shape transformation and polydispersity.

Small-angle neutron scattering (SANS) is an excellent technique for the study of micellar solutions [32,34]. The large difference in scattering lengths between ^{19}F and ^{1}H provides a strong contrast between the aqueous ($^{1}H_2O$) solvent and the micelles of the perfluorinated surfactant [32]. Herbst et al. [34b] observed that a sample of 30% w/w $C_8F_{17}COON(CH_3)_4$ in D_2O forms three lyotropic phases at different temperature ranges (see Sec. 7.4). The isotropic solution at $\geqslant$ 37°C close to the liquid crystal boundary contained disk-like micelles. The aggregation number of the micelles was about 150. The thickness of the micelles, including the counterions, was about 35 Å.

Berr and Jones [32] determined by SANS the charge and aggregation number for aqueous solutions of sodium perfluorooctanoate as a function of surfactant concentration above cmc. Their SANS data suggested a spherical micelle model with a water-free core of alkyl chains. The aggregation number increases linearly with increasing square root of the surfactant concentration. The cmc of sodium perfluorooctanoate is an order of magnitude smaller than the cmc of sodium octanoate. The aggregation number of sodium perfluoro-octanaote is larger than that of the sodium octanoate, in accord with the greater hydrophobicity of the perfluoroalkyl chains and the smaller headgroup charge of the fluorinated surfactant. The electronegative fluorine withdraws electron charge from the headgroup and the reduced Coulumbic repulsion between the headgroups allows a closer packing in the micelle.

Burkitt, Ottewill, Hayter and Ingram [35] investigated the size and shape of micelles of ammonium salts of octanoic, decanoic and perfluorooctanoic acids by SANS. Ammonium octanoate and ammonium decanoate formed spherical micelles having a micellar weight of 1640 and 12576, respectively. Ammonium perfluorooctanoate formed cylindrical micelles with the mean micellar weight of 17610. This corresponded to a mean association number of 43 at 0.12 M. Burkitt at al. proposed a cylindrical micelle in which the head groups are hexagonally close-packed and the fluorocarbon chains form helical rows (Fig. 7.3).

Berr and Jones [32] calculated the aggregation number for sodium perfluorooctanoate to be 28 at 0.12 M. The smaller aggregation number of sodium perflurooctanoate allows the the micelle to remain spherical whereas ammonium perfluorooctanoate with larger aggregation numbers [35] forms ellipsoidal micelles.

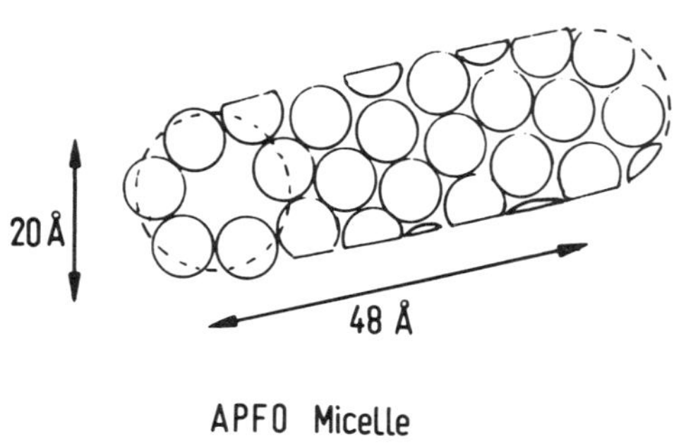

FIG. 7.3 A cylindrical micelle model for ammonium perfluorooctanoate. (From Ref. 35. Reproduced by permission of Academic Press, Inc.)

7.2 THEORY FOR MIXED MICELLES

A commercially produced fluorinated surfactant is almost always a mixture [36,37]. A typical fluorinated surfactant is a mixture of homologues which are difficult to separate. However, some commercial fluorinated surfactants contain hydrocarbon-type surfactants which have been blended with the fluorinated surfactant to improve its performance. A micellar solution containing a fluorinated surfactant and a selected hydrocarbon surfactant can exceed the surface activity predicted for an ideal solution in which the components do not interact. This synergism increases the effectiveness of the surfactant and, because of the lower cost of hydrocarbon-type surfactants, results in considerable savings. Since perfluorinated surfactants are not biodegradable (Chapter 10), the use of fluorinated surfactants at lower concentrations is also advantageous from the environmental point of view.

The synergism observed in solutions of mixed dissimilar surfactants is governed by monomer-micelle equilibria and interaction in micelles. The concentration of surfactant monomers below the cmc is so dilute that interactions between surfactant monomers are insignificant. The composition of mixed micelles has been determined from surface tension data, electric conductivity, NMR, self-diffusion measurements, fluorescence quenching, solubilization, ultrafiltration, and gel filtration. Complete nonmixing of a fluorinated surfactant and a hydrocarbon surfactant, partial mixing, and the existence of two types of mixed micelles or one mixed micelle has been reported.

In an ideal system of two similar surfactants of like charge the total monomer concentration is between the cmc values of the two surfactants, if the total surfactant concentration is at or above the mixture cmc (Fig. 7.4) [38]. At equilibrium [38]

$$C_M = \frac{\text{cmc}_A \text{cmc}_B}{y_A \text{cmc}_B + y_B \text{cmc}_A} \tag{3}$$

$$X_A = \frac{y_A C_M}{\text{cmc}_A} \tag{4}$$

where C_M is the monomer concentration, y_A and y_B are the mole fractions of monomers A and B, cmc_A and cmc_B are the cmc values of the monomers, and X_A is the micellar mole fraction of component A.

The ideal solution theory can predict the concentration of each surfactant in micelles or the monomeric state for mixed hydrocarbon surfactants of similar structure. However, hydrocarbon surfactant mixtures of dissimilar structures, such as ionic-nonionic or cationic-anionic surfactant mixtures, exhibit nonideal behavior. The deviation from the ideal relationship can be negative or positive (Fig. 7.4).

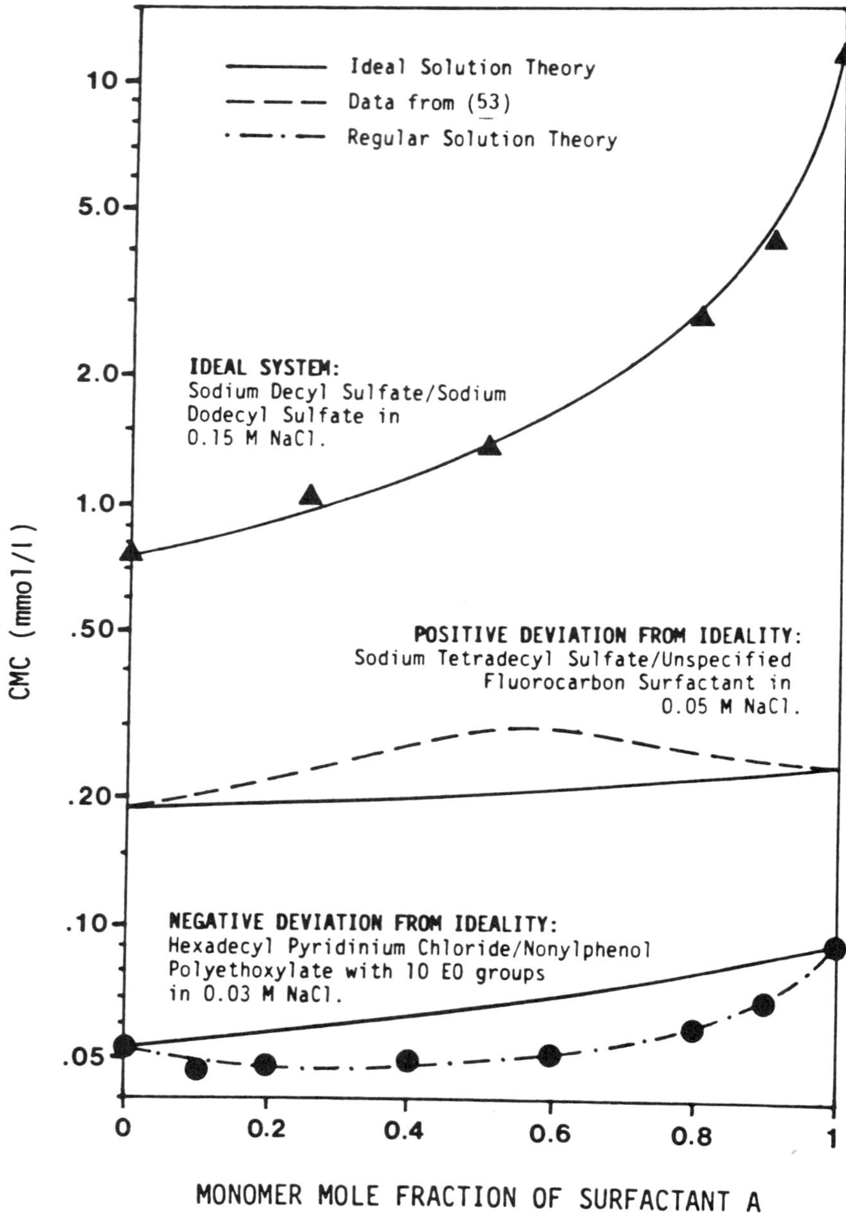

FIG. 7.4 Critical micelle concentrations for surfactant mixtures at 30°C (First surfactant listed is surfactant A.) (From Ref. 38. Reproduced by permission of American Chemical Society.)

Unlike mixed hydrocarbon chain surfactants of similar molecular structure, mixtures of fluorinated surfactants and hydrocarbon chain surfactants do not behave ideally, even when the surfactants have a similar hydrophilic group. Mixtures of anionic fluorinated surfactants with anionic hydrocarbon surfactants exhibit a positive deviation from the ideal relation (Fig. 7.4). In contrast, surfactant mixtures containing a nonionic surfactant or oppositely charged ionic surfactants exhibit a negative deviation from ideal predictions. The formation of mixed micelles is governed by hydrophobic interactions between hydrocarbon and fluorocarbon chains and electrostatic effects. Introduction of nonionic surfactants into micelles of anionic fluorinated surfactants reduces electrostatic repulsion between the ionic head groups. Apparently the resulting electrostatic effect overcomes the hydrophobic interaction between the fluorocarbon and hydrocarbon chains.

If the positive deviation from ideal mixing in the micelle is severe, the mixed micelles are thermodynamically unstable and separate into two coexisting micelle types of different composition. Hence surfactant mixtures containing anionic hydrocarbon and fluorocarbon surfactants form mixed micelles or two kinds of micelles, each rich in one of the components. The miscibility of hydrocarbon surfactants and fluorocarbon surfactants depends on the molecular structure of the surfactants, their concentrations, the composition of the solution, the temperature, and the pressure. Klevens and Raison [39] studied micellization of mixtures of sodium dodecyl sulfate and perfluorooctanoic acid. The phase diagram reveals mutual phobicity of hydrocarbon and perfluorocarbon chains of the surfactants (Fig. 7.5). The extent of the mutual phobia was examined by Mukerjee and Yang [40] with mixtures of sodium perfluorooctanoate (SPFO) and sodium decyl sulfate (SDeS) and with sodium laurate (SL). The cmc values of the mixed systems were calculated assuming (1) the extreme case of complete demixing or (2) ideal behavior. If the overall surfactant concentration is increased in a system of complete demixing, micelle formation begins when the concentration of one of the components exceeds its cmc value. If the surfactants are nonionic, the cmc versus mole fraction curve is given by the equation

$$\text{cmc} = \text{cmc}_0/X \tag{5}$$

where cmc_0 is the cmc value for the pure surfactant and X is the mole fraction in a mixture. For ionic surfactants, the common ion effect of the counterion has to be taken in account

$$\log \text{cmc} = A - B \log[\text{Na}^+] \tag{6}$$

where A and B are constants [41,42].

The cmc value of a mixed system is given by

$$\text{cmc} = \text{cmc}_0 \,\text{antilog}\{\log(1/X)/(1 + B)\} \tag{7}$$

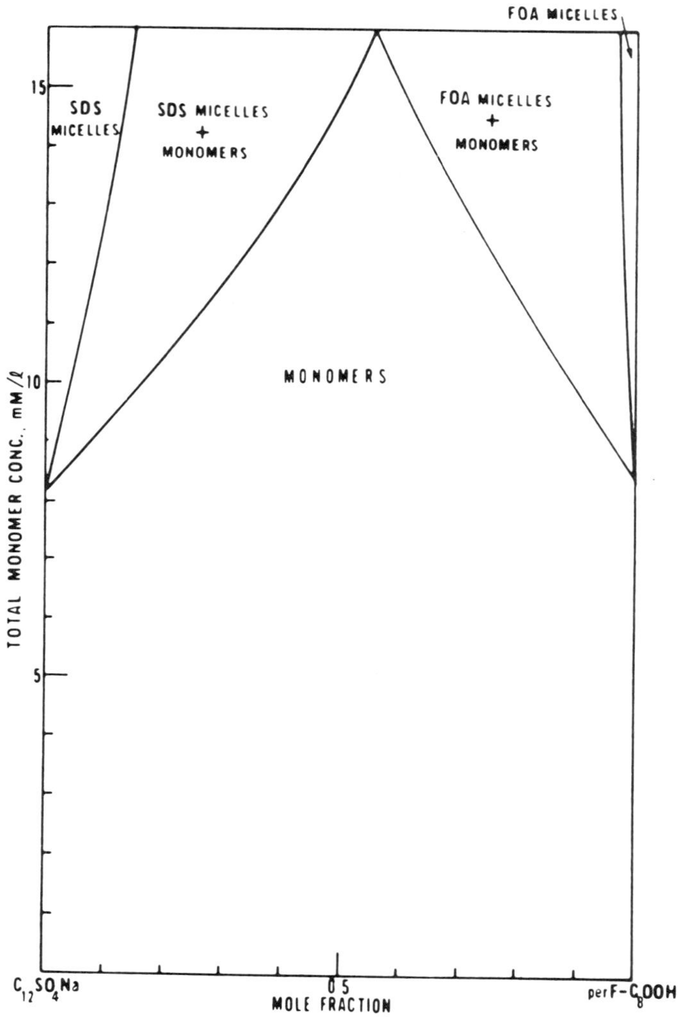

FIG. 7.5 Equilibrium compositions of micelles and monomers in the sodium dodecyl sulfate and perfluorooctanoic acid system. Two types of micelles are present with limited mutual solubility. (From Ref. 30. Reproduced by permission of American Chemical Society.)

The cmc values calculated for complete demixing and for ideal behavior are shown together with the observed cmc data in Fig. 7.6. The dashed lines show expected values for ideal mixing of micelles, assuming $B = 0.645$ for all three components. Curves 1, 2, 3, and 4 represent complete demixing of micelles (Eq. 7). Curve 1 was calculated for $B = 0.53$; for curves 2, 3, and 4 the value

$B = 0.645$ was used. The experimental data are close to the curves for complete demixing, indicating severe nonideality of mixing. The cmc data of Fig. 7.6 suggest that two micellar species may coexist, one rich in the fluorinated surfactant and the other micellar species consisting mainly of the hydrocarbon surfactant. Mukerjee and Yang [40] measured differential conductance

$$10^3 \frac{\Delta\kappa}{\Delta C} = 10^3 \frac{(\kappa_2 - \kappa_1)}{(C_2 - C_1)} \tag{8}$$

where κ_2 and κ_1 are the specific conductance at two adjacent concentrations C_2 and C_1. The differential conductance of mixtures of SPFO and sodium dodecyl sulfate (SDS) are plotted versus the mean concentration $(C_2 + C_1)/2$ in Fig. 7.7. Curves 1, 5, and 6 represent single surfactants, curves 2 and 4 show data for mixtures containing a fluorinated surfactant (SPFO), and curve 3 represents a mixture of hydrocarbon-type surfactants (SDS and SDeS). Curves 2 and 4 show two inflection points before flattening at higher concentrations. Mukerjee and Yang [40] proposed that micelles composed of SDS are formed first. As

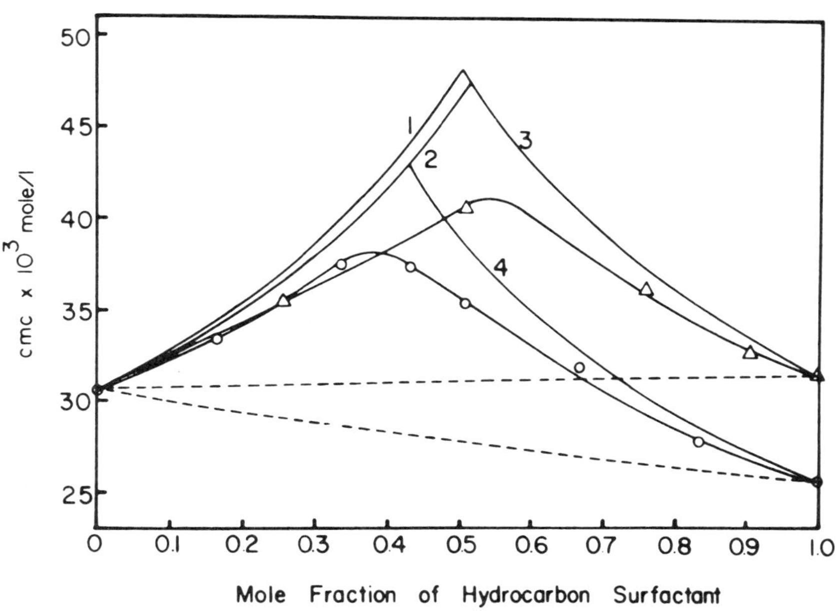

FIG. 7.6 Critical micelle concentrations for mixtures of sodium perfluorooctanoate with sodium decyl sulfate (△) and sodium laurate (○). Dashed lines represent expected values for ideal mixing, assuming $B = 0.645$ (Eq. 6.) Curves 1, 2, 3, and 4 show expected values for complete demixing of micelles (Eq. 7.) Curve 1 was calculated for $B = 0.53$; curves 2, 3, 4 were calculated assuming $B = 0.645$. (From Ref. 40. Reproduced by permission of American Chemical Society.)

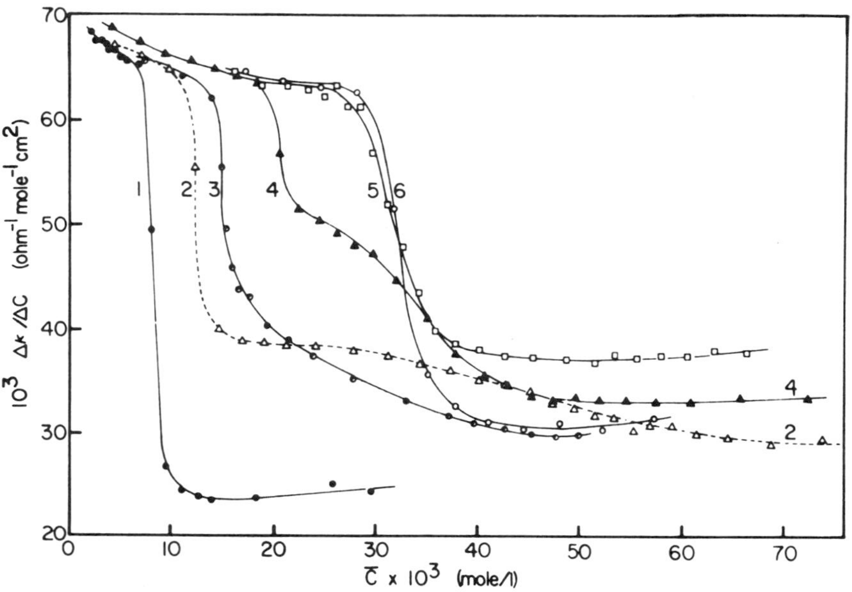

FIG. 7.7 Differential conductance, 10^3 $(\kappa_2 - \kappa_1)(C_2 - C_1)$ plotted against mean concentration $(C_2 + C_1)/2$: curve 1, SDS; curve 2, SDS + SPFO (SDS mole fraction 0.5); curve 3, SDS + SDeS (SDS mole fraction 0.29); curve 4, SDS + SPFO (SDS mole fraction 0.2); curve 5, SPFO; curve 6, SDeS. (From Ref. 40. Reproduced by permission of American Chemical Society.)

the total concentration increases the fraction of monomeric SPFO increases. At about the second inflection point, micelles form which consist mainly of SPFO. At higher concentrations two kinds of micelles coexist.

Advances in the theory of mixed micelle formation have made it possible to calculate the composition of mixed micelles formed by two or more surfactants. A thermodynamic treatment of micellar solutions of mixed surfactants is usually based on the *pseudophase separation theory* [38,43–46]. The pseudophase models developed for binary surfactant solutions assume ideal mixing of the surfactants in the micelle.

The pseudophase separation model of micellar solutions considers a micelle to be a pseudophase in a liquid state. Since the micelles are assumed to have a liquid-like core, the mutual solubility of a fluorinated surfactant and a hydrocarbon surfactant in mixed micelles is, according to the pseudophase model, governed by the miscibility of the fluorocarbon and hydrocarbon chain. For example, heptane and perfluoroheptane are immiscible at 25°C but above 50°C these liquids are miscible in all proportions [47]. A terminal substitution of a hydrophilic group depresses the enthalpy of mixing and makes the components miscible at 25°C.

The mutual solubility of the perfluoroalkanoic and alkanoic acids suggests that their salts should be miscible in micelles, at least partially. Depending on the chain length and the surfactant structure, the mixture may form a mixed micelle or two types of micelles, one rich in the fluorinated surfactant, the other rich in the hydrocarbon surfactant. Shinoda and Nomura [43] reported partial miscibility for the $C_8H_{17}COONa-C_{12}H_{25}SO_4Na$ system and miscibility in all proportions for the $C_7F_{15}COONa-C_{10}H_{21}SO_4Na$ mixed micelles at 25°C. NMR data indicated that $C_7F_{15}COOLi$ mixes with $C_{12}H_{25}SO_4Li$ in all proportions but $C_8F_{17}COOLi$, $C_9F_{19}COOLi$, and $C_{10}F_{21}COOLi$ do not mix completely with $C_{12}H_{25}SO_4Li$. The solubility and critical solution temperatures lead to the conclusion that a carbon chain longer than eight carbon atoms causes the liquid-liquid mixtures of alkanoic and perfluoroalkanoic acids to separate into two phases. For example, a 1:3 v/v mixture of $C_7F_{15}COOH$ and $C_7H_{15}COOH$ and an equimolar mixture of $C_8F_{17}COOH$ and $C_{11}H_{23}COOH$ form a single liquid phase above 19 and 47°C, respectively. Below these temperatures the respective perfluoroalkanoic acid precipitates. Studies have shown, however, that the interaction between hydrophilic groups can be more influential than the association of hydrophobic tails.

The pseudophase separation model has been fairly adequate in describing the behavior of binary nonionic or anionic surfactant mixtures, especially when the surfactants have the same hydrophilic group. When the hydrophilic groups of the mixed surfactants are different, significant deviations have been observed. The interactions between the different groups have been taken into account in a pseudophase separation model which assumes that the mixed micelle approximates a *regular solution* [38,44,45,48,49]. For nonideal surfactant mixtures, equations derived from the regular solution theory [50] can give the interaction parameter, W, and the micelle composition at a given monomer composition.

$$1 = \frac{y_A X_B \mathrm{cmc}_B \exp[W(X_A^2 - X_B^2)/RT]}{y_A X_A \mathrm{cmc}_A} \tag{9}$$

$$C_M = \frac{X_B \mathrm{cmc}_B \exp[WX_A^2/RT)}{y_A} \tag{10}$$

where R is the gas constant, T is the absolute temperature, cmc_A and cmc_B are the cmc values for the individual components, y_A and y_B are the monomer mole fractions of components A and B, respectively, and X_A and X_B are the micellar mole fractions of A and B, respectively.

Scamehorn [38] has emphasized, however, that although these equations have accurately predicted mixture cmc values and monomer-micelle equilibrium, they do not validate the regular solution theory for nonideal surfactant mixtures. Ample evidence contradicts the validity of the regular solution theory in describing the nonideal surfactant mixtures.

Rubingh (49], using the regular solution approximation, proposed an interaction parameter β for the treatment of the cmc of nonideal mixed surfactant solutions

$$\beta = \ln[(C_m\alpha)/(C_1X_1)]/(1 - X_1)^2 \tag{11}$$

where C_m is the cmc of mixed surfactants, C_1 is the cmc and X_1 the mole fraction of component 1, and α is mole fraction of surfactant 1 in the surfactant mixture present in the solution.

Osborne-Lee et al. [46] employed ultrafiltration to investigate binary mixtures of alkylbenzenesulfonates and oxyethylated nonylphenol. The regular solution approximation model was found to be inadequate for these mixtures. The model accurately represented cmc data but failed to predict correct micelle compositions. A new theory was proposed which included the effects of nonideal entropy variations resulting from nonrandom rearrangements in the mixed micelle and conformational entropy changes in the polyoxyethylene chain. The composition of mixed micelles was calculated using the regular solution theory and activity coefficients with a single empirical interaction coefficient. The solution of mixed micelles was treated as a classical multicomponent system.

Treiner and co-workers [51] measured partitioning of 1-pentanol to elucidate the micellar composition of two aqueous mixed anionic surfactant solutions (Na decyl sulfate + sodium perfluorooctanoate and sodium dodecyl sulfate + sodium perfluorooctanoate). The partition coefficient exhibited a maximum in hydrocarbon-rich mixtures in both mixed micellar solutions. The synergistic effect was attributed to hydrocarbon-fluorocarbon repulsive interaction and found to be in accord with the regular solution model using interaction parameters determined for binary surfactant solutions.

By applying the pseudophase model and considering the micelle as an ordinary solvent, the partition coefficient for the surfactant mixture can be expressed by the equation

$$\log P_m = X \log P_1 + (1 - X) \log P_2 + \beta X(1-X) \tag{12}$$

where P represents the partition coefficients of 1-pentanol in mixed (P_m) and pure surfactant solutions (P_1,P_2) and X is the mole fraction of the surfactant in the mixed micelles. The β values proposed by Asakawa et al. [52] were used in the calculations: 2.1 for Na decyl sulfate + sodium perfluorooctanoate and 1.8 for sodium dodecyl sulfate + sodium perfluorooctanoate. The agreement between the values predicted by Eq. 12 and the experimental values implies that the 1-pentanol molecules are distributed throughout the micelles and not confined near the palisade layer as usually assumed for dilute single micellar solutions. The results suggested that the mixed micelles are dynamic structures in a highly disordered state.

Treiner and co-workers [53] also determined the partition coefficient of 1-pentanol between water and mixed micelles of lithium dodecyl sulfate and lithium perfluorooctanoate. The solubilization of 1-pentanol in the mixed surfactant solution was calculated from the regular solution theory using the interaction parameter β. A positive deviation from ideal behavior was observed for the system as a consequence of the positive value of the interaction coefficient β. According to the regular solution theory, two types of mixed micelles should exist when the β value is above 2. The authors concluded, however, that the consequences of such a demixing effect on the solubilization profile depend on too many unknown parameters.

Kamrath and Franses [54] developed a single-micelle-size *mass-action model* for binary solutions of surfactants with the same hydrophilic group and counterion. The mass action model predicts micellar behavior more accurately than the pseudophase separation model [45] if the number of surfactant monomers in the mixed micelle is less than about 50.

The single-parameter nonideal mixed surfactant models are inadequate for treating the complex interactions between fluorinated surfactants and hydrocarbon surfactants. The possible involvement of water in micelle formation and electrostatic effects resulting from ionic interactions have to be taken into account. A more sophisticated model is therefore needed. Asakawa et al. [55] developed a *group contribution theory* to characterize binary mixtures of surfactants. The more complex model is based on the concept that a solution is composed of many functional groups [56]. A group interaction parameter can be used to calculate mixed cmc values when the cmc values, the micelle counterion binding parameter, and molecular structure data are given for the pure components.

The composition of micelles formed by mixed hydrocarbon and fluorocarbon surfactants is a controversial subject still being explored and debated. The miscibility of fluorocarbon surfactants and hydrocarbon surfactants in binary solutions depends on the structure of the mixed surfactants, the composition of the solution, and the temperature. At least some of the apparent incongruities have been caused by differences in the composition of the systems studied and physical conditions. Undoubtedly, interactions of fluorocarbons and hydrocarbons are also profoundly nonideal in micellar systems.

Mixtures of fluorinated and nonfluorinated surfactants may consist of anionic, nonionic, or cationic components. Most of the systems investigated so far have contained a fluorinated anionic surfactant and an anionic hydrocarbon surfactant. Anionic fluorinated surfactant mixtures with nonionic or cationic hydrocarbon-type surfactants have been investigated as well. The nonionic fluorinated-hydrocarbon surfactant mixtures and cationic fluorinated-hydrocarbon surfactant mixtures have been the subject of only a few studies.

Finally, most practical applications of surfactant mixtures involve adsorption at liquid-air, liquid-liquid, and solid-liquid interfaces. The theoretical understanding of the interactions in adsorbed surfactant mixtures is therefore of paramount importance (see Chapters 4 and 5).

7.3 SURFACTANT MIXTURES

Mixtures of Anionic Surfactants. Perfluoroalkanoates

The mutual, although limited, solubility of perfluoroalkanoic and alkanoic acids suggests that their salts should be miscible in micelles, at least partially [43]. The nature and the number of different types of mixed micelles are still being debated [57a]. However, at least some of the apparent incongruities are caused by differences in surfactant structure, composition of the solution, and temperature. The miscibility and critical solution temperature (cst) of binary systems depend on the chain length, the surfactant structures, and the composition of the solution. Above cst one kind of micelle exists in binary systems; below cst two kinds of micelles can coexist. Mukerjee and Yang [40] have reported that in a solution containing sodium perfluorooctanoate and sodium decyl sulfate two micellar species coexist, one rich in the fluorinated surfactant and the other micellar species consisting mainly of the hydrocarbon surfactant. On the other hand, studies made with pyrene as a fluorescence probe [2,17,57b] found only one mixed micellar species. Kalyanasundaram [57b] concluded that only one type of mixed Li perfluorononanoate (LiPFN)–sodium lauryl sulfate (NaLS) micelle may exist over a wide range, in spite of severe nonideality of mixing.

Burkitt et al. [58] studied mixtures of ammonium decanoate (AmDec) and ammonium perfluorooctanoate (APFO) by the SANS technique [35]. The proportions of the two surfactants in an ammonium chloride–ammonium hydroxide buffer at pH 8.8 and an ionic strength of 0.1 were varied. The mixed micelles formed were cylindrical in shape at APFO/AmDec ratios of 2:1, 1:1, and 1:2. The micelle appeared to have a diameter of 20 Å and a length of 150 Å. The authors proposed that the original ammonium perfluorooctanoate micelle, which also had a diameter of 20 Å, expanded along the lateral axis to accommodate the decanoate chains (Fig. 7.8). At a 1:9 APFO/AmDec ratio the micelles were found to be cylindrical but larger than those formed by ammonium decanoate alone.

Both the SANS results and surface tension measurements suggested that perfluorooctanoate and decanoate chains can mix to form a micelle but an incipient tendency toward phase separation exists and actual segregation may occur with longer hydrocarbon chains. The concept of a mixed micelle with intrami-

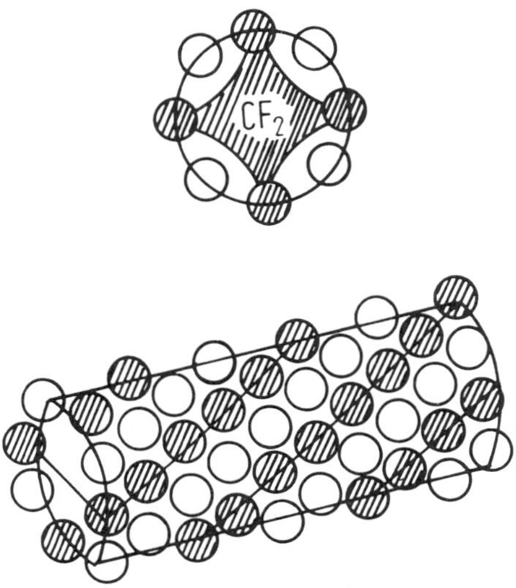

FIG. 7.8 Schematic mixed micelle model. Fluorocarbon chains (hatched) form intersecting bands which wind gradually through the structure. The more flexible hydrocarbon chains (open circles) fill the channels. (From Ref. 58. Reproduced by permission of Dr. Dietrich Steinkopf Verlag.)

cellar segregation between hydrocarbon and fluorocarbon chains is not consistent with the coexistence of two types of micelles. Carlfors and Stilbs [59], who based their conclusions on ^{19}F NMR self-diffusion measurements, proposed that alkanoate and perfluoroalkanoate mixtures form two kinds of micelles, one hydrocarbon rich and the other fluorocarbon rich. Burkitt et al. [58] explained the apparent contradiction by differences in the technique of measurement and suggested that the NMR technique measures temporal correlations, whereas small-angle neutron scattering measures spatial correlations.

The alkanoic fluorocarbon-hydrocarbon surfactant mixtures usually form a mixed micelle or two kinds of mixed micelles which contain a large amount of one component and a small amount of the other component. In contrast, potassium perfluorononanoylalaninate (PFNALA) and potassium acylalaninates are virtually immiscible [60]. Two types of micelles coexist, each micelle composed of only one surfactant. PFNALA is adsorbed on a water-air interface more strongly than potassium laurylalaninate.

Mixtures of Anionic Surfactants. Fluorinated Sulfonates

Mixtures of fluorinated surfactants featuring a sulfonate hydrophilic group and hydrocarbon surfactants are either immiscible, form two types of micelles, or form a mixed micelle. The miscibility limits depend on the molecular structure of the surfactants, the composition of the solution, and physical conditions of the system.

Micelles at equilibrium with surfactant monomers exhibit two relaxation times (see Chapter 9), one representing the exchange of monomers and the other the formation of new micelles. Bauernschmitt and Hoffmann [61] argued that if two different types of micelles exist in the micellar solution of mixed surfactants, the number of relaxation times should be four, instead of two. The observation that mixtures of $C_8F_{17}SO_3N(CH_3)_4/C_{14}H_{29}SO_4N(CH_3)_4$ and $C_8H_{17}SO_3Li/C_{12}H_{25}SO_4Li$ exhibited only two relaxation times appeared to be evidence for comicellization. The cmc values of the surfactant mixtures lent more support for mixed micelles. The cmc values of the surfactant mixtures $C_8F_{17}SO_3N(CH_3)_4/C_{14}H_{29}SO_4N(CH_3)_4$ and $C_8H_{17}SO_3Li/C_{12}H_{25}SO_4Li$ were found to be higher than the cmc values for the pure components. This result is consistent with mutual phobicity. However, the cmc values were lower than the cmc values expected for the mixture if no mixed micelles were formed. Hence the cmc data are in accord with the formation of mixed micelles. Kinetic data indicated that mixed micelles of the $C_8F_{17}SO_3NCH_3)_4-C_{14}H_{29}SO_4N(CH_3)_4$ mixture have lower aggregation numbers than the pure micelles. The residence time of the surfactant was found to be shorter in mixed micelles than in micelles formed by a single surfactant. The solubility of the fluorinated surfactant in hydrocarbon surfactant micelles is higher than vice versa.

The positive deviation of the lithium perfluorooctanesulfonate (LiFOS) and lithium dodecane sulfate (LiDS) mixture from ideal predictions indicates that both molecules tend to exclude each other. Ueno et al. [62] varied the mole fraction of lithium perfluorooctanesulfonate (LiFOS) and lithium dodecane sulfate (LiDS) and examined the micellar properties of the surfactant mixtures by surface tension, electroconductivity, and the spectral changes of pinacyanol chloride in the surfactant solution. The surface tension values plotted against the logarithm of the total surfactant concentrations (Fig. 7.9) exhibit a minimum near the cmc. The total concentrations of mixed cmc values obtained from surface tension curves, shown in Fig. 7.9, and from conductivity measurements are plotted against the mole fraction of LiFOS in Fig. 7.10. For comparison a curve is included which represents cmc values calculated for a hydrocarbon surfactant mixture by the Lange-Beck equation [48]. Both curves show a large positive deviation from ideal mixing and exhibit a maximum for an approximately equimolar mixture. The positive deviation of the cmc from ideal mixing was confirmed by spectral changes of pinacyanol chloride in single and mixed surfactant solutions.

The coexistence of two kinds of mixed micelles has been confirmed by gel filtration and ultrafiltration [63,64]. These techniques have elucidated the monomer-micelle equilibria above the cmc of the mixture and micelle demixing.

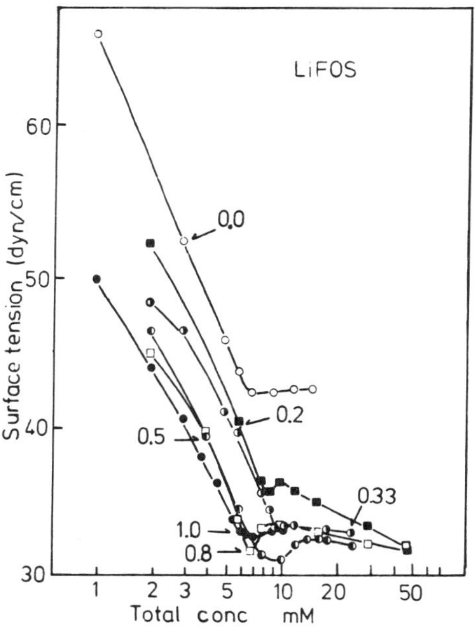

FIG. 7.9 Surface tension vs. total concentration of lithium perfluorooctanesulfonate–sodium decyl sulfate mixtures containing various mole fractions of lithium perfluorooctanesulfonate: (○) 0.0: (■) 0.2; (◑) 0.33; (◐) 0.5; (□) 0.8; (●) 1.0. (From Ref. 62. Reproduced by permission of Academic Press.)

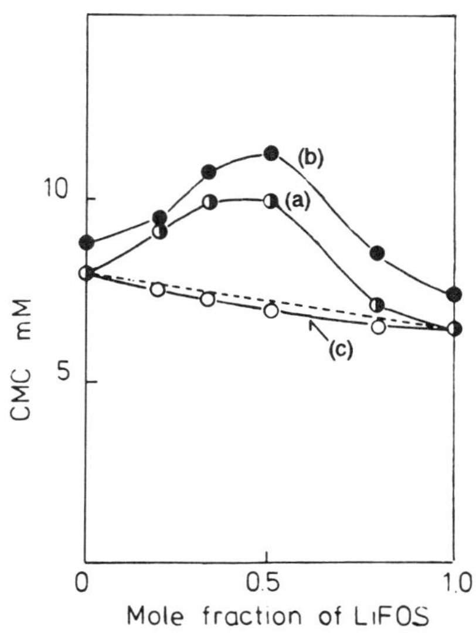

FIG. 7.10 Mixed cmc values vs. mole fraction of lithium perfluorooctanesulfonate [62]: (a) obtained from surface tension curves, (b) obtained from conductivity data, and (c) calculated by Lange's equation. (From Ref. 48. Reproduced by permission of Academic Press.)

Gel filtration [65–68] separates monomers from the micellar solution according to their molecular size. After the gel column has been equilibrated with a monomer solution, the sample solution is injected and eluted with the same monomer solution (sandwich method). At concentrations of 30 mM or above, Asakawa et al. [63] observed two peaks for the LiFOS-LiTS system but only one peak for the LiDS-LiTS system (Fig. 7.11). The first peak of the LiFOS-LiTS system corresponded to the LiTS-rich micelles and the second peak to the LiFOS-rich micelles. The LiFOS-rich peak diminished with decreasing total surfactant concentration and disappeared at low concentrations. The elution curves suggested that the LiTS-rich micelle exists above cmc, the second type of mixed micelle rich in LiFOS appears with increasing concentration, and the two types of micelles coexist at surfactant concentrations at or above 30 mM. The concentration needed to form the second type of mixed micelle was defined as the second cmc. The compositions of the mixed micelle remained constant in the demixing region, regardless of the mixing ratio.

Ultrafiltration has been used to measure cmc, monomer composition, and micelle composition of binary surfactant mixtures [46,64,69]. Asakawa et al.

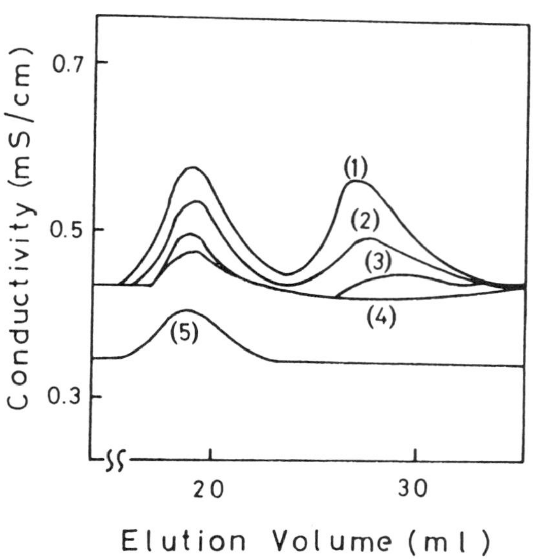

FIG. 7.11 Gel filtration elution curves of the equimolar LiFOS-LiTS system at various total surfactant concentrations using monomer solutions as the eluents. A 1-mL sample was introduced as a sandwich between monomer solutions. LiFOS-LiTS concentrations: (1) 50 mM , (2) 40 mM, (3) 30 mM, (4) 20 mM. Curve 5 30 mM 0.5 LiDS-LiTS solution. (From Ref. 63. Reproduced by permission of American Chemical Society.)

[64] used the ultrafiltration method to study SPFO-SDS systems and LiFOS-alkyl sulfate mixtures as a function of surfactant concentration. The alkyl sulfates had different chain lengths of 10 (LiDeS), 12 (LiDS), or 14 (LiTS) carbons. The micelle demixing region increased with increasing alkyl chain length. The "azeotropic point," where the micelle compositions are the same as the monomer surfactant composition, was found to be near the maximum of the mixture cmc curve for the LiFOS-LiTS and LiFOS-LiDS systems.

The coexistence of two kinds of mixed micelles in LiFOS-LiDS and LiFOS-LiTS systems is also supported by fluorescence probe analysis and changes in ^{1}H NMR and ^{13}C NMR spectra [70]. Abrupt changes far above the mixture cmc were attributed to the formation of a second type of mixed micelle at the second cmc.

A phase diagram for the lithium perfluorooctane sulfonate (LiFOS), lithium dodecylsulfate (LiDS), water system has been established by Tamori et al. [71] (see Sec. 7.4).

Mysels [72] introduced the concept of a critical demicellization concentration (cdc). Below cmc essentially no micelles are present; above cmc most of the surfactant has associated to form micelles (Fig. 7.12). When micelles in mixtures of surfactants have a limited solubility, a second cmc may exist [40]. As the surfactant concentration increases the monomers of one surfactant attain their cmc and form micelles, whereas the other surfactant is still in its monomeric state. At a higher concentration the second surfactant reaches its cmc and forms micelles. Above this cmc two kinds of micelles are present. As the surfactant concentration increases further, one kind of micelle may disappear at a concentration denoted as cdc.

The existence of cdc was demonstrated by Funasaki and Hada [73] with mixtures of an anionic fluorinated surfactant, Neos Ftergent (NF), $[(CF_3)_2CF]_2C{=}C(CF_3)OC_6H_4SO_3Na$ [sodium *p*-(perfluoro-3-isopropyl-4-methylpent-2-en-2-oxy)benzenesulfonate], and hydrocarbon surfactants. Funasaki and Hada found mixtures of fluorinated and hydrocarbon surfactants to contain under appropriate conditions one type of mixed micelle or two kinds of mixed micelles: one rich in one of the surfactants, and the other rich in the other surfactant. The coexistence of two kinds of mixed micelles in a solution depends on the nature of the surfactants, the molar ratio of the two surfactants, the total concentration of the surfactants, the temperature, and the concentration of added electrolyte [73–79].

The composition of mixed micelles in solutions containing NF and sodium tridecyl sulfate (STrS) above cmc was determined from the surface tension data [73]. The solutions contained 0.05 M NaCl to keep the surface tension value constant above cmc. A typical plot of surface tension versus the total surfactant concentration is shown in Fig. 7.13. The mole fraction NF-STrS was kept constant for each curve. The intersection of solid and dashed lines indicate the cmc.

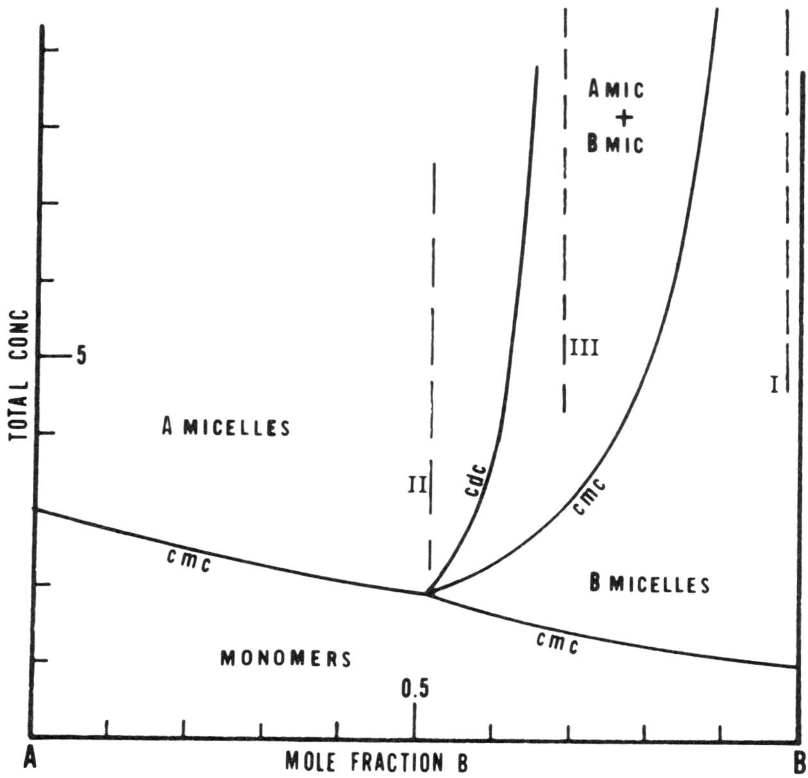

FIG. 7.12 Hypothetical micellization diagram for a pair of nonionic surfactants exhibiting a critical demicellization concentration (cdc.) The parameters used are $c_A = 3$; $c_B = 1$, and proportianility constants are $\sigma_A = 0.05$, $\sigma_B = 0.07$. (From Ref. 72. Reproduced by permission of Academic Press, Inc.)

In Fig. 7.14 the surface tension is plotted against the mole fractions of STrS in coexisting monomers (curve e) and micelles (curve f). The micellar composition was calculated assuming that (1) the surface tension of the pure surfactant remains constant above cmc and (2) the surface tension is a function of only the surfactant monomer concentrations regardless of coexisting micelles. In Fig. 7.14, the surface tension is constant at 18.73 mN/m between the mole fractions X_A and X_B. The compositions at both ends of this region, X_A and X_B, represent the mutual solubility in the mixed NF-STrS micelles. When the total concentration of the mixed surfactants of a fixed composition between X_A and X_B is increased, the number of mixed micelle types increases from zero to one, then to two, and finally decreases to one. The concentration at which the number of mixed micelle types begins to decrease is the cdc.

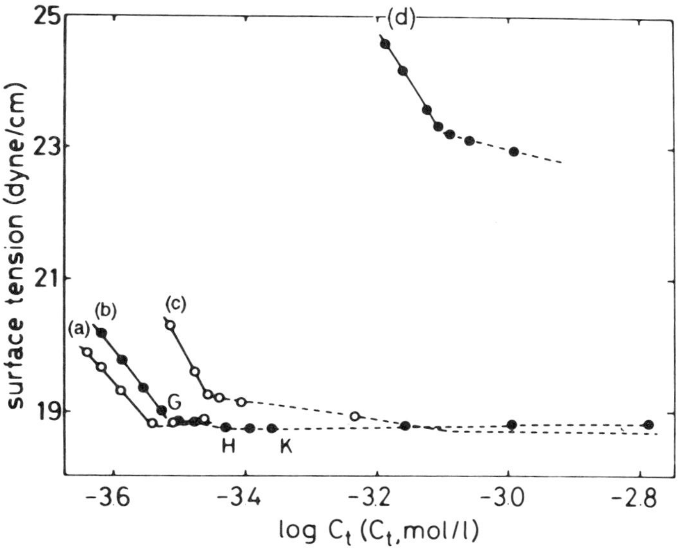

FIG. 7.13 Surface tension of the NF-STrS system vs. the logarithm of total surfactant concentration in 50 mM sodium chloride solution at 30°C. The overall mole fraction x_{STrS} is kept constant in each curve: (a) x_{STrS} = 0.094; (b) x_{STrS} = 0.249; (c) x_{STrS} = 0.540; (d) x_{STrS} = 0.903. (From Ref. 73. Reproduced by permission of Academic Press, Inc.)

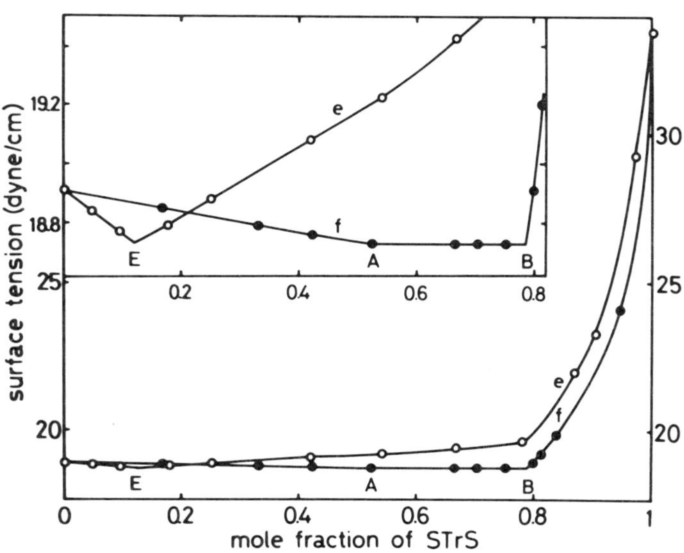

FIG. 7.14 Surface tension of the NF-STrS system plotted against the mole fractions of STrS in coexisting monomers (curve e) and micelles (curve f). (From Ref. 73. Reproduced by permission of Academic Press, Inc.)

Comicellization of NF and STrS is shown in Fig. 7.15 as a function of overall mole fraction of STrS. No micelles exist in region I, in region II micelles are NF rich, in region IV micelles are STrS rich, and in region III two kinds of mixed micelles coexist. The limited solubility of mixed micelles has been attributed to repulsion between hydrocarbon and fluorocarbon chains in the mixed micelle. To some extent, electrostatic interactions affect the miscibility of ionic fluorocarbon and hydrocarbon surfactants [74,75,80].

The fluorinated surfactant Neos Ftergent (NF) also forms two kinds of mixed micelles with sodium tetradecyl sulfate (STS) [74,75]. The plots of surface tension versus the logarithm of total surfactant concentration (Fig. 7.16), surface tension versus mole fraction of cmc (Fig. 7.17), and cmc versus the mole fraction of STS (Fig. 7.18) are similar to those of NF-STrS mixtures [73]. A comicellization diagram (Fig. 7.19) shows the relation between the total surfactant concentration and the overall mole fraction of STS. The area displayed is divided into four regions, I to IV, by solid lines. In region I there are no micelles. The micelles in region II are NF rich; in region IV the micelles are STS rich. In regions II and IV the compositions of coexisting monomers and micelles depend on the total concentration of the surfactants. In region III two kinds of micelles coexist with the monomers. The composition of the micelles is indicated by mole ratios X_A and X_B; the monomer ratio is given by the point E.

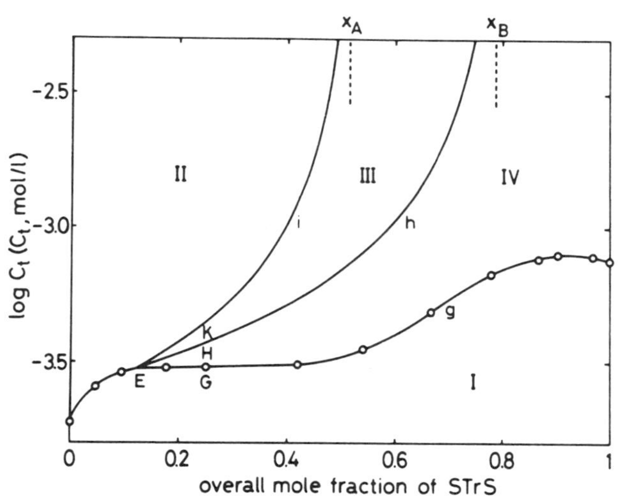

FIG. 7.15 Curves g and i show cmc and cdc of the NF-STrSD sytem. Region I, no micelles; region II, NF-rich micelles; region III, two kinds of mixed micelles; region IV, STrSD-rich micelles. (From Ref. 73. Reproduced by permission of Academic Press, Inc.)

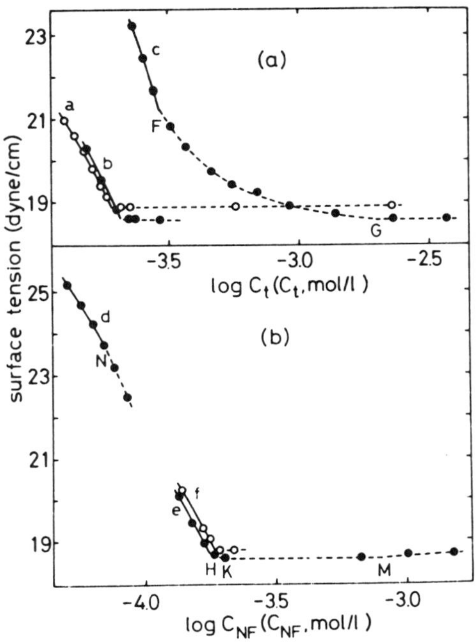

FIG. 7.16 (a) Surface tension plotted against the logarithm of total surfactant concentration at 30°C. The overall mole fraction, x_{STS}, of STS is kept constant: curve a, $x_{STS} = 0$; curve b, $x_{STS} = 0.097$; curve c, $x_{STS} = 0.649$. (b) Surface tension plotted against the logarithm of NF concentration at 30°C. The total STS concentration, C_{STS}, (mol/L), is kept constant: curve d, $C_{STS} = 2.01$; curve e, $C_{STS} = 5.75 \times 10^{-5}$; curve f, $C_{STS} = 7.60 \times 10^{-6}$. (From Ref. 74. Reproduced by permission of American Chemical Society.)

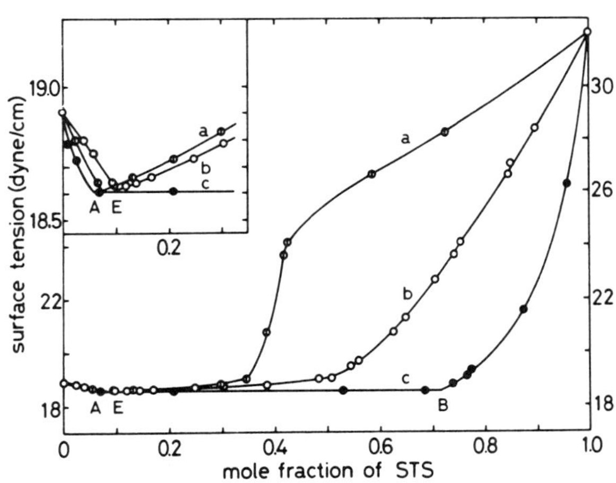

FIG. 7.17 Surface tension at the cmc (γ_{cmc}) plotted against the mole fraction of STS in adsorbed monolayers (curve a), in monomers (curve b), and in micelles (curve c). (From Ref. 74. Reproduced by permission of American Chemical Society.)

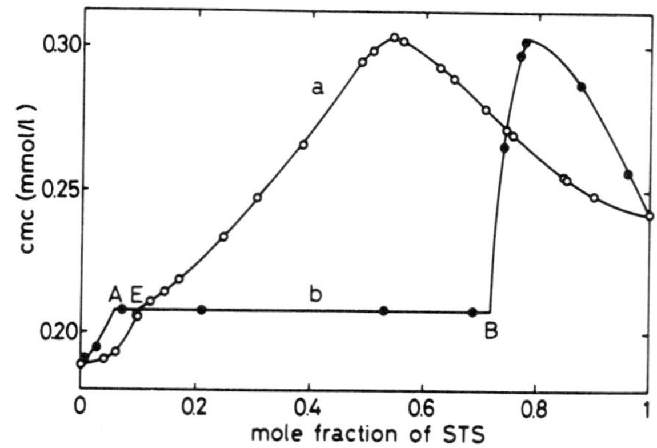

FIG. 7.18 Critical micelle concentration plotted against the mole fraction of STS in monomers (curve a) and in micelles (curve b). (From Ref. 74. Reproduced by permission of American Chemical Society.)

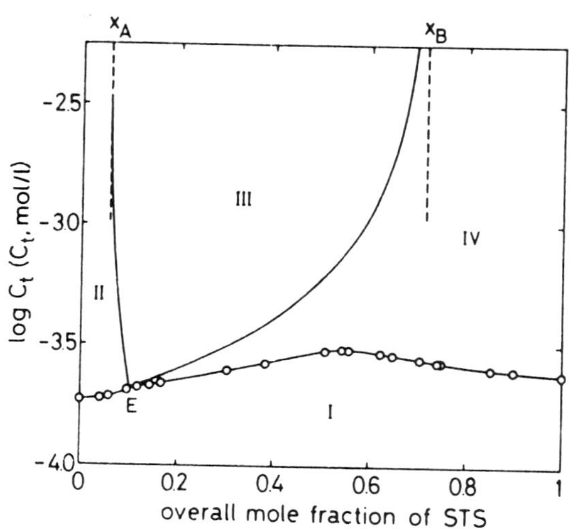

FIG. 7.19 Comicellization shown as a function of total concentration and overall mole fraction of STS in 0.05 M sodium chloride solution at 30°C: region I, no micelles; region II, NF-rich micelles; region III, two kinds of mixed micelles; region IV, SDTS-rich micelles. The dashed lines represent the asymptotes at infinity total concentration. (From Ref. 74. Reproduced by permission of American Chemical Society.)

The mutual solubility of fluorinated sulfonates and anionic hydrocarbon surfactants in micelles increases with increasing *temperature*. In the LiFOS-LiTS system the solubility of LiFOS increased substantially in the LiTS-rich micelle but only slightly in the LiFOS-rich micelle [63]. In the comicellar NF-STS system the temperature dependence of mutual solubility exhibited a critical solution temperature (cst). Above the cst only one kind of mixed micelle exists; below the cst two types of micelles can coexist. Above the critical demicellization concentration (cdc) only one kind of mixed micelle exists; below cdc two micellar species can coexist [75].

The miscibility of an anionic fluorinated surfactant with an anionic hydrocarbon surfactant is affected by the *electrolyte concentration* in the micellar solution [81]. Excess counterions increase binding of the counterions to ionic head groups and reduce electrostatic interactions between the head groups. Furthermore, the cst and the composition of the surfactant mixture at cst depend on the concentration of added electrolyte (Fig. 7.20). The mole fraction of NF in micelles at 30°C and the critical solution temperature increase when the NaCl concentration or the hydrocarbon chain length of sodium alkyl sulfate is increased [75].

The nonideal behavior of mixed hydrocarbon and fluorinated surfactants is evidenced by a *volume change* on comicellization [76,77]. According to the regular solution theory [50,82], the critical solution temperature and the composition of a fluorocarbon and hydrocarbon surfactant mixture should depend only on the solubility parameter and the molecular volume of the solutes. For

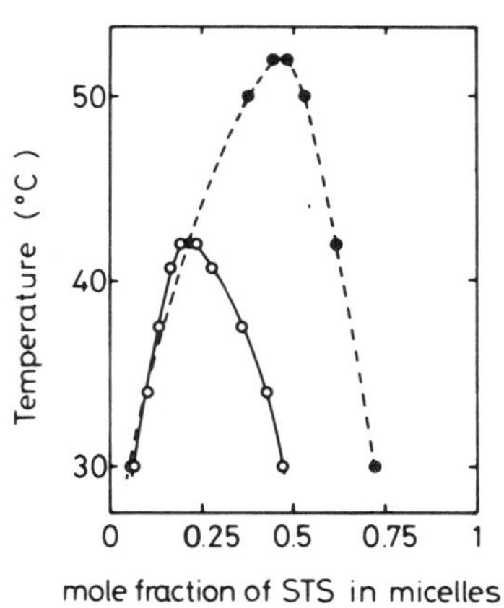

FIG. 7.20 The critical solution temperature and the composition of mixed NF-STS micelles in 10 mM (○) and 50 mM (●) sodium chloride solutions. (From Ref. 75. Reproduced by permission of Academic Press, Inc.)

an ideal mixture of two surfactants 1 and 2 in micelles, the molar volume of the mixed micelles is given by [77]

$$V = X_{m1}V_{m1} + X_{m2}V_{m2} \tag{13}$$

where V_{mi} is the molar volume of pure surfactant, i, micelles and x_{mi} the micellar molar fraction. For an ideal system the volume change on mixing is zero:

$$\Delta V = 0 \tag{14}$$

The volume change of mixing micelles of two nonionic surfactants with hydrocarbon chains, DE5 and DE7, is zero indeed (Fig. 7.21, curve b) (DE5 is pentaoxyethylene glycol dodecyl ether, DE7 is heptaoxyethylene glycol dodecyl ether). However, the molal volumes of the fluorinated surfactant NF and the hydrocarbon surfactant STS increase upon comicellization in the NF-STS system. ΔV is positive over the whole micellar composition range (curve a). A mixture of octaoxyethylene glycol dodecyl ether (DE8) and methyl *p*-hydroxybenzoate (MP) exhibited a negative molar volume change (curve c). The positive molar volume change of NTS and STS micelles is similar to a

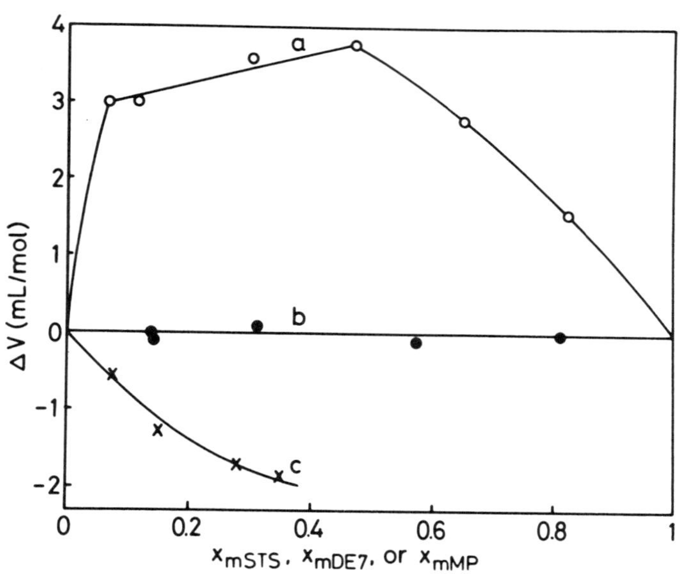

FIG. 7.21 Volume changes on mixing of micelles as a function of micellar composition x_{mSTS}, x_{mDE7}, or x_{mMP}: (○) NF-STS sytem; (●) DE5–DE7 system; (x) DE8-MP system; curve calculated from $\Delta V = 0.0089G_{ex}$ for the DE8-MP system, taking $\omega = -1.58RT$. (From Ref. 77. Reproduced by permission of American Chemical Society.)

volume increase when hydrocarbons and fluorocarbons are mixed [50]. The ΔV values for the NF-STS mixture appear to increase linearly with the mole fraction of STS in the range $0.065 < x_{mSTS} < 0.470$, and decrease at higher STS mole fractions, $x_{mSTS} > 0.470$, indicating the coexistence of two kinds of mixed micelles.

Mixtures of Anionic Fluorinated Surfactants

A mixed system of two fluorinated surfactants was studied by Yoda et al. [83]. Mixtures of LiPFO and LiFOS behaved almost like ideal systems, with an interaction parameter [44] β = -0.48 (Fig. 7.22). In contrast, the LiFOS and LiDS mixtures exhibited a positive deviation from ideal predictions. The β

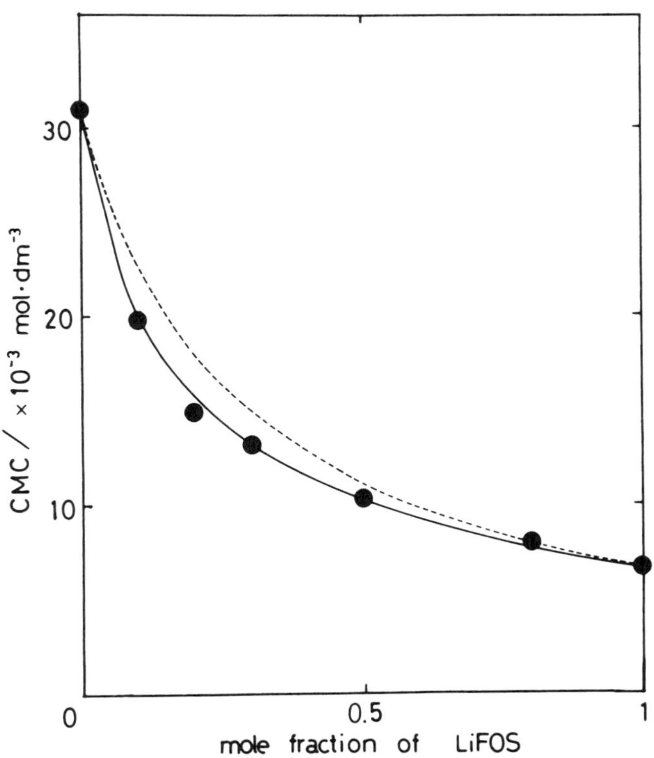

FIG. 7.22 Critical micelle concentrations for mixtures of LiPFO and LiFOS plotted against the mole fraction of LiFOS: solid line represents values calculated for a nonideal mixed micelle with $\beta = -0.48$, and dashed line represents cmc values calculated for ideal mixing. (From Ref. 83. Reproduced by permission of Academic Press, Inc.)

value of 1.36 was attributed to repulsive interactions between the hydrocarbon and fluorocarbon groups (Fig. 7.23).

Anionic-Nonionic Surfactant Mixtures

Binary hydrocarbon surfactant systems consisting of an anionic and a nonionic surfactant with almost equal cmc values generally exhibit a minimum cmc in the cmc versus composition curve. This negative deviation from the ideal cmc-composition relationship has been explained by electrostatic stabilization. Mixtures consisting of an anionic fluorinated surfactant and an anionic hydrocarbon surfactant exhibit a maximum in the cmc versus composition curve. The positive deviation from ideal predictions has been related to the mutual phobicity between fluorocarbon and hydrocarbon chains in the mixed micelle. In contrast, the deviation from ideal behavior is negative for mixtures consisting of an anionic fluorinated surfactant and a nonionic hydrocarbon surfactant. Apparently in the micelle containing a fluorinated anionic surfactant and a hydrocarbon chain nonionic surfactant, the reduction of electrostatic repulsion

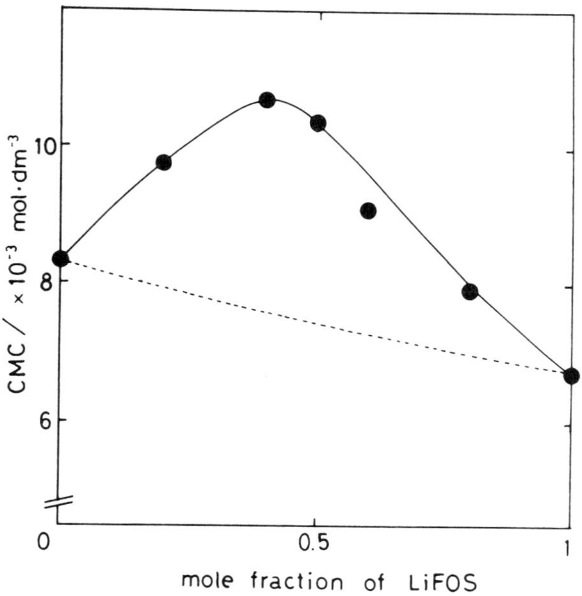

FIG. 7.23 Critical micelle concentrations for mixtures of LiDS and LiFOS plotted against the mole fraction of LiFOS: solid line represents values calculated for a nonideal mixed micelle with $\beta = +1.36$, and dashed line represents cmc values calculated for ideal mixing. (From Ref. 83. Reproduced by permission of Academic Press, Inc.)

between the hydrophilic groups overcomes the enthalpic destabilization resulting from the mixing of fluorocarbon and hydrocarbon chains [75]. A similar explanation was given by Yoda et al. [83] for mixtures of hexaoxyethylene dodecyl ether ($C_{12}E_6$) and LiFOS which deviated negatively from the ideal relationship with a β value of -4.8.

The surface tensions of mixtures containing Neos Ftergent, an anionic fluorinated surfactant, and dodecyl methyl sulfoxide (DMS) are plotted against the logarithm of total surfactant concentration C_t in Fig. 7.24 and against the mole fraction of DMS in Fig. 7.25. The mole fraction of curve c in Fig. 7.24 is equal to the composition of the intersection of curves a and b in Fig. 7.25. Funasaki and Hada [75] calculated micellar compositions from surface tension data assuming that the surface tension of a pure surfactant remains constant above cmc. This condition was met in the presence of an excess electrolyte. A plot of the cmc values against the mole fraction of DMS has a minimum and therefore shows a negative deviation from ideal behavior.

Surface tension curves of mixtures of LiFOS and hexaethyleneglycol *n*-dodecylether (6ED) are consistent with mixed micelle formation [84]. The surface tension curve of single LiFOS exhibited only one break corresponding to the cmc (Fig. 7.26). Addition of small amounts of 6ED lowered the surface tension of LiFOS from 33.0 mN/m above cmc to 25 mN/m. A minimum appeared near the cmc suggesting the existence of a mixed micelle.

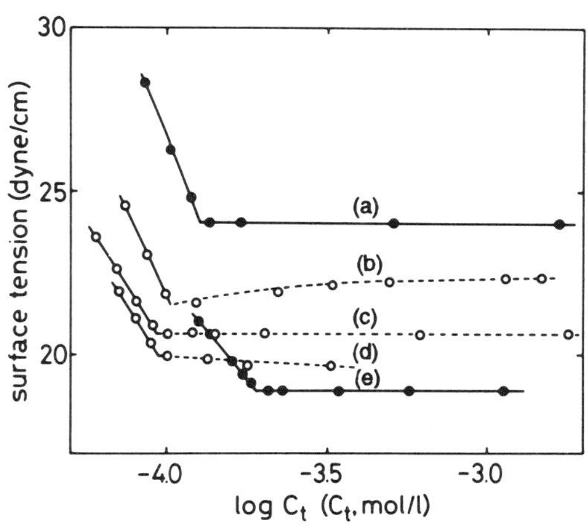

FIG. 7.24 Surface tension of the NF-DMS system vs. the logarithm of total surfactant concentration in 50 mM sodium chloride solution at 30°C. The overall mole fraction x_{DMS} is kept constant in each curve: (a) x_{DMS} = 1.0; (b) x_{DMS} = 0.961; (c) x_{DMS} = 0.818; (d) x_{DMS} = 0.572; (e) x_{DMS} = 0. (From Ref. 75. Reproduced by permission of Academic Press, Inc.)

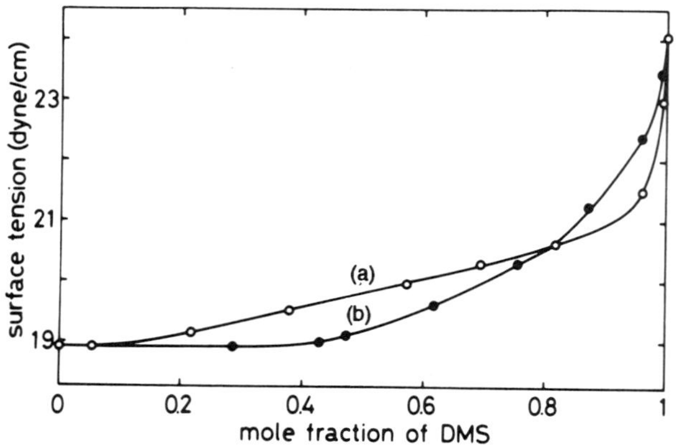

FIG. 7.25 Surface tension of the NF-DMS aqueous system plotted against the mole fractions of DMS in monomers (○) and in micelles (●). (From Ref. 75. Reproduced by permission of Academic Press, Inc.)

The surface tension of 6ED and that of its solutions containing small amounts of LiFOS are shown in Fig. 7.27. Above the cmc the surface tension of 6ED decreased with increasing amounts of LiFOS added. However, a minimum was not observed in the surface tension curves of 6ED and each curve exhibited a long plateau above cmc. Apparently LiFOS molecules added

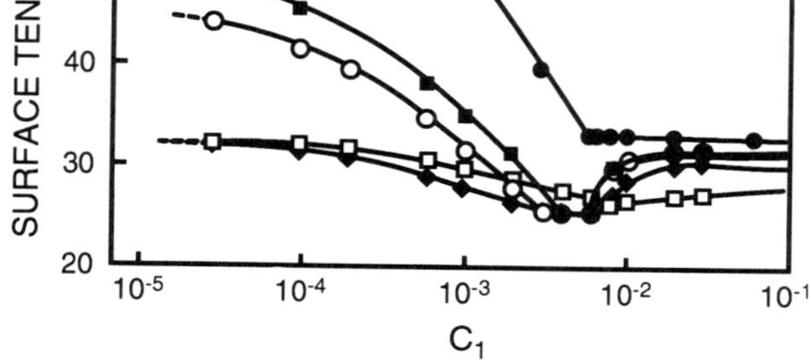

FIG. 7.26 Surface tension of LiFOS in the presence and absence of 6ED vs. the logarithm of LiFOS concentration, C_1. Additive concentration of 6ED (mM): (●), 0; (○), 0.005; (◆), 0.01; (■), 0.05; (□), 1.0. (From Ref. 84. Reproduced by permission of American Oil Chemists Society.)

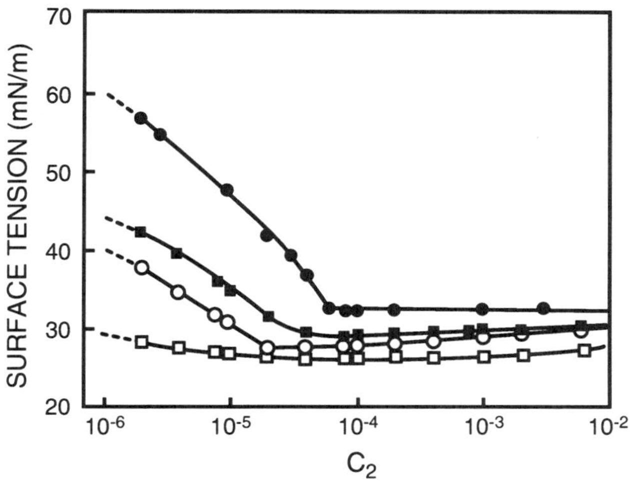

FIG. 7.27 Surface tension of 6ED in the presence and absence of LiFOS vs. the logarithm of 6ED concentration, C_2. Additive concentration of LiFOS (mM): (●), 0; (■), 0.5 ; (○), 1.0; (□), 7.0. (From Ref. 84. Reproduced by permission of American Oil Chemists Society.)

to the 6ED solution can replace some 6ED molecules adsorbed at the surface of the solution and some adsorbed LiFOS molecules can be interchanged with 6ED molecules. At a higher concentration of LiFOS, the surface excess of 6ED molecules is higher than in the single system. An addition of 6ED lowers the surface tension of LiFOS and a minimum in the curve (Fig. 7.26) suggests the formation of a mixed micelle.

The effect of LiFOS on micellization of 6ED was investigated by keto-enol tautomerism of benzoylacetanilide (BZAA) [85]. BZAA exists in the enol form in 6ED solutions above the cmc. The enolization in the hydrocarbon chain surfactant solution is similar to that in organic solvents. In aqueous solutions of LiFOS enolization was not observed. The enolic absorbance of BZAA at 315 nm above the cmc of 6ED decreases when LiFOS is added and the absorbance of the keto tautomer at about 250 nm increases. This result was explained by mixed micelle formation and the existence of a limited number of sites in the micelle which BZAA or LiFOS could occupy.

Meguro at al. [86] examined aqueous solutions of fluorocarbon and hydrocarbon surfactants by means of steady-state fluorescence of 8-anilino-1-naphthalenesulfonic acid ammonium salt (ANS). The fluorescence of ANS increases in the hydrocarbon environment. The fluorescence intensity of ANS in a hydrocarbon surfactant solution below the cmc is constant. However, above the cmc the fluorescence intensity of ANS in a solution of a hydrocarbon surfactant increases linearly with the surfactant concentration (Fig. 7.28). This indicates that ANS is solubilized by the hydrocarbon micelles. Unlike hydrocarbon surfactants, fluorinated surfactants do not solubilize ANS. These characteristics make ANS a useful probe for investigating mixed micellization. Three systems were studied: 6ED-LiFOS, 6ED-NFE7, and SDS-NFE7. The nonionic fluorinated surfactant NFE7 has the structure $[(CF_3)_2CF]_2C=C(CF_3)O(CH_2CH_2O)_7CH_3$. (The structure of the surfactant coded NF in Ref. 86 is different from the structure of the anionic surfactant NFtergent coded NF in Ref. 73 and is apparently related to the nonionic surfactant coded NFE in Ref. 79. The code NFE7 is used here to avoid possible confusion.)

The fluorescence data revealed that 6ED and LiFOS form mixed micelles over a wide concentration range. At constant 6ED concentrations above the cmc of 6ED, the fluorescence intensity of ANS decreases with increasing LiFOS concentration (Fig. 7.29), suggesting that mixed micelles are formed in which ANS is less soluble.

Unlike the 6ED and LiFOS mixture, the surfactants SDS and NFE7 are immiscible over the entire concentration range studied and form separate micelles.

Muto et al. [17] determined the aggregation number of lithium perfluorooctane sulfonate (LIFOS)-LiDS, hexaoxyethylene glycol dodecyl ether (C12E6)-LiFOS, octaoxyethylene glycol dodecyl ether (C12E8)-LiFOS, and

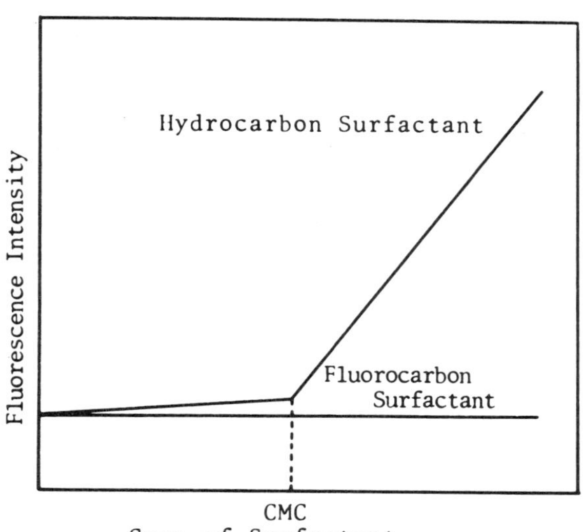

FIG. 7.28 Fluorescence intensity vs. surfactant concentration. (From Ref. 86. Reproduced by permission of American Chemical Society.)

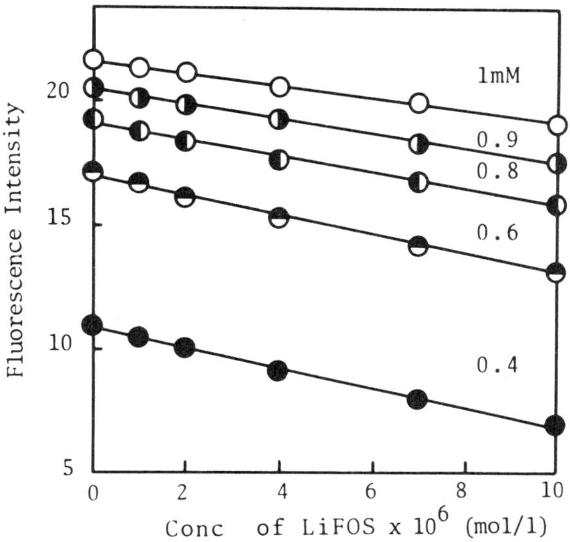

FIG. 7.29 The fluorescence of ANS as a function of LiFOS concentration in the 6ED-LiFOS mixed system. The numbers indicate fixed 6ED concentrations. (From Ref. 86. Reproduced by permission of American Chemical Society.)

C12E6-LiDS systems by fluorescence decay of micelle-solubilized pyrene in the presence of a micelle-solubilized quencher. The aggregation number increased considerably with the surfactant concentration. The measurements of the pyrene fluorescence lifetime and of the ratio of intensities of the first and third bands of the pyrene monomer fluorescence spectrum appeared to indicate that only one type of mixed micelle exists in the mixed system containing a fluorinated surfactant.

Muto et al. [87] studied interactions of anionic and nonionic surfactants by solubilization of a water-insoluble dye, Yellow OB. If the solubilizing power obeys the additivity rule, then the ideal solubilized amount, S_A, can be calculated as

$$S_A = S_1 C_{M1} + S_2 C_{M2} + S_W \qquad (15)$$

where S is the solubilizing power of each surfactant, denoted with the subscripts 1 and 2, C_M is the concentration of each surfactant in micelles, and S_W is the solubility of the solubilizate in water.

The solubilization efficiency was indicated by the ratio S_R:

$$S_R = \frac{S_A(\text{observed})}{S_A(\text{ideal})} \qquad (16)$$

where S_A(observed) and S_A(ideal) are the solubilized amounts measured and calculated for the same molar fraction of the nonionic surfactant in the mixed micelle.

When an anionic fluorocarbon surfactant (LiFOS) and a nonionic surfactant [hexaoxyethylene glycol dodecyl ether ($C_{12}E_6$) or octaoxyethylene glycol dodecyl ether ($C_{12}E_8$)] are mixed, the efficiency of solubilization decreases (Figs. 7.30 and 7.31). The incorporation of an anionic fluorinated surfactant into the nonionic micelle loosens the polyoxyethylene shell and causes desolubilization of the organic compound solubilized there [85,88,89]. The efficiency of solubilization decreases with increasing differences in the solubilization power of each individual component. The system containing LiFOS had a lower solubilization ratio than systems containing SDS or Aerosol OT.

The micellar size and charge in the mixed surfactant solutions of LiFOS-$C_{12}E_6$ and LiDS-$C_{12}E_6$ have been calculated using electric birefringence, interfacial tension, viscosity, dynamic light scattering, and electric conductivity data [90]. The component ratio at constant total surfactant concentration has a similar effect on the micellar shape and size for both systems (Fig.7.32). The length of the mixed rod-like micelles in the LiFOS-$C_{12}E_6$ system increases with increasing mole fraction of the anionic component and decreases after exhibiting a maximum value of 480 Å at a LiFOS mole fraction of 0.07 to 0.08. For the hydrocarbon LiDS-$C_{12}E_6$ system, the rod-like micelles have a maximum

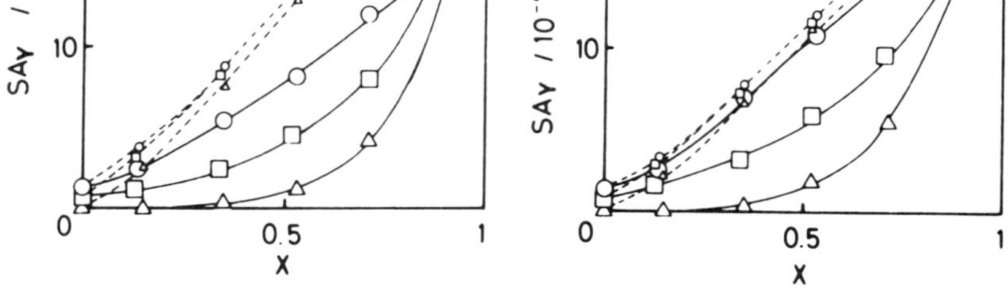

FIG. 7.30 The ideal (dashed line) and observed (solid line) amounts of solubilized Yellow OB (SA_y) as a function of the molar fraction of (a) $C_{12}E_6$ and (b) $C_{12}E_8$ in the mixed micelles. Total surfactant concentration 10 mmol/L. The symbols (○), (□), (△) refer to solutions containing SDS, AOT, and LiFOS, respectively. (From Ref. 87. Reproduced by permission of Academic Press, Inc.)

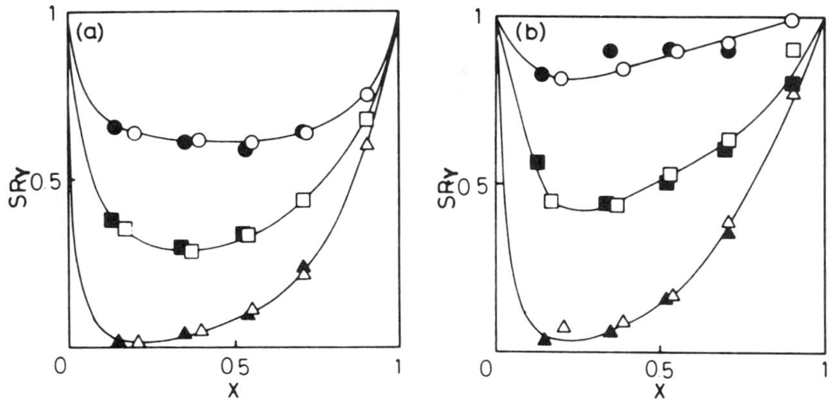

FIG. 7.31 The solubilized ratio (SR_y) of Yellow OB as a function of the molar fraction of (a) $C_{12}E_6$ and (b) $C_{12}E_8$ in the mixed micelles. Total surfactant concentration 10 mmol/L: (○), (□), (△); 20 mmol/L (·), (), (). The symbols ○, ● refer to solutions containing SDS; □, ■ to AOT; and △, ▲ to LiFOS, respectively. (From Ref. 87. Reproduced by permission of Academic Press, Inc.)

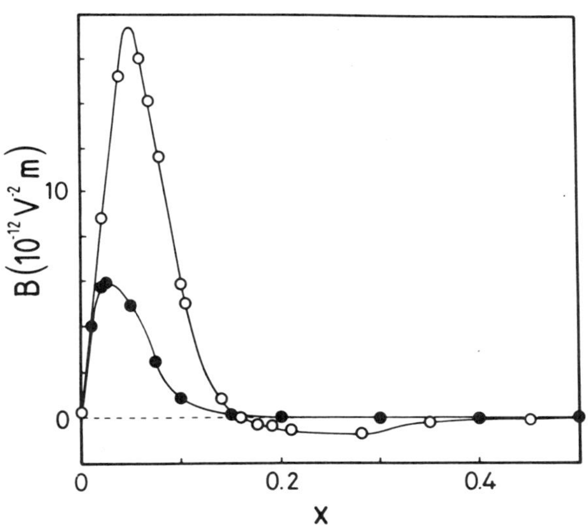

FIG. 7.32 The dependence of the Kerr constant, *K*, on the mole fraction of LiFOS in the mixture LiFOS-$C_{12}E_6$ (○) and the mole fraction of LiDS in LiDS-$C_{12}E_6$ (●). Total surfactant concentration 0.100 M, T = 25°C. (From Ref. 90. Reproduced by permission of Academic Press, Inc.)

value of 280 Å at an LiDS mole fraction of 0.025. A rod-to-sphere transition point corresponds to a mole fraction of 0.4 for the LiFOS-$C_{12}E_6$ system and 0.2 to 0.3 for the LiDS-$C_{12}E_6$ system. The mole fraction values determined by the electric birefringence method are in agreement with those obtained by viscosity and interfacial tension measurements.

The LiFOS-$C_{12}E_6$ micelles seem to have greater internal order than the LiDS-$C_{12}E_6$ micelles, probably because of the stiffer and shorter hydrophobe chain of the LiFOS surfactant than of the LiDS.

Sugihara et al. [91] investigated the effect of chain length on the *pressure* dependence of micellization in mixed surfactant solutions of nonyl-*N*-methylglucamine (MEGA9) and sodium perfluorooctanoate (SPFO). The cmc values of the mixed system are lower than the cmc values of the components (Fig. 7.33). The surface activity of the mixed system is higher than that of the components. The synergistic increase in surface activity leads to a decrease of the cmc [92,93]. The cmc curves (Fig. 7.33) indicate that the SPFO-MEGA9 system exhibits a negative deviation from ideal behavior, in contrast to the SPFO-SDeS system, which shows a positive deviation. The negative deviation implies that SPFO and MEGA9 are miscible in micelles and the repulsive forces between ionic head groups are reduced. The reduction in the repulsive forces probably results from the penetration of the nonionic surfactant into the space between the ionic head groups. The pressure dependence of this system is very small, probably because the volume change caused by micellization is small. The degree of counterion binding β, as well as the degree of head group ionization α, vary much less with pressure than α and β of anionic-anionic surfactant mixtures.

Esumi et al. [94,95] studied interactions between fluorocarbon-and hydrocarbon-type surfactants by their effect on monodispersed ferric hydrosols. The amounts of surfactants adsorbed showed that the formation of mixed bilayers of anionic fluorocarbon–nonionic hydrocarbon surfactants is more favorable than that of anionic fluorocarbon–anionic hydrocarbon surfactants (see Sec. 5.8).

Nonionic Surfactant Mixtures

The miscibility of anionic surfactant mixtures is affected by electrostatic effects which are absent in nonionic surfactant mixtures. This simplifies the interpretation of their solubility data. Furthermore, the miscibility of nonionic surfactants in micelles can be compared to their mutual solubility in the liquid state.

Funasaki and Hada [79] examined the mutual solubility of a fluorinated nonionic surfactant $[(CF_3)_2CF]_2C{=}C(CF_3)O(CH_2CH_2O)_nCH_3$ (NFE, average n = 18.4) and a nonionic hydrocarbon chain surfactant $CH_3(CH_2)_{11}O(CH_2CH_2O)_mH$ (DEm, m = 5,7,9, or 25). Curves of the surface tension,

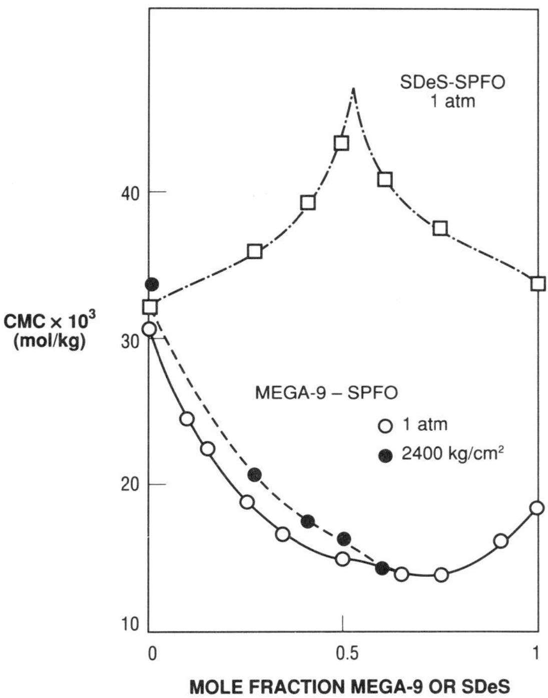

FIG. 7.33 The first cmc plotted vs. the mole fraction X_i for the MEGA9-SPFO system at 30°C under 1 atm and 2400 kg/cm^2 (lower curves.) The upper curve shows the first cmc values for the SDeS-SPFO mixed system. (From Ref. 91. Reproduced by permission of Plenum Publishing.)

plotted against the logarithm of total surfactant concentration for mixtures of NFE and DE7, are shown in Fig. 7.34. The constancy of surface tension beyond cmc (curve a) indicates that DE7 was highly pure, unlike NFE (curve f), which is difficult to purify and therefore contained impurities. The cmc values of NFE-DE7 mixtures exhibited a maximum at a mole fraction of 0.327 for DE7. The relation between the surface tension and the mole fraction of monomeric DE7 in the mixture of NFE and DE7 monomers is shown in Fig. 7.35 with open circles. The filled circles show the surface tension at about 10-fold concentrations above cmc as a function of DE7 mole fraction in the whole system. A plateau region in the micellar composition curve indicates the coexistence of two kinds of mutually saturated mixed micelles.

The surfactants NFE and DE7 mix partially in micelles and their mutual solubility increases with increasing temperature. The mutual solubility of NFE and DEm in micelles is shown in Fig. 7.36. Above the critical solution tem-

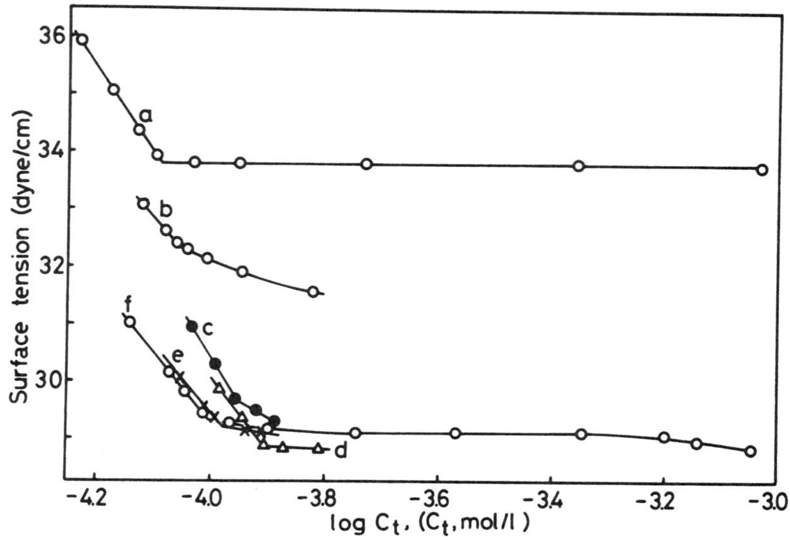

FIG. 7.34 Surface tension vs. log total surfactant concentration for mixtures of NFE and DE7 in water at 25°C. The overall mole fractions of DE7 are (a) 1.0; (b) 0.834; (c) 0.503; (d) 0.327; (e) 0.107; (f) 0. (From Ref. 79. Reproduced by permission of American Chemical Society.)

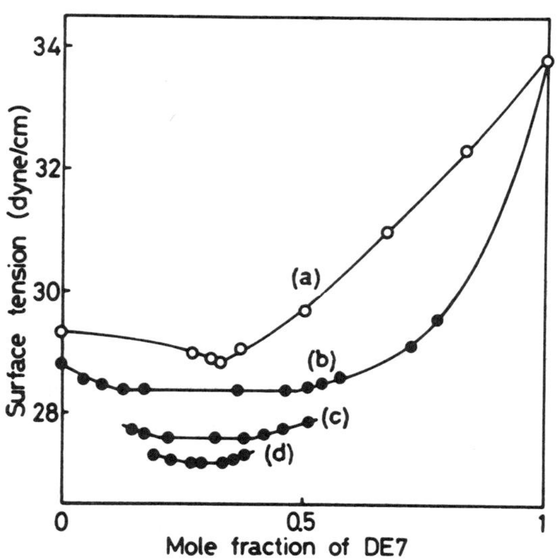

FIG. 7.35 Surface tension vs. mole fraction of DE7 in the system (●) and in monomers (○) for mixtures of NFE and DE7 in water at different temperatures: (a and b), 25°C; (c), 30°C; (d) 33°C. (From Ref. 79. Reproduced by permission of American Chemical Society.)

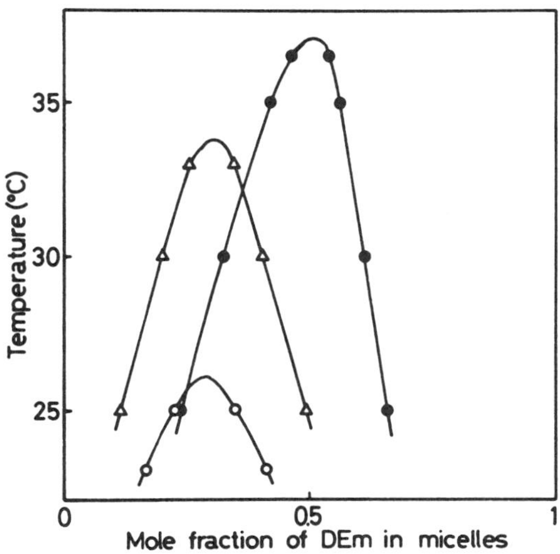

FIG. 7.36 Solubility-temperature diagram for the mixed NFE-DEm micelle: (●) NFE-DE5; (△) NFE-DE7; (●) NFE-DE9. (From Ref. 79. Reproduced by permission of American Chemical Society.)

perature (cst) only one kind of mixed micelle exists, whereas below cst two kinds of micelles can coexist, depending on the total concentration and chemical composition of the surfactants. The critical solution temperature and the critical mole fraction of DEm in the NFE-DEm system decrease with increasing number, *m*, of oxyethylene units. The critical solution temperature is higher for mixed NFE-DEm micelles than for liquid NFE-DEm mixtures, suggesting that the micellar core has a more extensively ordered structure than the liquid mixture.

Steady-state fluorescence studies by Meguro at al. [89] using ANS as the probe showed that the fluorinated surfactant $[(CH_3)_2CF]_2C=C[CF_3)O(CH_2CH_2)_7CH_3$ at concentrations below its cmc does not penetrate the micelles of the hydrocarbon surfactant $C_{12}H_{25}O(CH_2CH_2O)_6H$ (6ED). The surfactants form mixed micelles above the cmc of the fluorinated surfactant.

Oppositely Charged Surfactant Mixtures

Binary mixtures of an anionic hydrocarbon surfactant and a cationic fluorocarbon surfactant have a much lower surface tension and interfacial tension than the individual components [96] (Fig. 7.37). The same is true for mixtures of a cationic hydrocarbon surfactant and an anionic fluorocarbon surfactant. The

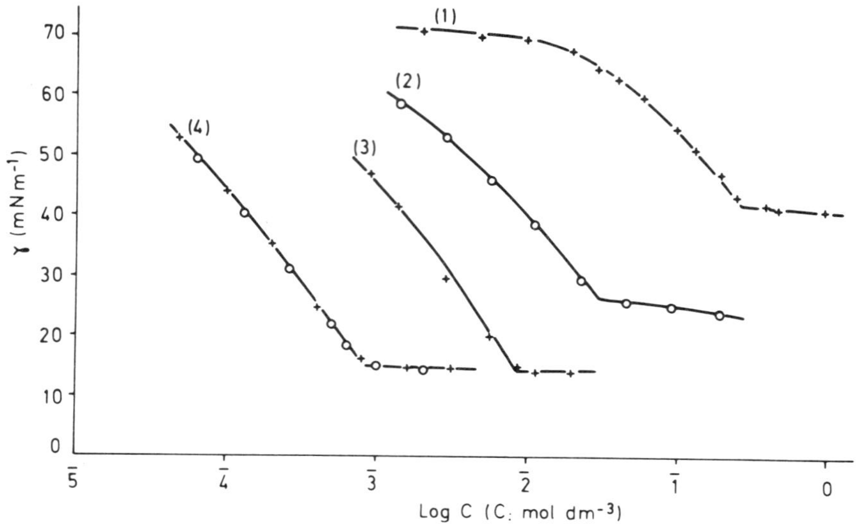

FIG. 7.37 Surface tension vs. log concentration curves for OTAB (1), SPFO (2), PFOA (3), and their 1:1 mixtures SPFO-OTAB (4,○) and PFOA-OTAB (4,+) at 30°C. (From Ref. 96. Reproduced by permission of Dr. Dietrich Steinkopf Verlag.)

mixtures investigated by Zhu and Zhao [96] included SPFO-OTAB, SPFO-DTAB, SPFO-HTAB, PFOA-OTAB, and FC3-SOS [OTAB, octyltrimethylammonium bromide; DTAB, dodecyltrimethylammonium bromide; HTAB, hexadecyltrimethylammonium bromide; SOS, sodium octylsulfate; PFOA, perfluorooctanoic acid; FC3, $C_3F_7OCF(CF_3)$ $CF_2OCF(CF_3CONH(CH_2H_5)_2$ $CH_3I]$.

The surface tension values of the mixtures SPFO-OTAB and PFOA-OTAB were identical (Fig. 7.37), indicating that the anion and cation interact so strongly that the different counterions of the individual components have no effect.

The mixed solutions of oppositely charged fluorocarbon and hydrocarbon surfactants have excellent spreading properties on oil. Hydrocarbon chains in the adsorption layer at a water-oil interface reduce the mutual phobicity between the fluorocarbon chain of the surfactant and the hydrocarbon chain of the oil and decrease the water-oil interfacial tension.

The mutual phobicity between hydrocarbon and fluorocarbon chains usually causes a positive deviation from ideal mixing with positive β values. In contrast, the strong interaction between anionic and cationic surfactants results in significant synergism with very large negative β values [97]. The interaction between oppositely charged ionic surfactants is much stronger than the interaction in mixtures of similarly charged ionic surfactants and nonionic surfactants.

7.4 MESOPHASES AND LIQUID CRYSTALS

A discussion of surfactant solutions usually involves the solid surfactant phase, dissolved surfactant monomers, and micelles in a solution above the cmc. Actually, a surfactant in a solvent can form several phases, depending on temperature and the surfactant concentration. A complete phase diagram of the surfactant-water system is essential for the understanding of the properties of the surfactant and its solutions [98].

Liquid crystals are thermotropic or lyotropic, depending on the mode of their formation. In solid crystals the ordering is three- or, in some exceptional cases, two-dimensional. When a solid is heated to its melting point, the crystal lattice of the solid collapses. When the melt is isotropic, molecules are randomly distributed and any orientational ordering is short range. In contrast, the melt of a thermotropic solid is anisotropic. The melt retains some its crystalline ordering, which is eliminated at a sufficiently higher temperature. At a second melting point the anisotropic melt converts to an isotropic fluid.

Lyotropic liquid crystals result from an interaction between a solid and a liquid. Hence lyotropic liquid crystals are binary systems. When a solid surfactant is brought into contact with water, the predominantly crystalline solid disintegrates. Molecular and micellar solutions are isotropic. Interactions and ordering in molecular solutions are only short range. However, the dissolution of the surfactant in water does not proceed directly to a micellar solution but involves transitions to intermediate phases. These mesophases (meso = in between) have retained some of the crystalline character of the solid. This dualism is expressed in the term liquid crystal [1,99,100].

Mesophases also form when the concentration of a surfactant in its micellar solution is increased. When the concentration of the surfactant above cmc is increased, the number of micelles and their size increase, in accord with the mass action model. Dilute micellar solutions are isotropic but at higher surfactant concentrations intermicellar interactions produce mesophases which are anisotropic and have one- or two-dimensional ordering.

The structure of mesophases has been investigated by polarizing microscopy, light scattering, x-ray scattering, and small-angle neutron scattering (see Chapter 9). The methods used in the past have been insufficient for the study of surfactant phases [101]. Recent studies have significantly benefited from NMR spectrography.

The structures of mesophases have been described as rod-like hexagonal phases (H_I and H_{II}), a lamellar phase (L_α), two sets of cubic phases (I_I, I_{II} and V_I, V_{II}), and nematic phases [1,100,102–104]. The liquid crystalline mesophases feature one-dimensional (lamellar), two-dimensional (hexagonal), or three-dimensional (cubic) translational periodicity. Nematic phases have only orientational order (Fig. 7.38).

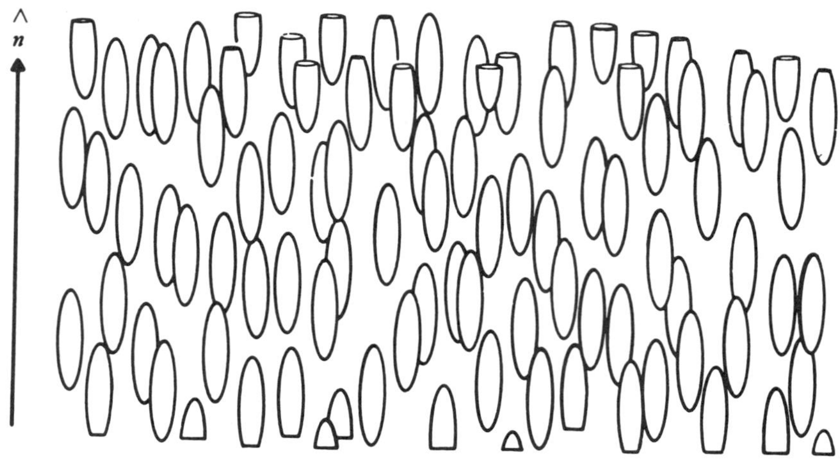

FIG. 7.38 Nematic liquid crystal structure. (From Ref. 99. Reproduced by permission of John Wiley & Sons.)

The hexagonal phases consist of rod-like micelles packed in a hexagonal assembly (Fig. 6.1). A lamellar phase consists of bilayer surfactant aggregates separated by water layers. Two types of cubic phases are known to exist in hydrocarbon-type surfactant systems. The first type occurs between micellar solutions and hexagonal phases, the second type between lamellar and hexagonal phases. The structure of the first type of cubic phase is believed to be a cubic arrangement of small micelles. The second type has been described as a regular three-dimensional network of surfactant aggregates. The cubic phases are isotropic.

In the hexagonal-lamellar phase transition region of hydrocarbon surfactants either a cubic phase exists, as with nonionic surfactants and short-chain ionic surfactants, or birefringent intermediate phases form, as for long-chain anionic surfactants and nonionic surfactants. The term intermediate phase refers to birefringent phases formed between hexagonal and lamellar phases [100,105]. This term used by Tiddy [100] is restricted to birefringent phases and excludes cubic phases. Since both anionic and nonionic surfactants form intermediate phases, intermicellar repulsion is not essential for the formation of an intermicellar phase. Hall and Tiddy [106] have postulated that the dominant factor which determines whether a cubic or an intermediate phase is formed is the balance between the type of the polar head and restrictions on alkyl chain conformational freedom and packing. The conformational restrictions increase with increasing chain length.

Nematic phases are solutions of orientationally ordered discoid (phase N_D) or columnar (N_C) micelles [107]. Nematic phases align in a magnetic field of adequate strength. The conditions for the existence of nematic phases are not completely understood [34]. Most of the binary surfactant systems at the transition from an isotropic solution to the liquid crystalline phase do not give a nematic phase but form a hexagonal or lamellar phase instead. One of the prerequisites for the formation of a nematic phase is the existence of anisotropic micelles in the isotropic phase. The magnitude of anisotropy and the size of micelles at the transition concentration must be within a narrow range favorable to the formation of the nematic phase.

Like surfactants with hydrocarbon chains, fluorinated surfactants can form liquid crystals. Liquid crystals of fluorinated surfactants are important because of industrial applications and their use as biological membrane models (see Chapter 8).

The mesomorphic phases formed by fluorinated surfactants have been the subject of numerous studies [22,33,34,71,102,108–125].

The phase behavior of fluorinated surfactants and hydrocarbon surfactants is remarkably similar. Tiddy and co-workers [108–111] observed that ammonium perfluorooctanoate and lithium perfluorooctanoate, like hydrocarbon surfactants, form a hexagonal phase, a lamellar phase, and an intermediate phase. A reversed hexagonal structure originally postulated for the intermediate phase [110,111] was found to be inconsistent with ^{19}F NMR observations and an alternative lamellar structure was proposed [112].

The question of whether fluorinated surfactants form cubic phases has not been answered unequivocally. Tiddy [100] has concluded that the preferential formation of intermediate phases instead of cubic phases depends on the balance between the polar head and the conformational restrictions of the hydrophobic chain. The fluorocarbon chains are stiffer than hydrocarbon chains and therefore have less conformational freedom. The stiffness of fluorinated alkyl chains is caused by a large energy difference between the gauche and trans conformations [100]. Hence fluorinated surfactants form intermediate phases rather than cubic phases. Although conformational restrictions decrease with decreasing chain length, even the shortest-chain fluorinated surfactant examined formed only intermediate phases [106]. Kekicheff and Tiddy [102] studied the structure of the intermediate phase using high-tensity, high-resolution x-rays from a synchrotron source. The intermediate phase was found to have a repeated layer structure closely related to the lamellar (L_α) phase. Kekicheff and Tiddy proposed that the intermediate phase is a lamellar phase where the layers have a regular array of holes through which water and ions can diffuse.

Guo et al. [22] studied fluorinated surfactants (perfluoroheptanoic acid, its salts, and ethoxylated amide of perfluoroheptanoic acid). The characteristics of ^{19}F NMR spectra and optical isotropy suggested that the aggregates of nonionic fluorinated surfactants (ethoxylated amides) may exist in a cubic phase. This conclusion is tentative and needs further investigation.

Boden et al. [119] showed that a thermodynamically independent nematic mesophase exists in the cesium perfluorooctanoate (CsPFO)-water system between 37 and 87% w/w ^2H_2O and 11 to 75°C. The nematic phase is intermediate to an isotropic micellar solution at higher temperatures and a smectic lamellar mesophase at lower temperatures. Microscopy and deuterium NMR spectroscopy indicate that the nematic phase is a solution of disk-shaped micelles in water. Their aggregation number initially increases with concentration, reaches a maximum, and then decreases [107]. The nematic phase of CsPFO-water is stable over a wide range of concentration without needing a cosurfactant or salt as a stabilizer. The disk-shaped micelles of the nematic phase orient with their unique axis parallel to the direction of an applied magnetic field.

Herbst et al. [34] have shown that a 30% w/w solution of tetramethylammonium perfluorononanoate in D_2O forms three lyotropic phases: an isotropic solution at $\geqslant$37°C, a nematic phase in the temperature range between 32 and 37°C, and a lamellar phase in the 25 to 32°C temperature range. The isotropic solution near the liquid crystal phase boundary contains disk-like micelles. The aggregation number of the micelles is about 150 and the thickness about 35 Å, including head groups and counterions.

Small-angle neutron scattering (SANS) indicated that the aligned nematic phase of tetramethylammonium perfluorononanoate maintains its alignment when it is cooled in a magnetic field to the temperature of the lamellar phase and the magnetic field is withdrawn.

The lamellar phase of tetramethylammonium perfluorononanoate consists of double layers of disk-like micelles alternating with layers of D_2O. The radius of the disks is probably about 59 Å. The polydispersity of the disk-like micelles is narrow. The layer does not have a two-dimensional long-range order but a long-range periodicity exists between the layers. In the nematic phase the long-range periodicity perpendicular to the layers is missing.

The *effect of additives* on mesophases of fluorinated surfactants has been studied by Tiddy and Wheeler [110] and Rosenblatt [123]. Tiddy and Wheeler described the effects of *n*-octanol on the ammonium perfluorooctanoate–water system with a three-component phase diagram (Fig. 7.39). The main differences between phase diagrams for this system and that for sodium octanoate-octanol-water were related to mutual phobicity between fluorocarbon and hydrocarbon chains. Octanol was found to be less soluble in the aqueous micellar phase of ammonium perfluorooctanoate than that of sodium octanoate. However, ammonium perfluorooctanoate is more soluble in octanol than sodium octanoate. This solubility difference is probably related to the effect of counterions, since ammonium salts are usually more soluble in octanol than sodium salts.

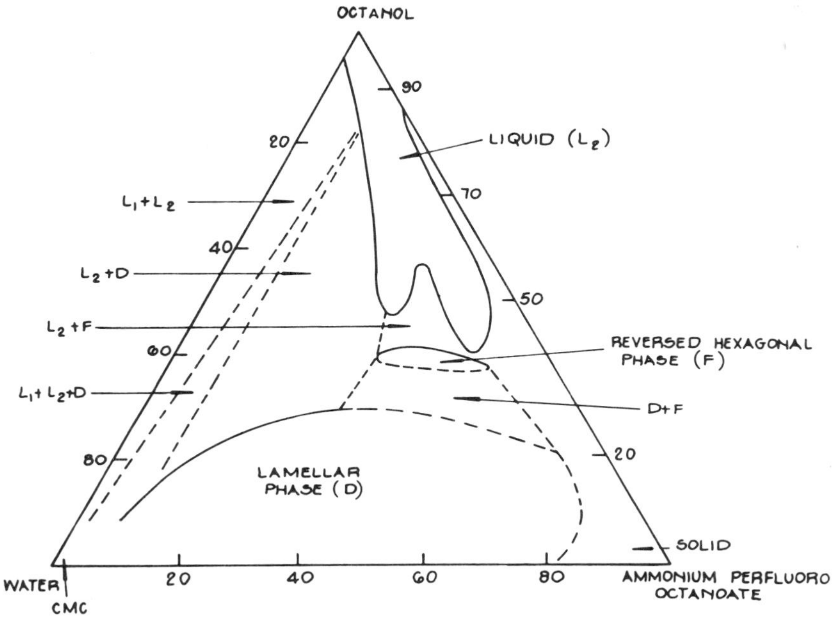

FIG. 7.39 Phase diagram of APFO-octanol-water system at 298 K. (---) Boundaries not accurately determined. (L_1) aqueous solution, (L_2) octanol solution, (D) lamellar phase consisting of lipid bylayers separated by water. (From Ref. 110. Reproduced by permission of Academic Press, Inc.)

Rosenblatt [123] found that the addition of perfluorooctanol, a cosurfactant, to the oblate cesium perfluorooctanoate–water micellar system raised the nematic-isotropic transition temperature. The micelles grew substantially when the cosurfactant was added.

The phase behavior of a *mixed surfactant system* consisting of lithium dodecyl sulfate (LiDS), lithium perfluorooctanesulfonate (LiFOS), and water was investigated by Tamori et al. [71]. The phase characteristics of the individual components, LiDS and LiFOS, are similar. The LiDS-water phase diagram, shown in Fig. 7.40, has three liquid crystal regions—hexagonal (H), bicontinuous cubic (V_I), and lamellar (Lam). The LiFOS-water phase diagram is similar (Fig. 7.41) and the same three liquid crystalline phases exist. In the mixed LiDS-LiFOS system (Fig.7.42), successive phase transitions, L_I-H-V_I-Lam-crystals, occur when the total surfactant concentration is increased. A phase region containing both Lam and H phases simultaneously has not been observed. This suggests that LiDS and LiFOS are mixed with each other, at least macroscopically.

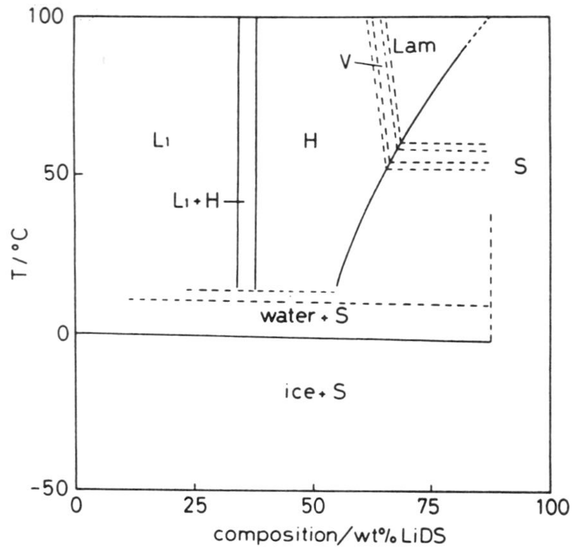

FIG. 7.40 The phase diagram of LIDS-water system; (L_1), aqueous surfactant solution; (H), hexagonal phase; (V), bicontinuous cubic phase; (Lam), lamellar phase; (S) indicates the presence of solid surfactant. (From Ref. 71. Reproduced by permission of Academic Press, Inc.)

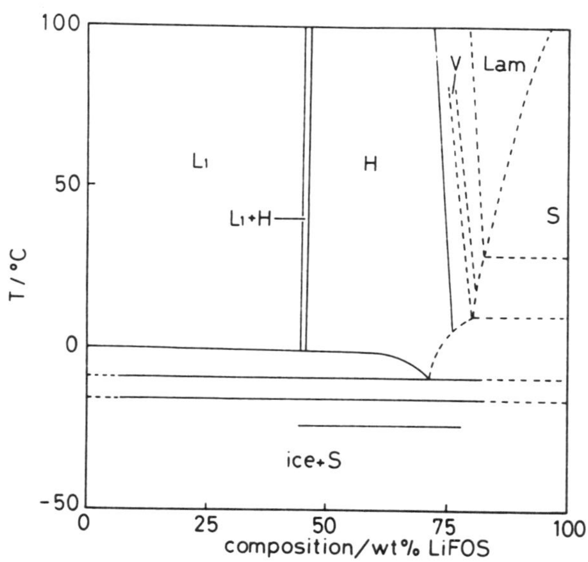

FIG. 7.41 Phase diagram of LIFOS-water system. Symbols as for Fig. 7.40. (From Ref. 71. Reproduced by permission of Academic Press, Inc.)

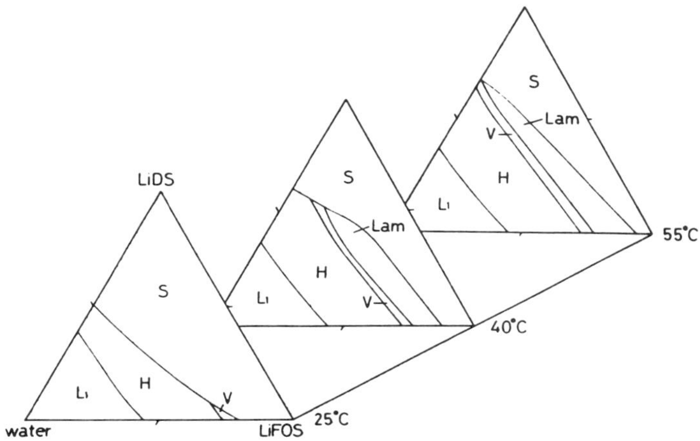

FIG. 7.42 Triangular phase diagram for the LiFOS-LiDS-water system. Symbols as for Fig. 7.40. (From Ref. 71. Reproduced by permission of Academic Press, Inc.)

The phase region of S becomes narrower with increasing temperature, whereas other boundary lines between each L_I, H, V_I, and Lam remain constant. This situation is related to the much lower entropy of the hydrocarbon chain in the crystal than in the other phases.

Tamori et al. [71] investigated the structure of the hexagonal phase by x-ray diffraction. The diameter of the rod-like micelles was estimated from the longest Bragg spacing and the molar volume. The pure LiDS micelles in the hexagonal phase had a diameter of 35 Å. The LiFOS micelle in the hexagonal phase had a diameter of 29 Å. The radius, 14.5 Å, is longer than the estimated length of the LiFOS molecule, 13 Å. This is in accord with the observations that perfluorocarbons prefer the trans form to the gauche form [49,125]. The diameter of the mixed hexagonal phase decreases linearly with increasing LiFOS mole fraction. This observation suggests that the LiFOS molecule does not reach to the center and the rod-like micelles have a core consisting of LiDS.

In concentrated solutions, mixed micelles exist in all LiDS-LiFOS mole fraction ranges with a positive interaction parameter. The mixed micelle has a LiDS core, similar to the mixed hexagonal phase.

7.5 CURRENT RESEARCH

Research on the micelle structure and interactions of fluorinated surfactants is continuing with the main focus on mixed surfactant systems. The interest in mixed surfactant systems has theoretical as well as practical reasons. The realization that surfactants are used most often as mixtures has shifted research from pure surfactants to mixed systems. Industrial fluorinated surfactants are usually mixtures because

1. They are synthesized as a mixture of isomers or telomers, or
2. Different surfactants are mixed to enhance the performance of the individual components by synergistic interactions.

Mixed systems containing fluorinated surfactants and hydrocarbon surfactants are currently investigated for several theoretical and practical reasons:

1. The increased efficiency of surfactant mixtures allows to reduce the amount of a fluorinated surfactant needed.

2. The fluorinated surfactant-hydrocarbon surfactant mixtures have unique properties. In two-phase systems of water and a hydrocarbon solvent, the fluorinated surfactant reduces surface tension and the hydrocarbon surfactant decreases the interfacial tension. For example, an aqueous foam of mixed surfactants spreads on a hydrocarbon solvent because the fluorinated surfactant adsorbs preferentially at the air-water interface while the hydrocarbon surfactant adsorbs at the water-oil interface (see Chapter 8, Fire-fighting foams).

3. The fluorinated surfactant-hydrocarbon surfactant mixtures are of great theoretical interest for studying the formation and structure of micelles. Fluorocarbon groups are not only hydrophobic but are oleophobic as well. The phobic interactions between fluorinated and hydrocarbon chains in surfactant mixtures cause nonideal behavior and in some systems, demixing. The formation of mixed micelles and the coexistence of two different kinds of micelles are being investigated to gain more insight into the behavior of micellar solutions.

Because of the extensive practical and theoretical interest in mixed fluorocarbon-hydrocarbon surfactants, it is not surprising that within a year after writing this chapter an excellent monograph on mixed surfactant systems [126] and several significant papers on fluorocarbon-hydrocarbon surfactant mixtures have appeared in print.

Abe et al. [127] have studied mixed surfactant systems consisting of a nonionic hydrocarbon surfactant ($C_{16}H_{33}O(C_2H_4O)_{20}H$) and an anionic fluorinated surfactant (ammonium perfluorooctanoate or ammonium perfluorodecanoate). Dynamic and static light scattering and fluorescence probe measurements revealed mixed micelle formation. Penetration of the anionic fluorinated surfactant into the nonionic micelle causes intermicellar repulsion and, with increasing mole fraction of the fluorinated surfactant, the micelle size (aggregation number) decreases.

Guo et al. [128] have studied mixtures of sodium perfluorooctanoate (SPFO) with a nonionic, an amphoteric, and a cationic hydrocarbon-type surfactant by NMR and surface tension measurements:

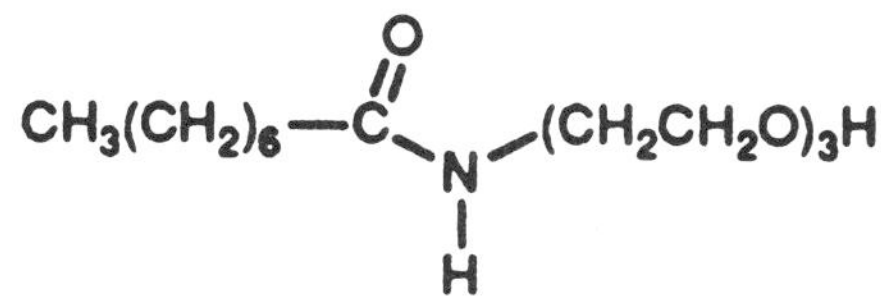

N-triethoxylated heptanylamide (HEA8-3)

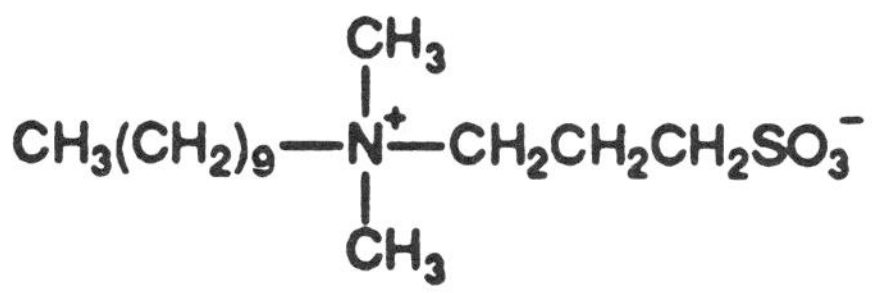

3-(decyldimethylammonio)-1-propanesulfonate (DEDIAP)

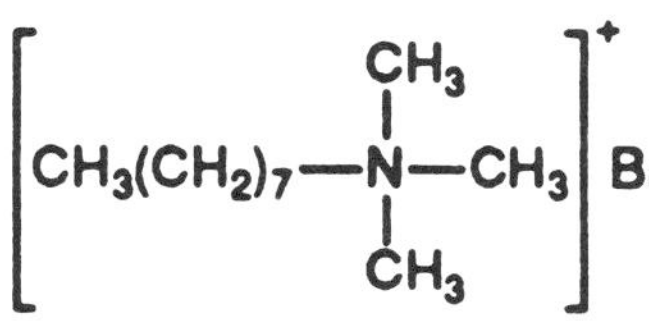

octyltrimethylammonium bromide (OCTAB)

The anionic fluorinated surfactant (SPFO) forms with the nonionic or the amphoteric hydrocarbon surfactant mixed micelles containing both types of surfactants. Both systems exhibit a negative deviation from ideality. Changes of the ^{19}F and ^{1}H chemical shifts of the two surfactants upon mixing are consistent with pseudo-phase diagrams, calculated from the cmc dependence on the fluorinated surfactant mole fraction. The interpretation of data was based a modified regular solution theory and the phase separation model. The anionic-cationic mixture forms instead of normal micelles larger aggregates, containing probably ion pairs.

The interaction of cationic fluorocarbon and hydrocarbon surfactants was studied by Tamori et al. [129]. The mixed cmc of diethanolheptadecafluoro-2-2-undecanolammonium chloride and dodecyltrimethylammonium chloride was determined by electric conductivity measurements. Partition coefficients of alcohols (methanol and $C_4F_7H_2OH$) and a fluorescent probe (pyrene-3-carboxaldehyde) between micelles and the bulk aqueous phase were determined. The data interpreted by a regular solution theory fitted an interaction parameter $\beta = 1$, indicating a much smaller repulsive interaction between the two cationic surfactants than that between an anionic fluorinated surfactant and an anionic hydrocarbon-type surfactant. The weak repulsion between the two cationic surfactants was explained by a large difference in their cmc values.

The miscibility of lithium tetradecyl sulfate (LiTS) and lithium perfluorooctanesulfonate (LiFOS) has been examined by Matsuki et al. [130] in aqueous solution and the adsorbed film. Surface tension measurements by the drop volume method indicated that the mixture of LiTS and LiFOS, like LiDS-LiFOS, forms a heterogeneous azeotrope (hetero-azeotrope). The LiTSD and LiFOS molecules are partially immiscible in the micellar state and form two kinds of micelles which coexist in equilibrium only at the composition of the hetero-azeotropic point. This situation is related to a strong repulsive interaction between the hydrocarbon and fluorocarbon chains in the micelle. A comparison of the results to those of the LiDS-LiFOS system suggested that the immiscible range of the composition of surfactants in the micelle increases with increasing chain length of the hydrocarbon surfactant. The hydrocarbon chain length affects as well the miscibility of the surfactants in the adsorbed film at the water/air interface.

Esumi and Ogawa [131] studied micellar solutions of an amphoteric fluorinated surfactant, [*N*-[3[[(Tridecafluorooctyl)sulfonyl]amino]propyl]-*N,N*-dimethylammonio]acetate [$C_6F_{13}C_2H_4SO_2NH(CH_2)_3N^+(CH_3)_2COO^-$, FOSAB], and its mixtures with LiFOS and LiDS by surface tension, fluorescence probing, and viscosity measurements. At cmc and at 25°C, the surface tension value for FOSAB alone is 20 mN/m. LiFOS and LiDS lower the surface tension of FOSAB to 14.5 and 16.5 mNm. This large reduction of surface tension results from electrostatic attraction between the cationic functionality of FOSAB and the anionic groups of LiFOS and LIDS.

In the FOSAB-LiDS system, the mixed cmc values coincided almost with those predicted for an ideal system. Esumi and Ogawa [131] explained this ideal behavior by a mutual compensation of two effects: (1) an increase of the mixed cmc, caused by a weaker interaction between hydrocarbon and fluorocarbon chains in the mixed micelle than between individual hydrocarbon and fluorocarbon chains, and (2) favorable mixed micelle formation because of electrostatic attraction between the cationic functionality of FOSAB and the anionic functionality of LiDS. At 0.8 molar fraction of FOSAB, the FOSAB-LiDS system exhibited strong shear dependence, suggesting the presence of rod-like micelles.

The cmc curves for the FOSAB-LiFOS system exhibited a large deviation from ideality, with an interaction parameter $\beta = -8.3$.

The solubilization of *n*-alkanols (1-butanol, 1-pentanol, and 1-hexanol) and fluorinated alkanols, ($C_nH_2F_{2n-1}OH$, $n = 3, 4,$ or 5) in single and mixed aqueous solutions of fluorinated surfactants LiDS, LiFOS, LiPFO and a nonionic hydrocarbon-type surfactant (hexaethyleneglycol dodecyl ether (6ED) was investigated by Yoda et al. [132]. In single surfactant solutions partition coefficients between micelles and water were higher for solutions in which the alkyl chains of the surfactant and of the alcohol were similar, than for solutions containing a surfactant and an alcohol with dissimilar alkyl chains. In the LiDS-LiFOS system with a positive β value, the partition coefficients of alkanols or fluorinated alkanols were higher than for solutions of single surfactants. The LiFOS-LiPFO system exhibited an almost ideal behavior. The partition coefficients of the fluorinated alkanols were higher than those of alkanols. In contrast, the β values for the 6ED-LiFOS system were negative and the partition coefficients of the alcohols (alkanols or fluorinated alkanols) were lower than in solutions of the single surfactants.

Mixtures of fluorinated surfactants and hydrocarbon surfactants with the same hydrophile form mixed micelles or two types of micelles, one rich in the fluorinated surfactant and the other rich in hydrocarbon surfactant. To circumvent the mutual solubility problem, surfactants have been synthesized with both hydrocarbon and fluorocarbon hydrophobes in the same molecule [133,134,135].

Guo et al. [135, 136] synthesized hybrid surfactants with a fluorocarbon group (m = 6–8) and a hydrocarbon group (n = 1–9)

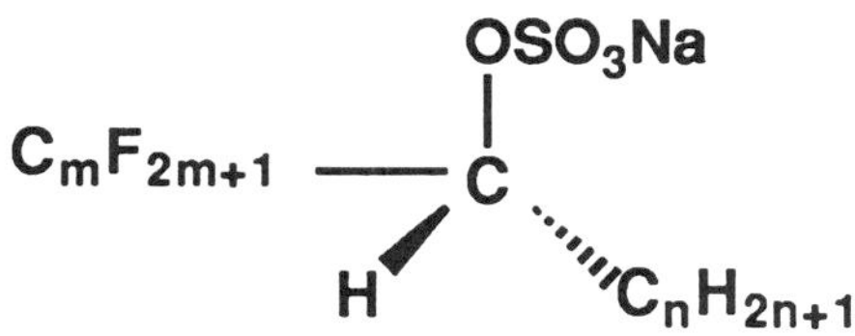

In aqueous solutions the surface tensions of these hybrid surfactants are about equal or slightly lower than that of the corresponding surfactant with a single hydrophobe. The cmc values are relatively low and are governed by the Kleven's equation. The increase of the hydrocarbon chain length CH_2 group decreases the cmc by about 35%. The increase of the fluorocarbon chain by each CF_2 group decreases the cmc by about 75%. Quantitative analysis of the F^{19} NMR spectra has revealed that the residence time of the fluorocarbon chain in the micelle is longer than that of single chain surfactants. The residence time of the CF_3 group increases monotonically with the increase in total surfactant concentration and is several times longer than that of the $\alpha-CF_2$ group. The residence time of the latter does not depend on the surfactant concentration.

Research on binary fluorinated surfactant-hydrocarbon surfactant mixtures is continuing with the main focus on the mutual solubility of mixed surfactants. It has been agreed [137] that an increase in the number of carbon atoms in either hydrophobe decreases the mutual solubility. An increase in the number of oxyethylene units in a nonionic hydrophile increases the mutual solubility of the mixed surfactants. An increase in temperature increases the mutual solubility. However, the mechanisms of fluorocarbon-hydrocarbon surfactant demixing and the formation of two types of coexisting micelles are still not completely understood. The gaps in knowledge are not caused by a lack of theories but are a result of experimental limitations. The development of more powerful analytical methods for the study of surfactants in micelles will advance the quantitative understanding of mixed surfactant systems. Finally, the advanced theoretical knowledge of micellar fluorinated surfactant systems will be applied to practical processes, such as adsorption, wetting, solubilization, emulsification, foaming, liquid crystal formation, and others.

REFERENCES

1. J. M. Corkill and J. F. Goodman, Adv. Colloid Interface Sci. 2, 297 (1969).
2. E. H. Lucassen-Reynders, ed., "Anionic Surfactants. Physical Chemistry of Surfactant Action," Surfactant Science Series, Vol. 11, Marcel Dekker, New York (1981).
3. R. Zana, ed., "Surfactant Solutions," Surfactant Science Series, Vol. 22, Marcel Dekker, New York (1987).
4. M. J. Schick, ed., "Nonionic Surfactants. Physical Chemistry," Surfactant Science Series Vol. 23, Marcel Dekker, New York (1987).
5. D. N. Rubingh and P. M. Holland, eds. "Cationic Surfactants. Physical Chemistry," Surfactant Science Series, Vol. 37, Marcel Dekker, New York (1991).
6. M. J. Rosen,"Surfactants and Interfacial Phenomena," 2nd ed., John Wiley & Sons, New York, NY (1989)
7. H. E. Gerry, P. T. Jacobs, and E. W. Anacker, J.Colloid Interface Sci. 62, 556 (1977).
8. J. D. Bernal, Trans. Faraday Soc. 42B, 197 (1946).
9. W. D. Harkins, R. W. Mattoon, and M. L. Corrin, J. Am. Chem. Soc. 68, 220 (1946).
10. R. W. Mattoon, R. S. Stearn, and W. D. Harkins, J. Chem. Phys. 16, 644 (1948).
11. P. A. Winsor, J. Phys. Chem. 56, 391 (1952).
12. F. Reiss-Husson and V. Luzzatti, J. Phys. Chem. 68, 3504 (1964).
13. F. Reiss-Husson and V. Luzzatti, J. Colloid Interface Sci. 21, 534 (1966).
14. R. Svens and B. Rosenholm, J. Colloid Interface Sci. 44, 495 (1973).
15. C. H. Arrington, Jr. and G. D. Patterson, J. Phys. Chem. 57, 247 (1953).
16. J. Turro and P. C. C. Lee, J. Phys. Chem. 86, 3367 (1982).
17. Y. Muto, K. Esumi, K. Meguro, and R. Zana, J. Colloid Interface Sci. 120, 162 (1987).
18. H. Hoffmann and B. Tagesson, Z. Phys. Chem. Neue Folge 110, 113 (1978).
19. H. Hoffmann and W. Ulbricht, Z. Phys. Chem. Neue Folge 106, 167 (1977).
20. J. Rassing and P. J. Sams, J. Chem. Soc. Faraday Trans. 2, 70, 1247 (1974).
21. H. Hoffmann, W. Ulbricht, and B. Tagesson, Z. Phys. Chem., Neue Folge 113, 17 (1978).
22. W. Guo, T. A. Brown, and B. M. Fung, J. Phys. Chem. 95, 1829 (1991).
23. N. Muller and R.H. Birkhahn, J. Phys. Chem. 71, 957 (1967).
24. N. Muller and R. H. Birkhahn, J. Phys. Chem. 72, 583 (1968).
25. N. Muller and T. W. Johnson, J. Phys. Chem. 73, 2042 (1969).
26. R. E. Bailey and G. H. Cady, J. Phys. Chem. 73, 1612 (1969).

27. N. Muller and F. E. Platko, J. Phys. Chem. 75, 547 (1971).
28. N. Muller and H. Simsohn, J. Phys. Chem. 75, 942 (1971).
29. N. Muller, J. H. Pellerin, and W. W. Chen, J. Phys. Chem. 76, 3012 (1972).
30. P. Mukerjee and K. J. Mysels, in "Colloidal Dispersions and Micellar Behavior," K. L. Mittal, ed., p. 239. ACS Symposium Series, No. 9, American Chemical Society, Washington, DC (1975).
31. U. Henrikson and L. Ödberg, J. Colloid Interface Sci. 46, 121 (1974).
32. S. S. Berr and R. R. M. Jones, J. Phys. Chem. 93, 2555 (1989).
33. E. H. Cordes and C. Gitler, in "Progress in Bioorganic Chemistry," E. T. Kaiser and F. J. Kézdy, eds., Vol. 2, p. 17, John Wiley & Sons, New York (1973).
34. (a). H. Hoffmann, J. Kalus, K. Reizlein, W. Ulbricht, and K. Ibel, Colloid Polym. Sci. 260, 435 (1982). (b). L. Herbst, H. Hoffmann, J. Kalus, K. Reizlein, U. Schmelzer, and K. Ibel, Ber. Bunsenges. Phys. Chem. 89, 1050 (1985).
35. S. J. Burkitt, R. H. Ottewill, J. B. Hayter, and B. T. Ingram, Colloid Polym. Sci. 265, 619 (1987).
36. M. J. Schwuger and R. Piorr, Tenside Deterg. 24, 70 (1987).
37. E. Jost, H. Leiter, M. J. Schwuger, Colloid Polym. Sci. 266, 554 (1988).
38. J. F. Scamehorn, in "Phenomena of Mixed Surfactant Systems," ACS Symposium Series 311, American Chemical Society, Washington, DC (1986).
39. H. B. Klevens and M. J. Raison, Chimie 51, 1 (1954).
40. P.Mukerjee and A. Y. S. Yang, J. Phys. Chem. 80, 1388 (1976).
41. K. Shinoda, T. Nakagawa, B. Tamamushi, and T. Isemura, "Colloidal Surfactants," Academic Press, New York (1963).
42. P. Mukerjee, K. J. Mysels, and P. Kapauan, J. Phys. Chem. 71, 4166 (1967).
43. K. Shinoda and T. Nomura, J. Phys. Chem. 84, 365 (1980).
44. P. M. Holland and D. N. Rubingh, J. Phys. Chem. 87, 1984 (1983).
45. R. F. Kamrath and E. I. Franses, Ing. Eng. Chem. Fundam. 22, 230 (1983).
46. I. W. Osborne-Lee, R. S. Schechter, W. H . Wade, and Y. Barakat, J. Colloid Interface Sci. 108, 60 (1985).
47. J. H. Hildebrand, B.B. Fisher, and H. A. Benesi, J. Am. Chem. Soc. 72, 7348 (1949).
48. H. Lange and K. H. Beck, Kolloid Z. Z. Polym. 251, 424 (1973).
49. D. N. Rubingh, in "Solution Chemistry of Surfactants," Vol. 1, p. 337, K. L. Mitall, ed., Plenum Publishing, New York (1979).
50. J. H. Hildebrand, J. M. Prausnitz, and R. L Scott, "Regular and Related Solutions," Van Nostrand, New York (1970).
51. C. Treiner, J. F. Bocquet, and C. Pommier, J. Phys. Chem. 90, 3052 (1986).
52. T. Asakawa, S. Myagishi, and M. Nishida, J. Colloid Interface Sci. 104, 279 (1985).
53. C. Treiner, A. A. Khodja, and M. Fromon, Langmuir 3, 729 (1987).
54. R. F. Kamrath and E. I. Franses, J. Phys. Chem. 88, 1642 (1984).
55. T. Asakawa, K. Johten, S. Miyagishi, and M. Nishida, Langmuir 1, 347 (1985).
56. Aa. Fredenslund, R. Johes, and J. M. Prausnitz, AIChE J. 21, 1086 (1975).
57a. E. Caponetti, D. C. Martino, M. A. Floriano, and R. Triolo, Langmuir 9, 1193 (1993).
57b. K. Kalyanasundaram, Langmuir 4, 942 (1988).

58. S. J. Burkitt, R. H. Ottewill, J. B. Hayter, and B. T. Ingram, Colloid Polym. Sci. 265, 628 (1987).
59. J. Carlfors and P. Stilbs, J. Phys. Chem. 88, 4410 (1984).
60. S. Miyagishi, M. Higashide, T. Asakawa, and M. Nishida, Langmuir 7, 51 (1991).
61. D. Bauernschmitt and H. Hoffmann, Makromol. Chem. 181, 2365 (1980).
62. M. Ueno, K. Shioya, T. Nakamura, and K. Meguro, in "Colloid and Interface Science," Vol. II, p. 411. Aerosols, Emulsions, and Surfactants, ed., M. Kerker, International Conference on Colloids and Surfaces, San Juan, Puerto Rico , Academic Press, Inc., New York (1976).
63. T. Asakawa, S. Miyagishi, and M. Nishida, Langmuir 3, 821 (1987).
64. T. Asakawa, K. Johten, S. Myagishi, and M. Nishida, Langmuir 4, 136 (1988).
65. F. Tokiwa, K. Ohti, I, Kokubo, Bull. Chem. Soc. Jpn. 41, 2845 (1968).
66. T. Nakagawa and H. Jizomoto, Kolloid Z. Z. Polym. 236, 79 (1969).
67. T. Nakagawa and H. Jizomoto, Kolloid Z. Z. Polym. 239, 606 (1970).
68. T. Nakagawa and H. Jizomoto, J. Am. Oil Chem. Soc. 48, 571 (1971).
69. I. W. Osborn-Lee, R. S. Schechter, and W. H. Wade, J. Colloid Interface Sci. 104, 279 (1983).
70. T. Asakawa, M. Mouri, S. Myagishi, and M. Nishida, Langmuir 5, 343 (1989).
71. K. Tamori, K, Esumi, and K. Meguro, J. Colloid Interface Sci. 142, 236 (1991).
72. K. J. Mysels, J. Colloid Interface Sci. 66, 331 (1978).
73. N. Funasaki and S. Hada, J. Colloid Interface Sci. 73, 425 (1980).
74. N. Funasaki and S. Hada, J. Phys. Chem. 84, 736 (1980).
75. N. Funasaki and S. Hada, J. Colloid Interface Sci. 78, 376 (1980).
76. N. Funasaki and S. Hada, J. Phys. Chem. 84, 1868 (1980).
77. N. Funasaki and S. Hada, J. Phys. Chem. 86, 2504 (1982).
78. N. Funasaki and S. Hada, Bull. Chem. Soc. Jpn. 56 (12) 3839 (1983).
79. N. Funasaki and S. Hada, J. Phys. Chem. 87, 342 (1983).
80. J. F. Rathman and J. F. Scamehorn, Langmuir 3, 372 (1987).
81. T. Asakawa, T. Fukiya, and S. Miyagishi, Langmuir 7, 2112 (1991).
82. J. H. Hildebrand and R. L. Scott, "The Solubility of Nonelectrolytes," Dover Publications, New York (1964).
83. K. Yoda, K. Tamori, K. Esumi, K. Meguro, J. Colloid Interface Sci. 131, 282 (1989).
84. T. Suzuki, M. Ueno, and K. Meguro, J. Am. Oil Chem. Soc. 58, 800 (1981).
85. T. Suzuki, K. Esumi. and K. Meguro, J. Colloid Interface Sci. 93, 205 (1983).
86. K. Meguro, Y. Muto, F. Sakurai, and K. Esumi, ACS Symposium Series, No. 311, p. 61 American Chemical Society, Washington, DC (1986).
87. Y. Muto, M. Asada, A. Takasawa, K. Esumi, and K. Meguro, J. Colloid Interface Sci. 124, 632 (1988).
88. K. Meguro, M. Ueno, and T. Suzuki, Jpn. Oil Chem. Soc. 31, 909 (1982).
89. K. Meguro, V. Muto, F. Sakurai, and K. Esumi, in "Phenomena in Mixed Surfactant Solutions," J. F. Scamehorn, ed., p. 61, ACS Symposium Series No. 311 (1986).
90. K. Tamori, K. Esumi, K. Meguro, and H. Hoffmann, J. Colloid Interface Sci. 147, 33 (1991).

91. G. Sugihara, M. Yamamoto, Y. Wada, Y. Murito, and Y. Ikawa, J. Solution Chem. 17, 225 (1988).
92. X. Y. Hua and M. J. Rosen, J. Colloid Interface Sci. 90, 212 (1983).
93. B. Y. Zhu and M. J. Rosen, J. Colloid Interface Sci. 99, 435 (1984).
94. K. Esumi, Y. Ono, M. Ishizuka, and K. Meguro, Colloids Surf. 32, 139 (1988).
95. K. Esumi, Y. Sakamoto, K. Yoshikawa, and K. Meguro, Colloids Surf. 36, 1 (1989).
96. Zhao Guo-Xi and Zhu Bu-Yao, Colloid Polym. Sci. 261, 89 (1983).
97. Guo-Xi Zhao and Bu-Yao Zhu, in "Phenomena in Mixed Surfactant Systems," J. F. Scamehorn, ed., ACS Symposium Series, No. 311, p. 184, American Chemical Society, Washington, DC (1986).
98. R. G. Laughlin, in "Advances in Liquid Crystals," Vol. 1, p. 41, G. H. Brown, ed., Academic Press, New York (1975).
99. D. B. DuPré, in "Kirk-Othmer, Chemical Encyclopedia," Vol. 14, p. 395, John Wiley & Sons, New York (1981).
100. G. J. Tiddy, in "Modern Trends of Colloid Science in Chemistry and Biology," H. F. Eicke, ed., p. 148, Birkhäuser, Basel (1985).
101. R. G. Laughlin, J. Am. Oil Chem. Soc. 67, 705 (1990).
102. P. Kekicheff and G.J.T. Tiddy, J. Phys. Chem. 93, 2520 (1989).
103. P. Ekwall, L. Mandell, and K. Fontell, in "Liquid Crystals," Vol. 2, p. 325, G. H. Brown, ed., Gordon & Breach Science, London (1969).
104. P. A. Winsor, Chem. Rev. 68, 1 (1968).
105. V. Luzzati, H. Mustacchi, A. Skoulios, and F. Husson, Acta Crystallogr. 13, 660 (1960).
106. C. Hall and G. J. T. Tiddy, 6th International Symposium on Surfactants in Solution, New Delhi, (August 1986).
107. M.C. Holmes, D.J. Reynolds, and N. Boden, J. Phys. Chem. 91, 5257 (1987).
108. G .J .T. Tiddy, Symp. Faraday Soc. 5, 150 (1971).
109. G. J. T. Tiddy, Trans. Faraday Soc. 68, 653 (1972).
110. G. J. T. Tiddy and B.A. Wheeler, J. Colloid Interface Sci. 47, 59 (1974).
111. E. Everiss, G. J. T. Tiddy, and B. A. Wheeler, J. Chem. Soc. Faraday Trans. 72, 1747 (1976).
112. G. J. T. Tiddy, Trans. Faraday Soc. I, 73, 1731 (1977).
113. P. G. Morris, P. Mansfield, and G.J.T. Tiddy, J. Chem. Soc. Faraday Symp. 13, 37 (1979).
114. G. J. T. Tiddy, M.F. Walsh, and E. Wyn-Jones, J. Chem. Soc. Faraday Trans. I, 78, 389 (1982).
115. K. Fontell and B. Lindman, J. Phys. Chem. 87, 3289 (1983).
116. K. Fontell, in "Surfactant Solutions," Vol. 1, K. Mittal and B. Lindman, eds., Plenum Publishing, New York (1984).
117. H. Hoffmann, G. Platz, H. Rehage, K. Reizlein, and W. Ulbricht, Makromol. Chem. 182, 451 (1981).
118. K. Reizlein and H. Hoffmann, Prog. Colloid Polym. Sci. 69, 83 (1984).
119. N. Boden, P. H. Jackson, K. McMullen, and M. C. Holmes, Chem. Phys. Lett. 65,476 (1979).

120. R. M. Hedge, R. T. Thomas, M. Mortimer, and J. W. White, J. Chem. Soc. Faraday Trans. I, 76, 236 (1980).
121. M. C. Holmes, D. J. Reynolds, and N. Boden, J. Phys. Chem. 91, 5257 (1987).
122. J. C. Ravey and M. J. Stebe (a) Comun. Jorn. Esp. deterg. 18, 271 (1987). (b) Progr. Colloid Polym. Sci. 82, 218 (1990).
123. C. Rosenblatt, J. Phys. Chem. 91, 3830 (1987).
124. R. Bongliovanni, E. Borgarello, C. Carniani, and C. Genova, Colloids Surf. 54, 75 (1991).
125. H. Hoffmann, Ber. Bunsenges. Phys. Chem. 88, 1078 (1984).
126. K. Ogino and M. Abe, eds., "Mixed Surfactant Systems," Surfactant Science Series, Vol. 46. Marcel Dekker, New York (1993).
127. M. Abe, T. Yamaguchi, Y. Shibata, H. Uchiyama, N. Yoshino, K. Ogino, and S. Christian, Colloids Surfaces 67, 29 (1992).
128. W. Guo, S. D. Heavin, Z. Li, B. M. Fung, and S. D. Christian, Langmuir 8, 2368 (1992).
129. K. Tamori, A. Ishikawa, K. Kihara, Y. Ishii, and K. Esumi, Colloids Surfaces 67, 1 (1992).
130. H. Matsuki, N. Ikeda, M. Aratono, S. Kaneshina, and K. Motomura, J. Colloid Interface Sci. 154, 454 (1992).
131. K. Esumi and M. Ogawa, Langmuir 9, 358 (1993).
132. K. Yoda, K. Tamori, K. Esumi, and K. Meguro, Colloids Surfaces 58, 87 (1991).
133. E. G. Shafrin and W. A. Zisman, J. Phys. Chem. 66, 740 (1962).
134. P. S. Abenin, F. Szönyi, and A. Cambon, J. Fluorine Chem. 55, 1 (1991).
135. W. Guo, Z. Lin., B. M. Fung, E. A. O'Rear, and J. H. Harwell, J. Phys. Chem. 96, 6738 (1992).
136. W. Guo, B. M. Fung, and E. A. O'Rear, J. Phys. Chem. 96, 10068 (1992).
137. N. Funasaki, in "Mixed Surfactant Systems," K. Ogino and M. Abe, eds., Surfactant Science Series, Vol. 46. Marcel Dekker, New York (1993).

8
Applications

8.1 PERFORMANCE CHARACTERISTICS OF FLUORINATED SURFACTANTS

The performance characteristics of fluorinated surfactants are related to the fluorocarbon segment in the surfactant molecule. The fluorocarbon segments are hydrophobic as well as oleophobic and the interaction between fluorocarbon chains is weak. Consequently, fluorinated surfactants can lower the surface tension of water more than hydrocarbon-based surfactants. Many fluorinated surfactants are powerful wetting agents. Fluorinated surfactants are needed for wetting surfaces that have a critical surface tension below 25 mN/m. For wetting surfaces of a higher critical surface tension, fluorinated surfactants can be useful because of their nonrewetting effect. Some fluorinated surfactants are strongly adsorbed on the surface with the fluorocarbon chains oriented toward the solution. As a result, the adsorbed fluorinated surfactant prevents or hinders rewetting of the surface by the solution. As an example, Zonyl FSP, an anionic surfactant, is a protective, anticorrosive agent for aluminum in acidic or alkaline media. Some fluorinated surfactants reduce the wettability of hydrocarbon surfaces, such as polyethylene.

A unique characteristic of fluorinated surfactants is the chemical stability of the fluorocarbon chain to strong acids, oxidizing agents, and concentrated alkalis. Hence fluorinated surfactants can be used in media where hydrocarbon-based surfactants would decompose.

Fluorinated surfactants are useful emulsifiers and dispersants for systems in which either the continuous phase or the disperse phase is a fluorocarbon. Mixtures of fluorinated and hydrocarbon surfactants can be more effective and less expensive than each component alone.

Leveling of paints and floor polishes is also based on the superior wetting power of fluorinated surfactants. Reduction of surface tension gradients in the paint film during drying is essential for a uniform coating. Very low concentrations, 50 to 150 ppm, of a fluorinated surfactant can effectively overcome streaking and beading.

The foaming properties of fluorinated surfactants vary widely (see Sec. 4.8). For example, amphoteric surfactants, Zonyl FSK and Zonyl FSC, are outstanding foaming agents. In contrast, the anionic fluorinated surfactant, Zonyl FSP, and the nonionic fluorinated surfactant, Zonyl FSN, are low foaming. In some systems, Zonyl FSP can function as an antifoam agent.

8.2 SELECTION OF FLUORINATED SURFACTANTS

Before using a fluorinated surfactant in a product or system, the following questions should be answered [1]:

1. What is the desired effect of the fluorinated surfactant?
2. Are the physical and chemical properties of the fluorinated surfactant suitable for the system?
3. Will the fluorinated surfactant cause problems for the product or system?
4. Is the fluorinated surfactant cost effective and are its benefits significant?

The relatively high price of fluorinated surfactants limits their use to applications where hydrocarbon-based surfactants are inadequate. Usually, fluorinated surfactants are cost effective because their high price is compensated by the low concentration needed. Sometimes as little as 50 to 150 ppm of the surfactant may be adequate. In some applications, a mixture of a fluorinated surfactant and a hydrocarbon-based surfactant is more cost effective or performs better than either one alone.

Fluorinated surfactants are available as liquids, pastes, or solids. Some are diluted with water or an organic solvent, some are sold in the 100% active ingredient form. If the system cannot tolerate water, an undiluted fluorinated surfactant or a surfactant formulated as an organic solution has to be used. Some nonionic fluorinated surfactants, for example, Zonyl FSN-100 and FSO-100, are soluble in several nonaqueous solvents. Generally, the surfactant must be soluble to be effective. Hence, solubility of the surfactant may limit its use in some systems.

Other physical properties—for example, the cloud point, cmc, HLB, pour point, and density—are important and have to be considered when selecting a fluorinated surfactant.

The surfactant has to be compatible with other surfactants or components present in the system. Some systems can react with the surfactant. Although the fluorinated segment of the surfactant is resistant to chemical attack, the functional groups attached to it, for example, the polyoxyethylene chain, may not withstand the intended chemical environment.

Usually, a fluorinated surfactant does not have all properties needed for a specific application. A fluorinated surfactant may dramatically improve wetting of a polyester or polyethylene sheet but not function as an effective emulsifier for oil in water. When selecting a fluorinated surfactant, the desirable and undesirable side effects of its use have to be considered. The selection should not be made merely on the basis of the tabulated properties supplied by the manufacturer. For example, surface tension reduction depends not only on the structure of the surfactant but also on the nature of the interface. A fluorinated surfactant should therefore be tested in the system or product before it is used on a large scale. The surfactant should be tested at various concentrations and the surface-active properties and side effects should be observed.

8.3 SPECIFIC APPLICATIONS

The use and application of fluorinated surfactants have been reviewed in several articles [2–7] and in trade literature [1,8]. This chapter is a more comprehensive review of various fluorinated surfactant applications, listed in alphabetical order.

Adhesives

Fluorinated surfactants are used in solvent-based adhesives, e.g., Zonyl FSN-100 and FSO-100, or in water-based adhesives, e.g., Zonyl FSA, FSP, and FSN. Fluorinated surfactants added to water-based adhesives facilitate wetting and penetration of the substrates being joined [1]. By improving leveling and spreading, fluorinated surfactants assure a complete contact between the joining surfaces and retard foaming.

Fluorinated surfactants should be evaluated at 0.001, 0.01, and 0.1% solids on weight of the formulated adhesive, since the effectiveness of fluorinated surfactants can vary significantly over this range. Anionic fluorinated surfactants, e.g., Zonyl FSA and FSP, should be evaluated in soft water. If hard water is used, a chelating agent should be added to reduce water hardness.

Foamable hot-melt adhesives containing polyamides and fluorinated or silicone surfactants are useful on coarse surfaces [9a].

A fluorinated surfactants added to rubber (SKF 26) allows adhesiveless bonding to steel [9b].

Antifogging

Fluorinated surfactants can form a durable antimist film on glass, metal, or plastic surfaces. Antimist formulations containing a fluorinated surfactant are very effective in preventing misting of glass surfaces exposed to humid atmospheres, such as mirrors in bathrooms, automobile windshields, and eyeglass lenses.

Anionic, nonionic, or amphoteric fluorinated surfactants prevent fogging of glass [10–12] and plastic cover sheets used in agriculture. Potassium perfluorooctanesulfonate and nonionic surfactants blended into transparent poly(vinyl chloride), polyethylene, or ethylene–vinyl acetate film reduce clouding caused by condensation of atmospheric moisture [13–17]. A weather-resistant agricultural cover film, made by blending poly(vinyl chloride), phthalic ester plasticizers, nonionic surfactants, and a phthalate-insoluble fluorinated surfactant (Zonyl FSN), displayed antifogging even after 2 months use in the field [18].

Ueno et al. [19] studied the antifogging effect of surfactants on transparent cellulose ester films. Nonionic fluorinated surfactants, applied as a monolayer, were more effective antifogging agents than the corresponding nonionic hydrocarbon surfactants. The surfaces of the thin film placed in a closed chamber filled with water vapor at 60°C remained transparent for a long time. The antifogging effect of nonionic fluorinated surfactants was attributed to a double structure consisting of fluorocarbon and poly(ethylene oxide) chains in the monomolecular layer.

Antistatic Agents

Antistats prevent the buildup of static electricity and dissipate the electric charge formed on the substrate.

Brueck [20] studied the electrostatic properties of a triboelectric series of polymers having polyethyleneimine at the positive end and a hexafluoropropylene-tetrafluoroethylene copolymer at the negative end. Various commercial polymers were compared to the polymers of the triboelectric series. The effects of different polar groups were in the order (from most positive to most negative) imino ≈ oxy, carbonate ≈ urethane ≈ aliphatic carboxylate ester, carboxyamide, hydroxy, cyano, aliphatic hydrocarbon ≈ aromatic hydrocarbon, arenecarboxylate ≈ carboxylic acid, nitrate ≈ chloro, fluoro. Surface treatments with fluorinated surfactants drastically altered the charging properties of polymers.

Amphoteric surfactants, such as carboxymethyl-3-nonadecafluorodecaneamidopropylammonium hydroxide inner salt, dissolved in an organic solvent, e.g., propanol-2, function as antistatic agents for magnetic tapes and phonograph records [21]. Anionic fluorinated surfactants, e.g. $C_6F_{13}SO_3Li$, have been used in antistatic rubber compositions [22]. Mixtures of anionic surfactants having C_{4-16} perfluorocarbon chains, e.g., $C_6F_{13}SO_3Li$, and nonionic surfactants having C_{4-16} perfluorocarbon and poly(oxyethylene) chains, e.g., $C_8F_{17}SO_2N(C_3H_7)(CH_2CH_2O)_{20}H$, have been claimed as antistatic agents for poly(vinyl chloride) [23]. A nonionic fluorinated surfactant,

$C_6F_{13}SO_2N[(CH_2)_5CH_3]CH_2CH_2(OCH_2CH_2)_nOH$, applied as a solution in isopropyl alcohol, reduced the surface charge of polyester film [24].

Nonionic fluorinated surfactants, e.g., Monflor 51 and Monflor 52, are effective internal antistats for low-density polyethylene.

Cement Additives

Fluorinated surfactants reduce shrinkage of cement [25]. Cement tiles pigmented with carbon black dispersions containing fluorinated surfactants are more weather resistant than tiles made with ligninsulfonate dispersants [26]. Fluorinated surfactants (Fluorad FC-340) improve primers used for coating cement mortar [27].

Cleaners for Hard Surfaces

Small amounts (0.2% w/w) of fluorinated surfactants included in hard-surface cleaners formulated with conventional surfactants can substantially enhance the cleaning power [28]. Fluorinated surfactants facilitate wetting of hard surfaces and aid cleaning of low-energy surfaces, such as polyethylene. They also promote a rapid runoff of rinse solutions. The fluorinated surfactant $C_{10}F_{21}CONH(CH_2)_5COONH_4$ in a detergent reduces wiping stripes and reflection glittering of cleaned glass [29].

Fluorinated surfactants, e.g., Zonyl FSN, Zonyl FSP, or Zonyl FSA, are used in cleaning formulations for removing calcium sulfate scale from reverse osmosis membranes [30].

The outstanding chemical stability allows fluorinated surfactants to be used in cleaners containing strong acids or alkali. A typical alkaline cleaner contains 5% sodium hydroxide and 0.01% an anionic fluorinated surfactant. A cationic fluorinated surfactant (about 0.01%) facilitates wetting and the removal of oily soil on concrete [28] and cleaning concrete or masonry with a phosphoric acid–hydrochloric acid mixture. Fluorinated surfactants are also used to facilitate cleaning of metal surfaces, for example, for cleaning the outside of airplanes, and degreasing of metals (see Metal Finishing).

Fluorinated surfactants in automotive waxes aid spreading and improve the resistance of the polish to water and oil. Fluorinated surfactants are used in other car care products as well: in cleaners for spray washing of automobiles [28] and shampoos for fabrics and vinyl surfaces.

Fluorinated surfactants, e.g., $C_8F_{17}SO_2N(C_3H_7)CH_2COOK$, in nonaqueous cleaning agents aid the removal of adhesives [31] and in dry cleaning of textiles or metal surfaces [32]. Cured epoxy resins are removed from integrated circuit modules by cyclic alcohols containing small amounts of a surfactant, e.g., $C_8F_{17}SO_2N(C_2H_5)CH_2COOK$ [33]. Machine parts, such as steel screws, are cleaned after nickel plating with trichlorotrifluoroethane containing a salt of perfluorooctanesulfonic acid [34].

Coatings

Fluorinated surfactants reduce pigment flotation, which is quite common when strong tinting pigments are used [35]. Fluorinated surfactants [1,8] improve the wetting and leveling of paints and prevent edge crawling, catering, and orange peeling in the dried paint film. Fluorinated surfactants [1,8] also improve gloss and antisoil properties and modify rheology, flow, and foaming. The active concentration of the fluorinated surfactant is about 0.05–0.1%, depending on the type of paint and pigment used. A fluorinated surfactant is more effective than silicon-based additives used for this purposes. Defects such as fisheyes and pinholes, sometimes observed with silicone-based compounds, do not occur with paints containing 0.05 to 1% fluorinated surfactant [35]. Overcoating is possible, unlike the case of paints formulated with silicone additives. Pigment dispersions containing a fluorinated surfactant, e.g., perfluoroalkyl phosphate, and an aluminum flake pigment are stable at high temperatures and can be used in automotive coatings applied by spraying and baking [36]. Fluorinated surfactants as paint additives improve the dust repellence of interior or exterior paints [37]. A dispersion containing acrylic and vinyl monomers, a fluorinated surfactant, and ammonium persulfate catalyst was mixed with TiO_2 pigment and hexakis(methoxymethyl)melamine sprayed onto a metal surface and cured at 150°C to effect polymerization of the coating. In the absence of the fluorinated surfactant similar coatings formed craters and cracks [38].

Fluorinated surfactants are used in floor finishes (see Polishes and Waxes).

Cosmetics

Fluorinated surfactants are used in cosmetics as emulsifiers, lubricants, or oleophobic agents.

Fluorinated surfactants in hair-conditioning formulations improve lubricity, facilitate wet combing, and render hair oleophobic. Fluorinated surfactants are added to hair creams and rinses to keep the hair from becoming oily. Examples are $(F(CF_2CF_2)_{3-8}P(O)(ONH_4)_2$ (Zonyl FSP); $(F(CF_2CF_2)_{3-8})_2P(O)(ONH_4)$ (Zonyl FSE) [39,40]; perfluorooctanoic acid [41]; $CF_3(CF_2)_6(CH_2)_yS(CH_2)_xCOOM$, where x = 1–20, y = 1–4, M = alkali metal or ammonium [42]; $C_8F_{17}(CH_2CH_2O)_8COC_{15}H_{31}$ [43]; $C_8F_{17}SO_2N(C_2H_5)CH_2CH_2OP(O)(OH)_2$, $C_8F_{17}SO_2N(C_2H_5)CH_2COOK$, or a cationic fluorinated surfactant $C_8F_{17}SO_2NH(CH_2)_3N^+(CH_3)_3I^-$ [44]; and $C_{10}F_{21}SO_2N(C_2H_5)CH_2CH_2OP(O)(OH)_2$ [45]. Only a small amount (<0.05%) of a fluorinated surfactant, e.g., Lodyne S-106B, Lodyne S-112B, Zonyl FSA, Zonyl FSC, and Zonyl FSN, is needed to enhance the effectiveness of cationic hair-conditioning agents [46].

Fluorinated surfactant oligomers have been patented [47] as hair-conditioning agents. Fluoroalkyl (meth)acrylate polymers [48] also are useful in hair care formulations.

Crystal Growth Regulators

The average size of Glauber salt ($Na_2SO_4 \cdot 10H_2O$) crystals formed in an aqueous solution in heat storage systems decreases when a fluorinated surfactant (potassium perfluorooctanesulfonate) is present [49].

Dispersions

Ferromagnetic metal oxide particles (magnetite) have been dispersed with a fluorinated surfactant [$C_8F_{17}SO_2N(C_2H_5)CH_2COOK$] in a fluorocarbon-type solvent to make magnetic fluids [50].

Poly(vinyl chloride) pipes manufactured from a polymer blend containing conductive carbon black dispersed with nonionic or cationic hydrocarbon-type or fluorinated surfactants have reduced surface electric resistance [51].

Fluorinated surfactants are used as dispersants in lubricating greases (see Greases and Lubricants; Dispersions).

Electroless Metallization

Fluorinated surfactants improve the quality of electroless plating of copper [52,53] and stabilize the coating bath to deposit nickel-boron layers without a dentritic layer structure [54].

Electronics

In the electronic industry fluorinated surfactants are used in a variety of applications. Insulators for electric wires and cables are made from polyethylene containing Fluorad FC-432, a fluorinated alkyl ester [55], or Saflon S-111, a perfluoroalkanoic acid salt [56]. Insulated wire with improved coiling propensity was prepared by coating the wire electrophoretically and treating the coated wire with a fluorinated surfactant solution before baking [57]. Insulating tapes have been prepared by impregnating glass fiber–reinforced mica tapes with a mixture of a fluorinated surfactant with an epoxy resin and curing [58,59]. A material for sealing electric circuits has been made by curing a mixture of polyamic acid, an epoxy compound derived from a silane, and a nonionic fluorinated surfactant in *N,N*-dimethylacetanilide (DMAC) [60].

Glass fiber–reinforced epoxy resin boards treated with a fluorinated surfactant (Fluorad FC-95 and FC-134) maintain their electric resistance when exposed to high humidity [61].

Fluorinated surfactants are added to zinc battery electrolyte to prevent the formation of dendrites [62]. A battery, having an improved charging-discharging cycle and increased service life, has a porous Zn anode, a nickel oxide cathode, and an electrolyte consisting of 0.001 to 0.1% of the nonionic fluorinated surfactant Zonyl FSN in 8 N KOH saturated with ZnO [63]. Alkaline manganese batteries have been made with MnO_2 cathodes containing carbon black treated with a fluorinated surfactant, e.g., potassium perfluoroalkylcarboxylate, and a hydrocarbon nonionic surfactant [64]. Nonionic fluorinated surfactants, $CF_3(CH_2)_mCH_2CH_2O(CH_2CH_2O)_nH$, where m is 5 to 9 and n is about 11, can make hydrophobic polymers, such as poly(tetrafluoroethylene) (PTFE), hydrophilic for use as microporous separators for electrolytic cells [65].

Fluorinated surfactants are also used as low-foaming noncorrosive wetting agents in solders for electronic parts and cleaning of electronic components. A fluorinated surfactant in a polymerized rosin provides a dull finish flux useful for automatic soldering [66].

Electroplating

The excellent chemical stability permits fluorinated surfactants to be used in electroplating baths where hydrocarbon-type surfactants would not survive [67–73]. Fluorinated surfactants are remarkably stable in a solution of chromic oxide in sulfuric acid at 50°C [74,75]. (See Chapter 3.1).

Fluorinated surfactants prevent mist formed by gas bubbles evolving at electrodes during electroplating of chromium [5,76–79]. Fluorinated surfactants reduce the size of gas bubbles by lowering the surface tension of the electrolyte solution. The fluorinated surfactant used for mist reduction has to meet several requirements. The surfactant should be readily soluble in the electrolyte, lower the surface tension rapidly, and have suitable foaming characteristics. Foam formed on the surface of the electrolyte solution acts like a barrier and retards the entrainment of the electrolyte. The formation of a layer of fine foam is advantageous but excessive foaming is undesirable.

Copper has been deposited from an acid copper sulfate solution containing cationic and amphoteric fluorinated surfactants, with the surfactant $C_6F_{13}SO_2N(CH_2CHOHCH_2SO_3Na)C_3H_6N^+(CH_3)_2C_2H_4OH]OH^-$ being the main component [72]. Fluorinated surfactants prevent haze of plated copper by regulating foam and improving the stability of plating baths. Brightness and adhesion are improved as well.

Fluorinated surfactants are used as nonfoaming surfactants in nickel plating baths to reduce the surface tension and increase the strength of nickel electroplate [80] by eliminating pinholes, cracks, and peeling [80,81].

Nickel has been codeposited with graphite fluoride, dispersed in a plating bath containing a cationic fluorinated surfactant derived from octanesulfonamide [82]. Fluorinated cationic, anionic, and nonionic surfactants were added to a suspension of graphite fluoride in a plating bath containing copper sulfate and sulfuric acid. The cationic surfactants were the most effective and the anionic surfactants were the least effective in increasing the deposition of graphite fluoride in the copper matrix [83].

A fluorinated surfactant added to a tin plating bath [73] produces a plate of uniform thickness.

Cationic or amphoteric fluorinated surfactants can impart a positive charge to fluoropolymer particles and aid electroplating of the polymer, e.g., PTFE, onto steel for surface protection [84].

Electropolishing

A fluorinated surfactant has been used for electropolishing gas turbine blades made from an Ni alloy [85].

Emulsions

Fluorinated surfactants are used to emulsify chloro- and fluorocarbons in emulsion polymerization [86–89] (see Polymerization). Fluorinated surfactants are used as emulsifiers in medicine (see Fluorochemical Oxygen Carriers in Sec. 10.4) and in cosmetics. Fluorinated surfactants improve leveling of acrylic emulsions, floor polishes, and shoe brighteners (see Polishes and Waxes).

Transportable aqueous crude petroleum emulsions have been prepared using a hydrocarbon surfactant and a fluorinated surfactant oligomer as emulsifiers [90].

Microemulsions of the oil-in-water type, consisting of perfluoropolyethers emulsified by fluorinated surfactants, are used as catholytes in electrochemical processes [91–93].

Certain fluorinated surfactants are used as emulsion breakers, for example, for breaking water-oil emulsions in oil wells [67].

Etching

Glass articles are polished and etched with solutions containing sulfuric acid, HF, and a fluorinated surfactant, for example, tetraethylammonium perfluorooctanesulfonate or potassium perfluorooctanesulfonate. Fluorinated surfactants increase the speed of etching, acid polishing, or frosting of flat (sheet, plate) glass with hydrofluoric acid. Preferred wetting agents are tetramethylammonium and tetraethylammonium perfluoroalkanesulfonates [67,94] or potassium perfluorooctanesulfonate [95].

Planar etching of fused silica [96] is effected by a mixture of HF, H_3PO_4, and a fluorinated surfactant, $[CF_3(CF_2)_{14}COO]_2Ca$. The fluorinated surfactant lowers the surface tension to facilitate wetting and the removal of etching byproducts.

Polyacetals are etched in solutions containing phosphoric and sulfuric acid and a fluorinated surfactant [97].

Fluorinated surfactants increase the wettability and adhesion of isotactic poly(propylene) when etched with a mixed solution of sulfuric and chromic acids.

A fluorinated surfactant, such as Fluorad FC-170, exhibits a maximum in the wettability–etching degree curve [98].

Fluorinated surfactants lower the surface tension and improve the performance of HF- and NH_4F-containing etchants used in the manufacture of integrated circuits. Fluorinated surfactants, e.g., $C_9F_{19}CH_2CH(OH)CH_2NHCH(CH_3)CH_2CH_3$ [99a], $H(CF_2)_6COOH$ [99b], and a perfluoroalkyl sulfonate [100], improve the wettability of semiconductor devices and inhibit dust collection. Copper can be removed from the circuit zone without underetching by including Zonyl FSP and Fluorad FC-135 in the coating [101].

Fluorinated surfactants improve the efficient life of alkali baths used in etching of aluminum. In mild steel etching, the effect of the fluorinated surfactant on the etching process depends on the structure of the surfactant. For example, Atsurf F-21 has given the best mist reduction, Atsurf F-35 the best surface. Fluorinated surfactants are also used for etching plastic preplates.

Fire-Fighting Foams and Powders

Fire-fighting foams [102] are formulated to float on flammable liquids and extinguish flames. The foams form a barrier for the vapors to escape and cool a hot surface to prevent reignition. Foams are arbitrarily divided into three groups [103], based on the expansion ratio of the foam. The expansion ratios are 20:1 for low-expansion foams, 20:1 to 100:1 for medium-expansion foams, and 200:1 to 1000:1 for high-expansion foams. High-expansion foams are used to fill large spaces with foam and extinguish flames without causing damage by water. The high-expansion foam is light, however, and is easily blown away outdoors. Medium- and low-expansion foams are used more often than the high-expansion foams.

Protein foam concentrates are produced by hydrolysis products of proteinaceous matter, such as hoof and horn meal, chicken feathers, or fish meal. *Fluoroprotein* concentrates are protein foam concentrates with fluorinated surfactants added. The fluorinated surfactants add to the cost of the protein-based foam agents but enhance fire-extinguishing efficiency by repelling hydrocarbon fuel when the foam is covered with fuel. Fluoroprotein foam introduced to a base of a burning fuel tank rises through the fuel and extinguishes the fire. The low surface tension allows fluoroprotein foam to move rapidly over a hydrocarbon fuel surface [104–109].

The *aqueous film-forming foaming* (AFFF) agents contain synthetic chemicals instead of protein-based materials. Fluorinated surfactants used in AFFF agents lower the surface tension of water and form a film on the fuel surface. AFFF agents used in unaspirating equipment like water hoses are effective mainly because of the aqueous film, and not the foam [102]. Alcohol-resistant AFFF agents contain a polymer for use on polar organic solvents. The polymer, usually a natural product, retards drainage and increases heat resistance of the foam [110,111].

An amine oxide containing a C_{4-20} alkyl group and two C_{1-14} alkyl groups improves the flashback resistance of a fire-extinguishing agent containing an amphoteric fluorinated surfactant [112]. Fluorinated surfactants, e.g., $C_8F_{14}\{OC_6H_4SO_2N[(CH_2)_2OSO_3NH(CH_2CH_2OH)]_2\}_2$, foam in seawater and produce stable foams for fire extinguishers used in a seawater environment [113].

Schuierer et al. [114] prepared fire-extinguishing foams from amphoteric fluorinated surfactants with and without phosphate ester groups. Fire-extinguishing agents containing a mixture of two amphoteric fluorinated surfactants, for example, $C_7F_{15}CF{=}CH_2N^+Me_2(CH_2)_2OSO_3^-$ and $C_7F_{15}CF{=}CH_2N^+Me_2C_2H_4OHCl^-$, a nonfluorinated alkyl sulfate, butyl glycol, and ethylene glycol have also been patented [115]. Cordes and Achilles [116] mixed amphoteric fluorinated surfactants, chlorofluorocarbons and a propylene oxide–ethylene oxide copolymer to prepare an emulsion which foams on mixing with water.

Mixtures of cationic and nonionic fluorinated surfactants have been patented as foamable fire-extinguishing agents to minimize the evaporation of flammable solvents and protect the solvent, e.g., hydrocarbon, from ignition [117,118].

Mixtures of fluorinated surfactants and silicone surfactants have been patented [119,120].

Fire-extinguishing foams containing mixtures of a cationic, an amphoteric, and a nonionic fluorinated surfactant [121] have been claimed to spread rapidly on hydrocarbon surfaces.

An AFFF agent for extinguishing or preventing fires has been formulated with a mixture of an amphoteric fluorinated surfactant, *N*-[(dimethylamino) propyl]-2(or 3)-(1,1,2,2-tetrahydroperfluoroalkylthio)succinamic acid, an anionic fluorinated surfactant, perfluoroalkanoic acid, and an ionic and a nonionic nonfluorinated surfactant [122].

Foam fire extinguishers containing potassium perfluorooctanoate, calcium chloride, and aluminum lactate are claimed to be powerful fire-extinguishing agents for hydrocarbon and polar organic solvents [123].

Numerous other patents have issued for fluorinated surfactant–containing fire-extinguishing foams.

Fluorinated surfactants are used in dry fire-extinguishing agents to make the powder nonwettable by hydrocarbons. The powder floats on the surface of the hydrocarbon and hinders the evaporation of the hydrocarbon. The hazard of reignition is thereby eliminated [124,125].

Flotation of Minerals

Fluorinated surfactants are used in ore flotation processes. Fluorinated surfactants are stable in concentrated acids and accelerate wetting of ore and the removal of the oxide layer. Perfluoroalkanoates, such as $CF_3(CF)_nCOOK$ (n = 2, 4, 6, 8) are more effective flotation agents than the corresponding hydrocarbon-type surfactants. The collecting power of the fluorinated surfactant in flotation of Al_2O_3 increases with increasing n (chain length) [126].

Vanadium compounds, such as NH_4VO_3, can be concentrated using a perfluorinated surfactant. At $\leqslant$300 mg V/L, 100% sorption has been achieved [127].

Fluorinated surfactants are useful in nitrogen flotation to recover uranium. The fluorinated surfactant improves separation of uranium contained in sodium carbonate and/or sodium bicarbonate solutions. In one example, Zonyl FSP yielded 95.5% recovery, as compared with 23.6% using sodium dodecyl sulfate [1].

Foams and Defoaming

Fire extinguishers—see Fire-Fighting Foams and Powders.

Fluorinated surfactants usually differ widely in their foaming power (see Sec. 4.9). Cationic fluorinated surfactants, for example, Zonyl FSC, and amphoteric fluorinated surfactants, for example, Zonyl FSK, are foaming agents. On the other hand, nonionics, for example, Zonyl FSN and Zonyl FSO, are low-foaming surfactants [1]. Endcapping reduces foaming.

Fluorinated surfactants are used to aid foaming of plastics and polymers, such as polyolefins [128], polyurethanes [129,130], poly(diethylene glycol diacrylate) [131], siloxanes [132], and foamable adhesives [133].

Anionic, cationic, or amphoteric fluorinated surfactants with a hydroxyl group on the α-carbon improve foam stability [134].

Fluorinated surfactants of the phosphoric ester type can function as antifoaming agents. A 175-mm-high foam produced by 1 g/L of a secondary alkanesulfonate collapsed during 1 s when a 0.02 g/L mixture of heptadecafluorooctylphosphonic acid and bis(heptadecafluorooctyl)phosphinic acid was added [135]. As little as 0.004% of the fluorinated surfactant Atsurf F-12 (a phosphoric ester–type anionic surfactant) lowered the initial foam of a typical nonionic surfactant from 80 to 38 mm and the 5-min reading of the Ross-Miles test from 38 to 5 mm [136]. An addition of 2.0% $H(CF_2)_{10}CH_2OPO(OH)_2$ to a detergent containing a coconut alcohol–ethylene oxide condensation product eliminated sudsing almost completely. A Dainippon Ink patent [137] claims fluorinated surfactants containing a C_{3-20} fluorinated aliphatic group and a $>C_{10}$ (preferably C_{12-36}) alkyl group. The addition of the fluorinated surfactant $C_8F_{17}SO_2NC_{18}H_{37}(CH_2CH_2O)_{10}H$ reduces foaming of Na dodecylbenzenesulfonate. The initial foam height of 230 mm and the height of 180 mm after 5 min decreased to a foam height of 10 mm initially and to zero foam height after 5 min.

Fluorinated alcohols, e.g., $CF_3(CH_2)_{5-11}C_2H_4OH$, reduce foaming of detergents used for mechanical washing [138,139].

Graphic Imaging

Printing inks are pigment dispersions or dye solutions which can be printed on a substrate and dried. Relief and lithographic inks are pigment dispersions in a viscous oil, and flexographic and gravure printing inks are liquids containing a resin in a volatile solvent. Ultraviolet and electron beam inks are cured and dried by a free radical polymerization mechanism. The radiation printing process produces very little solvent vapor.

Fluorinated surfactants reduce the surface tension and improve wetting properties of aqueous and oil-based inks [140]. The enhanced wetting is essential for printing on difficult-to-wet surfaces, such as plastics and metals. Fluorinated surfactants also aid pigment dispersion and control problems such as pigment flooding and flotation.

The success of gravure printing depends on the ability of the ink to wet the surface being printed and to keep the roll clean and free of contaminants. Adding 0.5% Zonyl FSN by weight of press-ready ink can improve the cylinder life 25 to 40%, depending on the ink and cylinder, and improve the quality of print by reducing "snowflaking" and streaking of certain inks and maintaining ink viscosity.

Fluorinated surfactants impart water resistance to water-based inks and improve the effective life of toner carrier materials.

Fluorinated surfactants are added to inks for ball-point pens, [141–144], marking pens [145–148], anticlogging jet recording inks [149], and printing inks for plastics [150,151] to improve leveling, wettability, a smooth flow, adhesion to the substrate, and water resistance of the print. Fluorinated surfactants in correction fluids repel ink and reduce bleeding of the print [152].

A cationic surfactant, a perfluoroalkyltrimethylammonium salt, in a developer for photosensitive lithographic plates facilitates the control of the development process [153].

Greases and Lubricants

Lubricating greases containing 15–40 wt % PTFE have been prepared by using a fluorinated surfactant as a dispersant [154].

Herbicides and Insecticides

The nonionic fluorinated surfactants tested by Sakakibara et al. [155] exhibited hardly any herbicidal activity. Hence fluorinated surfactants can be used safely as dispersants and adjuvants for agricultural chemicals. When compared to hydrocarbon surfactants, fluorinated surfactants are more powerful wetting agents for leaves, for example, wheat leaves [156].

Fluorinated surfactants are used in insecticide formulations to aid wetting and penetration of the insecticide into the insect. Insecticidal aerosols may contain an insecticide, solvent, and a fluorinated surfactant, e.g., Fluorad FC-128. An insecticide formulated with a fluorinated surfactant and dimethyl ether as the solvent is readily absorbed by insects [157]. Some fluorinated surfactants are insecticides in their own right, affecting the common housefly and the carmine mite [155]. The mechanism of insecticidal activity appears to be suffocation of the insect by the adsorbed fluorinated surfactant.

Leather

Fluorinated surfactants are used in various leather manufacturing processes and repellent treatments of tanned leather.

Fluorinated surfactants have been used in hydrating, bating, pickling, degreasing, and tanning processes. Fluorinated surfactants improve the efficiency of the process, reduce the processing time, and increase the quality of the product [158,159]. The use of fluorinated surfactants in leather tanning and dyeing processes has been investigated by Gratacos et al. [160]. In small amounts (0.025–0.05% on wt.) an anionic fluorinated surfactant increased the exhaustion of the chrome tanning agent and dyes, but at higher concentrations the fluorinated surfactant had the opposite effect. The distribution of Cr_2O_3 was more uniform when the skins were pretreated with a cationic fluorinated surfactant.

The techniques suitable for applying fluorinated surfactants [158,159] to leather after tanning are (1) tumbling in a drum, in which the leather sorbs the fluorinated surfactant from a emulsion, suspension, or solution; (2) spraying; and (3) cast coating. The water resistance of leather treated with fluorinated surfactants has been improved with cationic retention aids and by complexing with chromium and zirconium (see Repellency). The complexes provide oil repellency and are more water resistant than the parent fluorinated surfactants. The performance of monomeric fluorinated surfactants on leather has been substantially exceeded by designing polymeric fluorochemicals for repellent leather treatments.

Fluorinated surfactants improve the leveling of acrylic brightener emulsions on leather.

Liquid Crystals

Anionic fluorinated surfactants form liquid crystals with cationic fluorinated surfactants (see Chapter 7). Liquid crystals of fluorinated surfactants are used in various industrial applications and as biological membrane models.

As an example, stirring $C_9F_{19}CONH(CH_2)_3N^+(CH_2)_3\,I^-$, $C_9F_{19}COONH_4$, and H_2O at 60°C yields the $C_9F_{19}CONH(CH_2)_3N^+(CH_2)_3\,C_9F_{19}COO^-$ liquid crystal. The ion pair that forms the liquid crystal has a surface tension of 15.5 mN/m, compared to 18.3 mN/m for the cationic component and 21.0 for the anionic component alone [161,162]. Liquid crystal cells are less likely to become turbid in the absence of image-producing ultrasonic vibration or electric voltage when the inside surfaces of their glass support plates are coated with a fluorinated surfactant [163].

Medical and Dental Uses

Self-propelling aerosols of antiallergic, antibiotic, antitussive, or antianginal activities have been prepared using a fluorinated surfactant as the dispersant in Freon 114 and Freon 12 [164]. As an example, epinephrine bitartrate has been dispersed with FC-161 (perfluoro-*n*-octyl-*N*-ethylsulfonamidoethyl phosphate). The use of fluorinated surfactants in intravascular oxygen carriers and blood substitutes is discussed in Chapter 10.

A fluorinated surfactant in toothpastes containing potassium fluoride enhances fluoroapatite formation and inhibits caries [165]. A fluorinated surfactant formulation (1% Lodyne S-110), which consists of an amphoteric fluoroalkylaminocarboxylic acid and a nonionic fluoroalkylamide synergist, in the toothpaste increases enamel-fluoride interactions. A fluorinated surfactant in pharmaceutical formulations and in toothpaste must be nontoxic for the intended purpose.

Dispersions of cells are prepared in clinical laboratories to diagnose cell abnormalities. An anionic fluorinated surfactant, for example, Fluorad FC-128, facilitates the dispersion of cell aggregates from tissues in a saline solution [166].

Metal Finishing

Anionic, cationic, and nonionic fluorinated surfactants are used in various metal treatment processes. Metal surfaces are treated to prevent corrosion, reduce mechanical wear, or enhance the aesthetic appearance. Fluorinated surfactants are used in the phosphating process for aluminum and in bright dips for copper and brass. Some fluorinated surfactants, for example, the anionic surfactants Zonyl FSA and Zonyl FSP, are strongly adsorbed on metals and provide water and solvent repellency [1].

An effective surface treatment requires a clean surface. Metal surfaces are cleaned with an alkaline, neutral, or weakly acidic cleaner, an organic solvent, or by pickling with molten salt baths [5,67]. Fluorinated surfactants in a pickling and descaling bath disperse scum, speed runoff of acid when metal is removed from the bath, and increase bath life [167,168]. The fluorinated surfactant inhibits nascent hydrogen formation and therefore prevents embrittlement by hydrogen [169].

Some fluorinated surfactants function as corrosion inhibitors on steel [1,5,67,136,170–174]. For example, 0.01% Atsurf F-21 prevents corrosion of mild steel in 15% HCl for at least 20 days at ambient temperature [136]. Surface treatments with corrosion inhibitors containing fluorinated surfactants decrease the friction coefficient of magnetic audio- or videotapes [175–177].

Fluorinated surfactants promote the flow of metal coatings and prevent cracks in the coating during drying [178,179].

Some fluorinated surfactants are effective antiblocking agents for aluminum foil. For example, aluminum foil is coated with 0.025 g/m^2 (active ingredient) Monflor 91, applied as a 5% solution [35].

A nonionic fluorinated surfactant, Monflor 31, increases the penetration rate of penetrating oils by a factor of 3 [35].

Mild steel etching—see Etching.

Molding and Mold Release

Fluorinated surfactants are effective mold release agents because of their oleophobic and hydrophobic nature. Only small amounts of a fluorinated surfactant are needed, sometimes only 1/50 of the amount needed for hydrocarbon or silicone mold release agents. Because the amounts of fluorinated surfactants used are very small, the molding can be painted, metallized, or adhered to another surface without removing the molding agent. Fluorinated surfactants are used as mold release agents [180–184] for thermoplastics, polypropylene [180], epoxy resins [182], and polyurethane [183,184] elastomer foam moldings.

Fluorinated surfactants reduce autoadhesion and blocking [185,186] and prevent orange peel on film casting and coating [187].

Oil Containment

Oil spills on water can be contained and prevented from spreading by injecting into water a chemical barrier containing a fluorinated surfactant and a maleic anhydride–derived polymer featuring carboxylic acid and ester groups [188]. The fluorinated surfactant $C_8F_{17}SO_2N(C_3H_7)CH_2COOK$ prevents spreading of oils or gasoline on water [189].

Perlite or vermiculite, treated with a cationic fluorinated surfactant, is claimed to be hydrophobic and effective in cleaning oil spills [190].

Oil Wells

Fluorinated surfactants are used in oil well stimulation and for petroleum recovery by waterflooding [191] and in nonaqueous stimulation fluids for foaming hydrocarbon liquids, e.g., kerosine [192,193]. A foam, made by foaming a mixture of 1% $F(CF_2)_8CH_2CH_2O(CH_2CH_2O)_2H$ in methanol, stimulates underground formations in the production of petroleum [194].

Fluorinated surfactants lower the surface tension of stimulation fluids more than hydrocarbon surfactants and are stable in aqueous stimulation fluids at 100°C. Drainage of salt solution (2% KCl) from sandpacks is inversely related to the surface tension. Fluorinated surfactants enhance drainage more than hydrocarbon surfactants. Flow rates of fluids through sandpacks or low-permeability sandstone cores were unaffected by the fluorinated surfactant, indicating that formation blockage by a fluorinated surfactant is improbable. Aqueous fluorinated surfactant solutions do not form emulsions of kerosine.

Paper

Fluorinated surfactants function as oil and solvent repellents on paper and paperboard [158,195–197]. Monomeric fluorinated surfactants, their chromium or zirconium complexes, and polymeric fluorochemicals are used for repellent treatments.

Fluorinated surfactants can be added to the pulp slurry, applied to the paper surface, or included in pigmented coatings. The surface treatment process is the most efficient mode of fluorinated surfactant application and easier to control than the internal application process. In pigmented coatings on the outside of boxes or bags, fluorinated surfactants are used to prevent soiling and maintain the appearance of the package.

For internal application about 1.0 to 1.5% (based on the weight of dry fiber) of Zonyl RP, a fluoroalkyl phosphate, is needed for good oil repellency [195c]. Cationic retention aids are used with fluoroalkyl phosphates in internal application processes.

Materials treated with fluorinated surfactants include [195c]:

Liner board—for packaging machine parts, rope, twine, meat, etc.

Folding cartons—for snack foods, carry-out fast food, cake mixes, margarine, candy, bakery products, and pet foods. A repellent treatment with a fluorinated surfactant prevents fat and grease from seeping into the edges of stamped-out polyethylene- or polypropylene-lined cartons.

Multiwall bags—snack foods, cake mixes, pet food.

Flexible packaging—carry-out fast food, candy wrap.

Duplicator and reproduction paper—toluene holdout.

Support cards—candy and bakery products.

The treatment of paper or paperboard used for food and pet food packaging requires Food and Drug Administration (FDA) approval. This requirement excludes chromium and zirconium complexes of fluorinated surfactants. The monomeric fluorinated surfactants approved for the repellent treatment of paperboard and paper in direct contact with food are fluoroalkyl phosphates [158,195,197], for example, a mixture consisting of a mono-(fluoroalkyl) phosphate ester $(R_fCH_2CH_2O)P(O)(O_2^{2-}[H_2N^+(CH_2CH_2OH)_2]_2$ and a bis-(fluoroalkyl) phosphate ester $(R_fCH_2CH_2O)_2P(O)O^-H_2N^+(CH_2CH_2OH)_2$ (Zonyl RP) or $(C_8F_{17}SO_2N(C_2H_5)CH_2CH_2O)_nP(O)(ONH_4)_{3-n}$ (Scotchban). Studies by Du Pont have revealed that very low amounts of fluorinated surfactant are extracted from paperboard into solvents simulating food. Highly sensitive analytical methods had to be developed for the determination of trace amounts of organic fluorine in the extracts (see Chapter 9, Elemental Analysis).

Fluoroalkyl phosphates provide excellent oil and grease repellency. A moderate resistance to water penetration can be achieved with cationic retention aids and ketene dimer sizes or small amounts of alum. Excessive amounts of alum impede penetration of the fluorinated surfactant and reduce oil repellency. Optimum oil repellency with fluoroalkyl phosphates can be achieved by excluding alum from paper and paperboard or by using an alkaline sizing agent.

Emulsions containing a fluorinated surfactant and waxes and/or paraffins are release agents for paper coating compositions [198]. Cast-coated paper is produced by coating the paper with pigment-and adhesive-containing solutions, air drying, rewetting a polyethylene emulsion, and pressing the wet surfaces with a fluorinated surfactant–coated hot drum to give a paper with a high gloss [199].

Fluorinated surfactants are used in the manufacture of heat-sensitive recording paper [200–203] and ink-jet printing paper [204,205].

Photography

Fluorinated surfactants aid single-layer or multiple-layer, light-sensitive coating of photographic materials, such as films and papers, and function as wetting agents, emulsion additives, stabilizers, and antistats [8a,206–230]. Fluorinated surfactants impart antistatic and nontacky properties, prevent spot formation, and control edge uniformity in multilayer coatings. In combination with a nonionic surfactant or in a hydrophilic protective layer fluorinated surfactants prevent fogging and streaking caused by a static discharge. Fluorinated surfactants do not have undesirable effects on silver halide light-sensitive emulsions.

In a diffusion-transfer photographic process the photosensitive material and the image-accepting material are layered in a close contact to effect the diffusion transfer. When the photographic process is completed, the materials are peeled apart. Fluorinated surfactants in the timing layer of photographic diffusion transfer materials provide a good contact when wet or dry, so that rupture or peeling of the emulsion layer is prevented [231–235]. A low-surface-tension processing solution containing a fluorinated surfactant eliminates air bubbles that can cause failures in image transfer [236].

Plastics, Resins, and Films

See also Antifogging, Antistats, Mold Release, and Photography in this chapter.

Antiblocking agents for vulcanized or unvulcanized rubbers have been formulated with a nonionic fluorinated surfactant [237]. Films of poly(vinyl alcohol) or saponified ethylene–vinyl acetate copolymers were treated with a nonionic fluorinated surfactant [$C_8F_{17}SO_2NRCH_2CH_2O(CH_2CH_2O)_{10}H$ to reduce friction and blocking [238].

The water permeability of dialysis membranes containing fluorocarbon polymers can be increased by a surface treatment with a cationic fluorinated surfactant [239].

Partially fluorinated oligomers derived from hexafluoropropylene oxide, e.g., $F[C(CF_3)FCF_2O]_9C(CF_3)FCOOCH_2CH_2(OCH_2CH_2)_6OCH_3$, lower the critical surface tension of polymers on which they are adsorbed, for example, polystyrene, poly(methyl methacrylate), and a vinyl chloride–vinylidene copolymer [240].

An amphoteric fluorinated surfactant in silicone rubber sealants makes the seal soil resistant [241]. An anionic fluorinated surfactant prevents leakage of mineral oil around nitrile rubber seals [242].

Fluorinated surfactants facilitate coplating of polytetrafluoroethylene and metals onto a metal substrate [35]. A cationic fluorinated surfactant adsorbed onto the polymer particles imparts a positive charge and thus allows the polymer and metal to be electrolytically coplated.

Fluorinated surfactants improve wetting of fibers or fillers in composite resins and speed the escape of bubbles trapped in the viscous resin.

A fluorinated surfactant in a poly(oxymethylene)diacetate polymer–nylon copolymer reduces the frictional coefficient of the polymer [243].

Polishes and Waxes

Self-polishing liquid floor finishes depend on proper wetting and dry-down for complete coverage and shiny appearance. Some polymeric resin formulations do not wet floors completely and dry to a rough finish, especially on vinyl floors. As little as 50 ppm fluorinated surfactant added to the formulation can improve the appearance of the dried floor significantly by eliminating streaks and enhancing gloss. Fluorinated surfactants impart self-leveling properties to all types of polishes including styrene, acrylic, or wax-based floor polishes.

Fluorinated surfactants are used as leveling agents in cleaner-polishes for vinyl floors. The polishes contain usually a low-molecular-weight acrylic-methacrylic copolymer, a maleic anhydride–styrene copolymer or an acrylic-styrene copolymer, a fluorinated surfactant, a hydrocarbon-type surfactant, a fugitive plasticizer, a dimethylpolysiloxane antifoaming agent, sodium bicarbonate, aqueous ammonia, a fragrance, and other ingredients [244–251].

Polymerization

Fluorinated surfactants used as an emulsifier in emulsion polymerization of fluoropolymers improve physical properties of the polymer and increase the rate of polymerization. Since the fluorinated surfactants are more effective in emulsion polymerization than hydrocarbon surfactants alone, the total surfactant concentration can be reduced. For example, in emulsion polymerization of vinyl chloride, 160 ppm Monflor 31 can reduce the required concentration of sodium dodeylbenzenesulfonate by about 40%.

The emulsion-grade poly(tetrafluoroethylene) (PTFE) polymer is a dispersion of PTFE particles in an aqueous phase. PTFE is commercially produced by free radical polymerization of tetrafluoroethylene in water containing a fluorinated surfactant [87–89, 252–257]. The surfactants are usually alkali perfluoroalkanoates, for example, ammonium perfluorooctanoate or lithium perfluoroalkanoates $CF_3(CF_2)_nCOOLi$, $n = 5$–8. The surfactant solubilizes the monomer and stabilizes the PTFE particles formed. The polymerization rate, surfactant adsorption, and polymer particle morphology depend on the initial concentration of the surfactant [89,253]. Depending on the polymerization conditions, rod-shaped particles, spherical particles, small hexagons, or liquid crystalline suspensions of PTFE whiskers are formed. A correlation exists between the association state of the surfactant during the initial stages of polymerization and the PTFE particle morphology [253]. The particle size and shape change near the cmc of the fluorinated surfactant, suggesting that the particle nucleation is different in the two dominions. Below the cmc of the fluorinated surfactant the nucleation step of PTFE particles is homogeneous, while in micellar systems the nucleation step is heterogeneous. The micelles harbor the growing PTFE chains and as a tetrafluoroethylene reservoir provide a high concentration of the monomer. Copolymerization with a small amount of $CF_2{=}CFCF_3$ decreases the rod-shaped polymer formation [258].

Vinylidene fluoride has been polymerized in the presence of ammonium perfluorooctanoate [259], sodium perfluorooctanoate [260,261], ammonium perfluoroisooctanoate [262], or ammonium perfluorononanoate [263].

Ammonium ω-hydroperfluorononanoate [264], ammonium perfluoropelargonate [265], and sodium perfluoroheptanoate [266] function as emulsifiers for the polymerization of vinyl fluoride.

Nonionic fluorinated surfactants are used as emulsifiers for homo- or copolymerization of ethylene [267].

Kato et al. [268] studied emulsion polymerization of styrene in the presence of sodium dodecyl sulfate and lithium perfluorooctanesulfonate. The fluorinated surfactant decreases the polymerization rate and the molecular weight of the polymer formed.

Repellency

Fluorochemical repellents differ from silicone- or hydrocarbon-based repellents in several aspects, of which oil repellency is the most important [269]. Repellents with a fluorocarbon chain repel both water and oil, whereas repellents with silicone or hydrocarbon hydrophobes repel only water.

The resistance of a porous substrate or a textile fabric to wetting and penetration of a liquid, such as water or oil, depends on the chemical nature, geometry, and roughness of the surfaces and the capillary spacings in the substrate [270–275]. However, the initial repellency of a finish is not the only criterion for selecting a repellent. Durability to dry cleaning and laundering, resistance to abrasion and soiling, ease of application, and the cost of the repellent are important factors to be considered.

The relationship between repellency and the structure of the fluorocarbon segment is in accord with the critical surface tension concept developed by Zisman [276]. Shafrin and Zisman [277] determined the wettabilities and critical surface tensions of ω-perfluoroalkyl-substituted n-heptadecanoic acids synthesized by Brace [278]. Once the seven outmost carbon atoms are fully fluorinated ($x = 7$), the wettability of monolayers of the acids $F(CF_2)_x(CH_2)_{16}COOH$ approached that of the perfluorocarboxylic acid $F(CF_2)_xCOOH$ (Fig. 8.1).

This suggested that a terminal perfluoroalkyl chain of seven carbon atoms is sufficiently long to shield the nonfluorinated segment beneath the fluorinated segment. For fluorocarbon repellents on a fabric, approximately 10 fully fluorinated carbon atoms are needed in a normal alkane chain to achieve maximum repellency. Data of Grajeck and Petersen [279] suggest that oil repellency of a cotton fabric treated with poly(R_fCH_2O-acrylates) increases with increasing chain length of the perfuoroalkyl group R_f until the chain is about 12 carbon atoms long (Fig. 8.2).

The coverage of fiber surfaces by the fluorochemical repellent is a very important variable affecting repellency. Hence, the amount of a fluorochemical needed for maximum repellency depends on the fabric construction (Fig. 8.3) and the structure of the fluorochemical repellent.

At equal number of carbon atoms, normal perfluoroalkane chains are more effective than branched chains. Under otherwise identical conditions, a surface

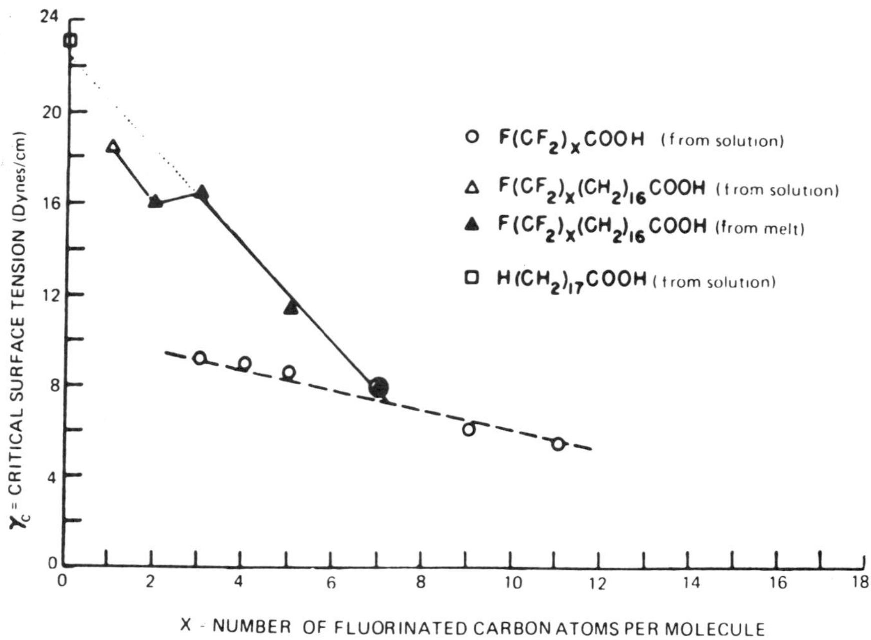

FIG. 8.1 The effect of fluorination of the adsorbed acid monolayer on the critical surface tension of wetting by *n*-alkanes. (From Ref. 277.)

composed of closely packed $-CF_3$ groups has the lowest surface energy and the highest repellency. Pittman et al. [280] concluded that polyacrylates containing a perfluoroisopropyl group have a lower critical surface tension than those containing an *n*-perfluoropropyl group. However, the comparison was made between poly(1*H*,1*H*-heptafluorobutyl acrylate) and poly(heptafluoroisopropyl acrylate) or poly(heptafluoroisopropyl methacrylate). Hence, structural features other than branching of the perfluoroalkane chain were also involved in the comparison.

Fluorochemical water- and oil-repellent finishes have been monomeric or polymeric. The first fluorochemical repellents used commercially were monomeric. In analogy to repellents with a hydrocarbon hydrophobe, perfluoroalkanoic acids have been applied as chromium complexes [281,282] or zirconium salts [283,284] or have been converted to cationic fluorinated surfactants:

$C_8F_{17}(CH_2)_5OCH_2N^+$ 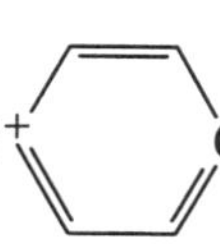 Cl^-

$F_3C(CF_2)_6CONH(CH_2)_3N^+(CH_3)_2C_2H_5Br^-$–

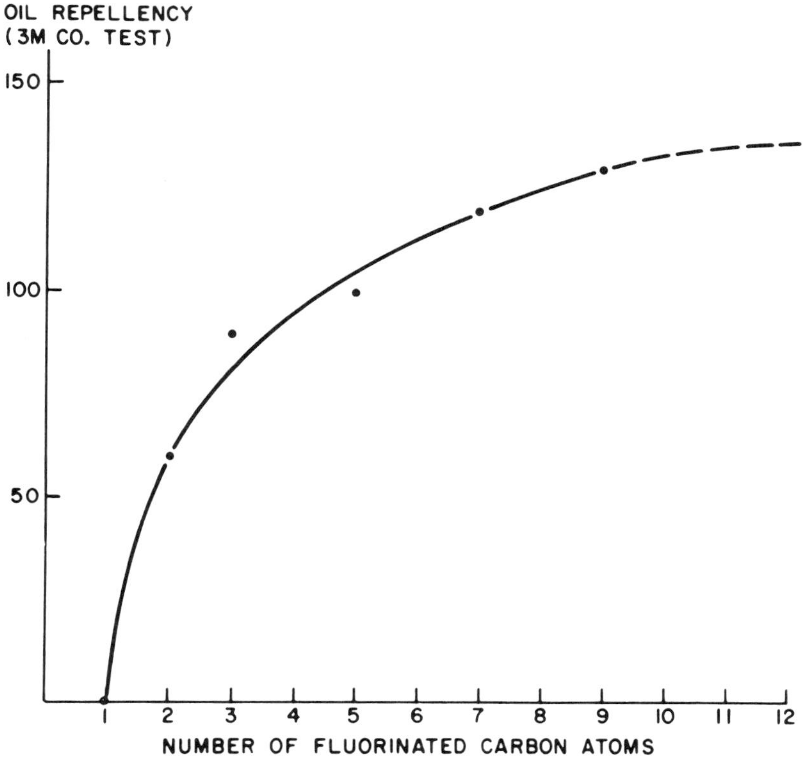

FIG. 8.2 The effect of fluorination of a hydrocarbon chain on oil repellency. (From Ref. 269.)

Since the repellent finishes produced with perfluorocarboxylic acid complexes were not durable to laundering [284], attempts were made to bond the repellent covalently to fibers. Benerito et al. [285] reacted cellulose with either perfluorooctanoyl chloride or perfluorobutanoic chloride in *N,N*-dimethylformamide [DMF] in the presence of a tertiary amine as the acid acceptor. The partial esters of cellulose thus produced exhibited good oil and water repellency but were hydrolyzed by alkaline detergents. Repellent finishes made by reacting glycidyl ethers of 1,1-dihydrofluoroalkanols with cellulose [286] were more durable but were probably not commercialized.

The low-molecular-weight fluorinated surfactants were superseded by polymeric fluorochemical repellents. However, the interest in monomeric fluorochemical repellents is increasing again. Modification of the fiber surface by fluorochemicals added to the polymer melt [287] or to the preformed fiber [288] is being explored and exploited commercially.

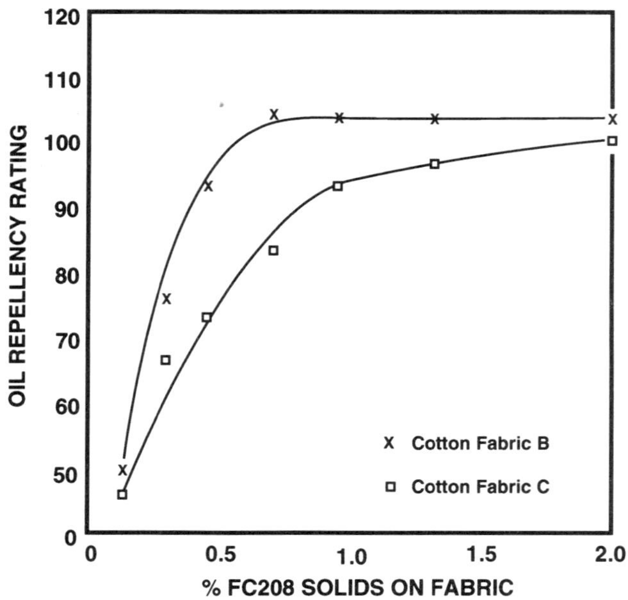

FIG. 8.3 Oil repellency as a function of fabric construction and the amount of fluoropolymer applied. (From Ref. 279.)

Monomeric as well as polymeric fluorinated surfactants are used to impart oil and fat repellency to paper or cardboard. See also "Paper".

Surface Treatment of Glass

See also Antifogging.

Optical glass lenses for cameras and optical instruments are made hydrophobic and oleophobic by a surface treatment with a cationic fluorinated surfactant [289,290]. The treated glass surface is more resistant to fingerprint soiling than the untreated surface. Glass can also be made oil repellent by coating with a methanolic solution of a 70:30 mixture of anionic fluorinated surfactants, $C_6F_{13}SO_3K$ and $C_6F_{13}SO_3NH_4$, and drying [291].

Fluorinated surfactants in windshield wiper fluids prevent icing of the windshield [292].

Textiles

Fluorinated surfactants impart oil and water repellency to textiles and paper (see Repellency) and increase surface lubricity. As an example, a size for polyester yarn containing C_{6-8} perfluoroalkanecarboxylic acid, poly(vinyl alcohol), and an acrylic polycarboxylate gave yarns easy to weave. When the fluorinated surfactant was replaced by potassium lauryl phosphate, the weavability was poor [293].

An anionic fluorinated surfactant, such as sodium 3-[3-perfluoromethylphenoxy]-1-propanesulfonate, has been claimed to increase the dye bath exhaustion of cationic dyes into acetate fibers [294].

In dry-cleaning formulations fluorinated surfactants improve soil suspension in perchloroethylene and reduce redeposition.

Vapor Barrier, Evaporation Retarders

Evaporation losses have been reduced by covering the liquid surface in petroleum product storage tanks with a floating layer of cereal grains, e.g., corn, wheat, or perlite treated with a fluorinated surfactant (Zonyl FSB) [295].

Evaporation of hydrocarbon fuel, e.g., gasoline, can be prevented with an aqueous layer containing anionic or amphoteric fluorinated surfactants [296,297]. Fluorinated surfactants are effective because of their low surface tension.

Katrizky et al. [298] examined the principles involved in vapor retardance. The postulated structural requirements for the optimum effect include a fluoroalkyl group, a rigid polar central section, and a lipophilic alkyl or arylalkyl substituent.

Wetting Agents

See also Herbicides and Insecticides.

Fluorinated surfactants are effective wetting agents in situations where conventional surfactants will fail, for example, in strongly alkaline or acid media. Fluorinated surfactants are used as stable wetting agents during ore treatment with concentrated acids. More rapid wetting of the ore, quicker breaking of the protective oxide layer, and reduction of caking can be achieved. In soldering fluids fluorinated surfactants function as a low-foaming wetting agent and reduce foaming encountered with silicone surfactants. Fluorinated surfactants aid wetting of sheepskins during desalination [35]. The addition of 0.1% fluorinated surfactant can reduce desalination time by seven times.

Mixtures of hydrocarbon surfactants and fluorinated surfactants are more effective wetting agents than either surfactant type used alone. A drop of water containing 0.1% trimethyl[3-(perfluorooctylsulfonylamino)-1-propyl]ammonium iodide and 1% sodium dodecyl sulfate spread on polyethylene and covered a circle with a 15-mm diameter. During the same time interval, drops containing 2% of either surfactant covered a circle having only a 6-mm diameter [299].

Tadros [156] studied wetting of wheat leaves with aqueous droplets containing fluorinated surfactants, hydrocarbon surfactants, or their mixtures. The

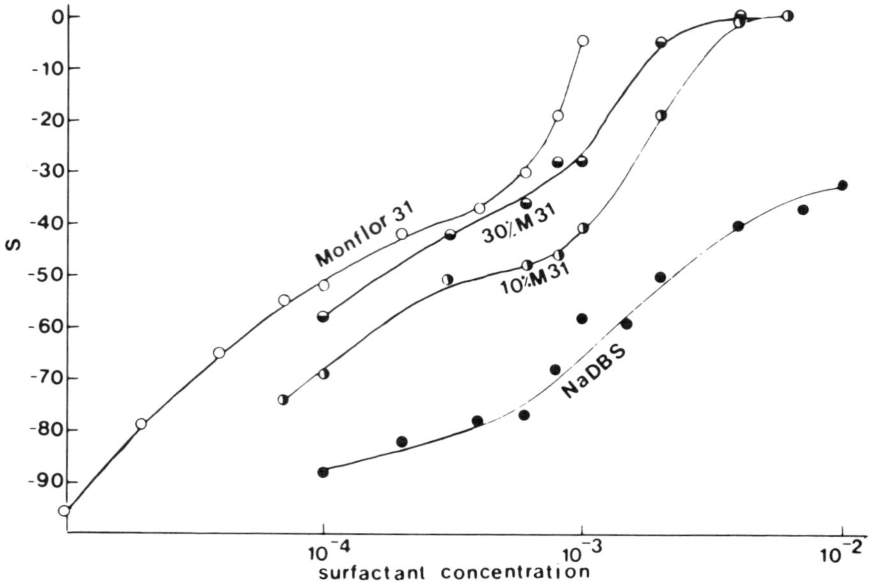

FIG. 8.4 Spreading coefficient S as a function of surfactant concentration, for Monflor 31 ($C_{10}F_{19}OC_6H_4SO_3Na$), sodium dodecylbenzenesulfonate, and their mixtures. (From Ref. 156.)

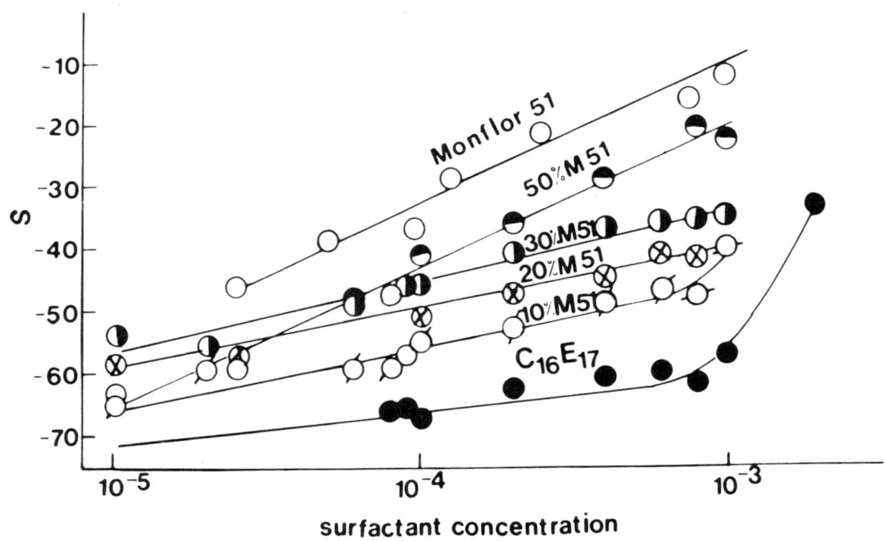

FIG. 8.5 Spreading coefficient S as a function of surfactant concentration, for Monflor 51, $C_{10}F_{19}O(CH_2CH_2O)_nH$, $C_{16}H_{31}O(CH_2CH_2O)_nH/C_{16}H_{33}O(CH_2CH_2O)_nH$ ($C_{16}E_{17}$), and their mixtures. (From Ref. 156.)

effect of surfactants on wetting was characterized by the spreading coefficient S and the retention factor F. The spreading coefficient S has been defined as

$$S = \gamma_{LA}(\cos\theta - 1) \tag{1}$$

where γ_{LA} is the liquid-air (vapor) interfacial tension and $\cos\theta$ the contact angle formed by the droplet on the leaf surface. The volume of liquid that will be retained on the sprayed surface is proportional to the retention factor F. Furmidge [300] derived the retention factor concept considering the sliding of drops on a tilted surface, the air-liquid surface tension, and the advancing and receding contact angles.

$$F = \theta_M \left[\frac{\gamma_{LA}(\cos\theta_R - \cos\theta_A)}{\rho} \right]^{0.5} \tag{2}$$

where θ_M is the mean of the receding contact angle θ_R and the advancing contact angle θ_A, and ρ is the density of the liquid.

Tadros [156] found the fluorinated surfactants to be more powerful wetting agents than hydrocarbon-type surfactants for wheat leaves. An order of magnitude lower concentration or even less of a fluorinated surfactant was needed to obtain the spreading coefficient exhibited by the corresponding hydrocarbon-type surfactant (Figs. 8.4 and 8.5). However, the retention coefficients for the fluorinated surfactants were lower than those for hydrocarbon surfactants (Figs. 8.6 and 8.7). To maintain satisfactory wetting and retention, Tadros proposed a mixture of fluorocarbon and hydrocarbon surfactants as a compromise. The use of mixed surfactants has a cost advantage and constitutes a lesser environmental burden as well.

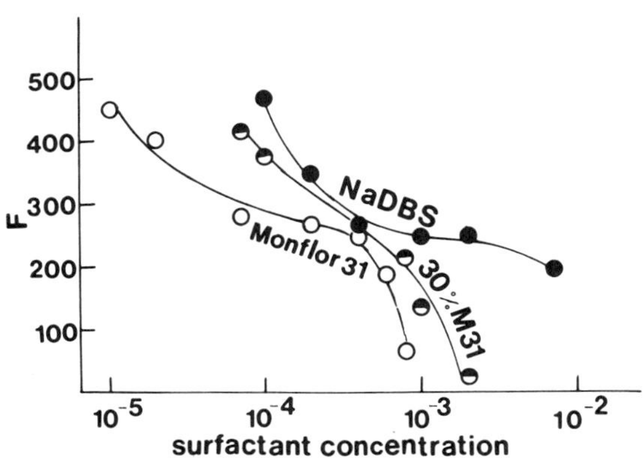

FIG. 8.6 Retention factor F as a function of surfactant concentration, for Monflor 31 ($C_{10}F_{19}OC_6H_4SO_3Na$), sodium dodecylbenzenesulfonate, and their mixtures. (From Ref. 156.)

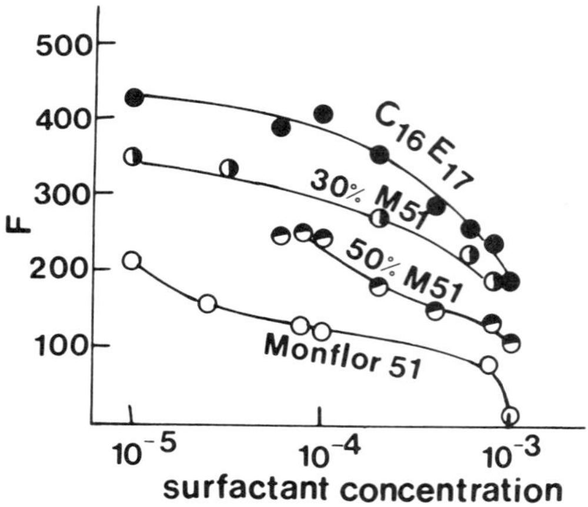

FIG. 8.7 Retention factor F as a function of surfactant concentration, for Monflor 51, $C_{10}F_{19}O(CH_2CH_2O)_nH$, $C_{16}H_{31}O(CH_2CH_2O)_nH/C_{16}H_{33}O(CH_2CH_2O)_nH$ ($C_{16}E_{17}$), and their mixtures. (From Ref. 156.)

REFERENCES

1. Du Pont Performance Products, Zonyl® Fluorosurfactants, (1987).
2. R. Jones, Fette Seifen Anstrichm. 77, 455 (1975).
3. H. Liebig and K. Ulm, Chem. Ztg. 99, 477 (1975).
4. E. Schuierer, Tenside 13, 1 (1976).
5. H. G. Klein, J. N. Meussdoerffer, and H. Niederprüm, Matalloberfläche 29, 559 (1975).
6. H. C. Fielding, in "Organofluorine Chemicals and Their Industrial Applications," R. E. Banks, ed., p. 214, Ellis Horwood, Chichester, England (1979).
7. J. N. Meussdoerffer and H. Niederprüm, Chem. Ztg. 104, 45 (1980).
8. (a) Bayer-Fluortenside, Bayer AG (1991). (b) Fluorad Fluorochemical Surfactants, 3M (1992). (c) Fluowet Fluorinated Surfactants, Hoechst (1980). (d) Forafac Fluorinated Surfactants, Atochem. (e) Lodyne High Performance Surfactants, Technical Bulletin, Ciba-Giegy Chem. Div. (1920). (f) Surflon Fluorinated Surfactants, Asahi Glass.

9a. Kao Corp., Jpn. Patent 58/74771 (1983).

9b. G. D. Semenov, V. S. Evchick, and L. A. Markova, Kauch. Rezina (7), 37 (1984).

10. Tohoku Hiryo Co., Jpn. Patent 56/41287 (1981).
11. Asahi Glass Co., Jpn. Patent 56/72071 (1981).
12. Mitsui Toatsu Chemicals, Jpn. Patent 59/145275 (1984)
13. Kokhoku Chemical Industries Co., Jpn. Patent 57/192445 (1982).
14. C. I. Kasei Co., Jpn. Patent 57/14648 (1982).
15. Achilles Foam Board Co., Jpn. Patent 58/32644 (1983).
16. C. I. Kasei Co., Jpn. Patent 59/217739 (1984).

17. Mitsubishi Monsanto Chemical Co., Jpn. Patent 59/102944 (1984).
18. Achilles Foam Board Co., Jpn. Patent 58/76440 (1983).
19. M. Ueno, Y. Ugajin, K. Horie, and T. Nişhimura, J. Appl. Polym. Sci. 39, 967 (1990).
20. R. Brueck, Kunststoffe 71(4), 234 (1981).
21. Asahi Glass Co., Jpn. Kokai Tokkyo Koho, JP 55/89375 (1980) and JP 55/144079 (1980).
22. Mitsubishi Metal Corp., Jpn. Patent 57/101302 (1982).
23. Tohoku Hiryo Co., Jpn. Patent 57/42741 (1982).
24. Dainippon Ink, Kawamura Phys. Chem. Inst., Jpn. Kokai Tokkyo Koho, JP 58001797 (1983).
25. Sanyo Chemical Industries and Nihon Cement Ltd., Jpn. Patent 59/97562 and 59/128240.
26. Brit. GB 1506464 (1978).
27. Aica Kogyo Co., Jpn. Patent 57/131263 (1982).
28. N. O. Vesterager, O. Michalsen, and T. Michalsen, Ger. Offen. DE 3,236,114 (1983).
29. Asahi Glass Co., Jpn. Patent 57/119,999 (1982).
30. J. Block, U.S. Patent 4,144,185 (1979).
31. Hitachi Ltd., Jpn. Patent 59/71398 and 59/217800.
32. Chem. Fabrik Kreussler und Co., Belg. Patent 861660 (1978).
33. P. Bakos, R. E. Darrow, J. P. Funari, and R. M. Poliak, Eur. Pat. Appl. EP 32179 (1981).
34. K. Heymann, R. Rolff, and W. Meyer, Ger. Offen. 2,636,993 (1978).
35. Monflor Fluorochemical Surface Active Agents, ICI United States, Inc.
36. J. C. Fang (Du Pont), U.S. Patent 3,839,254 (1974).
37. R. A. Bergfeld, S. P. Patel, and L. M. Schlickenrieder (NL Industries), U.S. Patent 4,208,496 (1980).
38. J. Mondt (Hoechst), Ger. Offen. DE 3128143 (1983).
39. J. C. Goebel and A. C. Lunn (American Cyanamid), U.S. Patent 4,183,367 (1980).
40. M. R. Lewis and S. A. Mitchell (Unilever Ltd.), Brit. GB 1598567 (1981).
41. J. A. Cella, R. A. Lukey, A. E. Fiebig, and F. J. Pum, U.S. Patent 3,993,745 (1976), CA 86, 47211.
42. J. A. Cella and A. E. Fiebig (Albert Culver Co.), U.S. Patent 3,993,744 (1976).
43. S. Omon, T. Deguchi, and K. Okamura (Daikin Kogyo Co.), Jpn. Kokai JP 53/31582 (1978).
44. M. Callingham and K. V. Curry (Unilever N.V.), Ger. Offen. 2816828 (1978).
45. J. A. Cella and A. E. Fiebig, U.S. Patent 4,176,176 (1979).
46. C. F. Iovanni and D. M. Biesel (Gillette Co.), U.S. Patent 4,765,975 (1988).
47. G. Vanlerberghe and H. Sebag (Oreal S.A.), FR 2521449 (1983).
48. (a) J. A. Keiner (Unilever N.V.), Ger. Offen. DE 2051523 (1971). (b) R. L. Parks and S. E. Builder (Procter and Gamble Co.), Ger. Offen. DE 2537374 (1976).
49. C. S. Herrick and F.F. Holub, U.S. Patent 4,267,879 (1981).
50. J. Shimoiizaka, Jpn. Patent 77/00783 (1977). CA 86,186129.

51. A. Naplava, L. Jahnatek, I. Krivosik, O. Kyslinger, and J. Oldrich, Czech. Patent 209,334 (1981).
52. H. Stefen, Ger. Offen. 2,409,251 (1975) and 2,462,435 (1977).
53. M. Oita, H. Hirohata, and N. Hamazaki, Jpn. Patent 51/135,836 (1976).
54. K. Stallmann and H. Speckhardt, Metall (Berlin) 39(3), 238 (1985).
55. Hitachi Cable, Ltd., Jpn. Kokai Tokkyo Koho JP 59/35315 (1984).
56. Hitachi Cable Ltd., Jpn. Kokai Tokkyo Koho JP 59/35314 (1984).
57. S. Masuda and M. Izumi (Sumimoto Electric Ind.), U.S. Patent 3,839,174 (1974).
58. M. Amano, K. Masahiko, and T. Kanashiro (Hitachi), Jpn. Kokai Tokkyo JP 54/65400 (1979).
59. T. Kaneshiro, M. Unno, and M. Sakai (Hitachi), Jpn. Kokai Tokkyo Koho JP 54/7200 (1979).
60. R. W. Angelo, R. M. Poliak, and J. R. Susko (IBM), Eur. Patent Appl. EP 6509 (1980).
61. G. Sprengling (Westinghouse), U.S. Patent 3,841,957 (1974).
62. C. D. Iacovangelo and F. G. Fritz, J. Electrochem. Soc. 132(4), 851 (1985).
63. K. Ohsawa, K. Fujiwara, Y. Takeda, and T. Yabumoto, Jpn. Patent 53/67834 (1978).
64. S. Suetsugu, A. Ota, A. Miura, Y. Nitta, and K. Yoshizawa (Matsushita), Jpn. Kokai Tokkyo Koho, JP 02114447 (1990).
65. C. A. Lazars, U. I. Keller, and E. H. Cook, Jr., Fr. Demande 2,477,162 (1981).
66. K. Aoyama, Jpn. Kokai JP 53/5052 (1978).
67. H. G. Klein, J. N. Meussdoerffer, H. Niederprüm, and M. Wechsberg, Tenside 15, 2 (1978).
68. (a) H. Brown (Udylite Corp.), U.S. Patent 2,913,377 (1959). (b) Societe Continentale Parker, Fr. Demande 2,241,542 (1975) and Fr. Demande 2,257,665 (1975).
69. H. G. Klein, J. N. Meussdoerffer, and H. Niederprüm, Galvanotechnik 68, 1045 (1977).
70. I. L. Knunyants, Ts. Adzhemyan, B. L Dyatkin, E. K. Tsirul, G. G. Kuznetsova, A. A. Krasilnikov, and L. M. Grubina, USSR. Patent 250131 (1977).
71. T. Umaba, T. Ito, and Y. Tashiro, Jpn. Patent 54/76443 (1979).
72. D. A. Bultman and M. H. Olson (3M), Brit. UK Patent Appl. GB 2,077,765 (1981).
73. Nippon Electric Co., Jpn. Patent 58/189386 (1983).
74. V. Glöckner, K. Lunkwitz, and D. Prescher, Tenside 26, 376 (1989).
75. I. G. Bondarenko, K. S. Burmistrov, and F. I. Danilov, Ukr. Khim. Zh. 47(2), 150 (1981).
76. K. H. Tillwich (3M), Ger. Patent 1,621,107 (1973).
77. M. A. Lombino, Bull. Doc. Cent. Inf. Chrome Dur. (July–August) 19–26 (1974). CA 83, 17579.
78. H. Enger and K. Szameitat, Metalloberfläche 37, 12 (1983).
79. D. Melzer, E. Tabel, and H. P. Biermann (Bayer), EP 299307 (1989).
80. S. Matsuura, Y. Mizutani, A. Mimura, and H. Yamamoto, Jpn. Patent 55/14869 (1980).
81. Nippon Electric Co., Jpn. Patent 57/200588 (1957).

82. F. Yamaguchi, S. Kurosaki, and N. Watanabe, Denki Kagaku Oyobi Kogyo Butsuri Kagaku, 43(1), 57 (1975). CA 83, 87129.
83. F. Yamaguchi, Y. Okamoto, S. Kurosaki, and N. Watanabe, Denki Kagaku Oyobi Kogyo Butsuri Kagaku, 43 (2) 106 (1975). CA 83, 154519.
84. C. Uemura, Jpn. Patent 59/182993 (1984).
85. V. Sova, Eur. Patent Appl. 139,958 (1985).
86. J. O. Hendricks, Ind. Eng. Chem. 45, 99 (1953).
87. (a) A. F. Benning and J. D. Park (Du Pont), U.S. Patent 2,396,076 (1946). (b) A. F. Benning (Du Pont), U.S. Patent 2,559,749 (1951). (c) A. F. Benning (Du Pont), U.S. Patent 2,597,702 (1952).
88. R. J. Plunkett (Du Pont), U.S. Patent 2,230,654 (1941).
89. (a) K. L. Berry (Du Pont), U.S. Patent 2,559,629 (1951). (b) K. L. Berry (Du Pont), U.S. Patent 2,559,752 (1951).
90. A. Karydas and J. Rodgers (Ciba-Geigy), EP 292427 (1988).
91. M. Visca and D. Lenti (Ausimont), EP 315078 (1989).
92. A. Chittofrati and D. Lenti (Ausimont), EP 315841 (1989).
93. A. Chittofrati, A. Tentorio, and M. Visca (Ausimont), EP 346933 (1989).
94. H. Niederpruem, H. G. Klein, and J. N. Meussdoerffer, Ger. Offen. DE 2535333 (1977) and DE 2,556,429 (1977).
95. E. Tabel (Bayer), Ger. Offen. DE 3038985 (1982).
96. C. J. Rossi, F. S. Kaveggia, K. H. Nelson, U.S. Patent 3,383,255 (1968).
97. Kokusan Kinzoku Kogyo Co., Jpn. Patent 58/5335 (1983), CA 99, 6646.
98. K. Yajima and T. Kanbe, Kinzoku Hyomen Gijutsu 25(3), 162 (1974).
99. (a) Daikin Kogyo Co., Jpn. Patent 59/56482 (1984), CA 102, 88540. (b) E. Naonori and T. Koji, Eur. Patent Appl. EP 133,584 (1985), CA 102, 213758.
100. R. J. Hopkins, E. G. Thomas, and H. J. Kieta (Allied Corp.), U.S. Patent 4,517,106 (1985), CA 103, 31250.
101. N. J. Nelson (PSI Star), Ger. Offen. DE 3420309 (1984).
102. F. Fitch, in "Foams: Physics, Chemistry and Structure," A. J. Wilson, ed., p. 207, Springer-Verlag, London (1989).
103. "Ullman's Encyclopedia of Industrial Chemistry," 5th ed., Vol. A11, p. 115, VCH Verlagsges., Weinheim (1988).
104. D. N. Meldrum and J. R. Williams (National Foam), U.S. Patent 3,475,333 (1969).
105. A. Rodriguez (Angus, George Co.), Ger. Offen. DE 1928556 (1969).
106. R. G. Hughes (ICI), Ger. Offen. DE 2240263 (1973).
107. Daikin Kogyo Co., Jpn. Kokai Tokkyo Koho JP 55/146172 (80/146172) (1980).
108. Neos Co., Jpn. Kokai Tokkyo Koho JP 57/72666 (82/72666) (1982).
109. A. Lantz, Fr. Demande FR 2,575,165 (1986).
110. E. Hervot and P. Schiffers (Sandoz), FR 2523456 (1983).
111. P. C. Clark and R. A. Falk (Ciba-Geigy), EP 311570 (1989).
112. P. Wirtz and R. Broell (Hoechst), Ger. Offen. DE 2732555 (1979).
113. R. D. Howard and W. I. Spurr (ICI), Ger. Offen. DE 2829594 (1979).
114. E. Schuierer, H. Bathelt and R. Broell (Hoechst), Ger. Offen. DE 2714013 (1978).
115. R. Broell, E. Schuierer, and H. Bathelt (Hoechst), Ger. Offen. DE 2 749 331 (1979).

116. H. Cordes and E. Achilles (Regeltechnische Geraete P. Luethge), Ger. Offen. DE 3409602 (1985).
117. L. Foulletier and R. Bertocchio (Ugine Kuhlmann), Ger. Offen. DE 2325855 (1973).
118. R. G. Hughes (ICI), Ger. Offen. DE 2240263 (1973). CA 78, 161183.
119. P. J. Chiesa, Jr. (Philadelphia Suburban Corp.), U.S. Patent 3,849,315 (1974), U.S. Patent 3,957,657 (1976); U.S. Patent 3,957,658 (1976).
120. Dainippon Ink and Chem., Jpn. Kokai Tokkyo Koho JP 58/50971 (1983).
121. R. Bertochio and L. Foulletier (Ugine Kuhlmann), Ger. Offen. DE 2 617 746 (1975).
122. R. A. Falk (Ciba-Geigy), U.S. Patent 4,042,522 (1977).
123. Nittan Co, Jpn. Kokai Tokkyo Koho JP 58/112565 (83/112565) (1983).
124. C. Raynolds and E. E. C. Cawood (Nu-Swift Intern.), GB 1208261 1970).
125. W. R. Warnock, D. V. Flatt, and J. R. Eastman (Ansul), U.S. Patent 3,553,127 (1971).
126. P. Somasundaram and R. D. Kulkarni, Inst. Mining Metall. Trans. Sect. C, 82 (Sept.), C164-C167 (1973).
127. V. G. Berezyuk, O. B. Dubrovina, N. A. Petrova, O. V. Evtyukhova, and A. M. Kasimov, Zh. Prikl. Khim. (Leningrad), 58(3), 669 (1985). CA 102, 209900.
128. Hitachi Chem. Co., Jpn. Kokai Tokkyo Koho JP 59/179635 (1984). CA 102, 114747].
129. Toshiba Corp., Jpn. Kokai Tokkyo Koho JP 57/125225 (1957). CA 98:55215.
130. Asahi-Olin Ltd., Jpn. Kokai Tokkyo Koho JP 58/76407 (1983). CA 100, 122184 (1983).
131. J. R. Allaway, D. R. Bloch, and M. P. Fischer, U.S. Patent 4,378,278 (1983).
132. Y. K. Kim, Chi-Long Lee, and G. M. Ronk, Ger. Offen. DE 2631267 (1977); CA 86:91107.
133. Kao Corp., Jpn. Kokai Tokkyo Koho JP 58/74771 (1983). CA 100, 35287.
134. S. Szönyi, R. Vandamme, and A. Cambon, J. Org. Chem. 30, 37 (1985).
135. C. Heid, D. Hoffmann, and J. Polster, Ger. Offenb. DE 2233941 (1974).
136. Atsurf F Fluorosurfactants, ICI Americas Inc. (1987).
137. Dainippon Ink and Chem., Jpn. Kokai Tokkyo Koho JP 58/70806 (1983).
138. F. J. Gohlke and W. Klinger (Hoechst), Brit. UK Patent Appl. GB 2004534 (1979). CA 91, 177016.
139. W. Klinger and E. Milewski (Hoechst), Eur. Patent Appl. EP 54894 (1982).
140. M. Hayek and R. J. Moody (Du Pont), U.S. Patent 3,948,668 (1976).
141. Mitsubishi Pencil Co., Jpn. Kokai Tokkyo Koho JP 56/55471 (1979).
142. Pilot Ink Co., Jpn. Kokai Tokkyo Koho JP 55/106273 (1979).
143. Shachihata Ind. Co., Jpn. Kokai Tokkyo Koho JP 59/86673 (1982).
144. Shachihata Ind. Co., Jpn. Kokai Tokkyo Koho JP 58/210972 (1983).
145. D. Sasage and T. Takashima (Pentel Co.), Jpn. Kokai Tokkyo Koho JP 53/128404 (1978).
146. Z. Kato (Tombow Pencil Co.), Jpn. Kokai Tokkyo Koho JP 54/4622 (1977).
147. Sailor Pen Co., Jpn. Kokai Tokkyo Koho JP 58/25375 (1981).
148. Carbon Paper Co., Jpn. Kokai Tokkyo Koho JP 60/13865 (1983).
149. Suwa Seikosha Co., Jpn. Kokai Tokkyo Koho JP 56/49771 (1979).

150. Pentel Co., Jpn. Kokai Tokkyo Koho JP 56/65060 (1981).
151. Sailor Pen Co., Jpn. Kokai Tokkyo Koho JP 59/59758 (1984).
152. Pentel Co., Jpn. Kokai Tokkyo Koho JP 58/76741 (1983).
153. M. Kyono, M. Uehara, M. Nakano, and A. Nogami (Konica), Jpn. Kokai Tokkyo Koho, JP 63175858 (1988).
154. G. Caporiccio, S. Soldini, and E. Strepparola, Eur. Patent Appl. EP 95825 (1983).
155. S. Sakakibara, R. Tatematsu, M. Kita, and H. Kimoto. Nagoya Kogyo Gijutsu Shikensho Hokoku 40(1), 20 (1991).
156. T. F. Tadros, in "Wetting, Spreading, Adhesion," J. F. Padday, ed., Academic Press, London (1978).
157. M. Matsumoto, Jpn. Kokai JP 51/35436 (1976). Kokai JP 51/35437 (1976).
158. W. V. Cohen (Du Pont), U.S. Patent 3,096,207 (1963).
159. (a) T. W. Berger (3M), Ger. Offen. 1952762 (1970). (b) R. D. Howells (3M), U.S. Patent 4,566,981 (1986). (c) C. M. Martinez and L. M. B. Rodriguez (3M), U.S. 5,098,446 (1992).
160. E. Gratacos, A. Marsal and M. Fort, J. Am. Leather Chem. Assoc. 85, 474 (1990).
161. T. Hayashi, S. Ono, and Y. Otoshi (Asahi Glass Co.), Jpn. Kokai Tokkyo Koho JP 55003428 (1980).
162. Asahi Glass Co., Jpn. Kokai Tokkyo Koho JP 56/98298 (1981).
163. Y. Arai, S. Kinoshita, and K. Kimura, U.S. Patent 3,910,682 (1975).
164. C. G. Thiel (3M), U.S. Patent 4,352,789 (1982). CA 98, 8181.
165. V. B. Caslavsky and P. Gron (Forsyth Dental Infirmary for Children), U.S. Patent 4,353,892 (1982).
166. D. Tokinaga and T. Kobayashi, Jpn. Kokai JP 52/105208 (1977).
167. H. W. Dettner, Ger. Offen. DE 1914759 (1970).
168. Nippon Steel Corp., Asahi Glass Co., Jpn. Kokai Tokkyo Koho, JP 57/198272 (1982).
169. Asahi Glass, Jpn. Kokai Tokkyo Koho, JP 59/226184 (1984).
170. Mobay Product Information, Fluoro Chemical Surfactant FT-248 (1981).
171. (a) 3M Company, Technical Information, 3M Brand Fluorochemical Surfactants (1963). (b) R. D. Burke, J. B. Kittredge, J. S. Spira, and M. L. Victor (3M), Mater. Protect. 4(7), 44 (1965), CA 63, 6720b. (c) R. Gritzky, Fette Seifen Anstrichm. 71, 971 (1969).
172. Nippon Paint Co., Jpn. Kokai Tokkyo Koho, JP 56/115397 (1981).
173. Dainippon Ink, Jpn.Kokai Tokkyo Koho, JP 58/213057 (1983).
174. I. A. Shelobasov, A. N. Kukushkin, V. A. Mikhailov, G. A. Saltanov, E. L. Fedorov, K. Schindler, and E. Chempick, Trenie Iznos. 5, 853 (1984); CA 102, 6597.
175. Matsushita Electric Ind. Co., Jpn. Kokai Tokkyo Koho, JP 599177 (1984).
176. Matsushita Electric Ind. Co., Jpn. Kokai Tokkyo Koho, JP 59/9178 (1984).
177. Matsushita Electric Ind. Co., Jpn. Kokai Tokkyo Koho, JP 59/16982 (1984).
178. S. Tanaka (Kansai Paint), Jpn. Kokai Tokkyo Koho, JP54/46235 (1979).
179. R. R. Alm, Proc. Water-Borne Higher Solids Coat. Symp., 8th(1), 51 (1981). CA 95, 188689.

180. R. Jones and L.E. Houghton (ICI), Ger. Offen. DE 2215387 (1972).
181. N. Horiuchi, S. Kai, and M. Shinjo, (Daikin Kogyo Co.), Ger. Offen. DE 2641898 (1972).
182. Toshiba Corp., Jpn. Kokai Tokkyo Koho JP 58/74743 (1983).
183. H. Franz (PPG Ind.), U.S. Patent 4,311,764 (1980).
184. Tokai Rubber Industries, Jpn. Kokai Tokkyo Koho JP 59/108040 (1984).
185. Sekisui Chem. Co., Jpn. Kokai Tokkyo Koho JP 58/217502 (1983).
186. H. Shirota and R. Sugae (Dainippon Ink and Chem.), Jpn. Kokai Tokkyo Koho JP 54/36342 (1979).
187. Hitachi Chem Co., Jpn. Kokai Tokkyo Koho JP 59/191760 (1984).
188. L. E. Houghton, R. I. Hancock, and P. L. Cornes, Brit. GB 1 545401 (1979).
189. T. Baba and T. Sato, Jpn. Kokai JP 50/22783 (1975).
190. P. Couderc, Vermiculite et la Perlite S.A., Fr. Demande FR 2333564 (1977).
191. T. Cooke (Ciba-Geigy), Ger. Offen DE 2 922 928 (1980).
192. H. B. Clark (3M), J. Pet. Technol. 32(10), 1695 (1980).
193. H. B. Clark, M. T. Pike, and G. L Rengel, J. Pet. Technol. 34(7), 1565 (1982).
194. G. S. Penny, J. E. Briscoe, and P. C. Harris (Halliburton), Ger. Offen. DE 3306593 (1983).
195. (a) N. O. Brace and A. K. Mackenzie (Du Pont), U.S. Patent 3,083,224 (1963). (b) Du Pont, Res. Discl. 175, 5–6 (1978). (c) Zonyl RP and Zonyl NF Paper Fluoridizers, Du Pont.
196. R. E. Brown, Paper Ind. 45, 205 (1963).
197. Asahi Glass Co. Ltd., Jpn. Kokai Tokkyo Koho, JP 60 64990 (1985).
198. H. Higuchi, H. Takahashi, and K. Hashimoto (Nippon Oils and Fats), Jpn. Kokai Tokkyo Koho, JP 215090/87 (1987).
199. T. Imai, K. Nojima, and M. Takahashi (Kanzaki Paper Mfg.), Jpn. Kokai Tokkyo Koho, JP 03106476 (1991).
200. Fuji Photo Film Co., Jpn. Kokai Tokkyo Koho, JP 59029194 (1983).
201. Y. Hotta, S. Kitamura, F. Ishi-i, T. Abe, K. Koshizuka, and M. Nakamura (Konishiroku Photo), Jpn. Kokai Tokkyo Koho, JP 94390/85 (1985).
202. H. Oike and M. Yanagimachi (Oike & Co.), Jpn. Kokai Tokkyo Koho, JP 199680/88 (1988).
203. N. Eto and T. Imai (Taiho Ind.), Jpn. Kokai, JP 11387/90 (1990).
204. K. Hasegawa, T. Mori, and M. Higuma (Canon), Jpn. Kokai Tokkyo Koho, JP 252779/88 (1988).
205. T. Sakai and T. Aono (Fuji Photo Film), Jpn. Kokai Tokkyo Koho, JP 146785/89 (1989).
206. M. Groh and A. Oetiker (Ciba), Ger. Offen. DE 1942665 (1970).
207. J. F. Willems, R. J. Pollet, F. J. Ville, R. F. Heylen, F. J. Sels, and A. J. Minne (Agfa-Gevaert), Ger. Offen. DE 1961638 (1970).
208. W. F. De Geesta and H. Vandenabeele (Agfa Gevaert), Ger. Offen. DE 2124262 (1971).
209. Imperial Chem. Ind. Ltd., Fr. Demande FR 2213514 (1974).
210. J. M. Babbitt and J. F. Houle (Eastman Kodak Co.), U.S. Publ. Patent Appl. B US 501379 (1976), Division of U.S. 3850640.
211. K. Hori and S. Nagamoto (Fuji), Ger. Offen. DE 2526970 (1976).

212. I. Horie, N. Sugimoto, M. Yoneyama, and N. Yamamoto (Fuji), Jpn. Kokai JP 51/7917 [76/7917] (1976).
213. M. Sato, H. Ishikawa, M. Ishihara, and K. Nakazato (Konishiroku), Ger. Offen. DE 2526481 (1976).
214. M. Yoneyama, T. Mikami, and N. Tsuji (Fuji), Ger. Offen. DE 2619248 (1976).
215. S. J. Cruikshank, Res. Discl. 175, 45 (1978).
216. Konishiroku Photo Industry Co., Ltd., Jpn. Kokai Tokkyo Koho JP 55/149938 [80/149938] (1980).
217. R. J. Pollet, H. L. Kokelenberg, R. P. Samijn, F. J. Sels, F. J. Ville, and N. C. De Jaeger (Agfa-Gevaert), Eur. Patent Appl. EP 15592 (1980).
218. C. E. Miller (Du Pont), Eur. Patent Appl. EP 30352 (1981). J. Minamizono, S. Kishimoto, and M. Yoneyama (Fuji), Ger. Offen, DE 3038818 (1981).
219. D. N. Miller and R. A. Kydd (Eastman Kodak), Belg. BE 891230 (1982).
220. N. V. Uvarova, V. F. Klyuchevich, and E. A. Zimkin, Zh. Nauchn. Prikl. Fotogr. Kinematogr. 27(5), 375 (1982).
221. J. F. Padday and A. R. Pitt (Eastman Kodak), WO 8300162 (1983); CA 98, 152754.
222. Fuji Photo Film Co. Ltd., Jpn. Kokai Tokkyo Koho JP 59/74554 A2 [84/74554] (1984).
223. Konishiroku Photo Industry Co. Ltd., Jpn. Kokai Tokkyo Koho JP 59/30536 [84/30356] (1984).
224. Konishiroku Photo Industry Co. Ltd., Jpn. Kokai Tokkyo Koho JP 59/206832 [84/206832] (1984).
225. N. Sugimoto, T. Kojima, and Y. Mukunoki (Fuji), Ger. Offen. DE 3327464 (1984).
226. V. F. Klyuchevich, N. V. Uvarova, and E. A. Zimkin, Zh. Nauchn. Prikl. Fotogr. Kinematogr. 30(1), 65 (1985).
227. Y. Kuraki and T. Inayama (Fuji), Jpn. Kokai Tokkyo Koho, JP 62173460 (1987).
228. K. Hesse and M. O. Oezelsel (Du Pont Deutschland), EU 19951 (1989).
229. T. Yamamoto, S. Goto, N. Suzuki, H. Tomiyasu, and Y. Kobayashi (Konica), Jpn. Kokai Tokkyo Koho, JP 01048849 (1989).
230. S. Yoneyama and J. Arakawa (Fuji), Jpn. Kokai Tokkyo Koho, JP 01263644 (1989).
231. Minnesota Mining and Manufg. Co., Fr. Demande FR 2191152 (1972).
232. E. Cavallo and F. Furlan (3M), Ger. Offen. DE 2337638 (1972).
233. K. J. Schreiber (Polaroid), U.S. Patent 4,267,255 (1981).
234. Mitsubishi Paper Mills, Ltd., Jpn. Kokai Tokkyo Koho JP 59/45441 [84/45441] (1984).
235. Konishiroku Photo Industry Co., Ltd., Jpn. Kokai Tokkyo Koho JP 59/72442 [84/72442] (1984).
236. Mitsubishi Paper Mills, Ltd., Jpn. Kokai Tokkyo Koho JP 59/37815 [84/37815] (1984).
237. Dainichi Nippon Cables, Jpn. Kokai Tokkyo Koho, JP 56/47476 (1981).
238. Japan Synthetic Chem., Jpn. Kokai Tokkyo Koho, JP 55/3445 (1980).

239. K. Motani, K. Yabuki, and S. Tomiguchi (Tokuyama Soda), Jpn. Kokai, JP51/1377 (1976).
240. M. K. Bernett, Proc. Chem. Inst. Symp. Polym. Interfaces: Focus Adhes. 135-52, Chem. Inst. Canada, Ottawa (1976).
241. Shin-Etsu Chem. Ind., Jpn. Kokai Tokkyo Koho, JP 58/167647 (1983).
242. D. G. Wharton, Adhesion 2, 167 (1978).
243. M. Hamada, I. Suzuki, H. Sakurai, N. Ohshima, and T. Yoshida (Asahi), Jpn. Kokai, JP 52/19751 (1977).
244. H. Kroll, A. R. Therrien, and A. L. Forchielli (P. A. Hunt, Chem. Corp.), U.S. Patent 3,654,206 (1972).
245. L. R. Parks (Procter & Gamble), U.S. Patent 3,741,914 (1973).
246. F. L. McCarthy, and H. T. Bach (Sterling Drug), U.S. Patent 4,299,749 (1981).
247. R. J. Posten (Airwick Ind.), BR 8103516 (1982).
248. H. Murahama, Jpn. Kokai Tokkyo Koho JP 58/52399 (1983).
249. F. Brandstetter, W. Ziegler, W. Dietsche, S. Weiss, and A. Hettche (BASF), Ger. Offen. DE 3420168 (1985).
250. O. M. Brown (S. C. Johnson and Son), EP 246928 (1987).
251. A. P. Jethwa (Sterling Drug), U.S. Patent 4,869,934 (1989).
252. F. J. Rahl, M. A. Evanco, R. J. Fredericks, and A. C. Reimschuessel, J. Polym. Sci. A-2, 10, 1337 (1972).
253. B. Luhmann and A. E. Feiring, Polymer 30, 1723 (1989).
254. D. Sianesi, G. Bernardi, G. Veroli (Mont Ed.), Ger. Offen. DE 1940293 (1970).
255. T. Yoshimura, K. Shimofukigoshi, T. Satokawa (Daikin Kogyo Co.), Japan JP 48/34836 (1973).
256. M. B. Mueller, P. P. Salatiello, and D. L. Sawhney (Allied), Ger. Offen. DE 2157170 (1972).
257. S. V. Gangal (Du Pont), U.S. Appl. 916844 (1978).
258. T. Shimizu, S. Yamaguchi, and H. Koizumi, Kobunshi Kako 30(10), 473 (1981).
259. J. P. Stallings (Diamond Shamrock), Ger. Offen. DE 2213135 (1972).
260. Y. Toyoda and M. Shirai (Kureha), Japan JP 47/51233 (1972).
261. K. Kido, H. Wakamori, F. Suzuki, and G. Asai (Kureha), Jpn. Kokai Tokkyo Koho JP 54/85290 (1979).
262. J. E. Dohany (Pennwalt), U.S. Patent 4360652 (1982).
263. S. S. Ivanchev, V. P. Budtov, A. I. Andreeva, G. A. Otradina, and Yu. A. Zaichenko, Vysokomol. Soedin. Ser. A 25(11), 2335 (1983).
264. M. Tatemoto and S. Sakata (Daikin), Japan JP 49/43386 (1974).
265. T. S. Sirlibaev, V. G. Kalyadin, I. Tirkashev, and Kh. U. Usmanov, Dokl. Akad. Nauk. UzSSR (8), 38 (1982).
266. M. Petruschke, K. H. Goebel, W. Jaeger, G. Reinisch, D. Prescher, H. Kaltwasser, I. Richter, and H. J. Wolf, Ger (East) DD 159079 (1983).
267. E. Jones and J. Walker (ICI), Ger. Offen. DE 2501239 (1975).
268. K. Kato, K. Esumi, and K. Meguro, Bull. Chem. Soc. Jpn. 59, 249 (1988).
269. E. Kissa, Repellent Finishes, in "Handbook of Fiber Science and Technology," Vol. II, p. 143, Part B, M. Lewin and S. B. Sello, eds., Marcel Dekker, New York (1984).
270. S. Baxter and A. B. D. Cassie, J. Text. Inst. 36, T67 (1945)]

271. A. B. D. Cassie and S. Baxter, Trans. Faraday Soc. 40, 546 (1944).
272. G. H. Segall, Textile Res. J. 22, 736 (1952).
273. C. A. Davis, Am. Dyest. Rep. 56, P555 (1967).
274. M. Kärrholm and G. Kärrholm, Text. Res. J. 20, 215 (1950).
275. A. M. Sookne, F. W. Minor, J. E. Simpson, and M. Harris, Am. Dyest. Rep. 35, 295 (1946).
276. W. A. Zisman, in "Contact Angle, Wettability, and Adhesion," R. F. Good, ed., Adv. Chem. Ser., Vol. 43, p. 1, American Chemical Society, Washington, DC (1964).
277. E. G. Shafrin and W. A. Zisman, J. Phys. Chem. 66, 740 (1962).
278. N. O. Brace, J. Org. Chem. 27, 4491 (1962).
279. E. J. Grajeck and W. H. Petersen, Text. Res. J. 32, 320 (1962).
280. A. G. Pittman, D. L. Sharp, and B. A. Ludwig, J. Polym. Sci. PT. A-1, 6, 1729 (1968).
281. T. S. Reid (3M), U.S. Patent 2,662,835 (1953).
282. F. J. Philips, L. Segal, and L. Loeb, Text. Res. J. 27, 369 (1957).
283. V. W. Tripp, R. L. Clayton, and B. R. Porter, Text. Res. J. 27, 340 (1957).
284. L. Segal, F. J. Philips, L. Loeb, and R. L. Clayton, Text. Res. J. 28, 233 (1958).
285. R. R. Benerito, R. J. Berni, and T. F. Fagley, Text. Res. J. 30, 393 (1960).
286. R. R. Benerito, R. J. Berni, and T. F. Fagley, Text. Res. J. 30, 576 (1960).
287. F. Mares and B. C. Oxenrider, Text. Res. J. 47, 551 (1977).
288. F. Mares and B. C. Oxenrider, Text. Res. J. 48, 218 (1978).
289. J. F. Paddey and T. D. Blake (Kodak), Res. Discl. 180, 138 (1979)
290. Brit. GB 1588962 (1981); CA 95, 152499.
291. Dainippon Ink and Kawamura Phys. Chem. Res. Institute, Jpn. Kokai Tokkyo Koho, JP 58/213057 (1983).
292. O. Wack and H. Schmid, Ger. Offen. DE 3208219 (1983).
293. M. Ueda, Jpn. Kokai Tokkyo Koho, JP 54131094 (1979).
294. H. Moriga, JP 47043155 (1972).
295. P.R. Scott, W. D. Johnston, and J. L. Kyrish (Shell), U.S. Patent 4,035,149 (1977).
296. Dainippon Ink and Chemicals, Inc., Kawamura Rikagaku Kenkyoshu, Jpn. Kokai Tokkyo Koho JP 55/145780 (1980).
297. Dainippon Ink and Chemicals, Inc., Kawamura Rikagaku Kenkyoshu, Jpn. Kokai Tokkyo Koho JP 57/78473 (1982).
298. A. R. Katrizky, T. L. Davis, G. W. Rewcastle, G. O. Rubel, and M. T. Pike, Langmuir 4, 732 (1988).
299. M. J. Owen and J. Thompson, Brit. GB 1337467 (1973).
300. C. G. L. Furmidge, J. Colloid Sci. 17, 309 (1962).

9
Analysis of Fluorinated Surfactants

9.1 DETERMINATION AND CHARACTERIZATION OF FLUORINATED SURFACTANTS

Analytical techniques are employed to determine the purity or the concentration of a fluorinated surfactant and to characterize a fluorinated surfactant and its solutions. Since most fluorinated surfactants are mixtures of homologues, the term purity has to be redefined for each particular case. In most cases the determination of purity begins with the analysis of intermediates used to synthesize the surfactant. Usually the intermediates can be readily analyzed by chromatography and the homologue distribution determined. Gas chromatography has only a limited value for the analysis of fluorinated surfactants proper since most fluorinated surfactants are not sufficiently volatile for gas chromatography.

In general, the concentration of a fluorinated surfactant in solution can be determined by conventional volumetric or spectroscopic methods used for hydrocarbon-type surfactants [1–5]. In addition to the functional groups utilized for the analysis of hydrocarbon-type surfactants, the fluorine content is a unique feature useful for the determination of fluorinated surfactants. If the fluorinated surfactant is the only fluorine-containing species in a solution or a substrate, then the fluorine content indicates the concentration of the fluorinated surfactant.

9.2 ELEMENTAL ANALYSIS

Elemental analysis is more important for fluorinated surfactants than for hydrocarbon-type surfactants because the fluorine content can indicate the concentration of a fluorinated surfactant in admixture with hydrocarbon-type surfactants or nonfluorinated chemicals. Hence the concentration of a fluorinated surfactant can be determined without having to resort to complicated separation schemes.

Fluorine in an organic substance can be determined by nondestructive methods or by destruction of the organic matter by combustion or fusion. Nondestructive methods include neutron activation [6] and x-ray fluorescence. Nondestructive methods for elemental fluorine analysis are rapid but require unusual equipment or are not adequately accurate, sensitive, or versatile.

Fluorine in organic compounds is usually determined by converting organic fluorine to an inorganic fluoride. Various combustion methods are routinely used for this purpose. However, the carbon-fluorine bond is exceptionally strong and extremely vigorous conditions are needed for a quantitative mineralization. Conventional combustion conditions used for the determination of carbon and hydrogen in nonfluorinated organic compounds are not adequate for a quantitative analysis of fluorinated surfactants.

The most vigorous analytical technique for the determination of fluorine in organic fluorochemicals is combustion in an oxyhydrogen flame. The original torch designed by Wickbold [7] used an oxygen-hydrocarbon gas mixture. The Wickbold torch was modified in a Du Pont laboratory by Sweetser [8] (Fig. 9.1), who replaced the hydrocarbon with hydrogen. Dobratz [9], in Jackson Laboratory of Du Pont, provided the Sweetser apparatus with a bypass system to allow continuous operation during introduction of samples and collection of

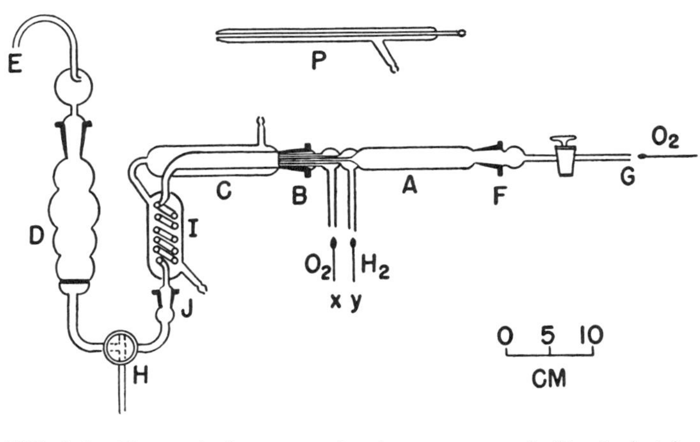

FIG. 9.1 The oxyhydrogen combustion apparatus. A, Pyrolysis tube; B, oxyhydrogen torch; C, flame chamber; D, absorber; E, spray trap; F, removable joint; G, sweep oxygen inlet; H, 3-way stopcock; I, spiral condenser; J, joint; P, probe burner; x, oxygen inlet; y, hydrogen inlet. (From Ref. 8. Reproduced by permission of American Chemical Chemical Society.)

analyte without disrupting the operation of the oxyhydrogen flame. The temperature of the combustion chamber was raised by cooling it with air instead of water. Some modifications of the apparatus were made by Kissa [10] and procedures for the determination of fluorine in biological samples and metal-containing compounds were developed. The sample is pyrolyzed or vaporized in a quartz apparatus and the vapors or pyrolysis products are swept into an oxygen-hydrogen flame burning at 2000°C. Organic fluorine is converted to hydrofluoric acid, which is absorbed in water or in aqueous solutions of sodium carbonate or sodium hydroxide. The fluoride ion collected is determined with a fluoride ion–selective electrode [11].

Two units are routinely used in Jackson Laboratory of Du Pont, and during their 30-year history no serious incidents have occurred. Although the mixture of oxygen and hydrogen is potentially hazardous, elaborate safety devices, such as automatic shut-down valves built into the apparatus, assure a safe operation.

Other versions of the oxyhydrogen torch have been developed [12]. A combustion apparatus is commercially available from Heraeus [13].

Combustion in an oxygen Parr bomb [14,15], although less vigorous than combustion in a oxyhydrogen flame, has given quantitative results for perfluorooctanoic acid and its salts. However, the method is not suitable for volatile organic fluorine compounds. Aqueous samples, for example, blood, have to be dried and pelletized.

Combustion in an oxygen flask [16,17], although convenient, tends to give low results. The combustion converts fluorine to the fluoride ion, which is determined titrimetrically [18–25] or with a fluoride ion–selective electrode [11,26–29]. Satisfactory results have been obtained for perfluorodecalin and perfluorotripropylamine [30]. However, the fluorine recovery of samples containing trifluorobenzoic acid and *p*-fluorobenzoic acid was found to be low (82–87%) by the oxygen flask combustion method [31]. Although the method is simple and easy to use, it is unsuitable for low fluorinated surfactant concentrations because the sample size is limited to 50 mg. Like the oxygen Parr bomb, the oxygen flask is not suitable for aqueous samples.

Ashing [31] or fusion with metallic potassium [32] or with sodium biphenyl [33–35] may give low results caused by either losses of fluorine or incomplete mineralization.

The limitations of various combustion and fusion methods leave the oxyhydrogen flame as the most powerful technique for quantitative mineralization of a fluorinated surfactant.

9.3 VOLUMETRIC METHODS AND ION-PAIR SPECTROSCOPY

Volumetric methods used for hydrocarbon-type surfactants [1] are applicable to fluorinated surfactants, unless the solubility of the fluorinated surfactant imposes some limitations. Anionic fluorinated surfactants can be titrated potentiometrically with benzethonium chloride (Hyamine 1622), using a surfactant-selective [36] or a nitrate-selective electrode (Fig. 9.2). Cationic surfactants can be titrated with sodium dodecyl sulfate. Alternatively, turbidity of the titration medium can be used as an end-point indication [10].

Two-phase titration methods [1] are less convenient and not applicable to a fluorinated surfactant, if a suitable water-immiscible solvent cannot be found for the ion pair formed by the fluorinated surfactant.

Anionic hydrocarbon-type surfactants can be determined spectrophotometrically by forming an ion pair with a cationic dye [1]. The ion pair is extracted into a water-immiscible solvent, for example, chloroform, and determined spectrophotometrically. Jones [37] suggested the use of methylene blue for the determination of anionic surfactants, but his original method occasionally gave erroneous results. Numerous modifications of the method have been published

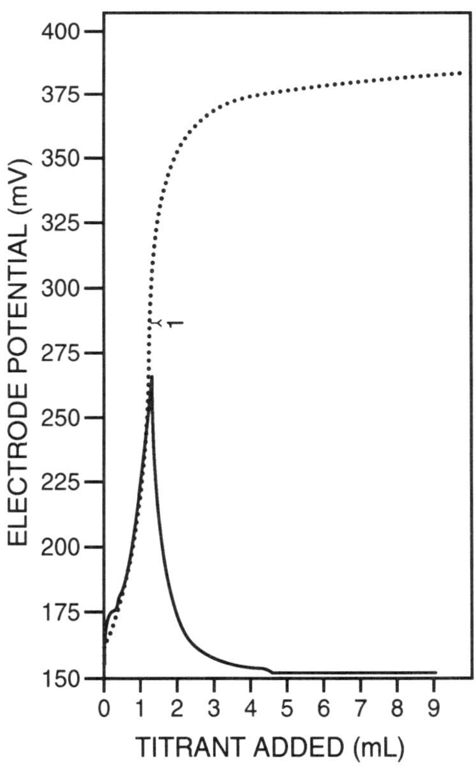

FIG. 9.2 Titration of Zonyl FSA with 0.05 N Hyamine 1622. Metrohm model 670 titrator, Orion model 9342BN surfactant-selective electrode, and model 90–02 double junction reference electrode [10].

[1,38]. Sharma et al. [39] used a methylene blue method to determine perfluorinated carboxylic acids with a 7–10-carbon-long perfluorinated alkane chain. Mixtures of perfluorinated carboxylic acids and alkyl sulfates were analyzed by extracting the ion pairs from a citrate buffer and from 0.25 N H_2SO_4. In a medium of high pH, both surfactants, sodium dodecyl sulfate and sodium perfluorooctanoate, form a colored ion pair. At a low pH (sulfuric acid) the sodium salt of perfluorooctanoic acid is present as the free acid, which is extracted as a colorless species. Hence the difference between the two absorbances represent the perfluorooctanoic acid concentration.

9.4 CHROMATOGRAPHY

Gas chromatography is the main tool for analyzing intermediates for fluorinated surfactants. Most fluorinated surfactants are not sufficiently volatile for gas chromatography. Fluorinated surfactants suitable for gas chromatography include perfluoroalkanecarboxylic acids, perfluoropolyether surfactants [40], and semifluorinated alkanes [41]. Perfluorooctanoic acid has been derivatized with diazomethane and determined as its methyl ester by gas chromatography [42].

Supercritical fluid chromatography, with CO_2 as the mobile phase, can determine the telomer and homologue distribution in nonionic fluorinated surfactants (Fig. 9.3). Supercritical fluid chromatography with an octadecylsilane bonded stationary phase has been utilized in studies of semifluorinated alkanes [43].

High-performance liquid chromatography (HPLC) has been employed to analyze fluorinated surfactants. Mixtures of fluorinated surfactant and hydrocarbon-type surfactant monomers, separated from their micellar solutions by gel permeation [44] or ultrafiltration, have been analyzed by HPLC [45]. The nonionic fluorinated surfactant Fluowet OTN has been analyzed by reversed phase HPLC/MS/MS with a thermospray interface [46], using a 15-cm, 3.9-mm inner diameter C_{18} column with 5-μm spheres. The methanol-water eluent gradient was from 80:20 to 20:80 in 10 min. After separation of the fluorinated surfactant, 0.10 mol/L ammonium acetate was injected on line.

Gel filtration [47–50] is a partition chromatography technique that separates surfactant monomers from their micellar solution according to their molecular size. After the gel column has been equilibrated with a monomer solution, the sample solution is injected and eluted with the same monomer solution (sandwich method). Nakagawa and Jizomoto [50] developed a gel filtration method for binary surfactant solutions. Asakawa et al. [51] developed a new simulation technique based on the group contribution method [52] for the micelle-monomer equilibrium. The gel permeation column is considered to consist of a series of plates, each of which consists of a mobile phase and a sta-

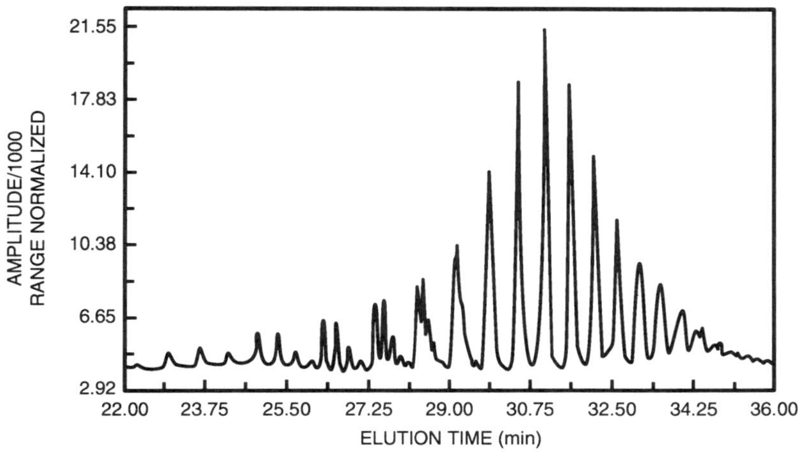

FIG. 9.3 Supercritical fluid chromatography of a nonionic fluorinated surfactant $R_fCH_2CH_2O(CH_2CH_2O)_nH$. A 50-μm-ID capillary column, 10 m long, coated with a 0.25-μm dimethylpolysiloxane film. (Courtesy of J. J. Kozlowski, Du Pont.)

tionary phase. The surfactants in the mobile phase move to the neighboring lower phase and the two phases are assumed to reach equilibrium immediately. The monomer exchange between the micelle and bulk phase is rapid compared to the elution rate. Hence the kinetics of the monomer-micelle exchange or the micelle formation and dissociation are not taken in account.

9.5 ULTRAVIOLET AND INFRARED SPECTROSCOPY

Aliphatic carboxylic acids and their anions are known to absorb in the ultraviolet region as a result of n-π^* and π-π^* transitions. Mukerjee at al. [53] found that long-chain perfluorocarboxylates, such as perfluorooctanoate ($\epsilon = 344$ L mol^{-1} cm^{-1} at 25°C), have higher molar absorptivities in the 205–230-nm region than perfluoroacetate ($\epsilon = 57$ L mol^{-1} cm^{-1} at 25°C). The absorptivity of perfluoroalkanoates is sufficient for a quantitative determination of the fluorinated surfactant down to the 10^{-5} M concentration range using a 10-cm cell. Mukerjee at al. [53] observed that below the critical micelle concentration (cmc) perfluoroheptanoate and perfluorooctanoate solutions obeyed the Beer-Lambert law within 1%. A somewhat better linear relationship was obtained by relating absorbance data to the fluorinated surfactant concentration by the equation

$$A = a_1 + b_1 c \tag{1}$$

where A is absorbance, a_1 and b_1 are constants, and c is the fluorinated surfactant concentration. At the critical micelle concentration (cmc) the absorptivity increases markedly. The increase (Fig. 9.4) is large enough to permit the determination of the cmc from ultraviolet (UV) absorption data.

Anionic fluorinated surfactants of the structure $R_fCH_2CH_2SCH_2CH_2COOLi$ do not absorb in UV, although they have a perfluoroalkyl chain and a carboxylate function.

The application of UV spectroscopy to the analysis of nonionic hydrocarbon-type surfactants is limited to nonionics which contain functional groups which absorb in the UV region, such as aromatic nuclei [54]. The main functional group of nonionics, the oxyethylene ether linkage, does not absorb in the UV region. In spite of this limitation, UV spectroscopy can be useful for determining impurities in nonionic fluorinated surfactants.

Infrared (IR) spectroscopy [55] is used mainly for identification and characterization of fluorinated surfactants. A beam of infrared radiation is passed through the sample and focused at a monochromator, which disperses the radiation into a spectrum. An infrared spectrum is recorded by plotting the percent transmission of the sample as a function of frequency or wavelength. IR spectra can be used to identify functional groups of the sample or to identify a compound by comparing its spectra to reference spectra of a known pure compound. Conventional dispersive IR spectroscopy is not very sensitive; the detection limit of a component in a mixture is above 0.2–1.0%. Quantitative

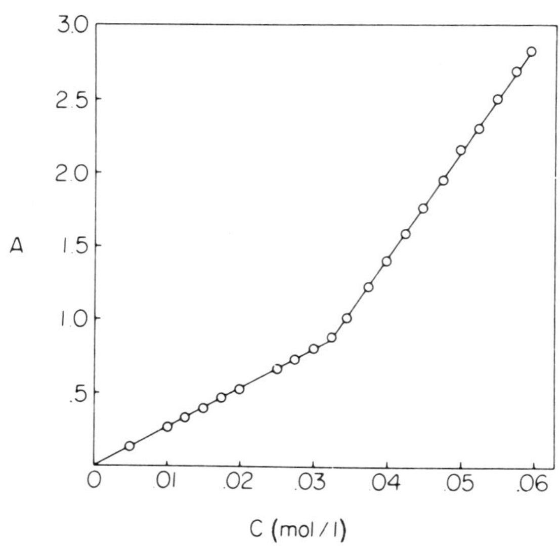

FIG. 9.4 Plot of absorbance (A) at 230 nm versus perfluoroheptanoic acid concentration. (From Ref. 53. Reproduced by permission of American Chemical Society.)

analysis of surfactants by infrared spectroscopy was less important until Fourier transform spectroscopy was developed [55]. In Fourier transform spectroscopy the entire frequency range of interest is transmitted through an interferometer. The output signal is recorded as a function of interference, and the resulting interferogram is converted to a spectrum using a Fourier transform and a computer. Fourier transform infrared spectroscopy (FTIR) has increased the accuracy of absorption data by accumulating repetitive spectra and combining digitalized data electronically. The precision of IR spectroscopy is limited mainly by the signal/noise ratio. Since noise is random, cumulative collection of absorption data by Fourier transform spectroscopy increases the precision and sensitivity of quantitative IR spectroscopy.

Fluorinated surfactants exhibit absorption bands arising from CF stretching and CF_2 vibration modes (Fig. 9.5).

Raman spectroscopy analyzes frequency changes in scattered monochromatic radiation. Light passing through a material medium is transmitted, absorbed, and/or scattered. When scattering involves only a direction change, the scattered light has the same frequency as the unscattered light. However, when light interacts with matter and various transitions are involved, the scattered light will have gained or lost energy. The resulting change in frequency is characteristic of the material studied. Since the intensity of Raman scattering is low, Raman spectroscopy was almost dormant until the development of the laser provided a high-density monochromatic light source.

Raman spectroscopy has been very useful for conformation studies of hydrocarbon-type surfactants in aqueous systems but has only limited value for

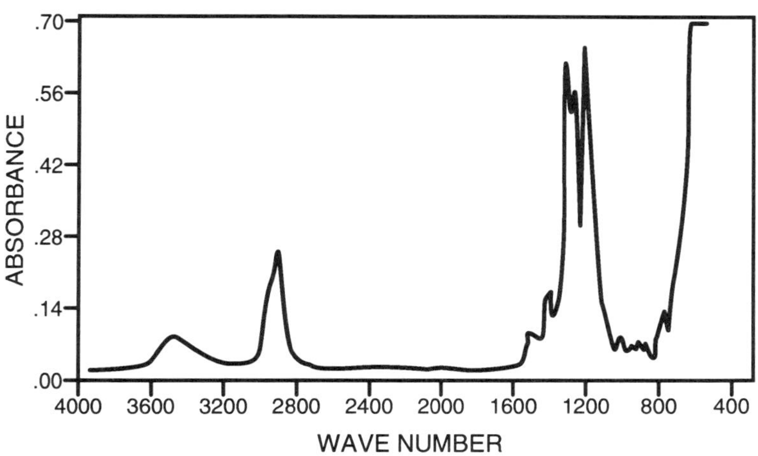

FIG. 9.5 Infrared spectra of a nonionic fluorinated surfactant consisting of homologues $R_fCH_2CH_2O(CH_2CH_2O)_nH$. (Courtesy of J. T. Cronin, Du Pont.)

fluorinated surfactants. Unlike a strong IR absorption band arising from the CF stretching mode, the intensity of Raman bands is low for these vibrations. In contrast to the characteristic CH stretching mode, the CF and CF_2 modes are in a region where other molecular modes occur and complicate absorption patterns. Amorim da Costa and Santos [56] have nevertheless been able to show Raman spectroscopy to be useful for structure and conformation analysis of fluorinated surfactants (Fig. 9.6).

Developments such as Fourier Raman spectrometry and the charge-coupled device (CCD) detector will enhance the usefulness of Raman spectroscopy for fluorinated surfactants.

9.6 MASS SPECTROMETRY

Mass spectrometry [57–64] involves four steps: (1) isolation of the component of interest, (2) ionization, (3) separation of the ions in a combination of electric and magnetic fields according to their mass/charge (m/Z) ratio, and (4) detection. The molecular ions and ionic fragments are detected by an electrometer and their relative abundances are recorded in the mass spectra. The sensitivity of detection can be increased with an electron multiplier.

The first step in mass spectrometry isolates a component of the sample by (1) vaporization using a direct insertion probe to heat the sample to about 200–300°C, (2) flash desorption at a very rapid heating rate to minimize thermal degradation, or (3) chromatography. Mass spectrometry is made more powerful by adding a chromatographic "front end" to separate the components of the sample before they enter the mass spectrometer. Most fluorinated surfactants, even when derivatized, are not sufficiently volatile to be analyzed by gas chromatography–mass spectrometry (GC-MS), the most commonly used "hyphenated mass spectrometer". High-performance liquid chromatography (HPLC) is more useful for fluorinated surfactants. Several techniques have been developed for interfacing liquid chromatography with a mass spectrometer [64]. Over a dozen LC-MS interfaces are commercially available, including a transport interface using a belt to transport the eluent through a desolvation chamber to the ionization source, direct liquid introduction into the ion source, particle beam, thermospray, electrospray, and others [64].

The thermospray technique [65–68] uses a heated vaporizer from which the HPLC eluent containing the dissolved electrolyte is sprayed as a jet into a heated chamber. A sampling orifice is positioned normal to the axis of the vaporizer probe. The ions and molecules are pumped through the sampling orifice into the mass spectrometer. Electron impact or collision-activated ionization, although optional, provides structural information.

Schröder [69,70] analyzed fluorinated surfactants in water and wastewater using HPLC coupled by a thermospray interface to a tandem mass spectrome-

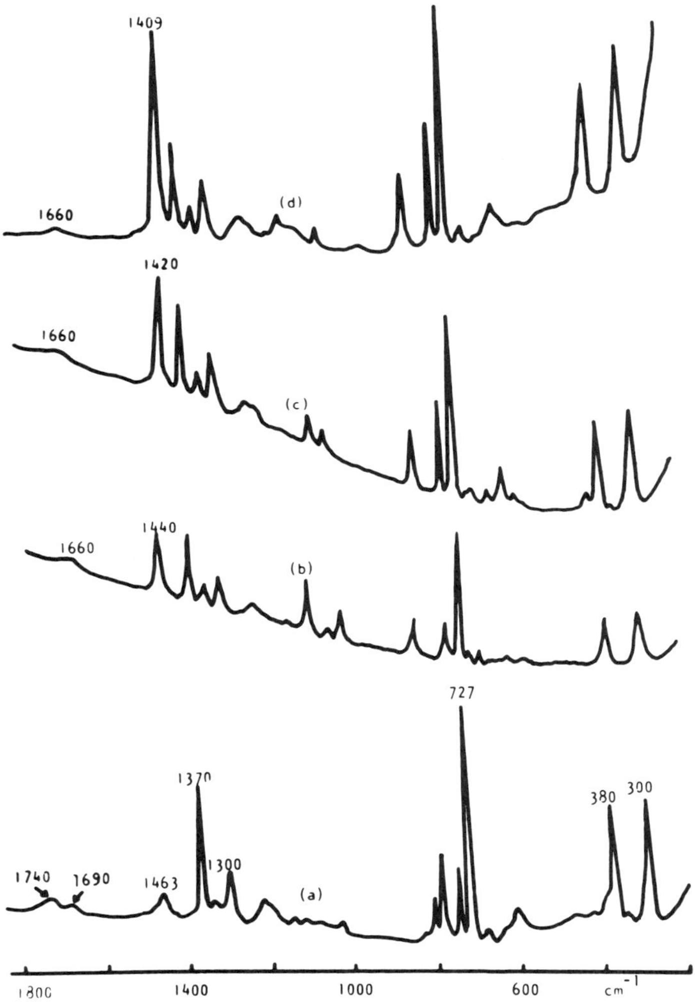

FIG. 9.6 The 200–1800 cm^{-1} Raman spectra of perfluorodecanoic acid (a) and its lithium (b), sodium (c), and ammonium (d) salts at 22°C. (From Ref. 56. Reproduced by permission of Academic Press, Inc.)

ter (MS/MS). Alternatively, the chromatographic column was bypassed and the analyte was injected into the mass spectrometer (FIA, flow injection analysis).

Supercritical fluid chromatography using CO_2 as the mobile phase eliminates the problems associated with the evaporation of a liquid eluent and is therefore more compatible than liquid chromatography with MS.

The second step in mass spectrometry, ionization of the sample, is accomplished by one of several techniques, some of which include sampling as well.

Usually the sample is bombarded with a beam of electrons or energetic particles. Electron impact ionization employs electrons from a heated filament to ionize a gas phase sample. The energy of the commonly used 70-eV electrons is in excess of the energy required for removing an electron from a molecule to produce a molecular ion. Hence the electron impact ionization causes fragmentation of the sample and provides a fragmentation pattern which gives useful structural information. In field ionization a volatile sample passes through a strong electric field (10^7–10^8 V/cm) which generates molecular ions with little fragmentation. Both ionization techniques, electron impact ionization and field ionization, require a vaporizable sample, limiting their application to volatile surfactants or degradation products of nonvolatile surfactants.

Modern "soft" ionization techniques have overcome the sample volatility requirement by combining the first two steps in mass spectrometry, sampling and ionization. The soft ionization techniques used for the analysis of surfactants include fast atom bombardment (FAB), field desorption (FD), desorption chemical ionization (DCI, also called direct chemical ionization), secondary-ion mass spectrometry (SIMS), and laser desorption methods.

FAB directs a beam of energetic inert gas atoms onto a sample in a viscous liquid matrix, usually in glycerol or triethanolamine. A flow of the sample in the solution replenishes the sample on the surface and limits degradation by the particle beam. The surfactant suppresses ionization of glycerol and by providing a cleaner background increases the sensitivity of the FAB method [71,72]. The FAB techniques usually provide quasimolecular ions with only a few fragment ions. Electrolytes added to the sample solution facilitate the formation of quasimolecular ions, for example, adduct ions with alkali metal cations.

Heller et al. [73] observed large cluster ions of perfluoroalkanesulfonates desorbed under FAB conditions. Cesium perfluorohexanesulfonate formed clusters containing as many as 29 anions and 30 cations. The abundance of these high-mass ions produced by perfluoroalkanesulfonates was much higher than that of the ions formed by CsI. For example, the abundance of the cluster $Cs(C_6F_{13}SO_3Cs)_{19}^+$ at *m/z* 10240.4 was eight times greater than the abundance of the $Cs_{40}I_{39}^+$ at 10265.5. Discontinuities in surface tension of cesium perfluorohexanesulfonate and in the slope of cluster abundance occurred at the same surfactant concentration range in tetraglyme solutions, suggesting that the cluster formation and aggregation in solution are related.

Lyon et al. [74] characterized fluroalkanesulfonates by FAB ionization combined with tandem MS/MS spectrometry. The samples were dissolved in glycerol or triethylamine, placed on the copper target of the FAB probe, and bombarded with 8-keV xenon atoms. The ions formed were accelerated into the analyzer of the mass spectrometer. Normal spectra were recorded by scanning the first spectrometer, MS-I, and leaving the second mass spectrometer, MS-II, fixed to pass all ions. Tandem mass spectrometry was used to enhance the FAB technique by collision-activated dissociation (CAD). An appropriate ion selected with MS-I was subjected to collisions with helium atoms in the collision cell and the CAD spectra were recorded by the MS-II unit. Examples of the FAB and CAD spectra are shown in Figs. 9.7, 9.8, and 9.9.

The fragmentation of perfluoroalkanesulfonates involved the cleavage of the C−C bond with the loss of a $C_nF_{2n+1}\cdot$, followed by the detachment of tetrafluoroethylene:

$$CF_3(CF_2)_n-CF_2CF_2CF_2(CF_2)_mSO_3^- \rightarrow CF_3(CF_2)_n\cdot + \cdot CF_2CF_2CF_2(CF_2)_mSO_3^- \rightarrow$$

$$CF_2=CF_2 + \cdot CF_2(CF_2)_mSO_3^-$$

The fragmentation reaction sequence is analogous to the thermal decomposition mechanism of poly(tetrafluoroethylene).

Substitution of hydrogen for one of the terminal fluorine atoms changes the fragmentation mechanism. Formation of HF then becomes the main reaction in the fragmentation process.

An amphoteric fluorinated surfactant, Du Pont's Zonyl FSB, has been used as a calibration standard for high-resolution FAB-MS measurements [75].

In field desorption (FD) the sample is deposited directly onto carbon dendrites serving on the anode as activated emitters. For hydrocarbon-type anionic, cationic, and nonionic surfactants, field desorption usually produces molecular or quasimolecular ions free of fragmentation. For amphoteric nonfluorinated surfactants, molecular ions have been obtained together with fragment ions providing structural information [76–78] which showed that perfluoroalkanesulfonates are desorbed as high-mass clusters under FD conditions.

Desorption chemical ionization (DCI) places the sample onto a direct-insertion probe located within the chemical ionization plasma [79–81]. Cationic surfactants produce molecular ions and decomposition ions useful for quantitative analysis [72,82]. The DCI technique is less informative for anionic or nonionic surfactants.

9.7 NUCLEAR MAGNETIC RESONANCE SPECTROSCOPY

Nuclear magnetic resonance (NMR) is a very powerful tool for investigating surfactant systems. The theory of nuclear magnetic resonance spectroscopy has been described in several books [83–92] and will not be discussed here in detail. The applications of NMR to surfactant systems have been reviewed by Lindman et al. [93].

NMR spectroscopy is based on the allowed orientation (Zeeman energy levels) of nuclei with nonzero angular momentum when the sample is placed into a magnetic field. The nuclei can be realigned by varying the external magnetic

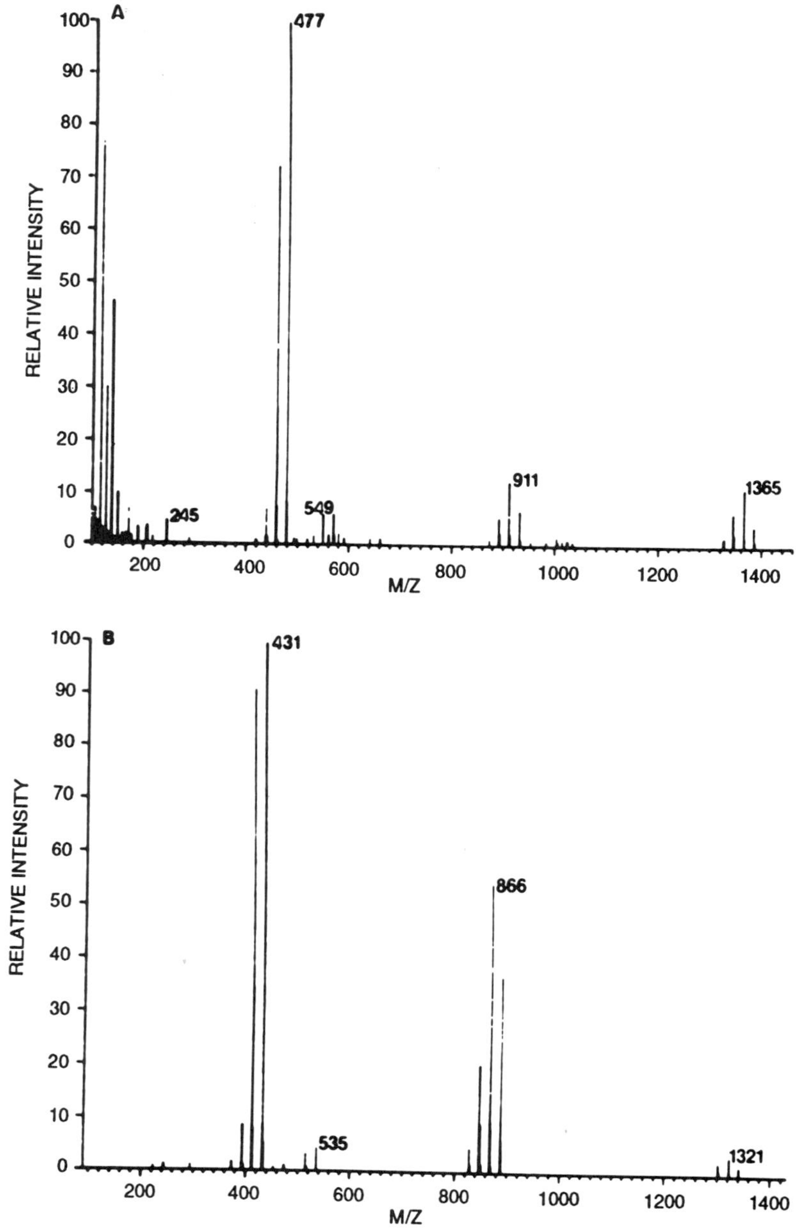

FIG. 9.7 FAB spectrum from Na 2-hydroperfluoroheptanesulfonate: (A) positive ions, (B) negative ions. (From Ref. 74. Reproduced by permission of American Chemical Society.)

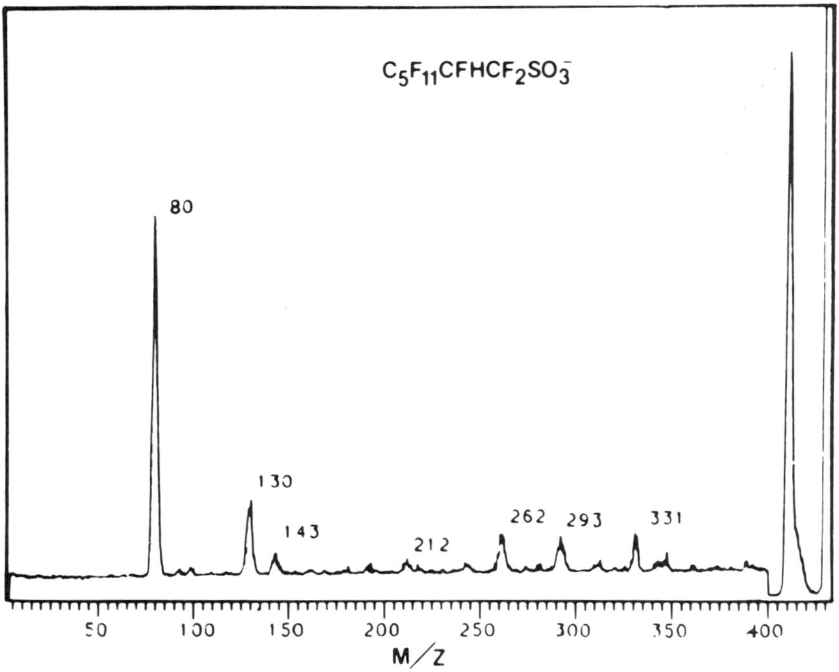

FIG. 9.8 CAD spectrum of negative ions from *m/z* 431, the 2-hydroperfluoroheptanesulfonate anion. (From Ref. 74. Reproduced by permission of American Chemical Society.)

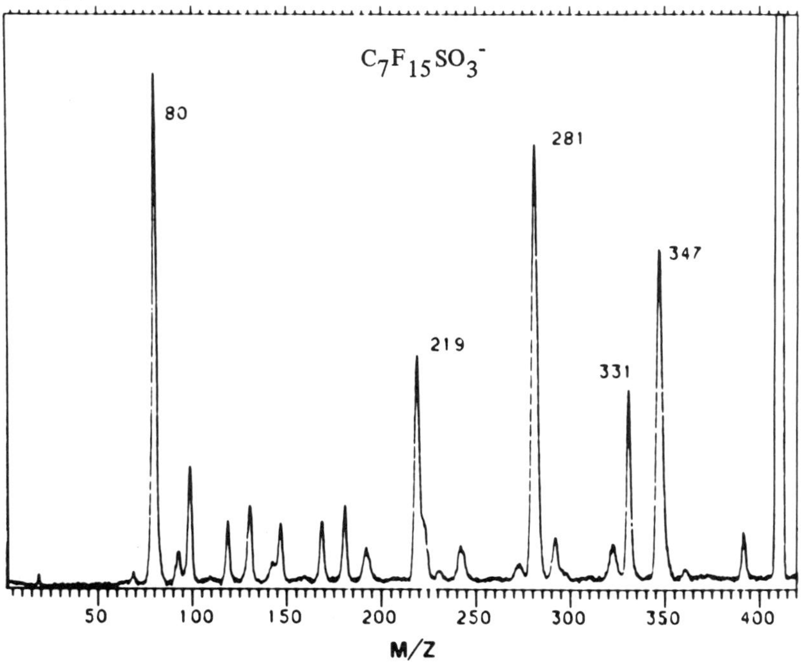

FIG. 9.9 CAD spectrum of negative ions from *m/z* 411, the perfluoroheptanesulfonate anion. (From Ref. 74. Reproduced by permission of American Chemical Society.)

field or by radio-frequency irradiation. When the applied energy matches the energy required for the transition between Zeeman levels, resonance results.

Nuclei do not all have the same resonance frequency because their chemical environment can vary the applied magnetic field. As a result of differences in shielding, nuclei in functional groups have characteristic resonance frequencies. The difference in the resonance frequencies of two chemically and/or magnetically unequal nuclei indicates the chemical shift, expressed in ppm. To calculate the chemical shift, the difference between the resonance frequency of the sample peak and the resonance frequency of the reference peak is divided by the resonance frequency of the reference peak or by the "observed frequency" given by the instrument manufacturer [91]. The chemical shifts observed by NMR depend on the concentration of the species and the solvent [94]. The solvent effect has been used to investigate the environment of atoms within the fluorinated surfactant micelle [95–98].

NMR spectroscopy yields structural information on surfactants and their micelles, values of the free energy of micellization, ΔG^o_M, and the corresponding enthalpy and entropy changes, ΔH^o_M and ΔS^o_M. For the analyses of fluorinated surfactants, ^{1}H, ^{13}C, and ^{19}F NMR spectroscopies have been employed.

^{1}H NMR spectroscopy can provide information on the environment of the fluorinated surfactant in the micelle. The high sensitivity of the ^{1}H nucleus is a definite advantage. Monduzzi et al. [99] utilized the Fourier transform pulsed-gradient spin-echo (FTPGSE) ^{1}H NMR technique to determine the self-diffusion coefficients of water in W/O microemulsions containing perfluoropolyether (PFPE) oils and an anionic surfactant with a PFPE hydrophobe. The self-diffusion data provided quantitative information on the amount of water in the composition range where continuous water coexists with water in droplets.

^{13}C NMR spectroscopy, because of its high resolution and wide chemical shift range [100–103], can give qualitative information on molecular conformation of fluorinated surfactants in solution and quantitative information on cmc values. However, the ^{13}C signal is relatively weak, for two reasons. The abundance of the ^{13}C isotope in carbon-containing substances is about 1.1% of carbon atoms. Furthermore, the ^{13}C has a lower magnetogyric ratio (lower magnetic strength) than ^{1}H and this reduces the sensitivity further to a total factor of about 1:5800. Several techniques have been developed to overcome the loss of sensitivity. Instead of sweeping the resonance frequencies successively, the Fourier transform method uses a radio-frequency pulse to excite all resonance frequencies at once and the signal is enhanced by repeated pulses and signal averaging.

For NMR studies of fluorinated surfactants the most useful nucleus is ^{19}F, in addition to ^{13}C and ^{1}H nuclei. Changes in the ^{19}F chemical shift at cmc are larger than changes in the proton chemical shifts and therefore provide more information on fluorinated surfactants and their micellar structures. ^{19}F NMR spectra have been recorded for structural characterization of perfluorononanoic acid [104] and perfluoropolyether surfactants [105]. Micelle formation in solutions of fluorinated surfactants has been studied by measuring the ^{19}F chemical shift [95–98,106–109] (Fig. 9.10). Muller and co-workers studied the effect of the environment on fluorine atoms in a surfactant micelle by ^{19}F NMR [95–98] (see Sec. 7.1 for the results and conclusions of their studies). Carlfors and Stilbs [110] used the Fourier transform NMR pulsed-gradient spin-echo (FTPGSE) method [111–113] for the determination of multicomponent self-diffusion coefficients in micellar solutions of sodium perfluorooctanoate and sodium perfluorooctanoate–sodium decanoate. Partition coefficients were calculated from the self-diffusion data for a homologous series of *n*-alkanols, benzene, and benzyl alcohol.

Palepu and Rainsborough [114] measured the ^{19}F chemical shift changes for 1:1 mixtures of sodium perfluorooctanoate with α- and β-cyclodextrins. In α-cyclodextrin mixtures the shifts for terminal fluorine atoms changed more than those for fluorine atoms in the middle of the chain. In β-cyclodextrin mixtures,

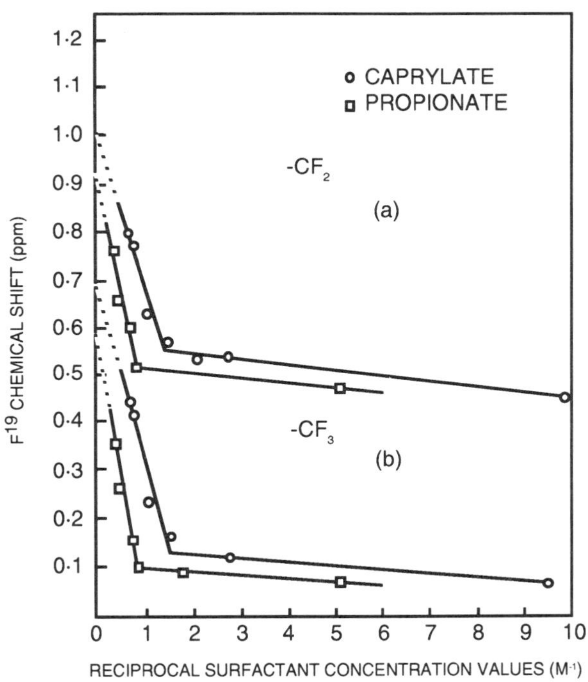

FIG. 9.10 Plot of ^{19}F chemical shift against the inverse concentration of sodium perfluorocaprylate and sodium perfluoropropionate (a), CF_2; (b), CF_3. (From Ref. 106. Reproduced by permission of American Chemical Society.)

the fluorine atoms in the middle of the chain were affected more than the terminal ones by cyclodextrin. Guo et al. [115] employed ^{19}F NMR to investigate the association of α-, β-, and γ-cyclodextrins with sodium perfluorobutanoate, sodium perfluoroheptanoate, sodium perfluorooctanoate, and sodium perfluorononanoate. Trifluoroacetic acid was used as the external reference and the difference between the ^{19}F chemical shift for the mixed system and that for the solution containing only the fluorinated surfactant was measured. The results of this systematic study showed a weak association of α-cyclodextrin with the fluorinated surfactants. Fluorinated surfactants with a short chain formed a 1:1 complex with β-cyclodextrin. Fluorinated surfactants having a longer chain form a 2:1 complex, especially at higher β-cyclodextrin concentrations. γ-Cyclodextrin forms a 1:1 complex with the fluorinated surfactants. The association complexes for the 1:1 complexes were calculated from the ^{19}F chemical shifts measured for various cyclodextrin concentrations. Guo et al. [115] explained the results by the cavity size of the host cyclodextrins. The cavity of α-cyclodextrin is apparently too small to accommodate fluorinated surfactants and form an inclusion complex.

The dynamic parameters of fluorinated surfactant solutions have been studied by NMR relaxation methods. The theory of NMR relaxation has been discussed in detail by Henriksson and Ödberg [116] and reviewed by Lindman et al. [93]. Spin-lattice relaxation transfers energy from the higher energy level to the lattice as thermal energy. (The term "lattice" is used here to denote molecules other than the fluorinated surfactant in the sample.) Since the resulting temperature change is too small to be detected, the relaxation time of recovery of the absorption signal from saturation is measured. Fluorine relaxation times can be measured both in water and in deuterium oxide and the different magnetic properties of protons and deuterons can provide information about the environment of the fluorocarbon segments.

Henriksson and Ödberg [116] determined ^{19}F spin-lattice relaxation time for heptafluorobutyric acid by the $\pi/2$, $\pi/2$ pulse method and concluded that fluorocarbon chains in the heptafluorobutyric acid micelles are to some extent exposed to water. Ulmius and Lindman [117] measured ^{19}F spin-lattice relaxation time for various perfluorinated or partially fluorinated carboxylic acids and concluded that the fluorocarbon chains come in contact with water only at the micellar surface. Since the deuterium relaxation time is affected by the state of water (bound or free), Burger-Guerisi et al. [118a] studied phase transitions in fluorinated microemulsions by measuring ^{2}H NMR relaxation times. Serratrice et al. [118b] investigated the influence of water on ^{13}C chemical shifts and relaxation times of nonionic fluorinated surfactants dissolved in a fluorocarbon. The observed chemical shift variations were attributed to the hydration of the hydrophilic chain. In another study [118c], relaxation data indicated similar flexibilities of the fluorinated chain in the various nonionic fluorinated surfactant molecules studied [118c]. Tiddy [118d] measured ^{19}F spin-lattice and spin-spin relaxation times for the lamellar phase of the ammonium perfluorooctanoate–water system. The ^{19}F relaxation rates were found to be qualitatively similar to the relaxation rates of protons in analogous hydrocarbon surfactant systems. The spin-lattice relaxation times indicated that the CF_2 groups at 298 K rotate about the long axis of the hydrophobic chain more slowly than the CH_2 groups in hydrocarbon systems. The activation energy of rotation is similar to or smaller than the activation energy for analogous hydrocarbon systems.

9.8 CHEMICAL RELAXATION METHODS

Chemical relaxation methods (CRM) observe a mixture of reactants and reaction products in thermodynamic equilibrium and perturb this equilibrium by generating a rapid but very small change in one of the parameters affecting equilibrium, such as pressure or temperature [119]. As a result of the perturbance, the system shifts to another equilibrium, commensurate to the change in a parameter. The change of equilibrium is characterized by one or several time constants, the relaxation times, which are related to the rate constants of the chemical reactions studied.

CRM has been very useful in studies of micellization kinetics, based on the theory of Aniansson and Wall [120–122], modified by Kahlweit and coworkers [123–127]. Chemical relaxation techniques have been described in several articles and books [128–133] and reviewed by Lang and Zana [119].

The chemical relaxation methods usually employed are the temperature jump, pressure jump, shock tube, ultrasonic absorption, and stopped flow methods. The temperature jump method utilizes rapid heating techniques, such as heating by a microwave pulse, discharge of a charged coaxial cable in the solution, discharge of a capacitor, and heating by a pulse of laser light. The pressure jump method utilizes an autoclave with a thin metal diaphragm which bursts and allows the pressure of the autoclave to drop very rapidly to atmospheric pressure. A pressure jump apparatus with conductivity detection and twin cell arrangement is shown in Fig. 9.11. One cell contains the sample investigated and the other cell contains an electrolyte of similar conductivity but no relaxation.

The shock tube technique is somewhat similar. The bursting of the diaphragm generates a pressure drop which propagates through a tube half filled with water or ethanol. Reflections of the pressure jump at the bottom and the top of the tube cause addition and subtraction of the incident and reflected pressure waves. As a result, the equilibrium is shifted by a rectangular change in pressure.

The ultrasonic absorption method shifts the equilibrium periodically by harmonic changes of pressure and temperature caused by the propagation of ultra-

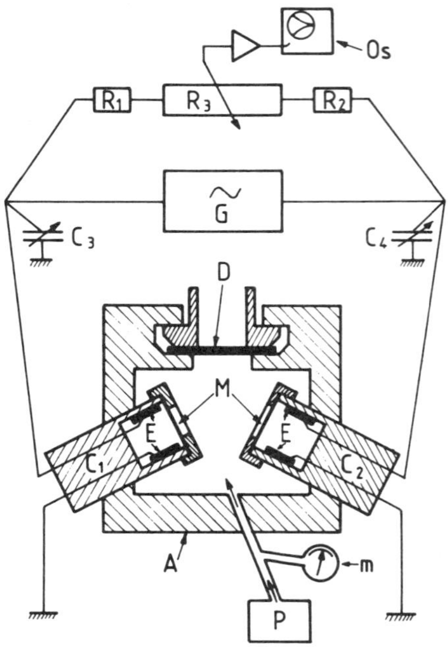

FIG. 9.11 Pressure jump apparatus. A, Autoclave; C_1, C_2, conductivity cells; E, electrodes; M, elastic membrane; D, metal diaphragm; P, pressure pump; m, manometer; G, 40-kHz generator for the conductivity bridge; C_3, C_4, tunable capacitors; R_1 and R_2, helipot resistors; R_3, potentiometer; Os, oscilloscope. (From Ref. 119.)

sonic waves in fluids. The stopped flow method involves rapid mixing of two solutions in less than a millisecond. Since the mixture observed is not in equilibrium, the stopped flow method is not truly a chemical relaxation method. The stopped flow method is useful, nevertheless, for observing perturbations by composition jumps.

Dilute micellar solutions of surfactants are characterized by two well-separated relaxation times. The molecular origin of the fast relaxation time has been related to a monomer-micelle exchange [134–137]. It was realized later that the relaxation spectra of micellar solutions really exhibit two relaxation times. The theory of Aniansson and Wall [120,138] assumes a stepwise aggregation of surfactant monomers to form micelles [139]. The fast relaxation time is attributed to the exchange of monomeric surfactants between the micelles and the intermicellar solution. The slow relaxation time is attributed to micelle formation and breakdown. The theory and its modifications by Kahlweit and co-workers [123–127] have been the basis for most interpretations of chemical relaxation times and provided valuable information on kinetics of micellization.

Pressure jump and a shock wave method with conductivity detection have been used by Hoffmann and co-workers in their studies of micelles formed by perfluorinated surfactants [140–146]. The pressure jump and shock wave techniques were utilized in micellization studies on cationic surfactants with perfluorinated counterions as well [144].

The temperature jump relaxation technique [128] has been used by Hoffmann and Ulbricht [141] with optical detection, utilizing a pH indicator (thymol blue) to observe relaxation processes of a 1:1 mixture of perfluorooctanoic acid and its sodium salt.

Ultrasonic absorption has been measured to determine relaxation times in surfactant solutions [147–158].

The kinetics of micelle formation have been investigated using the ultrasonic relaxation method for alkali metal salts of perfluorooctanoic acid [159]. The periodic fluctuations in temperature and pressure caused by the acoustic wave are several magnitudes less than the temperature or pressure perturbations of jump techniques. Rassing et al. [159] suggested that the ultrasonic and jump methods measure different modes of micelle formation whose relaxation times differ by several orders of magnitude. Ultrasonic absorption techniques [157] have also been used to measure relaxation spectra of cesium perfluorooctanoate in water and in deuterium [158].

9.9 SMALL-ANGLE SCATTERING METHODS

Small-angle scattering allows us to measure distances in the range of 0.5 to 50 nm. Small-angle scattering methods differ in principle from imaging methods, such as microscopy. Imaging methods collect and focus radiation scattered by the objects being studied and reconstruct their image. In contrast, small-angle scattering methods produce an interference pattern of the radiation scattered by the objects under study. The interference pattern can be converted to reconstruct an average image and interpreted to obtain basic information on surfactant micelles. It is important to keep in mind that small-angle scattering techniques provide only an average image in the space of correlation functions. Averaging limits severely the amount of information obtainable because thermal agitation in surfactant solutions produces large fluctuations. However, this limitation has the advantage that all structures are described by several averaged parameters and, in that sense, averaging facilitates interpretation of data [160].

Small-angle and wide-angle x-ray diffraction techniques have been reviewed in several articles and books [160–164].

Fontell and Lindman [165] investigated phase equilibria of two-component systems consisting of water and perfluorononanoic acid or its salts. Small-angle x-ray scattering (SAXS) showed liquid crystalline regions besides regions of micellar solutions. The thickness of fluorocarbon layers in the liquid crystalline region and the area per polar head group were estimated.

Small-angle neutron scattering (SANS) methods have been described in the literature [160,166–167].

Hoffmann et al. [168] measured small-angle neutron-scattering (SANS) of lithium perfluorooctanoate, diethylammonium perfluorononanoate, and tetraethylammonium perfluorooctanesulfonate micelles in D_2O or in mixtures of D_2O and H_2O. The radii, micelle concentrations, and aggregation numbers were calculated.

Ravey and Stebe [169] studied SANS of nonionic fluorinated surfactant gels. SANS spectra of systems containing a fluorocarbon, nonionic fluorinated surfactant, and large amounts of water were interpreted in terms of water-in-(water-in-oil microemulsions) emulsions.

Mathis et al. [170] determined the aggregation number and droplet size of microemulsions prepared from a ternary mixture of water, a fluorocarbon, and a nonionic fluorinated perfluoroalkylpoly(oxyethylene) surfactant. The solutions were diluted with an H_2O-D_2O mixture in the variable contrast method that allows varying the scattering length [171] of the solvent. Scattering lengths, b_i, where i = FC, S, A, B, or P, for the fluorocarbon (FC) and the surfactant (S) molecules, for the hydrophilic and hydrophobic moieties (A and B) of the surfactant, and for the whole particle (P) were computed from atomic coherent scattering lengths. Scattering densities were calculated from the scattering lengths, b_i, and the corresponding molecular volumes

$$Q_i = b_i N_{av} / V_i \tag{2}$$

where N_{av} is the Avogadro number and V_i are the molar volumes.

The number of scattering particles per unit volume, N_p, has been calculate assuming an average aggregation number (N) and neglecting the monomeric surfactant molecules (the cmc values were low)

$$N_p = \frac{N_{av} Z_s}{M_s V N} \tag{3}$$

where Z_s and M_s are the mass fraction and molecular weight of the surfactant and V is the specific volume of the solution.

Assuming that interparticle effects are negligible, the scattered intensity, $I(q)$, for a sample containing monodisperse particles can be written as [172]

$$I(q) = K N_p (Q_p - Q_w)^2 V_p^2 P_N(q) \tag{4}$$

where K is a calibration constant, Q_w is the scattering density of the H_2O-D_2O mixture, Q_p is the scattering length density of particles, and P_N (q) is the scattering form factor.

Burkitt et al. [173,174] used SANS to examine the size and shape of micelles in solutions containing ammonium perfluorooctanoate or mixtures of ammonium perfluorooctanoate with ammonium decanoate. The SANS measurements were made using a neutron diffractometer and a computer to process the basic data to give the intensity of scattering $I(Q)$ as a function of Q relative to water. The scattering vector Q for elastic scattering is defined by

$$Q = \left(\frac{4\pi}{\lambda}\right) \sin\left(\frac{\theta}{2}\right) \qquad (5)$$

where λ is the wavelength of the radiation (neutrons) in the medium and θ the scattering angle.

Burkitt et al. [173,174] concluded that SANS is an excellent method for the determination of micellar weights of ammonium octanoate, ammonium decanoate, and ammonium perfluorooctanoate. The scattering data suggested that the ammonium perfluorooctanoate micelles are cylindrical.

9.10 LIGHT SCATTERING

Light scattering has provided a wealth of information on the shape, size, polydispersity, and micellization of surfactant solutions and microemulsions. A large number of papers have been published and the theory of light scattering has been reviewed in several books [175–178].

The theoretical aspects of light scattering are in several ways similar to those of small-angle scattering. However, important differences exist in the status of experimental techniques. Dynamic light scattering is now in routine use, dynamic neutron scattering is a recently developed technique, but the practical feasibility of dynamic x-ray scattering is uncertain [178].

The use of light-scattering methods for studying micellar structures of fluorinated surfactants is limited to partially fluorinated surfactants. The micelles of perfluorinated surfactants are very weak light scatterers. The refractive index–concentration slope, *dn*/*dc*, is 10 to 100 times smaller than that for hydrocarbon chain surfactants [145]. However, partially fluorinated surfactant solutions are amenable to light scattering measurements. Micellar aggregation numbers of fluorinated surfactants with a terminal $-CF_3$ group (12,12,12-trifluorododecyltrimethylammonium bromide) have been determined by light scattering [179].

9.11 LUMINESCENCE PROBING METHODS

Luminescence probes are molecules or ions which upon photoexcitation emit light having characteristics sensitive to the immediate environment of the probe

[180]. The characteristics of emitted light serve to characterize the environment of the luminescence probe.

Luminescent probes can be divided into two groups: fluorescence probes and phosphorescence probes. Fluorescence is an emission of light associated with the transition from excited single states to the ground state. Phosphorescence is emission of light associated with the transition from the lowest triplet state to the ground state (Fig. 9.12) [181]. Radiative lifetimes of fluorescence generally range from 10^5 sec^{-1} to 10^9 sec^{-1} and radiative lifetimes of phosphorescence range from 10^{-1} sec^{-1} to 10^3 sec^{-1} [182].

Some molecules or ions can function as quenchers and inhibit luminescence. Quenching, caused by interactions between the luminescent probe and a quencher, may be reversible or irreversible. Excimer formation is a case of reversible quenching. Some luminescent probes react, in the excited state, with an identical molecule in the ground state and form an excimer:

$$^1P^* + P \underset{k_{-E}}{\overset{k_E}{\Longleftrightarrow}} {}^1P^*_2$$

where k_E and k_{-E} are the rate constants for excimer formation and dissociation [180]. A probe in the excited state may associate in an analogous manner with a dissimilar molecule in the ground state and form an exciplex. The fluorescence characteristics of excimers and exciplexes differ from those of the

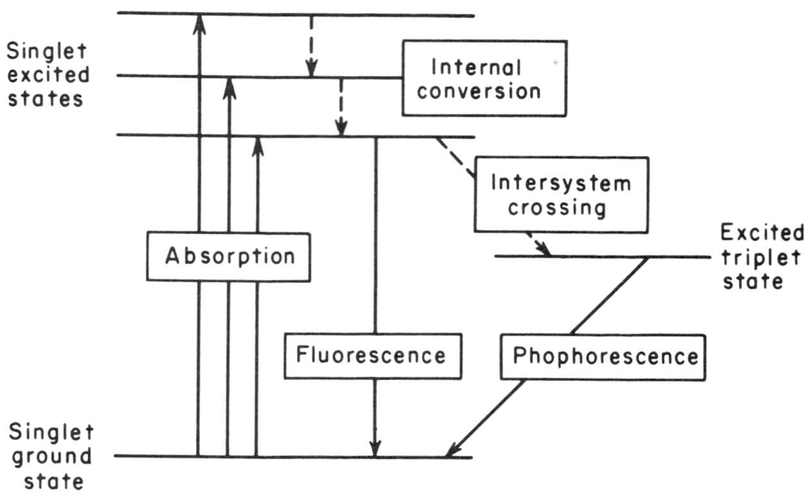

FIG. 9.12 Molecular energy levels involved in photochemical processes. (From Ref. 181. Reproduced by permission of Prentice-Hall.)

monomeric probe. The excimer or exciplex formation is sensitive to the viscosity of the environment around the probe and therefore provides useful information on the structure of surfactant solutions.

Another reversible quenching technique used in micellar studies is energy transfer from the probe in the excited state to an energy acceptor molecule.

The luminescence methods involve solubilization of the probe in micelles and the determination of fluorescent spectra and fluorescence polarization. Various steady-state and transient-state fluorescence methods have been employed. Experimental details can be found in the literature [183–187]. A careful selection of the probe is essential for obtaining meaningful results. Luminescence methods are based on the assumption that the probe does not affect the fundamental nature of the solution and the micelles. This assumption must be validated for the system being studied.

The fluorescence intensity of ammonium 1-anilinonaphthaline-8-sulfonate (ANS) in a solution of a hydrocarbon-type surfactant is constant below the cmc of the surfactant but increases linearly with increasing surfactant concentration above the cmc. The concentration dependence of fluorescence intensity indicates that the ANS probe is solubilized in the micelles of the hydrocarbon-type surfactant. In contrast, fluorinated surfactants do not solubilize ANS [188]. The ANS probe is therefore useful for investigating fluorinated surfactant and hydrocarbon-type surfactant mixtures (see Sec. 7. 3). Asakawa et al. [100] studied the micellar environment of mixed fluorinated surfactants and hydrocarbon-type surfactants by fluorescence intensities of ANS, auramine, and pyrene. The ANS fluorescence intensity is shown in Fig. 7.28 as a function of total surfactant concentration. The ANS fluorescence intensity increased with increase in hydrocarbon-type surfactant concentration (LiDS = lithium dodecyl sulfate). Because the ANS probe was not incorporated in LiFOS micelles, the fluorescence intensity increased very little with increasing fluorinated surfactant (LiFOS = lithium perfluorooctane sulfonate) concentration.

Muto et al. [189] measured pyrene fluorescence lifetime τ_0 and the ratio I_1/I_3 of the intensities of the first vibronic and the third vibronic band of the monomeric pyrene. The pyrene fluorescence data revealed the existence of a single type of mixed micelle in solutions of LiDS-LiFOS, LiFOS-hexaoxyethylene glycol dodecyl ether, or LiFOS-octaoxyethylene glycol dodecyl ether mixtures. The lifetime and the intensity ratio of vibronic peaks have been used to determine the cmc of fluorinated surfactant micelles [190]. However, the solubility of pyrene in micelles of fluorinated surfactants is not adequate for determining the micelle aggregation number [190,191].

The I_1/I_3 ratio is very sensitive to the polarity of the medium sensed by the pyrene probe. The pyrene fluorescence technique has therefore been utilized for the characterization of adsorbed layers of hydrocarbon surfactants and fluorinated surfactants on alumina [192,193].

9.12 X-RAY PHOTOELECTRON SPECTROSCOPY

X-ray photoelectron spectroscopy (XPS, commonly termed ESCA as an abbreviation for electron spectroscopy for chemical analysis) is eminently suited to the study of surfactant adsorption. The XPS method is highly sensitive to the surface composition and can characterize adsorbed surfactant layers without elaborate sample preparation.

The theory and praxis of XPS have been reviewed in several monographs and journal articles [194–200]. The sample is placed into a chamber and positioned for analysis. The chamber is evacuated to a high vacuum of $<10^{-7}$ torr and the sample is irradiated with soft x-rays, usually from an MgK_α (1253.6 eV) or AlK_α (1486.6 eV) source. The x-ray irradiation generates photoelectrons which are emitted with kinetic energy, E_K, governed by the energy of the exciting radiation, $h\nu$, and the binding energy, E_B, of the electron

$$E_K = h\nu - E_B - \phi \tag{7}$$

The "work function" ϕ depends on the sample and the spectrometer used for measuring photoelectron emission.

The binding energies of the electrons are characteristic of the element and the environment of the atom in the molecule. Hence XPS can characterize the composition and the chemical state of the near-surface region.

XPS spectra are strongly affected by the orientation of the sample, the source, and the spectrometer. Almost all (about 95%) of the signal emerges from the distance 3λ within the solid, where λ is the inelastic mean free path of the electron, also called the attenuation length of the emerging electron. The sampling depth, d, of the subsurface analyzed by XPS is given by

$$d = 3\lambda \sin\alpha \tag{8}$$

where α is the exit angle of the emitted electron, relative to the sample surface. The mean free path depends on the kinetic energy of the photoelectron, which in turn is affected by the energy of the radiation source. The sampling depth has a maximum when $\alpha = 90$ and is usually below 50 Å..

Although the method is considered to be nondestructive, sample damage and evaporative losses have been of concern [201]. The fluorine-to-carbon photoelectron peak intensity ratios, F(1s)/C(1s), have been found to decrease during an x-ray exposure of several minutes, depending on experimental conditions [202].

XPS can yield qualitative and quantitative information on adsorbed surfactant layers. The overlayer on the substrate decreases the intensity, I_e, of a photoelectron peak, originating from a component in the substrate, by a factor

$$\frac{I_c}{I_e} = \exp\left(\frac{-\delta}{\Lambda_e \sin\alpha}\right) \tag{9}$$

where δ is the thickness of the overlayer sampled, I_c is the intensity of the photoelectron peak originating from the covered substrate, Λ_e is the inelastic mean free path (IMFP) of the electron in the overlayer, and α is the electron takeoff angle relative to the sample surface [203]. IMFP, the average distance a photoelectron travels before an inelastic collision, depends on the binding energy of the photoelectron and the composition of the sample.

Angle-dependent XPS (variable takeoff angle) can confirm the surfactant overlayer thickness and determine the continuity of the surfactant overlayer [203] (Fig. 9.13). The effective sampling depth of XPS analysis is a function of the electron IMFP and the takeoff angle. By decreasing the takeoff angle, the signal intensity contributions to the photoelectron spectrum from the top surface region can be selectively enhanced. This relationship serves as the basis for angle-dependent depth profiling. The angle-dependent ratio of overlayer to substrate, R, can be calculated using a simplified expression given by Fadley [204]:

$$R\left(\frac{\text{overlayer}}{\text{substrate}}\right) = K\left[\exp\left(\frac{\tau}{\sin\alpha}\right) - 1\right] \tag{10}$$

where K is a function of atom density, instrument response, the kinetic energies of the substrate and overlayer atoms within the measured levels, and the effective cross sections of the atoms. The effective overlayer thickness, τ, is given by

$$\tau = \delta/\Lambda_e \tag{11}$$

where δ is the actual overlayer thickness.

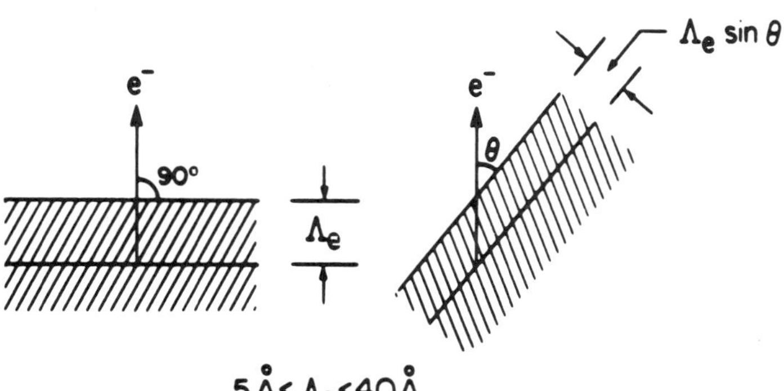

FIG. 9.13 The principle of angle-dependent XPS, where Λ_e is the IMFP of the electron being analyzed and Θ is the angle between the sample surface and the emitted electrons. (From Ref. 203. Reproduced by permission of American Chemical Society.)

The orientation of an adsorbed surfactant can be determined by measuring the intensity of the peak for an atom on one end of the surfactant molecule relative to the intensity of a peak for an atom on the opposite end. Fluorine bound to carbon in $CF_3(CF_2)_n$- groups induces a chemical shift to a higher electron binding energy. The resulting peak is readiliy distinguished in the C (1s) spectrum from the peak for the carbon in the nonfluorinated portion of the molecule. If the surfactant molecule prefers a certain orientation, one peak is enhanced relative to the other for a given takeoff angle [203].

Gerenser et al. [203,205] studied adsorption of Zonyl FSC, a cationic fluorinated surfactant, on SiO_2 and poly(ethylene terephthalate). Samples of adsorbed Zonyl FSC were irradiated with monochromatic AlK_α x-rays. Use of a monochromatic x-ray source minimized radiation damage of the sample. The pressure in the spectrometer was typically 5×10^{-9} torr. Gerenser et al. [203] found that the thickness of the adsorbed surfactant layer calculated by the angle-dependent method (Eq. 10) is always larger than the thickness value calculated by the substrate attenuation method (Eq. 11). The discrepancy was explained by orientation of the surfactant, which affects only the angle-dependent method and has no effect on the attenuation method. The results of their study are discussed in Sec. 5.5.

Batts [202] used XPS to study surface chemistry of dried gelatin layers which contained a cationic fluorinated surfactant FC134. The samples were irradiated by x-rays from a magnesium target (MgK_α, 1253.6 eV). Conditions were chosen to attain a theoretical resolution of 1 eV and to minimize sample exposure. Errors caused by sample damage were minimized by recording the spectra for each takeoff angle and surfactant concentration on fresh surfaces. Contact angle and XPS data indicated that the progressive adsorption of a fluorinated surfactant at the aqueous gelatin-air interface can be monitored by analyzing the corresponding dried layers. The surface chemical composition determined by XPS was in accord with wettability data (see Sec. 5.5).

Claesson et al. [206] studied the adsorbed monolayers of a cationic, double-chained fluorinated surfactant on mica. The XPS spectrometer was equipped with an AlK_α x-ray source. The known number of exchangeable potassium and sodium ions on the mica basal plane served as the internal standard for the quantitative determination of adsorbed surfactant. The surfactant oriented preferentially with both nitrogen atoms or only the quaternary ammonium group toward the surface, depending on the deposition method.

9.13 ELECTROCHEMICAL METHODS

Electric Conductivity

Electric conductivity provides highly useful information on the association of surfactants in solution. Usually the specific conductivity is plotted against the surfactant concentration or the equivalent electric conductivity is plotted against the square root of surfactant concentration [207]. The break in the curve indicates the cmc. Polarization is avoided by using AC or applying short pulses of opposing polarity. Dörfler and Müller [208] determined the cmc value for tetraethylammonium perfluorosulfonate from $\kappa/\sqrt{c}$ plots (cmc = 1.7×10^{-3}) (Fig. 9.14) and by AC polarography from the differential capacity-potential curves (cmc = 1.2×10^{-3}) (Fig. 9.15).

Transient Electric Birefringence

Electric birefringence can give useful information on the shape and size of micelles and fluorinated surfactant aggregates, especially when complemented by other physical methods [180,209–219]. Electric birefringence has been successfully used to determine the shape and size of polymers, polyelectrolytes, and surfactant micelles. Colloidal particles or molecular aggregates, which have a permanent dipole moment or are polarized anisotropically, orient in an electric field. The colloidal solution becomes optically anisotropic and exhibits electric birefringence termed the Kerr effect [183,220,221].

An apparatus for electric birefringence measurements is shown schematically in Fig. 9.16 [217]. Rectangular high-voltage pulses of short duration are applied to the solution and the buildup and decay of electric birefringence are measured. The beam of an He-Ne laser is polarized by a Glan prism set at 45 degrees with respect to the electric field applied across the Kerr cell. The

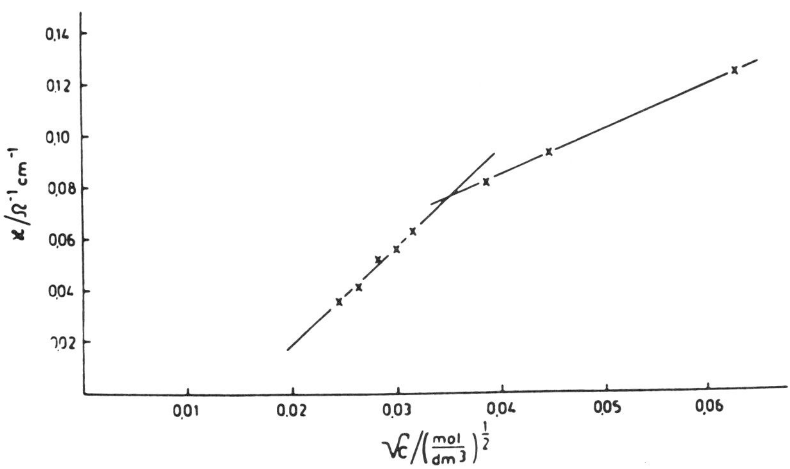

FIG. 9.14 The specific conductivity of tetraethylammonium perfluorooctanesulfonate plotted against the square root of its concentration. (From Ref. 208. Reproduced by permission of Carl Hanser Verlag.)

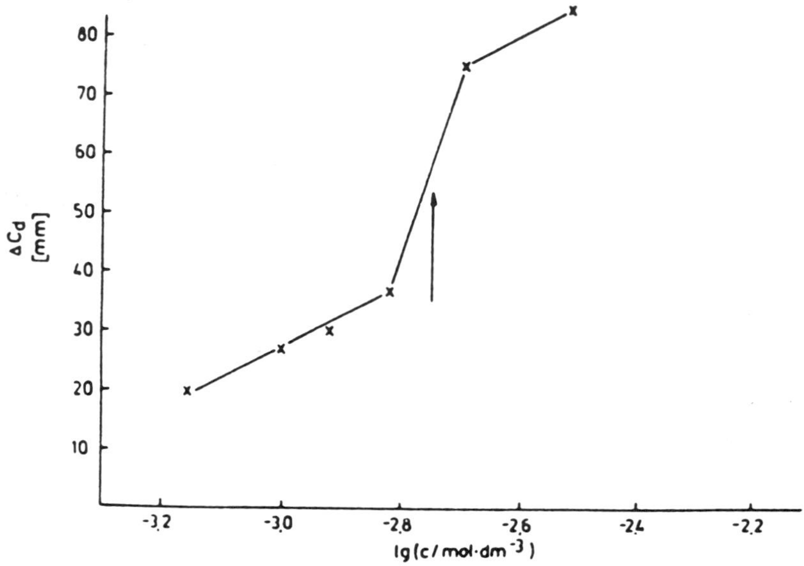

FIG. 9.15 The relative depression of the AC polarographic differential capacity, ΔC_d, plotted against the log concentration of tetraethylammonium perfluorooctanesulfonate. (From Ref. 208. Reproduced by permission of Carl Hanser Verlag.)

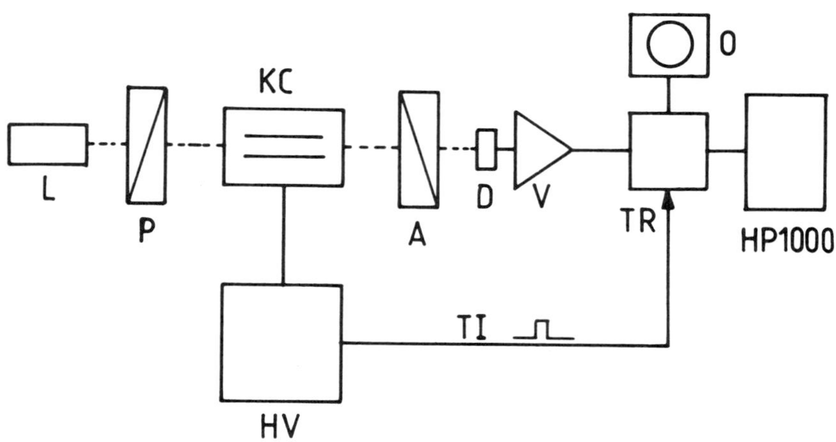

FIG. 9.16 Apparatus for electric birefringence measurements. L, He-Ne laser; P, polarizer; KC, Kerr cell; A, analyzer; D, photodiode-detector; V, amplifier; TR, transient recorder; O, oscilloscope; HP, HP1000 computer; HV, high-voltage pulse generator; TI, trigger impulse to start the recording system. (From Ref. 217. Reproduced by permission of American Chemical Society.)

polarized light traverses the Kerr cell containing the sample and passes through the analyzer to the photodiode detector. The analyzer and the polarizer are in a crossed position. The signal of the detector is digitized by a recorder interfaced with a computer.

The electric birefringence, $\Delta n = n_{\parallel} - n_{\perp}$, is the difference in refractive indexes parallel and perpendicular to the direction of the applied electric field. Electric birefringence is related to the optical retardation or phase shift, δ, by the equation

$$\Delta n = \delta\lambda/(2\pi l) \tag{12}$$

where l is the path length of the Kerr cell and λ is the vacuum wavelength [212]. The Kerr constant, B, is given by

$$B = \lim_{E \to 0} \Delta n/(\lambda E^2) \tag{13}$$

where E is the field strength. The Kerr constant depends on the temperature and the surfactant concentration (Fig. 9.17) [212]. Below cmc the surfactant does not exhibit birefringence and only the solvent birefringence is observed. The Kerr constant of ionic surfactants can assume positive or negative values, depending on temperature and the counterion [212].

The rise and decay of the electric birefringence of a fluorinated surfactant are shown in Fig. 9.18 [212]. The birefringence relaxation time, τ_B, is related to the rotational diffusion constant, D_R,

$$\tau_B = (6D_R)^{-1} \tag{14}$$

The rotational diffusion constant, D_R, is proportional to L^{-3}, where L is the length of a rod-like micelle or the diameter of a disk-like spheroid. Hence τ_B values cannot distinguish between disks or rods, unless complemented with Kerr constant values or data obtained by other methods.

Tamori et al. [219] used electric birefringence to estimate the micellar size and shape in mixed surfactant solutions containing hexaoxyethylene glycol dodecyl ether and lithium perfluorooctanesulfonate or lithium dodecyl sulfate.

The time dependence of birefringence is affected by intermicellar interaction, electrolytes, and polydispersity. If the aggregates are polydisperse, the time dependence of birefringence deviates from a single exponential relationship. The size distribution function must be known, because the deviation depends on the width of the distribution function [180]. In spite of these limitations, Shorr and Hoffmann [212,217] conclude that electric birefringence measurements are useful for the determination of the dimensions of anisotropic surfactant aggregates. Although electric birefringence data alone are not sufficient to define the type of surfactant aggregates, the electric birefringence method complements other physical methods.

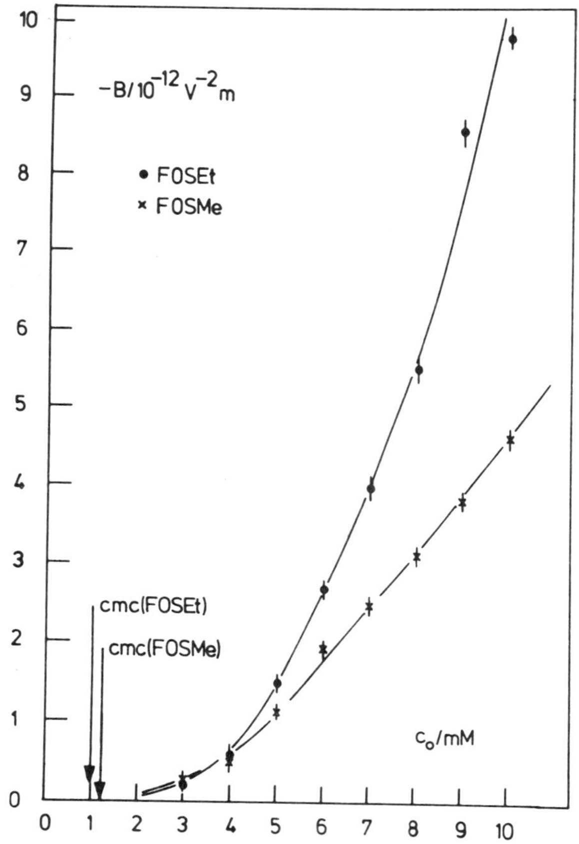

FIG. 9.17 Kerr constant *B* of tetraethylammonium perfluorooctanesulfonate (FOSET) and tetramethylammonium perfluorooctanesulfonate (FOSMe) as a function of total concentration c_0 for $T = 20°C$. (From Ref. 212. Reproduced by permission of American Chemical Society.)

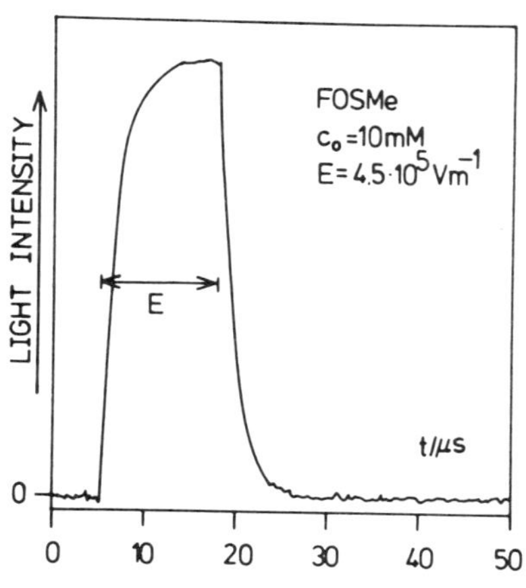

FIG. 9.18 The rise and decay of the electric birefringence of FOSMe. The duration of the applied electric field pulse is indicated with the arrow. (From Ref. 212. Reproduced by permission of American Chemical Society.)

9.14 ULTRAFILTRATION

Ultrafiltration techniques [222–224] have been used to separate surfactant monomers from their micelles. Asakawa et al. [225] used the ultrafiltration method to study fluorinated surfactant and hydrocarbon-type surfactant mixtures. The method is based on the assumption that surfactant monomers pass through an ultrafiltration membrane which has pores sufficiently small to prevent the passage of micelles. The membrane used (YC-05, Amicon Corp.) can exclude molecules having a molecular weight larger than 500. The surfactant solution was forced through the membrane by applying pressurized nitrogen gas. The concentrations of the filtrand and the filtrate were determined by HPLC or isotachophoresis, a high-resolution electrophoretic method [226]. Ultrafiltration experiments were conducted as a function of initial surfactant concentration. Below cmc the surfactant concentrations in the filtrand and filtrate were equal. Above cmc the surfactant concentration in the filtrate became constant and was equal to cmc, in accord with the postulate that the filtrate contains monomeric surfactant molecules in equilibrium with micelles which did not pass the filter.

9.15 SURFACE TENSION

It is perhaps needless to state that surface tension is the most important physical property of a surfactant to be determined. Methods for surface and interfacial tension measurement have been the subject of numerous papers and review articles [227–242].

In spite of the apparent simplicity of surface tension measurement, correct and reproducible values are not always readily obtainable. In addition to the specific limitations of each technique, time dependence of surface tension of surfactant solutions can be a major complication. Surface tension depends on the adsorption and orientation of molecules at the liquid-air interface. Whereas in pure liquids only microseconds are needed for equilibrium orientation, in surfactant solutions hours or even days may be needed to attain equilibrium surfactant adsorption at a freshly created surface. Adsorption and orientation kinetics are especially critical when measuring surface tension of surfactant mixtures.

Surface tension methods measure either static or dynamic surface tension. Static methods measure surface tension at equilibrium, if sufficient time is allowed for the measurement, and characterize the system. Dynamic surface tension methods provide information on adsorption kinetics of surfactants at the air-liquid interface or at a liquid-liquid interface. Dynamic surface tension can be measured in a time scale ranging from a few milliseconds to several minutes [243]. However, a demarkation line between static and dynamic methods is not very sharp because surfactant adsorption kinetics can also affect the results obtained by static methods. It has been argued [244] that in many industrial processes, sufficient time is not available for the surfactant molecules to attain equilibrium. In such situations, dynamic surface tension, dependent on the rate of interface formation, is more meaningful than the equilibrium surface tension. For example, peaked alcohol ethoxylates, because they are more water soluble, do not lower surface tension under static conditions as much as the conventional alcohol ethoxylates. Under dynamic conditions, however, peaked ethoxylates are equally or more effective than conventional ethoxylates in lowering surface tension [245].

Most techniques stretch the liquid-air surface at the moment of measurement. For example, the drop weight method [246] and the ring method [247–250] stretch the surface during detachment. However, instruments are now available which measure surface tension without detaching the ring from the liquid, for example, the Krüss Tensiometer K12.

The surface tension methods measure a force, pressure, or drop size (volume, weight, or dimensions). Examples of methods which measure a force are the ring method [247–250] and the plate method [251,252]. Capillary height [253–256,264] and the maximum-bubble-pressure method [257–264] measure pressure. Pendant drop [256,265–267], sessile drop [256,268], drop volume [269–271], drop weight [246,264,272–275], and spinning drop methods [276–278] measure the size or the dimensions of a drop. Special techniques [243] for measuring dynamic surface tension include the oscillating jet [279], dynamic drop volume [243,280], inclined plate [281], strip [282], free falling [283], pulsed drop [284], dynamic maximum-bubble-pressure [244,259–262,285–287], and dynamic capillary [243] methods.

Modern tensiometers are interfaced with computers to increase the accuracy of the measurement and obtain dynamic surface tension readings within short but accurately measured time intervals.

The *ring method* [247–250] is one of the most frequently used techniques for surface tension measurement. A platinum ring, attached to a vertical wire, is placed horizontally into the liquid (Fig. 9.19). The force, P, needed to pull the ring through the interface is measured. If one assumes that the ring supports a cylinder of liquid,

$$P = 4\pi\gamma R \tag{15}$$

where R is the radius of the ring. Actually, the liquid column lifted by the ring is not a cylinder (Fig. 9.19) and a correction factor [247–250,288] is needed. The accuracy of surface tension values obtained by the ring method is limited by the accuracy of the correction factor. Lunkenheimer [289] considers Huh and Mason correction factors to be sufficiently accurate for a ring with a two-point suspension and suggests that the correction factors of Harkins and Jordan

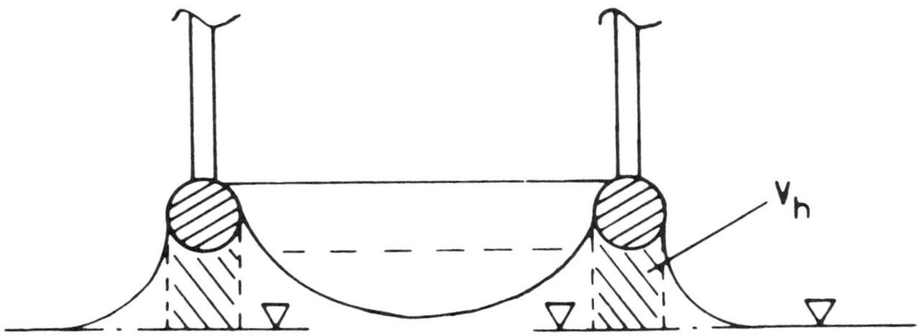

FIG. 9.19 The ring method. (Reproduced by permission of Krüss USA.)

should not be used. To simplify calculations, modern tensiometers are calibrated to make the correction factor for water at 20°C equal to one. Consequently, the deviation of the obtained value from the actual value is reduced.

The number of papers being published on the ring method indicates, however, that this technique still has problems [250,289–293]. The ring method gives reproducible and accurate values only if certain precautions are taken [289,294]:

The surface of the liquid should be sucked to remove impurities in the surface layer.
Complete wetting of the ring by the liquid is a prerequisite.
The ring should perpendicular to the suspension and planar.
The ring should not be detached from the liquid during the measurement.
Sufficient time should be allowed for the surfactant to attain equilibrium.
The vessel should provide a liquid surface much larger than the area covered by the ring.

It has been argued that the ring method is suitable only for measuring the surface tension of pure liquids. The applicability of the method for the measurement of surface tension of surfactants has been debated [252,295–297].

The wire loop method [298,299] is similar to the ring method.

The *Wilhelmy plate method* [251,252], the sessile drop method [256,268], and the capillary height method [253–256] measure equilibrium surface tension, if sufficient time is allowed for the adsorption of surfactant molecules at the surface to attain the state of equilibrium. The Wilhelmy plate method measures the force exerted on a vertical plate partially immersed in the liquid (Fig 9.20). If wetting of the plate is complete, the force, F, is proportional to the surface tension, γ, and the circumference, L, of the plate

$$F = \gamma \times L \tag{16}$$

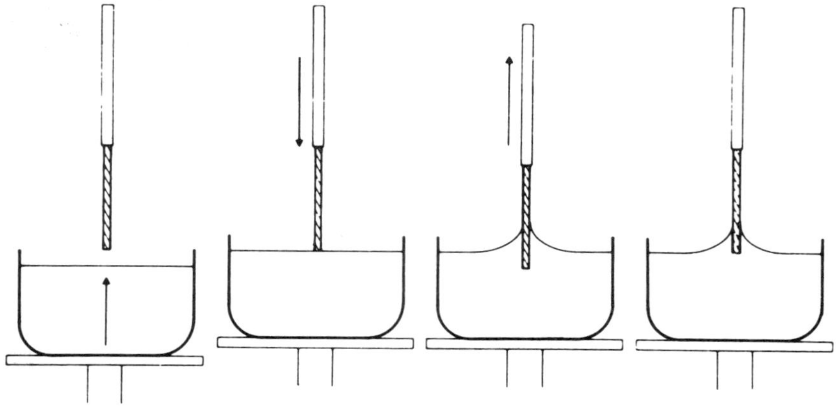

FIG. 9.20 The Wilhelmy plate method. (Reproduced by permission of Krüss USA.)

The Wilhelmy plate method has the advantages of measuring strictly static surface tension and being less sensitive to vibrations of the vessel or a slight deformation of the plate. The prerequisite is complete wetting of the plate, indicated by a zero contact angle. Significant contact angles in the wetting of the plate by some liquid systems have been observed [300]. Wetting is facilitated by using a roughened plate or a platinized platinum plate from which the liquid does not recede. Tadros [301] used a glass plate for measuring the surface tension of fluorinated surfactants by the Wilhelmy method. Hirt et al. [244] used a platinum wire instead of a plate for surface tension measurements of fluorinated surfactants.

The principle of the Wilhelmy method has also been employed for the measurement of wettability of single fibers [302].

The *pendant drop method* has been described by Andreas at al. [303] and others [256,265–267]. The apparatus is simple but the technique requires skill for forming the drop and maintaining its size and shape during the measurement of the diameters d_1 and d_2 (Fig 9.21). The measurement is complicated by optical effects (diffraction and dispersion). The detrimental effects of vibration and evaporation have to be considered. A correction table has been compiled by Fordham [304].

The measurements of the pendant drop are usually determined from a photograph of the drop [267]. The laborious calculations have been greatly facilitated by using video equipment interfaced with a computer [267].

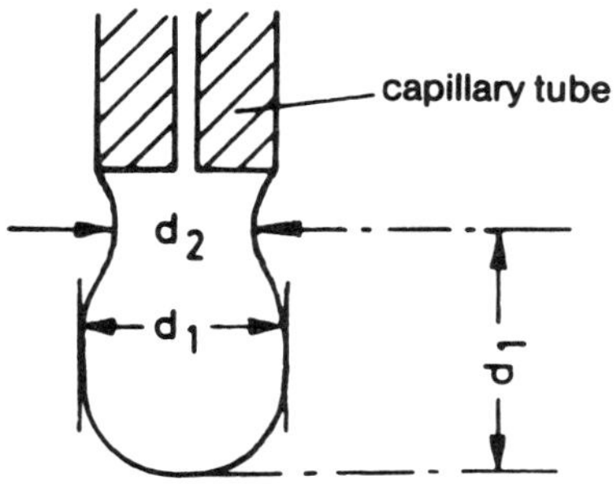

FIG. 9.21 The pendant drop method. (Reproduced by permission of Krüss USA.)

The *drop volume method* [269–271] requires only a buret or a syringe (Fig. 9.22). Either the volume required to form the drop, V, is measured or the number of drops formed by a measured volume of liquid is counted.

$$\gamma = \frac{V\rho g}{2\pi r} \tag{17}$$

The dripping radius, r, has to be equal to the capillary radius or be known. The method is not very precise and demands a careful manipulative skill. Semiautomatic [305] and automatic drop volume methods [306] have been developed. The calculation of interfacial tensions corrected for transport processes inside the growing drop is simplified by interfacing the tensiometer with a computer.

The *drop weight method* measures the weight of a drop (or several drops) emerging from a capillary of known dimensions [246,264,272–275]. Slight vacuum is applied to the apparatus through a tubing until the drop, forming at the outlet of the capillary, assumes almost its full size. The drop is then allowed to detach itself from the capillary. Surface tension is calculated from the equation

$$\gamma = \left(\frac{mg}{r}\right)F \tag{18}$$

where m is the mass of the drop, g the acceleration due to gravity, r the capillary radius, and F a factor dependent on the drop volume and r [307]. The drop volume can be calculated from the drop weight and the density of the liquid. With proper corrections [308], the method is quite accurate [309,310].

Maximum-bubble-pressure method [242,244,257–264] measures pressure in a bubble formed at the end of a capillary when a gas, e.g., air, is blown through the capillary into the liquid. The pressure increases when the bubble grows and attains its maximum value when the bubble has obtained the shape

FIG. 9.22 The drop volume method. (Reproduced by permission of Krüss USA.)

of a hemisphere (Fig. 9.23). The pressure decreases when the bubble grows further and finally bursts. Maximum-bubble-pressure methods have been compared [311] and equipment for automated surface tension determination by maximum-bubble-pressure measurement has been developed [312–315].

Dynamic surface tension measurements by Hirt et al. [244], based on the maximum-bubble-pressure method, revealed large differences between equilibrium and dynamic surface tension values of fluorinated surfactants (see Sec. 4.4). The surface tension transition from equilibrium values to dynamic diffusion-limited values depended on the surfactant type, concentration, and bubble generation rate.

The *spinning drop method* [276–278] is used to determine interfacial tensions between two liquids. A capillary tube is mounted in a chamber leaving the ends open (Fig. 9.24). The chamber and the tube are filled with the heavier of the two liquids and the capillary is rotated at a high speed (about 2000 rpm). A drop of the other liquid having a lower density is placed into the capillary. The drop moves into the center of the capillary tube and usually assumes the

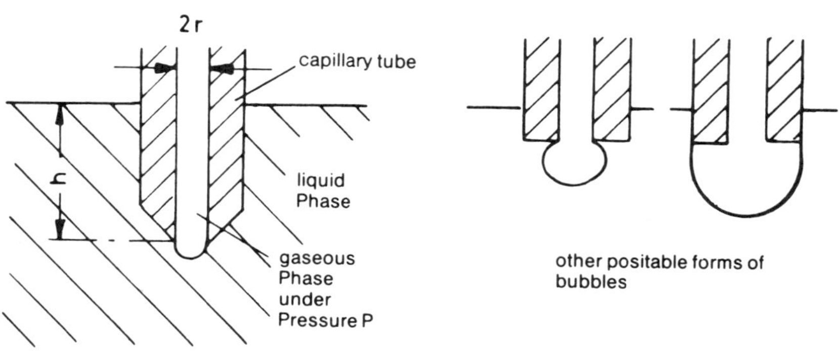

FIG. 9.23 Maximum bubble pressure method. (Reproduced by permission of Krüss USA.)

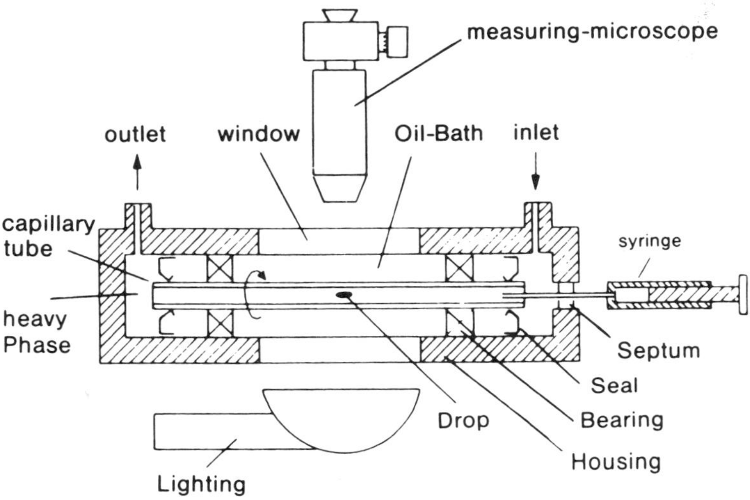

FIG. 9.24 Spinning drop tensiometer. (From Ref. 230. Reproduced by permission of Krüss USA.)

shape of a cylinder with curved edges. The radius of the drop is measured using a camera or a microscope.

$$\gamma = \tfrac{1}{4} r^3 \, \Delta\rho \, \omega^2 \tag{19}$$

where r is the radius of the drop, $\Delta\rho$ the density difference, and ω the angular velocity.

The accuracy of the method depends on the technique used for measuring the radius of the drop. A camera interfaced with a computer allows the image to be frozen for an accurate measurement [316,317].

Examples of methods used to measure surface tensions of fluorinated surfactants are shown in Table 9.1.

9.16 FLUORINATED SURFACTANTS IN BIOLOGICAL SYSTEMS

The discovery of two types of fluorine, organic and inorganic, in human blood [329–333] intensified the interest in the absorption and retention of fluorochemicals in biological systems. The conjecture that organic fluorine in blood originated from exposure to fluorinated surfactants prompted analyses of blood and other biological samples for fluorinated surfactants. A study of the exposure to fluorinated surfactants [334] found fluorine concentrations ranging from 1 to 71 ppm in the blood of workers handling ammonium perfluorooctane.

TABLE 9.1 Examples of Surface Tension Methods Used for Fluorinated Surfactants

Method	Author	Reference
Ring	Bernett and Zisman (1959)	318
	Caporiccio et al. (1984)	105
	La Mesa and Sesta (1987)	319
	Glöckner et al. (1989)	320a
Wireloop	Thoai (1977)	299
Plate	Funasaki and Hada (1979)	321
	Tadros (1980)	301
	Jost et al. (1988)	322
Pendant drop	Jarvis and Zisman (1959)	323
	Motomura et al. (1989)	324a
Drop volume	Zhao Guo-Xi and Zhu Bu-Yao (1983)	325a
Drop weight	Shinoda and Nakayama (1963)	309
	Mukerjee and Handa (1981)	246, 326
	Nishikido et al. (1989)	327
Maximum bubble pressure	Scholberg et al. (1953)	328
	Hirt et al. (1990)	244

The methods used for the determination of fluorinated surfactants in biological samples can be divided into two groups: (1) the determination of organic fluorine, which represents the concentration of a fluorinated surfactant if other fluorochemicals are absent, and (2) a specific method for the fluorinated surfactant of interest.

The determination of organic fluorine in biological samples, such as whole blood, serum, and plasma, involves destruction of organic matter by combustion or ashing to convert organic fluorine to inorganic fluoride and the determination of fluoride in the sample. The organic fluorine is calculated as the difference between fluoride found in the combusted sample minus inorganic fluoride present in the uncombusted sample. Ashing in the presence of magnesium carbonate [331], magnesium oxide [333], or calcium phosphate [335,336] leads to low results and has been superseded by combustion in a closed system with oxygen. Venkateswarlu [337] and Belisle and Hagen [15] have employed an oxygen Parr bomb for the determination of fluorine in whole blood or serum. Since liquid blood cannot be combusted in an oxygen Parr bomb, Belisle and Hagen [15] removed water from blood in vacuo at ambient temperature. The solids were pelletized after mixing with a preweighed amount of benzoic acid. The pellet was combusted in an oxygen Parr bomb which contained 0.025 N NaOH to absorb fluoride. The reaction mixture was acidified with perchloric acid and fluoride extracted into *m*-xylene containing triethylsilanol. The fluoride, converted to triethylfluorosilane, was determined by gas chromatography. The method gives accurate results for perfluorooctanoic acid in blood, but volatile fluorochemicals in the sample evaporate together with water when the blood sample is dried.

Kissa [10] combusted biological samples, including whole blood, serum, and various organic tissues, in an oxyhydrogen flame. A 0.5- to 1.0-g sample of whole blood, serum, or plasma was introduced into the combustion apparatus. Hydrofluoric acid formed during the combustion is collected in water and determined with a fluoride ion–selective electrode [11]. To calculate the organic fluorine concentration, the inorganic fluoride concentration of the sample was determined with an analyte addition method using a fluoride ion–selective electrode [338] and deducted from the total fluorine concentration. The method has the advantage of being applicable to liquid samples without conversion of the liquid sample to a solid. The oxyhydrogen combustion method can determine volatile fluorochemicals which would be lost if a liquid sample had to be dried. A large number of samples of blood and organic tissues have been analyzed routinely by the oxyhydrogen combustion method in a Du Pont laboratory [10].

The direct determination of fluorinated surfactants is possible if the fluorinated surfactant is amenable to chromatography or spectroscopy. Belisle and Hagen [42] determined perfluorooctanoic acid in blood, urine, and liver tissue. Perfluorooctanoic acid was extracted from blood or other biological samples with hexane in the presence of hydrochloric acid and converted to its methyl ester with diazomethane. The recovery of known amounts of perfluorooctanoic acid added to human plasma was essentially quantitative. The precision of the method was inferior to that of the determination of perfluorooctanoic acid by elemental fluorine analysis but could probably be improved by using a capillary chromatographic column instead of the packed column used by the authors.

9.17 FLUORINATED SURFACTANTS IN THE ENVIRONMENT

The presence of fluorinated surfactants in the environment is of concern in air and in water or wastewater.

Fluorinated surfactants can enter air as vapor if volatile or as a liquid or solid aerosol. To protect the health of workers, fluorinated surfactants have to be monitored in air at industrial sites where fluorinated surfactants are produced or used.

The determination of fluorochemicals in *air* usually involves two steps: collection of an air sample and determination of the fluorochemical in the sample collected. Sampling techniques used have included grab sampling, concentration by cryogenic techniques, trapping in a solvent, and adsorption on a solid adsorbent. The latter method is convenient for collecting a sample. However, the common adsorbents hold fluorinated surfactants firmly and desorption of the collected species for its determination is difficult and frequently not quantitative. Kissa [339] therefore introduced the concept of the total organic fluorine concentration in air. The fluorinated surfactant in air is collected on a solid adsorbent, such as activated carbon, graphitized carbon, silica, or Tenax. The adsorbent is combusted in an oxyhydrogen torch. Hydrogen fluoride, formed by combustion of organic fluorine on the adsorbent, is collected in water and determined with a fluoride ion–selective electrode [11]. When only one fluorine-containing species is present in air, the fluorine content represents its concentration. When several fluorochemicals are present, the fluorine concentration indicates the maximum concentration a fluorinated surfactant may possibly have in the mixture of fluorochemical air contaminants. If this maximum concentration value indicated by the total fluorine concentration is below the acceptable limit, a specific analytical method is not needed.

Fluorinated surfactants present in air as solid aerosols, such as dust, are collected on filters made of mixed cellulose esters and combusted together with the filter in the oxyhydrogen torch [339]. Volatile fluorochemicals can be collected on a solid adsorbent in an adsorption tube connected to the outlet of the filter.

The determination of fluorinated surfactants in *water* and *wastewater* is essential for (1) the detection of pollution by fluorinated surfactants, (2) study of biodegradation, and (3) determining the effect of fluorinated surfactants on aquatic life. If a specific method is not needed, the oxyhydrogen combustion method is the most effective [10]. By introducing a 10-mL water sample into the oxyhydrogen torch in several portions, as little as 20–40 ppb fluorinated surfactant can be detected without the need to concentrate the sample before combustion.

When other fluorochemicals are present, the fluorinated surfactant has to be separated and determined by a specific analytical method. Some of the conventional methods for the analysis of hydrocarbon-type surfactants [1] are also applicable for the determination of fluorinated surfactants (see Secs. 9.3 and 9.4). However, when the fluorinated surfactant has to be identified and structural information is needed, mass spectrometry [340] is the method of choice. Schröder [46,69,70] employed a tandem mass spectrometer (MS/MS) to analyze fluorinated surfactants in water and wastewater. High-performance liquid chromatography (HPLC) was coupled by a thermospray interface to a tandem mass spectrometer (MS/MS). A nonionic fluorinated surfactant (Fluowet OTN) was separated by reverse phase HPLC using a 15-cm, 3.9-mm-ID C_{18} column with 5-μm spheres. The gradient of the methanol-water eluent was from 80:20 to 20:80 in 10 min. 0.10 mol/L ammonium acetate was injected on line after separation. Alternatively, the chromatographic separation was bypassed and the analyte injected on line by the flow injection analysis (FIA) method.

An anionic fluorinated surfactant (Fluowet PL 80) and a cationic fluorinated surfactant (Fluowet L 3658-1) were quantitatively analyzed in water. However, analyses of these fluorinated surfactants in wastewater containing sludge were complicated by strong adsorption of the surfactants on the sludge. Extraction of the surfactants with acidified methanol was incomplete. The determination of the anionic surfactant (Fluowet PL 80) by combustion in an oxyhydrogen flame gave a quantitative result, although only 41% of the fluorine in the nonionic surfactant (Fluowet OTN) was recovered.

REFERENCES

1. T. M. Schmitt, "Analysis of Surfactants," Surfactant Science Series, Vol. 40, Marcel Dekker, New York (1992).
2. Martin J. Schick, ed., "Nonionic Surfactants," Surfactant Science Series, Vol. 1, Marcel Dekker, New York (1967).
3. E. Jungermann, ed., "Cationic Surfactants," Surfactant Science Series, Vol. 4, Marcel Dekker, New York (1970).
4. J. Cross, ed., "Anionic Surfactants—Chemical Analysis," Surfactant Science Series, Vol. 8, Marcel Dekker, New York (1977).
5. Martin J. Schick, ed., "Nonionic Surfactants. Physical Chemistry," Surfactant Science Series, Vol. 23, Marcel Dekker, New York (1987).
6. E. A. M. England, J. B. Hornsby, W. T. Jones, and D.R. Terrey, Anal. Chim. Acta 40, 365 (1968).
7. R. Wickbold, Angew. Chem. 64, 133 (1952).
8. P. B. Sweetser, Anal. Chem. 28, 1766 (1956).
9. I. Dobratz, unpublished Du Pont report.
10. E. Kissa, to be published.
11. E. Kissa, Anal. Chem. 55, 1445 (1983).
12. H.J. Friese and A. Pukropski, Fr. Demande FR 2592162 (1987)
13. Heraeus PCL-B-40, Verbrennungsapparatur V5 (Heraeus Quartzschmelze GmbH, Hanau, Germany), Tenside 23, 271 (1986).
14. J. J. Bailey and D. G. Gehring, Anal. Chem. 33, 1760 (1961).
15. J. Belisle and D. F. Hagen, Anal. Biochem. 87, 545 (1978).
16. W. Schöniger, Mikrochim. Acta 1955, 123 (1955).
17. T. S. Ma and R. C. Rittner, "Modern Organic Elemental Analysis," Marcel Dekker, New York (1979).
18. N. H. Gelman and L. M. Kiparenko, Zh. Anal. Khim. 20, 229 (1965).
19. F. H. Oliver, Analyst 91, 771 (1966).
20. A. Steyermark, R. R. Kaup, D. A. Petras, and E. A. Pass, Microchem. J. 4, 55 (1960).
21. H. Trutnovsky, Mikrochim. Acta 1963, 498 (1963).

22. J. Horacek and V. Pechanec, Mikrochim. Acta 1966, 17 (1966).
23. T. S. Light and R. F. Mannion, Anal. Chem. 41, 107 (1969).
24. F. W. Cheng, Mikrochim. Acta 1970, 841 (1970).
25. W. Selig, Z. Anal. Chem. 249, 30 (1970).
26. H. J. Francis, Jr., J. H. Donarine, and D. D. Pesing, Microchem. J. 14, 580 (1969).
27. D. A. Shearer and G. F. Morris, Microchem. J. 15, 199 (1970).
28. J. Pavel, R. Kuebler, and H. Wagner, Microchem. J. 15, 192 (1970).
29. Y. Inoue and K. Yokoyama, Iyakuhin Kenkyu 14, 25 (1983).
30. Y. Asano and Y. Shimada, Koku Eisei Gakkai Zasshi 35, 526 (1985).
31. J. L. Stuart, Analyst 95, 1032 (1970).
32. T. S. Ma and J. Gwirtsman, Anal. Chem. 29, 140 (1957).
33. W. K. R. Muskgrave, Analyst 86, 842 (1961).
34. P. P. Wheeler and M. J. Fauth, Anal. Chem. 38, 1970 (1966).
35. M. Noshiro, T. Yarita, and S. Yonemori, Bunseki Kagaku, 32(12), E403 (1983).
36. Model 93–42 Surfactant Electrode Instruction Manual, Orion Research (1988).
37. J. H. Jones, J. Assoc. Offic. Agric. Chem. 28, 398 (1945).
38. J. Waters and C. G. Taylor, in "Anionic Surfactants—Chemical Analysis," J. Cross, ed., Surfactant Science Series, Vol. 9, Marcel Dekker, New York (1977).
39. R. Sharma, R. Pyter, and P. Mukerjee, Anal. Lett. 22(4),999 (1989).
40. N. Ishikawa and M. Sasabe, J. Fluorine Chem. 25, 241 (1984).
41. J. F. Rabolt, T. P. Russell, and R. J. Twieg, Macromolecules 17, 2786 (1984).
42. J. Belisle and D. F. Hagen, Anal. Biochem. 101, 369 (1980).
43. M. P. Turberg and J. E. Brady, J. Am. Chem. Soc. 110, 7797 (1988).
44. T. Asakawa, S. Miyagishi, and M. Nishida, Langmuir 3, 821 (1988).
45. T. Asakawa, S. Miyagishi, and M. Nishida, Langmuir 4, 136 (1988).
46. H. Fr. Schröder, Vom Wasser 78, 211 (1992).
47. F. Tokiwa, K. Ohti, I, Kokubo, Bull. Chem. Soc. Jpn. 41, 2845 (1968).
48. T. Nakagawa and H. Jizomoto, Kolloid Z. Z. Polym. 236, 79 (1969).
49. T. Nakagawa and H. Jizomoto, Kolloid Z. Z. Polym. 239, 606 (1970).
50. T. Nakagawa and H. Jizomoto, J. Am. Oil. Chem. Soc. 48, 571 (1971).
51. T. Asakawa, S. Miyagishi, and M. Nishida, Langmuir 3, 821 (1987).
52. T. Asakawa, S. Miyagishi, and M. Nishida, Langmuir 1, 347 (1985).
53. P. Mukerjee, M. J. Gumkowski, C. C. Chan, and R. Sharma, J. Phys. Chem. 94, 8832 (1990).
54. H. Nadeau and S. Siggia, in "Nonionic Surfactants," p. 825, M. J. Schick, ed., Surfactant Science Series, Vol. 1. Marcel Dekker, New York (1967)
55. (a) P. R. Griffits and J. A. de Haseth, "Fourier Transform Infrared Spectrometry," Chemical Analysis Vol. 83, J. Wiley & Sons, New York (1986). (b) D. Lin-Vien, N. B. Colthup, W. G. Fateley, and J. G. Grasselli, "The Handbook of Infrared and Raman Characteristic Frequencies of Organic Molecules," Academic Press, San Diego, CA (1991).
56. A. M. Amorim da Costa and E. B. H. Santos, Colloid Polym. Sci. 261, 58 (1983).
57. (a) K. D. Cook, in "Encyclopedia of Polymer Science and Engineering," Vol. 9, p. 319, John Wiley & Sons, New York (1987). (b). M. B. Bursey, in "McGraw-

Hill Encyclopedia of Science and Technology," Vol. 10, p. 492, McGraw-Hill, New York (1987).
58. K. Levsen, "Fundamental Aspects of Organic Mass Spectrometry," Verlag Chemie, Weinheim (1978).
59. M. E. Rose and R. A. W. Johnstone, "Mass Spectrometry for Chemists and Biochemists," Cambridge University Press, New York (1982).
60. J. T. Watson, "Introduction to Mass Spectrometry," Raven Press, New York (1982).
61. F. W. McLafferty, ed., "Tandem Mass Spectroscopy," John Wiley & Sons, New York (1983).
62. J. R. Chapman, "Practical Organic Mass Spectrometry," John Wiley & Sons, London (1986).
63. J. B. Lambert, H. F. Shurvell, D. Lightner, and R. G. Cooks, "Introduction to Organic Spectroscopy," Macmillan, New York (1987).
64. W. M. A. Niessen and J. van der Greef, "Liquid Chromatography-Mass Spectrometry," Chromatographic Sci. Ser., Vol. 58, Marcel Dekker, New York (1992).
65. J. V. Iribarne and B. A. Thomson, J. Chem. Phys. 64, 2287 (1976).
66. B. A. Thomson and J. V. Iribarne, J. Chem. Phys. 71, 4451 (1979).
67. C. R. Blakely and M. L. Vestal, Anal. Chem. 55, 750 (1983).
68. M. L. Vestal and G. J. Fergusson, Anal. Chem. 57, 2373 (1985).
69. H. Fr. Schröder, Vom Wasser 73, 111 (1989).
70. H. Fr. Schröder, Vom Wasser 77, 277 (1991).
71. K. H. Ott, Fette Seifen Anstrichm. 87, 377 (1985).
72. E. Schneider, K. Levsen, P. Daehling, F. W. Roellgen, Fresenius Z. Anal. Chem. 316, 488 (1983).
73. D. N. Heller, C. Fenselau, J. Yergey, R. J. Cotter, and D. Larkin, Anal. Chem. 56, 2274 (1984).
74. P. A. Lyon, K. B. Tomer, and M. L. Gross, Anal. Chem. 57, 2984 (1985).
75. J. M. Gilliam, P. W. Landis, and J. L. Occolowitz, Anal. Chem. 55, 1531 (1983)].
76. J. Kawase, K. Tsuji, Y. Yasuda, and K. Yashima, J. Chromatogr. 267, 133 (1983).
77. J. Kawase, K. Tsuji, and Y. Yasuda, J. Chromatogr. 267, 149 (1983).
78. G. D. Roberts and E. V. White, Biomed. Mass Spectrom. 11, 273 (1984)
79. M. A. Baldwin and F. W. McLafferty, Org. Mass Spectrom. 7, 1353 (1973).
80. R. J. Cotter, Anal. Chem. 52, 1589A (1980).
81. C. D. Daves, Jr., Accounts Res. Chem. 12, 359 (1979).
82. R. J. Cotter, G. Hansen, and T. R. Jones, Anal. Chim. Acta 136, 135 (1982)
83. A. Abragam, "Principles of Nuclear Magnetism," Oxford University Press, Oxford (1961).
84. R. K. Harris, "Nuclear Magnetic Resonance Spectroscopy: A Physicochemical Approach," Pitman, London (1983).
85. M. L. Martin, J.J. Delpuech, and G. J. Martin, "Practical NMR Spectroscopy," Heyden, London (1983).

86. A. Rahman, "One and Two Dimensional NMR Spectroscopy," Elsevier, New York (1989).
87. C. P. Sichter, "Principles of Magnetic Resonance," 3rd ed., Springer-Verlag, New York (1989).
88. R. Kitamaru, "Nuclear Magnetic Resonance: Principles and Theory," Elsevier, Amsterdam (1990).
89. H.-O. Kalinowski, S. Berger, and S. Braun, "Carbon C-13 NMR Spectroscopy," John Wiley & Sons, New York (1991).
90. R. Ernst, G. Bodenhausen, and A. Wokaun, "Principles of Nuclear Magnetic Resonance in One or Two Dimensions," Oxford Science Publications, Oxford University Press, New York (1991).
91. H. Friebolin, "Basic One- and Two-Dimensional NMR Spectroscopy," VCH Publishers, New York (1991).
92. J. K. M. Saunders and B. K. Hunter, "Modern NMR Spectroscopy, a Guide for Chemists," Oxford University Press, New York (1991).
93. B. Lindman, O. Söderman, and H. Wennerström, in "Surfactant Solutions," R. Zana, ed., Vol. 22, p. 295, of Surfactant Science Series, Marcel Dekker, New York (1987).
94. P. Mukerjee and K. Mysels, Pap. Symp. 1974, ACS Symp. Ser. 9, 239 (1975).
95. N. Muller and R. H. Birkhahn, J. Phys. Chem. 71, 957 (1967).
96. N. Muller and R. H. Birkhahn, J. Phys. Chem. 72, 583 (1968).
97. N. Muller and T. W. Johnson, J. Phys. Chem. 73, 2042 (1971).
98. N. Muller and H. Simsohn, J. Phys. Chem. 75, 942 (1971).
99. M. Monduzzi, A. Chittofrati, and M. Visca, Langmuir 8, 1278 (1992).
100. T. Asakawa, M. Muori, S. Miyagishi, and M. Nishida, Langmuir 5, 343 (1989).
101. B.-O. Persson, T. Drakenberg, and B. Lindman, J. Phys. Chem. 80, 2124 (1976).
102. R. J. E. M. de Weerd, J. W. de Haan, J. M. van de Ven, M. Achten, and H. M. Buck, J. Phys. Chem. 86, 2523 (1982).
103. D. L. Chang, H. L. Rosano, and A. E. Woodward, Langmuir 1, 669 (1985).
104. K. Fontell and B. Lindman, J. Phys. Chem. 87, 3289 (1983)
105. G. Caporiccio, F. Burzio, G. Carniselli, and V. Biancardi, J. Colloid Interface Sci. 98, 202 (1984).
106. R. Haque, J. Phys. Chem. 72, 3056 (1968).
107. R. E. Bailey and G. H. Cady, J. Phys. Chem. 73, 1612 (1969).
108. N. Muller and F. E. Platko, J. Phys. Chem. 75, 547 (1971).
109. N. Muller, J. H. Pellerin, and W. W. Chen, J. Phys. Chem. 76, 3012 (1972).
110. J. Carlfors and P. Stilbs, J. Colloid Interface Sci. 103, 332 (1985).
111. E. L. Hahn, Phys. Rev. 80, 580 (1950).
112. E. O. Stejskal and J. E. Tanner, J. Chem. Phys. 42, 288 (1965).
113. P. Stilbs, J. Colloid Interface Sci. 87, 385 (1982).
114. R. Palepu and V. C. Rainsborough, Can. J. Chem. 67, 1550 (1989).
115. W. Guo, B. M. Fung, and S. D. Christian, Langmuir 8, 446 (1992).
116. U. Henriksson and L. Ödberg, J. Colloid Interface Sci. 46, 212 (1974).
117. J. Ulmius and B. Lindman, J. Phys. Chem. 85, 4131 (1981).

118. (a) C. Burger-Guerisi, C. Tondre, and D. Canet, J. Phys. Chem. 92, 4974 (1988). (b) G. Serratrice, M. J. Stebe, and J. J. Delpuech, J. Chim. Phys. Phys.-Chim. Biol. 85, 875 (1988). (c) G. Serratrice, M. J. Stebe, and J. J. Delpuech, J. Fluorine Chem. 25, 275 (1984). (d) G. J. T. Tiddy, Symp. Faraday Soc. No. 5, 150 (1971).
119. J. Lang and R. Zana, in "Surfactant Solutions," R. Zana, ed., p. 405, Vol. 22 of Surfactant Science Series, Marcel Dekker, New York (1987).
120. E. A. G. Aniansson and S. N. Wall, J. Phys. Chem. 78, 1024 (1974).
121. E. A. G. Aniansson and S. N. Wall, J. Phys. Chem. 79, 857 (1975).
122. E. A. G. Aniansson, S. N. Wall, J. Lang, C. Tondre, R. Zana, H. Hoffmann, I. Kielmann, and W. Ulbricht, J. Phys. Chem. 80, 905 (1976).
123. M. Kahlweit and M. Teubner, Adv. Colloid Interface Sci. 13, 1 (1980).
124. E. Lessner, M. Teubner, and M. Kahlweit, J. Phys. Chem. 85, 1529. (1981).
125. E. Lessner, M. Teubner, and M. Kahlweit, J. Phys. Chem. 85, 3167 (1981).
126. M. Kahlweit, J. Colloid Interface Sci. 90, 92 (1982).
127. M. Kahlweit, Pure Appl. Chem. 53, 2069 (1981).
128. M. Eigen and L. de Maeyer, in "Technique of Organic Chemistry," Vol. VIII, p. 895, S. L. Friess, E. S. Lewis, and A. Weissberger, eds., Wiley-Interscience, New York (1963).
129. D. Hague, "Fast Reactions," Wiley-Interscience, New York, (1971).
130. G. C. Hammes, ed., "Technique of Organic Chemistry," Vol. VI, Part II: "Investigation of Elementary Reaction Steps in Solution and Very Fast Reactions," Wiley-Interscience, New York, (1974).
131. E. Wyn-Jones, ed., "Chemical and Biological Applications of Relaxation Spectrometry," D. Reidel, Dordrecht (1975).
132. (a) C. F. Bernasconi, "Relaxation Kinetics," Academic Press, New York (1976). (b) K. L. Mittal, ed., "Micellization, Solubilization, and Microemulsions," Vol. I, Plenum, New York (1977).
133. W. J. Gettins and E. Wyn-Jones, eds., "Techniques and Applications of Fast Reactions in Solution," D. Reidel, Dordrecht (1979).
134. E. Graber, J. Lang, and R. Zana, Kolloid Z. 238, 470 (1970).
135. E. Graber and R. Zana, Kolloid Z. 238, 479 (1970).
136. P. J. Sams, J. E. Rassing, and E. Wyn-Jones, Chem. Phys. Lett. 13, 233 (1972).
137. P. J. Sams, J. E. Rassing, and E. Wyn-Jones, J. Chem. Soc. Faraday Trans II 70, 1247 (1974).
138. E. A. G. Aniansson and S. N. Wall, J. Phys. Chem. 82, 2805 (1978).
139. G. E. A. Aniansson, in "Aggregation Processes in Solution," E. Wyn-Jones and J. Gormally, eds., Studies in Physical and Theoretical Chemistry, Vol. 26, Elsevier, Amsterdam (1983).
140. G. Platz and H. Hoffmann, Ber. Bunsenges. Phys. Chem. 76, 491 (1972).
141. H. Hoffmann and W. Ulbricht, Z. Phys. Chem. Neue Folge 106, 167 (1977).
142. H. Hoffmann and B. Tagesson, Z. Phys. Chem. Neue Folge 110, 113 (1978).
143. H. Hoffmann, W. Ulbricht, and B. Tagesson, Z. Phys. Chem. Neue Folge 113, 17 (1978).
144. H. Hoffmann, B. Tagesson, and W. Ulbricht, Ber. Bunsenges. Phys. Chem. 83, 148 (1979).

145. H. Hoffmann, G. Platz, H. Rehage, K. Reizlein, and W. Ulbricht, Makromol. Chem. 182, 451 (1981).
146. H. Hoffmann, G. Platz, and W. Ulbricht, J. Phys. Chem. 85, 1418 (1981).
147. T. Yasunaga, H. Oguri and M. Miura, J. Colloid Interface Sci. 23, 352 (1967).
148. T. Yasunaga, S. Fujii and M. Miura, J. Colloid Interface Sci. 30, 399 (1969).
149. R. Zana and J. Lang, Compt. Rend. C, 266, 893 (1968).
150. R. Zana and J. Lang, Compt. Rend. C, 266, 1377 (1968).
151. J. Lang and R. Zana, in "Surfactant Solutions", R. Zana, ed., p. 424, Surfactant Science Series Vol. 22, Marcel Dekker, New York (1987).
152. A. Mathieson, Molecular Acoustics, Wiley-Interface, London (1970).
153. P. J. Sams, E. Wyn-Jones, and J. Rassing, Chem. Phys. Lett. 13, 233 (1972).
154. P. J. Sams, E. Wyn-Jones, and J. Rassing, J. Chem. Soc. Faraday Trans 2 69,180 (1973).
155. J. E. Rassing and E. Wyn-Jones, Chem. Phys. Lett. 21, 93 (1973).
156. J. E. Rassing, P. J. Sams, and E. Wyn-Jones, J. Chem. Soc. Faraday Trans. 2, 70(7), 1247 (1974).
157. G. T. Tiddy, M. F. Walsh, and E. Wyn-Jones, J. Chem. Soc. Commun. 1979, 252.
158. J. Gettins, P. L. Jobling, M. F. Walsh, and E. Wyn-Jones, J. Chem. Soc. Faraday Trans 2 76, 794 (1980).
159. J. E. Rassing, P. J. Sams, and E. Wyn-Jones, J. Chem. Soc. Faraday Trans. 2, 70(7), 1247 (1974).
160. B. Cabane, in "Surfactant Solutions," R. Zana, ed., p. 57, Surfactant Science Series, Vol. 22, Marcel Dekker, New York (1987).
161. A. Guinier and G. Fournet, "Small Angle Scattering of X Rays," Wiley-Interscience, New York (1955).
162. K. Fontell, in "Liquid Crystals and Plastic Crystals," G. W. Gray and P. A. Winsor, eds., Vol. 2, p. 80, Horwood, Chichester (1974).
163. O. Glatter and O. Kratky, "Small Angle X Ray Scattering," Academic Press, London (1982).
164. K. Fontell, in "Surfactant Solutions," K. L. Mittal and B. Lindman, eds., Vol. 1, p. 69, Plenum, New York (1984).
165. K. Fontell and B. Lindman, J. Phys. Chem. 87, 3289 (1983).
166. (a) B. Jacrot, Rep. Prog. Phys. 39, 911 (1976). (b) E. Caponetti, D. C. Martino, M. A. Floriano, and R. Triolo, Langmuir 9, 1193 (1993).
167. J. B. Hayte, in "Physics of Amphiphiles: Micelles, Vesicles, and Microemulsions," V. Degiorgio and M. Corti, eds., North-Holland, Amsterdam (1985).
168. (a). H. Hoffmann, J. Kalus, K. Reizlein, W. Ulbricht, and K. Ibel, Colloid Polym. Sci. 260, 435 (1982). (b). H. Hoffmann, J. Kalus, and H. Thurn, Colloid Polym. Sci. 261, 1043 (1983).
169. J. C. Ravey and M. J. Stebe, Physica B (Amsterdam) 156–157, 394–7 (1989).
170. G. Mathis, P. Leempoel, J.-C. Ravey, C. Selve, and J.-J Delpuech, J. Am. Chem. Soc. 106, 6162 (1984).
171. C. G. Windsor, in "Chemical Applications of Thermal Neutron Scattering," p. 1. B. T. M. Willis, ed., Oxford University Press, Oxford (1973).
172. J. P. Cotton, D. Dekker, H. Benoit, B. Farnoux, J. Higgins, G. Jannink, R. Ober, C. Picot, and J. De Cloizeaux, J. Macromol. 7, 863 (1974).

173. S. J. Burkitt, R. H. Ottewill, H. B. Hayter, and B. T. Ingram, Colloid Polym. Sci. 265, 619 (1987).
174. S. J. Burkitt, R. H. Ottewill, H. B. Hayter, and B. T. Ingram, Colloid Polym. Sci. 265, 628 (1987).
175. M. Kerker, "The Scattering of Light and Other Electromagnetic Radiation," Academic Press, New York (1969).
176. B. J. Berne and R. Pecora, "Dynamic Light Scattering," Wiley-Interscience, New York (1976).
177. V. Degiorgio, M. Corti, and M. Giglio, eds., "Light Scattering in Liquids and Macromolecular Solutions," Plenum, New York (1980).
178. S. J. Candau, in "Surfactant Solutions," R. Zana. ed., p. 147, Surfactant Science Series, Vol. 22, Marcel Dekker, New York (1987).
179. H. E. Gerry, P.T. Jacobs, and E. W. Anacker, J. Colloid Interface Sci. 62, 556 (1977).
180. R. Zana, in "Surfactant Solutions," R. Zana, ed., p. 241, Surfactant Science Series, Vol. 22, Marcel Dekker, New York (1987).
181. W. J. Moore, "Physical Chemistry," 4th ed., Prentice-Hall, Englewood Cliffs, NJ (1972).
182. N. J. Turro, M. W. Geiger, R.R. Hautala, and N. E. Schore, in "Micellization, Solubilization, and Microemulsions," K. L. Mittal, ed., Vol. 1, p. 75, Plenum, New York (1977).
183. R. Zana, ed. "Surfactant Solutions," Surfactant Science Series, Vol. 22, Marcel Dekker, New York (1987).
184. C. A. Parker, "Photoluminescence of Solutions," Elsevier, Amsterdam (1968).
185. S. Schulman, "Fluorescence and Phosphorescence Spectroscopy: Physicochemical Principles and Practice," Pergamon Press, Oxford (1976).
186. N. J. Turro, "Modern Molecular Photochemistry," Benjamin/Cummings, Menlo Park, CA (1978).
187. J. Lakowicz, "Principles of Fluorescence Spectroscopy," Plenum, New York (1983).
188. K. Meguro, Y. Muto, F. Sakurai, and K. Esumi, ACS Symp. Ser. 1986, 311, p. 61 (1986).
189. Y. Muto, K. Esumi, K. Meguro, and R. Zana, J. Colloid Interface Sci. 120, 162 (1987)
190. K. Kalyanasundaram, Langmuir 4, 942 (1988).
191. Y. Muto, K. Esumi, K. Meguro, and R. Zana, Langmuir 5, 885 (1989).
192. K. Esumi, H. Otsuka, and K. Meguro, J. Colloid Interface Sci. 142, 582 (1991).
193. K. Esumi, H. Otsuka, and K. Meguro, Langmuir 7, 2313 (1991)
194. C. R. Brundle and A. D. Baker, eds., "Electron Spectroscopy, Theory, Techniques and Applications," Vols. 1–4 (1977–1981), Academic Press.
195. D. W. Dwight, T. J. Fabish, and H. R. Thomas, eds., "Photon, Electron, and Ion Probes of Polymer Structure and Properties," ACS Symposium Series 162, Washington, DC, (1981).
196. H. Siegbahn and L. Karlsson, "Photoelectron Spectroscopy," Handbuch der Physik, W. Mehlhorn, ed., Vol. 31, Springer-Verlag, Berlin (1982).
197. K. Siegbahn, J. Electron Spectrosc. Relat. Phenom. 51, 11 (1990).
198. H. Agren, Int. J. Quantum Chem. 39, 455 (1991).

199. D. Briggs and M. P. Seah, eds., "Practical Surface Analysis," Vol. 1, "Auger and X-ray Photoelectron Spectroscopy," 2nd ed., John Wiley & Sons, Chichester (1990).
200. N. H. Turner and J. A. Schreifels, Anal. Chem. 64, 302R (1992).
201. R. W. Phillips and R. G. Dettre, J. Colloid Interface Sci. 56, 251 (1976).
202. G. N. Batts, Colloids Surf. 22, 133 (1987).
203. L. J. Gerenser, J. M. Pochan, M. G. Mason, and J. F. Elman, Langmuir 1, 305 (1985).
204. C. S. Fadley, Prog. Solid State Chem. 2, 265 (1976).
205. L. J. Gerenser, J. M. Pochan, J. F. Elman, and M. G. Mason, Langmuir 2, 765 (1986).
206. P. M. Claesson, P. C. Herder, J. M. Berg, and H. K. Christenson, J. Colloid Interface Sci. 136, 541 (1990)
207. J. W. McBain, "Colloid Science," D. C. Heath, Boston (1950).
208. H.-D. Dörfler and E. Müller, Tenside 26, 381 (1989).
209. H. Hoffmann, G. Platz, H. Rehage, and W. Schorr, Ber. Bunsenges. Phys. Chem. 85, 255 (1981).
210. H. Hoffmann, G. Platz, H. Rehage, and W. Schorr, Ber. Bunsenges. Phys. Chem. 85, 877 (1981).
211. H. Hoffmann, G. Platz, and W. Ulbricht, J. Phys. Chem. 85, 1418 (1981).
212. W. Schorr and H. Hoffman, J. Phys. Chem. 85, 3160 (1981).
213. H. Hoffmann, H. Rehage, G. Platz, W. Schorr, H.Thurn, and W. Ulbricht, Colloid Polym. Sci. 260, 1042 (1982).
214. H. Hoffmann, G. Platz, H. Rehage, and W. Schorr, Adv. Colloid Interface Sci. 17, 275 (1982).
215. H. Hofmann, Ber. Bunsenges. Phys. Chem. 88, 1078 (1984).
216. H. Hoffmann, H. Rehage, W. Schorr, and H. Thurn, in "Surfactants in Solution," Vol. 1, K. Mittal and B. Lindman, eds., p. 425, Plenum, New York (1984).
217. W. Schorr and H. Hoffmann, in "Physics of Amphiphiles: Micelles, Vesicles, and Microemulsion," pp. 160–180, V. Degiorgio and M. Corti, eds., North-Holland, Amsterdam (1985).
218. M. Angel and H. Hoffmann, Z. Phys. Chem. Neue Folge 139, 153 (1984).
219. K. Tamori, K. Esumi, K. Meguro, and H. Hoffmann, J. Colloid Interface Sci. 147, 33 (1991).
220. C. T. O'Konski, ed., "Molecular Electrooptics," Marcel Dekker, New York (1976).
221. B. R. Jennings, ed., "Electro-Optics and Dielectrics of Macromolecules and Colloids," Plenum, New York (1979).
222. H. Scott, J. Phys. Chem. 68, 3612 (1964).
223. I. W. Osborne-Lee, R. S. Schechter, and W. H. Wade, J. Colloid Interface Sci. 94, 179 (1983).
224. I. W. Osborne-Lee, R. S. Schechter, W. H. Wade, and Y. Barakat, J. Colloid Interface Sci. 108, 60 (1985).
225. T. Asakawa, M. Mouri, S. Myagishi, and M. Nishida, Langmuir 4, 136 (1988).

226. T. Asakawa, S. Sunazaki, and S. Miyagishi, Colloid Polym. Sci. 270, 259 (1992).
227. J. J. Bikerman, "Surface Chemistry," 2nd ed., Academic Press, New York (1958).
228. P.Becher "Emulsions: Theory and Practice," 2nd ed., pp. 381–400, Reinhold, New York (1965).
229. R. Heusch, Fette Seifen Anstrichm. 72, 969 (1970).
230. C. Weser, GIT Fachzeitschrift für das Laboratorium, 24, 642-848, 734-742 (1980).
231. H.J. König and G. Hartmannn, Textiltechnik 37, 259 (1987).
232. C. C. Addison, J. Chem. Soc. 535 (1943).
233. C. C. Addison, J. Chem. Soc. 252 (1944).
234. C. C. Addison, J. Chem. Soc. 477 (1944).
235. C. C. Addison, J. Chem. Soc. 98 (1945).
236. C. C. Addison, J. Chem. Soc. 354 (1945).
237. A. F. H. Ward and L. Torday, J. Chem. Phys. 14, 453 (1946).
238. E. J. Burcik, J. Colloid Sci. 5, 421 (1950).
239. E. J. Burcik, J. Colloid Sci. 8, 520 (1953).
240. E. J. Burcik and R. Newman, J. Colloid Sci. 9, 498 (1954).
241. H. Lange, J. Colloid Sci. 20, 50 (1965).
242. X. Y. Hua and M. J. Rosen, J. Colloid Interface Sci. 124, 652 (1988).
243. J. van Hunsel and P. Joos, Colloid Polym. Sci. 267, 1026 (1989).
244. D. E. Hirt, R. K. Prudhomme, B. Miller, and L. Rebenfeld, Colloids Surf. 44, 101 (1990).
245. M. F. Cox, AOCS/CSMA Conference, Hershey, PA, Oct. 29 (1989).
246. P. Mukerjee and T. Handa, J. Phys. Chem. 85, 2298 (1981).
247. W. D. Harkins, T. F. Young, and L. H. Cheng, Science 64, 333 (1926).
248. W. D. Harkins and H. F. Jordan, J. Am. Chem. Soc. 52, 1751 (1930).
249. B. B. Freud and H. Z. Freud, J. Am. Chem. Soc. 52, 1772 (1930).
250. C. Huh and S. G. Mason, Colloid Polym. Sci. 253, 566 (1975).
251. L. Wilhelmy, Ann. Phys. (Leipzig) 119, 177 (1863).
252. J. F. Padday and D. R. Russell, J. Colloid Sci. 15, 503 (1960).
253. T. W. Richards and L. B. Coombs, J. Am. Chem. Soc. 37, 1656 (1915).
254. T. W. Richards and L. B. Coombs, J. Am. Chem. Soc. 43, 827 (1921).
255. S. Sugden, J. Chem. Soc. 1483 (1921).
256. C. D. Holcomb and J. A. Zollweg, J. Colloid Interface Sci. 134, 41 (1990).
257. P. Z. Rehbinder, Z. Phys. Chem. 111, 447 (1924).
258. J. Kloubek, Tenside 5, 317 (1968).
259. J. Kloubek, J. Colloid Interface Sci. 41, 1 (1972).
260. J. Kloubek, J. Colloid Interface Sci. 41, 7 (1972) .
261. J. Kloubek, J. Colloid Interface Sci. 41, 17 (1972).
262. K. Lunkenheimer, R. Miller, and J. Becht, Colloid Polym. Sci. 260, 1145 (1982).
263. K. J. Mysels, Langmuir 2, 428 (1986).
264. C. D. Holcomb and J. A. Zollweg, J. Colloid Interface Sci. 134, 41 (1990).

265. Y. Rotenberg, L. Boruvka, and A. W. Neumann, J. Colloid Interface Sci. 93, 169 (1983).
266. S. H. Anastasiadis, J. K. Chen, J. T. Koberstein, A. F. Siegel, J. E. Sohn, and J. E. Emerson, J. Colloid Interface Sci. 119, 55 (1987).
267. F. K. Hansen and G. Rødsrud, J. Colloid Interface Sci. 141, 1 (1991).
268. J. E. Verschaffelt, Proc. Konikl. Akad. Wetenschap. Amsterdam 21, 357 (1919).
269. H. C. Parreira, J. Colloid Sci. 20, 44 (1964).
270. E. A. Hauser, H. E. Edgerton, B. M. Holt and J. T. Cox, Jr., J. Phys. Chem. 40, 973 (1936).
271. H. Dunken, Ann. Phys. 41, 567 (1942).
272. W. D. Harkins and F. E. Brown, J. Am. Chem. Soc. 41, 499 (1919).
273. J. L. Lando and H. T. Oakley, J. Colloid Interface Sci. 25, 526 (1967).
274. R. Kumar and S. G. T. Bhat, Tenside 24, 86 (1987).
275. L. Holysz and E. Chibowski, Tenside 25, 377 (1988).
276. B. Vonnegut, Rev. Sci. Instrum. 13, 6 (1942).
277. H. T. Patterson. K. H. Hu, and T. H. Grindstaff, J. Polymer Sci. 34, 31 (1971).
278. J. L. Cayias, R. S. Schechter, and W. H. Wade, in "Adsorption at Interfaces," K. L. Mittal, ed., p. 234, ACS Symposium Series 8, American Chemical Society, Washington, DC (1975).
279. R. Defay and G. Pètrè, in Surface Colloid Sci., 3, 27 E. Matijevic, ed., (1971), Wiley-Interscience.
280. A. P. Brady and A. G. Brown, in "Monomolecular Layers," H. Sobotka, ed., Symp. Amer. Assoc. Advan. Sci., p. 33, AAAS, Washington, DC (1951).
281. R. Van den Bogaert and P. Joos, J. Phys. Chem. 83, 2244 (1979).
282. E. Rillaerts and P. Joos, J. Colloid Interface Sci. 88, 1 (1982).
283. J. Van Havenbergh and P. Joos, J. Colloid Interface Sci. 95, 172 (1983).
284. J. H. Clint, E. L. Neustadter, and T. Jones, Dev. Pet. Sci. 13, 135 (1981).
285. S. J. Sugden, J. Chem. Soc. 121, 858 (1921).
286. S. J. Sugden, J. Chem. Soc. 125, 27 (1924).
287. R. L. Bendure, J. Colloid Interface Sci. 35, 238 (1971).
288. W. D. Harkins, T. F. Young, and L. H. Cheng, Science 64, 333 (1926).
289. K. Lunkenheimer, Tenside 19, 272 (1982).
290. P. J. Cram and J. M. Haynes, J. Colloid Interface Sci. 35, 706 (1971).
291. W. A. Gifford, J. Colloid Interface Sci. 64, 588 (1978).
292. K. Lunkenheimer and K.D. Wantke, Colloid Polym. Sci. 259, 354 (1981).
293. K. Lunkenheimer, J. Colloid Interface Sci. 131, 580 (1989).
294. E. Bartholomé and K. Schäfer, Melliand Textilber. 31, 487 (1950).
295. E. A. Boucher, T. M. Grinchuk and A. C. Zettlemoyer, J. Colloid Interface Sci. 23, 600 (1967).
296. A. M. Mankowich, J. Colloid Interface Sci. 25, 590 (1968).
297. A. C. Zettlemoyer, V. V. Subba Rao, J. Colloid Interface Sci. 29, 172 (1969).
298. P. Lenard, R.v.Dallwitz-Wegener, and E. Zachmann, Ann. Phys. [4] 74, 381 (1924).
299. N. Thoai, J. Colloid Interface Sci. 62, 222 (1977).
300. A. H. Ellison and W. A. Zisman, J. Phys. Chem. 60, 416 (1956).

301. Th. F. Tadros, J. Colloid Interface Sci. 74, 196 (1980).
302. W. Asche, Seifen Oele Fette Wachse 112, 543 (1986).
303. J. M. Andreas, E. A. Hauser, and W. B. Tucker, J. Phys. Chem. 42, 1001 (1938)
304. S. Fordham, Proc. R. Soc.(London) 194, 1 (1948).
305. K. Prochaska, Z. Górski, and J. Szymanowski, Tenside 27, 233 (1990).
306. R. Miller and K.-H. Schano, Tenside 27, 238 (1990).
307. A. E. Alexander and J. B. Hayter, in "Physical Methods of Chemistry", Vol. I, Part V, p. 501, eds. A. Weissberger and B. W. Rossiter, Wiley-Interscience, New York (1971).
308. W. D. Harkins and F.E. Brown, J. Am. Chem. Soc. 41, 519 (1919).
309. K. Shinoda and H. Nakayama, J. Colloid Sci. 18, 705 (1963).
310. K. Shinoda, M. Hato, and T. Hayashi, J. Phys. Chem. 76, 909 (1972).
311. S. W. Morrall, A. D. Clauss, and T. J. Adams, AOCS Annual Meeting, Toronto, May 10–14 (1992).
312. M. J. Rosen and X. Y. Hua, J. Colloid Interface Sci. 139, 397 (1990).
313. X. Y. Hua and M. J. Rosen, J. Colloid Interface Sci. 141, 180 (1991).
314. P. Joos, J. P. Fang, and G. Serrien, J. Colloid Interface Sci. 151, 144 (1992).
315. A. Mehreteab, G. Broze, and J. Rouse, AOCS Annual Meeting, Toronto, May 10–14 (1992).
316. A. Müller and M. Albrecht, Tenside 27, 399 (1990).
317. J. K. Borchardt and C. W. Yates, AOCS Annual Meeting, Toronto, May 10–14 (1992).
318. M. K. Bernett and W. A. Zisman, J. Phys. Chem. 63, 1911 (1959).
319. C. La Mesa and B. Sesta, J. Phys. Chem. 91, 1450 (1987)
320. (a) V. Glöckner, K. Lunkwitz, and D. Prescher, Tenside 26, 376 (1989). (b) DIN 53914 (1978).
321. (a) N. Funasaki and S. Hada, J. Colloid Interface Sci. 73, 425 (1980). (b) N. Funasaki and S. Hada, Chem. Lett. 717 (1979). (c) N. Funasaki and S. Hada, Bull. Chem. Soc. Jpn. 49, 2899 (1976).
322. F. Jost, H. Leiter, and M. J. Schwuger, Colloid Polym. Sci. 266, 554 (1988).
323. N. L. Jarvis and W. A. Zisman, J. Phys. Chem. 63, 727 (1959).
324. (a) K. Motomura, I. Kajwara, N. Ikeda, and M. Aratono, Colloids Surf. 38, 61 (1989). (b) N. Matubayasi, K. Motomura, S. Kaneshina, M. Nakamura, and R. Matuura, Bull. Chem. Soc. Jpn. 50, 523 (1977).
325. (a) Zhao Guo-Xi and Zhu Bu-Yao, Colloid Polymer Sci. 261, 89 (1983). (b) Zhu Bu-Yao and Zhao Guo-Xi, Hua Xue Tong Bao (Chem.) No. 6, p. 341 (1981).
326. T. Handa and P. Mukerjee, J. Phys. Chem. 85, 3916 (1981).
327. N. Nishikido, W. Mahler, and P. Mukerjee, Langmuir 5, 227 (1989).
328. H. M. Scholberg, R. A. Guenther, and R. I. Coon, J. Phys. Chem. 57, 923 (1953).
329. D. R. Taves, Nature 211, 192 (1966).
330. D. R. Taves, Nature 200, 582 (1968).
331. D. R. Taves, Nature 217, 1050 (1968).
332. D. R. Taves, Talanta 15, 1015 (1968).

333. W. S. Guy, D. R. Taves, and W. S. Brey, Am. Chem. Soc. Symp. Ser. 28, 117 (1976).
334. F. A. Ubel, S. D. Sorenson, and D. E. Roach, Am. Ind. Hyg. Assoc. J. 41, 584 (1980).
335. P. Venkateswarlu, L. Singer, and W.D. Armstrong, Anal. Biochem. 42, 350 (1971).
336. L. Singer and R. H. Ophaug, Anal. Chem. 49, 38 (1977).
337. P. Venkateswarlu, Anal. Biochem. 68, 512 (1975).
338. E. Kissa, Clin. Chem. 33, 253 (1987).
339. E. Kissa, Environ. Sci. Technol. 20, 1254 (1986).
340. D. Barceló, Anal. Chim. Acta 263, 1 (1992).

10
Toxicology and Environmental Aspects

10.1 TOXICOLOGY

The toxicology of organic fluorine compounds is unusual in the sense that most fluorine compounds are harmless, whereas some are among the most toxic nonproteins known. The structural differences between a toxic and a nontoxic compound are usually not extensive [1]. Organic fluorine compounds therefore have to be treated with caution unless proven to be nontoxic by toxicological testing.

The toxicity of organic fluorine compounds depends on their biochemical stability and the toxicity of metabolites. The carbon-fluorine bond is highly resistant to biochemical degradation and, as a consequence, perfluorocarbons are physiologically inactive. However, if a fluorochemical is metabolized, the toxicity of the metabolized fluorocompound depends on the structure of the *metabolites*. Monofluorinated alkanoic acids are toxic when the β-oxidation mechanism can produce monofluoroacetic acid [2]. Steric hindrance and branching of the alkane chain reduce toxicity (Table 10.1).

Monofluoroacetic acid is one of the most toxic compounds known. Its toxicity is related to blocking the citric acid cycle by hindering the enzyme acotinase. Compounds which can readily be hydrolyzed or oxidized to fluoroacetic acids, e.g., esters of fluoroacetic acid and fluoroethanol, are also highly toxic.

In addition to metabolism, *chemical reactions*, such as pyrolysis, can convert a nontoxic fluorine compound to a toxic one. Pyrolysis of some organic fluorine compounds can produce perfluoroisobutylene, $(CF_3)_2C{=}CF_2$, probably one of the most highly toxic fluorocompounds known [3].

TABLE 10.1 Toxicity of Monofluorocarboxylic Acids

Compound	LD_{50} male mouse (mg/kg)
$F(CH_2CH_2)_3CH_2COOH$	0.64
CH_3 $F(CH_2CH_2)_3CHCOOH$	1.1
CH_3CH_2 $F(CH_2CH_2)_3CHCOOH$	5.5
$CH_3CH_2CH_2$ $F(CH_2CH_2)_3CHCOOH$	75
CH_3CHCH_3 $F(CH_2CH_2)_3CHCOOH$	100

Source: Ref. 1.

The toxicity of *intermediates* used for synthesizing fluorinated surfactants also has to be considered. The toxicity of perfluoroalkylethyl iodides depends on the length of the perfluoroalkyl group. The inhalation toxicity LC_{50} value of $C_2F_5CH_2CH_2I$ is 400 to 500 ppm, that of $C_4F_9CH_2CH_2I$ 4000 ppm [4].

The biochemical effects of *fluorinated surfactants* are not completely understood. Perfluorocarbons are used as oxygen carriers in blood without toxic effects [5,6]. The toxicity of some fluorinated surfactants is so low that they have been tested in vivo as emulsifiers in blood substitutes and biomedical oxygen carriers (see Sec. 10.4). However, some fluorinated surfactants are considerably toxic although they are not metabolized or, if metabolized, produce presumably a nontoxic fluorocompound.

A discussion of the toxicity of fluorinated surfactants has to distinguish between the toxicity of (1) a pure surfactant, (2) a surfactant containing impurities, and (3) a commercial product consisting of a fluorinated surfactant and a solvent. The exposure to the fluorinated surfactant may be acute or chronic to small amounts for a long period of time. Both systemic effects and local effects have to be considered. Systemic effects result from absorption of a fluorinated surfactants by oral intake, inhalation, or skin penetration. Local effects are observed at the location of contact with a surfactant. Irritation of skin, eye, or nasal mucous membranes and sensitization of skin are typical local effects.

The intrinsic toxicity of some fluorinated surfactants has been related to their exceptionally *high surface activity*. The acute oral toxicity of tetraethylammonium perfluorooctanesulfonate has been reported to be considerable [LD_{50} = 190 mg/kg (Wistar rat)] although the surfactant does not irritate mucous membranes and is not a bactericide. At low concentrations the surfactant is not toxic to fish. However, when the concentration of the surfactant in water is increased, toxic effects appear at a concentration where the surface tension starts to drop markedly (see Aquatic Toxicity in Sec. 10.2) [7]. Surfactants are adsorbed on interfaces and interact with biological membranes, proteins, and enzymes [8]. Anionic surfactants form ionic adsorption complexes with proteins, whereas the interaction of nonionic surfactants is weak. This is in accord with the observed toxicity, which is much higher for anionic than for nonionic fluorinated surfactants.

The *lipophobicity* of fluorinated surfactants also may contribute to their toxicity. It is plausible that the lipophobicity of fluorinated surfactants in conjunction with a high surface activity and consequent strong adsorption may interfere with normal functioning of cells.

The interaction of fluorinated surfactants with cells is evidenced by the effect of fluorination on hemolytic activity of the surfactant. Fluorinated surfactants containing both fluorocarbon and hydrocarbon segments are less hemolytic than their hydrocarbon analogues. The hemolytic activity decreases with increasing fluorocarbon/hydrocarbon ratio (see Sec. 10.4). The undoubtedly complex mechanism of hemolysis is not fully understood [9]. It is usually believed that a hydrocarbon surfactant is adsorbed onto the erythrocyte membrane before it penetrates the cell and causes its disintegration. The weaker hemolytic activity of fluorinated surfactants has been explained by a low affinity of fluorocarbon chains for nonfluorinated materials and weak adhesion to the cells [9]. In contrast to this generally accepted view, it may be argued that the decrease in hemolytic activity is related not to weak adsorption but to strong adsorption. The fluorocarbon tails of the adsorbed fluorinated surfactant may render the blood cell lipophobic as well as hydrophobic and hinder penetration of the cell.

Although the lipophobicity of the surfactant appears to be beneficial for reducing hemolytic activity, lipophobicity of fluorinated surfactants may be detrimental to normal diffusion processes in cells.

Since most fluorinated surfactants are commercial products containing several components, the toxicity of *impurities* in fluorinated surfactants has to be considered. Commercial fluorinated surfactants are usually sold as solutions in an aqueous solvent [10]. In some cases, the solvent may cause more systemic or local toxic effects than the surfactant itself. The solvent and volatile impurities may dominate the toxic effects produced by inhalation. Nonionic surfactants with a poly(oxyethylene) hydrophilic chain may contain 1,4-dioxane, which has shown carcinogenic activity in some animal tests. 1,4-Dioxane is a by-product found in nonionic surfactants, regardless of whether the surfactants are fluorinated. However, the concentration of 1,4-dioxane in nonionic surfactants is carefully controlled and is usually very low (about 0.1% or less). Air monitoring has indicated that at a workplace where there are nonionic fluorinated surfactants containing about 0.1% dioxane, the 1,4-dioxane concentration in air would be below 1 ppm, well below the acceptable concentration limit of 25 ppm, the ACGIH threshold limit value (TLV) [11].

The acute toxicities of some *commercial fluorinated surfactants* marketed as solutions in an aqueous solvent are given in Table 10.2 [10].

A 10-dose oral subacute test in rats has been conducted for Zonyl TBS [11]. Zonyl TBS was administered by intragastric intubation to three groups of male rats, 10 rats per group, five times a week for 2 weeks in repeated dose levels of either 10, 100, or 1000 mg/kg. Half of the rats were sacrificed after the last dose, the other half after a 14-day recovery period. For the 10 and 100 mg/kg doses the mortality was 0/10. For the 1000 mg/kg dose the mortality was 9/10; only one rat survived the recovery period. The organic fluorine content of blood was analyzed by combustion in the oxyhydrogen flame. Kissa [12] found the organofluorine content of blood to increase linearly with the square root of the dose level (Fig. 10.1). A deviation from linearity at the highest dose level may have been related to the high mortality rate. Substantial amounts of organofluorine remained in the blood after the 14-day recovery period, indicating a relatively slow elimination rate, typical of anionic fluorinated surfactants. The sorption and elimination kinetics of fluorine in blood for a subchronic inhalation study of Zonyl TBS are discussed in Sec. 10.3.

TABLE 10.2 Toxicity of Zonyl Fluorinated Surfactants

			LD_{50} (g/kg)	
Name	Structure[a]	Solvent	Acute oral (rat)	Skin (rabbit)
Zonyl FSA	Anionic (25)(38)	Propanol-2	4.7	12.9
Zonyl TBS	Anionic (33)(3)	Acetic acid	1.9	2.1
Zonyl FSE	Anionic (14)(24)	Ethylene glycol	4.0	>22
Zonyl FSP	Anionic (35)(20)	Propanol-2	4.7	12.9
Zonyl FSC	Cationic (50)(25)	Propanol-2	6.1	22
Zonyl FSN	Nonionic (40)(30)	Propanol-2	13.8	12.9
Zonyl FSO	Nonionic (50)(25)	Ethylene glycol	>17	>22
Zonyl FSK	Amphoteric (47)(53)	Acetic acid	3.3	1.1

[a] The first number in parentheses indicates the percent solids, the second number the weight percent of solvent in the product. The difference between the sum of the two numbers and 100 is the percent water in the product.
Source: Ref. 10.

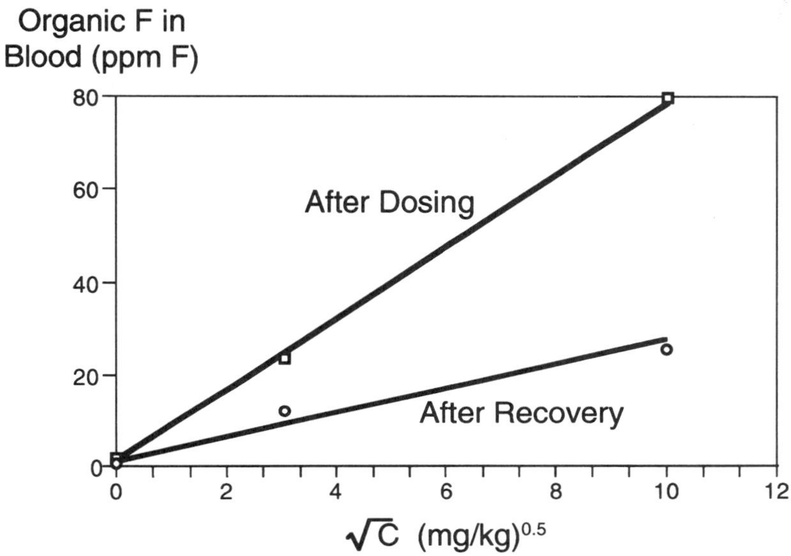

FIG. 10.1 Concentration of organic fluorine in rat blood as a function of the dose in a 10-dose oral subacute test. (□) 4-h recovery; (○) 14-day recovery. (From Ref. 12.)

Zonyl TBS has a low acute toxicity by inhalation. Groups of six male rats were exposed to dust atmospheres of Zonyl TBS in the free acid form for a single, 4-h period. All rats survived the 14-day postexposure period after the exposure to up to and including 2100 mg/m^3 [11].

Zonyl TBS is a moderate skin irritant. In contrast, the anionic fluorinated surfactant Zonyl FSA with a $-CH_2CH_2COOLi$ group is neither a skin irritant nor an eye irritant and is not a skin sensitizer. However, it is a good practice to avoid skin or eye contact with commercial grade fluorinated surfactants and flush the contaminated skin area with water, should a contact occur.

From a practical viewpoint the important question is, how does the toxicity of fluorinated surfactants compare to that of their hydrocarbon analogues? A general statement cannot be made because the toxicity of fluorinated surfactants varies greatly with their structure. Some fluorinated surfactants derived from natural products, such as carbohydrates or lipids, are sufficiently biocompatible to be considered for intravascular use (see Sec. 10.4). Clearly, a fluorinated surfactant per se is not necessarily more toxic than a nonfluorinated surfactant. The toxicities of surfactants with corresponding structures have to be compared.

Most fluorinated surfactants have been tested as commercial products containing a solvent and toxicity data for pure, well-characterized fluorinated surfactants are scarce. The acute oral toxicities of solvent-free fluorinated surfactants are listed with those of hydrocarbon surfactants in Table 10.3. The acute oral toxicity of the nonionic fluorinated surfactants listed is low and comparable to that of their hydrocarbon analogues. However, the data suggest that anionic surfactants when ingested are more toxic than their hydrocarbon counterparts. The toxicity of sodium carboxylates, e.g., sodium myristate, is relatively low. In contrast, ammonium perfluorooctanoate is moderately toxic when ingested and toxic when inhaled, with a 4-h ALC (approximate lethal concentration) in the rat of 800 mg/m^3 [11].

The local toxicity of anionic fluorinated surfactants depends on their structure, as can be expected. The surfactant FT-248, tetraethylammonium perfluorooctanesulfonate, is not a skin irritant or a eye irritant [7b]. Sodium perfluorooctanoate is a moderate skin and eye irritant [11,14].

It is important to recognize that fluorinated surfactants are used only in small quantities, usually 10–100 times smaller than the amounts of hydrocarbon-type surfactants. Most fluorinated surfactant applications are in industry, where the handling and disposal of fluorinated surfactants can be controlled. Hence a consumer is exposed much less to fluorinated surfactants than to hydrocarbon surfactants. It is reasonable to conclude that the toxicity of most fluorinated surfactants should not prohibit their proper use. Under usual application conditions an exposure to fluorinated surfactants can be kept well within acceptable limits.

TABLE 10.3 Acute Oral Toxicity of Surfactants with a Fluorocarbon or Hydrocarbon Hydrophobe

Surfactant	LD_{50} (rat) (g/kg)	Reference
Nonionic		
$R_fCH_2CH_2O(CH_2CH_2O)_nH$	13.8	11
$R_fCH_2CH_2O(CH_2CH_2O)_mH$	17.0	11
Alkyl polyoxyethylene ethers	0.87–>25	13
Anionic		
Ammonium perfluorooctanoate $C_7F_{15}COONH_4$	0.47, 0.54	11, 14
Sodium myristate $C_{13}H_{27}COONa$	>10	15
Tetraethylammonium perfluorooctanesulfonate $C_8F_{17}SO_3(C_2H_5)_4$	0.19	7
Sodium octanesulfonate $C_8H_{17}SO_3Na$	2.0 (mouse)	15

10.2 ENVIRONMENTAL ASPECTS

Release to the Environment

Fluorinated surfactants can affect the environment by their occurrence in air or water. Some fluorinated surfactants are sufficiently volatile to be present in air, some are introduced into the air as dust particles. It is believed that the presence of organic fluorine in blood (see Sec. 10.3) originates from the inhaled vapors or dust of fluorinated surfactants.

Fluorinated surfactants are used in small quantities and for this reason do not constitute a heavy burden on the environment. The largest concentrations of fluorinated surfactants can be found in the industrial environment, where they can be more readily contained. The population at large is not exposed to fluorinated surfactants in the same way as to hydrocarbon-type surfactants. Unlike hydrocarbon-type surfactants, fluorinated surfactants are not used in household detergents. Hence fluorinated surfactants do not come in frequent contact with the skin and are not inevitably ingested like small amounts of hydrocarbon-type surfactants.

Aquatic Toxicity

The aquatic toxicity of a chemical depends not only on its intrinsic toxicity but also on its biodegration rate. Chemicals which are toxic but are degraded at a rapid rate may not affect aquatic life. However, most fluorinated surfactants are biochemically stable and their aquatic toxicity is not reduced by degradation.

The aquatic toxicity of fluorinated surfactants has been studied by Knaack and Walther [16]. The biochemical oxygen demand of three fluorinated surfactants was measured by the Warburg method, using a mixed bacterial culture. The fluorinated surfactants tested were

(A) $R_fCF{=}CFSO_3Na$, where $R_f = C_6$ to C_9, average C_8

(B) $R_fCF{=}CF(OCH_2CH_2)_nOH$, where $R_f = C_6$ to C_9, average C_8; $n = 7$

(C) $R_fCON(CH_2CH_2OH)_2$, where $R_f = C_7$

Fluorinated surfactant (A) in concentrations below 100 mg/L did not affect the oxygen consumption of the bacterial culture significantly. Above the concentration of 100 mg/L the biochemical oxygen demand decreased drastically (Fig. 10.2). Fluorinated surfactants (B) and (C) were less toxic to the bacteria. The oxygen consumption was not inhibited until complete inhibition occurred at about 7500 mg/L (Fig. 10.3). The median inhibitory concentration values (IC_{50}) for bacterial cultures of the three fluorinated surfactants tested decreased in the order of increasing toxicity: (C) 6750 mg/L $<$ (B) 5650 mg/L $<$ (A) 110 mg/L. The anionic fluorinated surfactant, like anionic hydrocarbon-type surfactants, exhibited higher antibacterial activity than the nonionic surfactants tested. The high IC_{50} values of the two nonionic fluori-

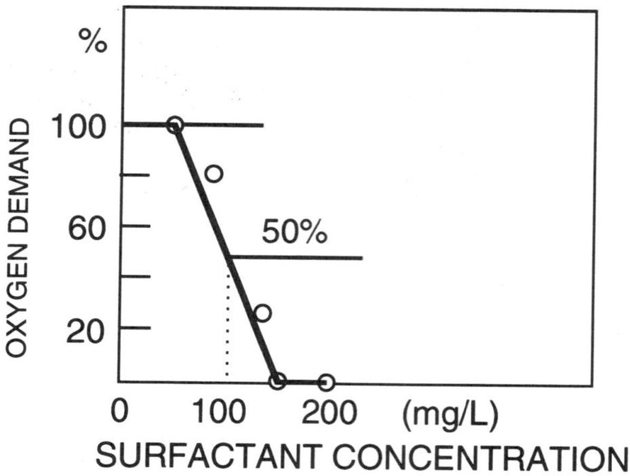

FIG. 10.2 Biochemical oxygen demand of the anionic fluorinated surfactant A. (From Ref. 16.)

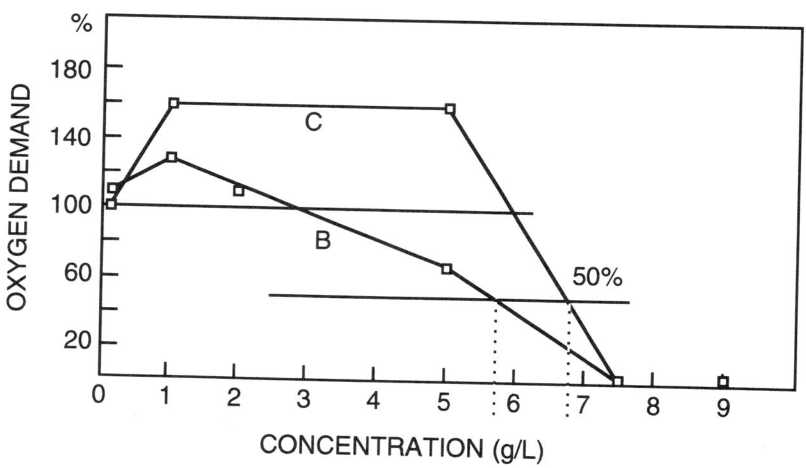

FIG. 10.3 Biochemical oxygen demand of the nonionic fluorinated surfactants B and C. (From Ref. 16.)

nated surfactants indicate that these surfactants in normal use should not affect aquatic bacterial flora.

Prescher et al. [17] studied the aquatic toxicities of four fluorinated surfactants to guppy (*Poecilia reticulata)* and green algae (*Monoraphidium griffithii*). The structures of the four fluorinated surfactants were

(I) $CF_3(CF_2)_nCF{=}CF(OCH_2CH_2)_7OH$, where $n = 4$ to 7

(II) $CF_3(CF_2)_nCF{=}CFCF_2SO_3Na$, where $n = 6$ to 7

(III) $CF_3(CF_2)_nCFHCOONa$, where $n = 4$ to 10

(IV) $CF_3(CF_2)_nCOONa$, where $n = 7$ to 10

Median lethal concentration (LC_{50}/96 h) values for *Poecilia reticulata* were 88, 11, 9, and 4 mg/L, respectively. Median inhibitory concentrations (IC_{50}) for *Monoraphidium griffithii* were 418, 42, 15, and 30 mg/L, respectively. The suggested maximum permissible concentrations are $\leqslant 0.1$, 0.2, $\leqslant 0.05$, and 0.05 mg/L, respectively.

With the exception of fluorinated surfactant (III), the fluorinated surfactants tested are less toxic to algae than to fish. Fluorinated surfactant (I) is less toxic to algae and to fish than the corresponding hydrocarbon-type nonionic surfactants (oxyethylated *n*-tetradecanol, with 10 to 15 EO units). The no-effect concentrations and the IC_{50} values for fluorinated surfactant (II) and for its nonfluorinated counterpart (sodium pentadecanesulfonate) are about the same, but the LC_{50} value for the hydrocarbon-type surfactant is about 10 times lower.

The aquatic toxicity of fluorinated surfactants to fish has been related to their high surface tension. The very sensitive orfes (*Leuciscus idus melanotus*) have tolerated as much as 20 mg/L of tetraethylammonium perfluorooctanesulfonate for 70 h without noticeable effects. However, when the concentration of the surfactant was increased to a level where the surface tension started to drop markedly, toxic effects were immediately observed [7].

Nonaquatic Biological Activity

The biological activity of nonionic fluorinated surfactants as herbicides, fungicides, and insecticides was studied by Sakakibara et al. [18]. The nonionic fluorinated surfactants were derived from alcohols of the structure $F(CF_2CF_2)_3CH_2OH$ or $H(CF_2CF_2)_{2\text{-}4}CH_2OH$ by adding 1 to 19 oxyethylene units. The nonionic ethylenediamine derivatives had the structure $CF_3(CF_2)_{2 \text{ or } 6}CONHCH_2CH_2NH_{2\text{-}0}(CH_2CH_2OH)_{0\text{-}2}$.

The 18 nonionic fluorinated surfactants studied exhibited hardly any significant herbicidal activity. Foliar spray application had only weak effects on plant life and the herbicidal activity was even weaker for fluorinated surfactants with a longer oxyethylene chain. None of the surfactants tested were strong fungicides. Based on these results, the authors concluded that the fluorinated surfactants tested can be used safely as adjuvants, such as emulsifiers and dispersants, for agricultural chemicals.

Some of the fluorinated surfactants tested were found to be insecticides, affecting especially the common housefly (*Musca domestica*) and carmine mite (*Tetranychus cinnabarius*). The mechanism of the insecticidal activity appeared to be suffocation of the insect, attributable to adhesion of the surfactant to the cuticle of the insect.

Biodegradation

The oxygen demand curves obtained by Knaack and Walther [16] (Figs. 10.2 and 10.3) suggested that surfactants (B) and (C) were at least partially biodegraded. The oxygen demand for these surfactants exceeded 100%, suggesting that the nonfluorinated hydrophile of these surfactants was degraded, unless the oxygen consumption was caused by impurities. Since surfactant (A) consisted of a perfluorinated group and a sulfonate hydrophile, the surfactant can be expected to be nonbiodegrable. However, Knaack and Walther could not verify this experimentally. The high antibacterial activity of surfactant (A) permitted biodegradation tests only at low surfactant concentrations, where accuracy is somewhat uncertain.

Prescher et al. [19] found no evidence of significant biodegradation. Their study included one nonionic and three anionic fluorinated surfactants (I, II, III, and IV; see structures above). The conclusion was based on biological oxygen demand (BOD), $KMnO_4$ consumption, ignition loss, and dehydrogenase activity. Prescher et al. [19] classified the fluorinated surfactants as biochemically not degradable or difficult-to-degrade substances.

Schröder [20] studied the biodegradation of an anionic, a cationic, and a nonionic surfactant. The anionic fluorinated surfactant, Fluowet PL 80, was found to be a mixture of a phosphonic acid, $C_nF_{2n+1}PO(OH)_2$, and a phosphinic acid, $(C_nF_{2n+1})_2PO(OH)$. The phosphinic component was adsorbed on activated sludge, whereas the phosphonic component of the fluorinated surfactant remained in the aqueous phase. Biodegradation of the perfluoroalkane chain did not occur, and no metabolites were detected [20a].

The cationic surfactant Fluowet L 3658 has the structure $C_nF_{2n+1}CHFCF_2N^+(CH_2CH_2OH)_3\ X^-$. In wastewater containing activated sludge, the cationic surfactant was adsorbed on the sludge. Deactivation of the sludge with sodium azide had no effect on the removal of the cationic surfactant from the aqueous phase, indicating that the sorption of the surfactant on the sludge did not involve biodegradation. In the absence of sludge, the cationic surfactant remained in the aqueous phase but biodegradation was not evident [20a].

Schröder [20b] used HPLC/MS/MS with a thermospray interface [20c] to detect, identify, and quantify metabolites of Fluowet OTN, a nonionic fluorinated surfactant with the structure $R_nF_{2n+1}CH_2CH_2(OCH_2CH_2)_mOH$. The biodegradation was limited to the poly(oxyethylene) hydrophile [20b]. The absence of fluoride ions indicated that the perfluorocarbon chain was not degraded.

The biodegradation of FC-171, a nonionic fluorinated surfactant, was measured by the modified ISO Standard Aerobic Shake Flask Test [21]. The test uses nutrient salt in a brine (seawater) solution in which the fluorinated surfactant is the only organic solute. No significant biodegradation was observed after 9 days, but after 57 days dissolved organic carbon (DOC) was reduced from 14.5 to 6.0 mg/L. Since the perfluoro chain was probably not degraded, the results indicate a 90% degradation of the oxyethylated part of the surfactant molecule.

Removal

Since fluorinated surfactants are not biodegradable, their removal from wastewater is an important practical problem. The results obtained by Schröder [20a] indicate that ionic fluorinated surfactants are strongly adsorbed on sludges and sediments and their concentrations in wastewater are probably low.

Prescher et al. [22] investigated the removal of four fluorinated surfactants by chemical and physical means. The structures of the fluorinated surfactants were the same as in their previous study [17]. The study showed that the nonionic surfactant can be removed by flocculation with aluminum oxychloride, by chlorination, or by adsorption on activated carbon. For the anionic fluorinated surfactant, adsorption on activated carbon or on a resin, such as Wofatit EA_{60}, is the most effective removal method.

A treatment of synthetic sewage containing 6.0 mg/L FC-171, a nonionic fluorinated surfactant, with activated sludge reduced the toxicity of the surfactant to water flea (*Daphnia magna*). The 48-h LC_{50} was increased from 0.26 to 2.4 mg/L (14 days of continuous operation before sampling).

Staude et al. [23] evaluated the removal of tetraethylammonium perfluorooctanesulfonate from rinse solutions of the electrochemical plating industry by hyperfiltration.

Fluorinated surfactants have unique properties and are therefore indispensable. A potential effect on the environment can be reduced by (1) using synergism with hydrocarbon-type surfactants to minimize the concentration of fluorinated surfactant where feasible and (2) removing fluorinated surfactants from wastewater at industrial sites by adsorption or converting the surfactant by partial biodegradation to physiologically inert substances.

10.3 PHYSIOLOGY: SORPTION, METABOLISM, AND EXCRETION

Taves et al. [24,25] observed that human blood serum contains organic fluorine, in addition to inorganic fluoride. The presence of organic fluorine was related to fluorinated surfactants which are significantly volatile and absorbed in the body by inhalation. Perfluorooctanoic acid is known to be adsorbed on protein in blood [26,27]. The blood of workers handling ammonium perfluorooctanoate was found to contain from 1 to 71 ppm organic fluorine [28]. The long retention times of fluorinated surfactants in the body have inspired efforts to minimize exposure to fluorinated surfactants by reducing their volatility. However, the elimination rate of fluorinated surfactants from the body by expiration is also related to volatility. Hence, the main benefit of using less volatile fluorinated surfactants may be a decrease in sorption but not necessarily a shorter retention time of organic fluorine. It should be kept in mind that fluorinated surfactants are not the only fluorine-containing substances which may be present in air.

The sorption and elimination kinetics of a fluorinated surfactant in blood has been investigated by Kissa and Kinney [29]. An inhalation subchronic study on Zonyl TBS, a partially neutralized ammonium salt of telomer sulfonic acids, used rats as the test animals. Airborne Zonyl TBS was formed using two-stage dust generators. Zonyl TBS powder was agitated in the reservoirs of the generator and swept through the elutriators into the rat exposure chambers. The total fluorine content of rat blood was determined by combustion in an oxyhydrogen torch [30–32] and analysis of the combusted analyte by the fluoride ion–selective electrode [33]. Inorganic fluoride in blood was determined by an analyte addition method using a fluoride ion–selective electrode [34].

The concentration of organic fluorine in blood, C_b, increased exponentially with increasing concentration, C_a, of the fluorinated surfactant in the air inhaled (Fig. 10.4):

$$C_b = K_s C_a^n \tag{1}$$

where K_s is the sorption coefficient. The value of n was found to be in the 0.5 to 0.6 range. A plot of the residual organic fluorine concentration in blood versus the concentration of the fluorinated surfactant in air is reasonably linear (Fig. 10.5).

The elimination of fluorocompounds from the body is usually assumed to be a first-order kinetic process. The half-life of a fluorocompound in the blood or body is used to characterize retention [5,35,36]. However, the first-order elimination plot for Zonyl TBS exhibited curvature, suggesting a departure from uncomplicated kinetics (Fig. 10.6). The elimination of fluorinated surfactants from blood can be viewed as a first-order process with a postexposure time, t, dependent rate constant, K_t.

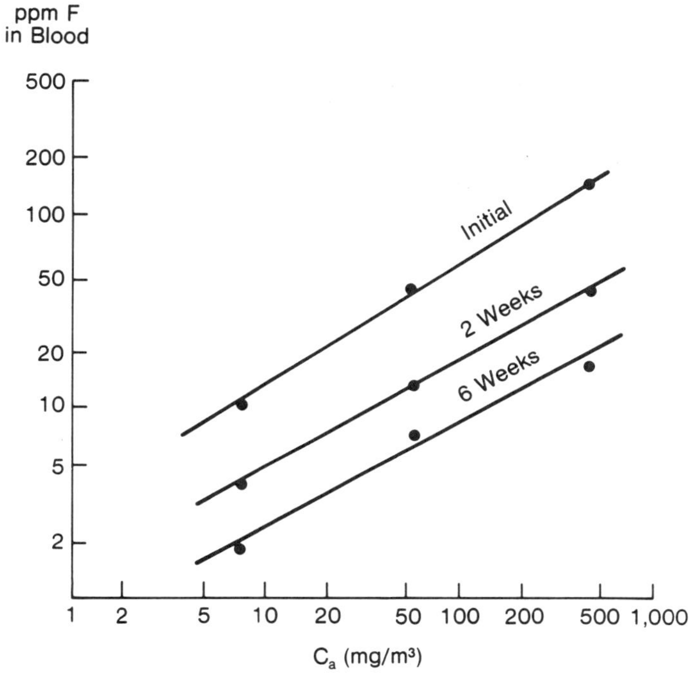

FIG. 10.4 Organofluorine concentration in rat blood as a logarithmic function of fluorinated surfactant (Zonyl TBS) concentration in air inhaled during exposure. 0, 2, 6 weeks of postexposure time. (From Ref. 29. Reproduced by permission of Carl Hanser Verlag.)

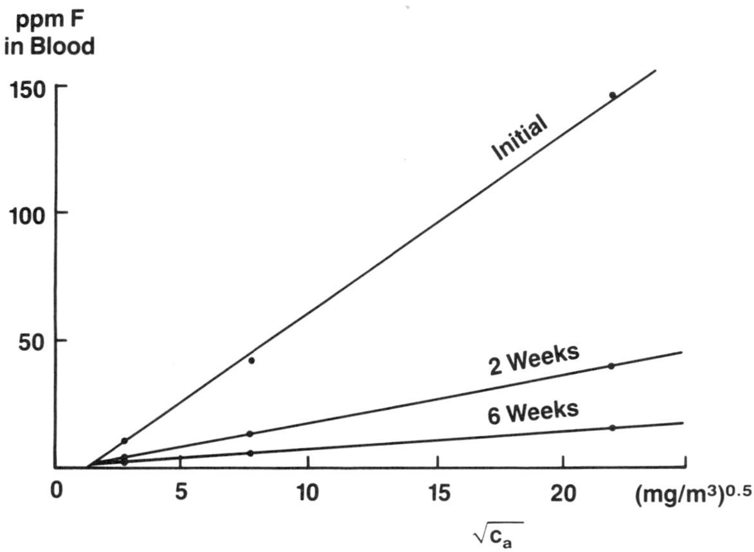

FIG. 10.5 Organofluorine concentration in rat blood versus the square root of fluorinated surfactant (Zonyl TBS) concentration in air inhaled during exposure. 0, 2, 6 weeks of postexposure time. (From Ref. 29. Reproduced by permission of Carl Hanser Verlag.)

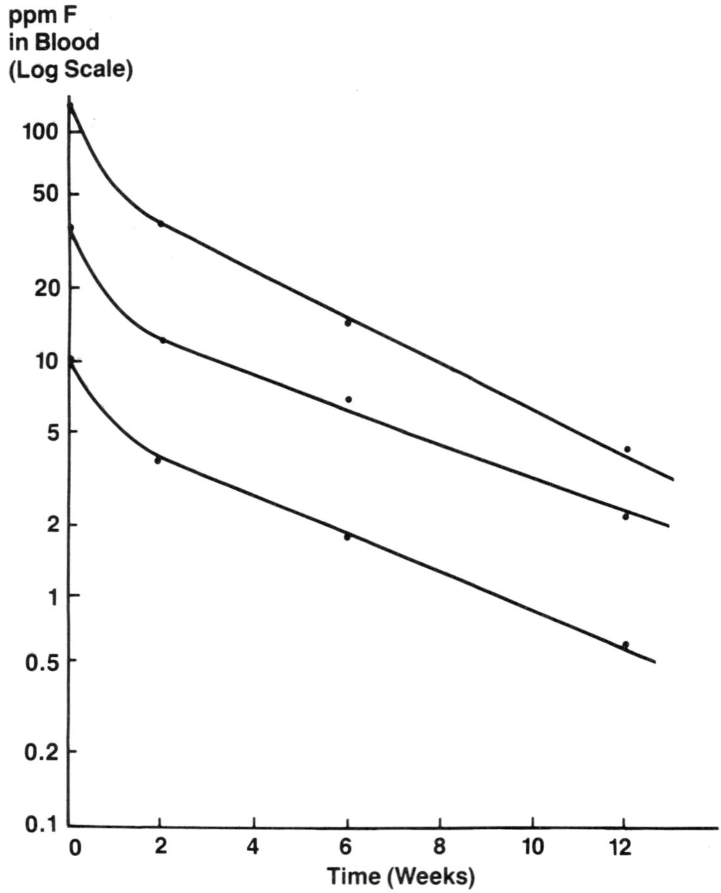

FIG. 10.6 A first-order plot of fluorinated surfactant (Zonyl TBS) elimination from blood. (From Ref. 29. Reproduced by permission of Carl Hanser Verlag.)

$$\log(C_b/C_0) = K_t t \qquad (2)$$

where

$$K_t = K_e/t^{0.5} \qquad (3)$$

and K_e is the elimination rate constant for the first-order process (Fig. 10.7).

Kissa and Kinney explained the deviation from simple first-order kinetics by adsorptive site-dependent variation of adsorptive bonding, heterogeneity of the telomer-derived fluorinated surfactant, and involvement of other organs.

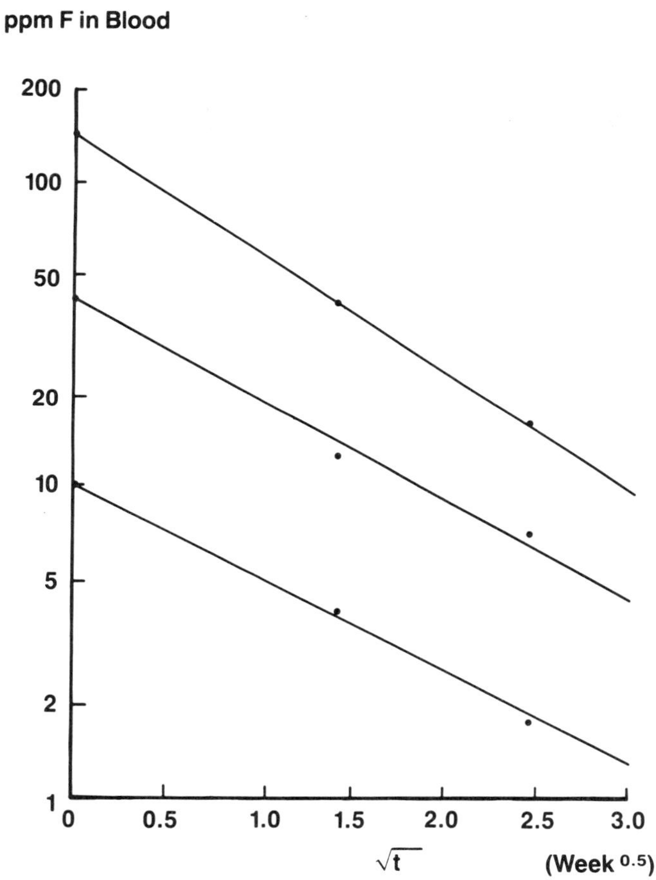

FIG. 10.7 A plot of the logarithm of organofluorine concentration in rat blood versus the square root of postexposure time. (From Ref. 29. Reproduced by permission of Carl Hanser Verlag.)

Perfluorooctanoic acid adsorbs on protein in blood [26]. It is reasonable to assume that Zonyl TBS, a partially neutralized fluorinated alkanesulfonic acid, may also be adsorbed on protein in the blood. The strength of the adsorptive bond may depend on the particular adsorption site and the less strongly held fluorinated surfactant molecules may be eliminated faster than the more strongly held ones.

The elimination kinetics may be complicated by the heterogeneous composition of the fluorinated surfactant. Zonyl TBS is a mixture of fluorinated surfactants differing in their chain length and conceivably in their diffusion rate. The elimination kinetics of fluorinated surfactants from blood are also complicated by the involvement of other organs, mainly the liver and the spleen, in the storage and transport of the fluorinated surfactant.

MacNicol and Robertson [37] have cautioned that perfluorochemicals cannot be assumed to be inert in biological systems. Kissa and Kinney [29] found, however, that the presence of fluorinated surfactant–related organic fluorine in blood does not affect the inorganic fluoride concentration significantly (Table 10.4). The increase in inorganic fluoride concentration in blood was barely significant, even when the organic fluorine concentration in blood exceeded 2000 mg/L after an exposure to very high fluorinated surfactant concentrations in air. The very slightly elevated inorganic fluoride concentration in blood returned to its initial level shortly after the exposure.

Biological inertness and biocompatibility are essential when considering the use of fluorinated surfactants in biomedical oxygen-carrying emulsions (Sec. 10.4).

10.4 FLUOROCHEMICAL EMULSIONS FOR BIOMEDICAL OXYGEN TRANSPORT

Blood Substitutes and Biomedical Oxygen Carriers

Transfusion with whole blood has several limitations and problems [38]. The storage stability of donated whole blood is limited. Blood must be stored at 4°C or frozen, but even under favorable conditions the lifetime of refrigerated blood is only 5–6 weeks. Therefore, a blood reserve must be replenished con-

TABLE 10.4 Inorganic (Ionic) Fluoride in Rat Blood

Exposure: Zonyl TBSA concentration in air (mg/m^3)	Post-exposure time (weeks)	Organofluorine concentration in blood (ppm)	Inorganic F concentration in blood (ppm)
0	0	1.4	0.016
0	2	0.7	0.007
0	2	0.7	0.025
0	2	1.1	0.019
480	0	112	0.034
480	0	140	0.055
480	0	168	0.063
480	2	41	0.020

Source: Ref. 29. Reproduced by permission of Carl Hanser Verlag.

tinuously. It is not practical to build up a large blood reserve for an unforeseeable natural disaster or war because most of the stored blood would have to be discarded from time to time.

The transfusion with whole donated blood is complicated by the intergroup incompatibility of red blood cells. About 20 antigenic systems have been identified and are considered for blood type matching. Crossmatching of blood types requires time but hypovolemia must be treated as rapidly as possibly because the time a patient remains hypovolemic is critical for recovery. The incompatibility of blood groups complicates the supply and storage of donated blood as well. In a large-scale emergency situation, such as a natural disaster or war, large volumes of blood may be needed and shortages of particular donor blood types may develop. A significant number of patients has so many antibodies that they cannot receive foreign red cells [38].

An additional problem with transfusions of donated blood is the potential risk of transmitting bacterial or viral diseases, such as hepatitis or the human immunodeficiency (HIV) virus. Blood donors have to be carefully screened for their past and present diseases to reduce the risk of disease transmittal.

Blood substitutes are needed for patients who refuse blood transfusion for religious reasons, in spite of their critical condition.

Because of the limitations and complications with transfusions of whole blood, a biocompatible synthetic blood substitute is an important research objective. The blood substitutes, used to supplement blood, can be divided into two classes: (1) plasma substitutes used to restore the blood volume and (2) oxygen carriers which not only can correct blood volume deficits but can transport respiratory gases as well. Although hypovolemia is usually the life-threatening condition caused by a massive hemorrhage, a supply of oxygen is essential in case of continuing hemorrhage. The administration of a colloidal blood plasma substitute can restore the normal blood volume but is insufficient when oxygen supply has to be restored as well. Oxygen-carrying resuscitation fluids have therefore been developed. Hemoglobin solutions [38] and emulsions of fluorocarbons [5,6,38–44] have been investigated.

There are differences and similarities between the two categories of blood substitutes. Hemoglobin or synthetic metal chelates combine oxygen chemically, whereas fluorocarbons dissolve oxygen without a chemical reaction. Both hemoglobin and perfluorocarbons can transport oxygen to the tissues but blood substitutes derived from hemoglobin or modified hemoglobin have oxygen sorption and desorption characteristics of whole blood [28]. However, blood substitutes based on modified hemoglobins are mainly of interest for acute blood volume restoration, because of their relatively short residence times in blood [45].

Fluorochemical blood substitutes differ from blood in several important aspects. Oxygen uptake in fluorochemical emulsions increases linearly with the partial oxygen pressure pO_2, unlike the S-shaped oxyhemoglobin dissociation curve (Fig. 10.8) [42]. Desorption of oxygen is rapid, because oxygen is not chemically bonded but is dissolved in fluorochemical emulsions. Fluorochemical particles in properly prepared emulsions have a diameter smaller than 0.3 μm, about 70 times smaller than red blood cells. The fluorochemical particles can pass through fine capillaries and emboli [46–49] which are impermeable for the much larger red blood cells. Because of the small particle size, the surface area of the fluorochemical particles in the emulsion is large. The rapid desorption of oxygen from the emulsion and the large surface area of the fluorochemical particles facilitate the delivery of oxygen to tissues [50]. Fluorochemical emulsions are therefore of considerable interest for cytotoxic treatments and diagnostic procedures.

Development of more concentrated injectable perfluorochemical emulsions has extended their diagnostic and therapeutic applications in medicine. Fluorochemical oxygen carriers have a considerable potential for treating myocardial and cerebral ischemia. Fluorochemical emulsions enhance the effectiveness of radiation therapy of cancer by delivering oxygen to ischemic areas of tumors [51]. Fluorochemical oxygen carriers containing bromine are radiopaque. 1-Bromoperfluorooctane has been tested as a contrast agent in x-ray tomography and for magnetic resonance imaging (MRI) [52–56]. A very promising poten-

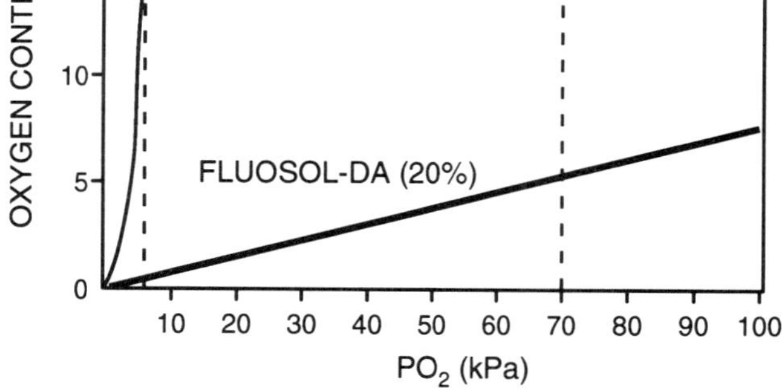

FIG. 10.8 The oxygen content of normal human blood and Fluosol-DA 20% as a function of oxygen partial pressure. (From Ref. 61. Reproduced by permission of Ellis Horwood Ltd.)

tial application of fluorochemical emulsions is in the perfusion and preservation of donated organs [39,57–61].

Fluorochemical Oxygen Carriers

Fluorochemicals (Table 10.5) are attractive as biomedical oxygen carriers for several reasons [5,6,39,42,43]. Perfluorocarbons are inert, available in large quantities, and pathogen free. The solubility of oxygen and carbon dioxide in fluorochemicals is related to weak intermolecular forces. Within a homologue series, the solubilities of oxygen and carbon dioxide decrease with increasing molecular weight and molal volume of the perfluorocarbon (Fig. 10.9) [42,62]. For a given molecular weight the linear fluorocarbons dissolve more oxygen than fluorocarbons with a cyclic structure. This is contrary to earlier beliefs that cyclic fluorocarbons are to be preferred.

The solubility of oxygen in dilute emulsions parallels the solubility of oxygen in neat fluorocarbons [63]. Perfluorocarbons dissolve oxygen, about 30 to 50 mL O_2/100 mL at 760 torr and 37°C. However, solubility of oxygen in the commercial perfluorocarbon emulsions is only about 0.8 to 8 mL $O_{2/100}$ mL

TABLE 10.5 Examples of Fluorocarbons Used in Biomedical Oxygen-Carrying Emulsions

	$N(C_3F_7)_3$	$N(C_4F_9)_3$
Mol. wt. 462 F-decalin (FDC) |PPS|	Mol.wt. 521 F-tripropylamine (FTPA)	Mol. wt. 671 F-tributylamine (FTBA) |FC-43|
Fluosol-DA : FDC/FTPA 70/30		
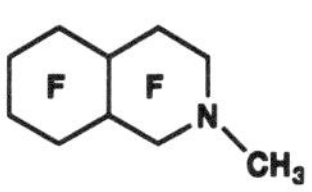	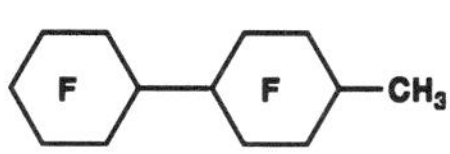	
Mol. wt. 495 F-*N*-methylisoquinoline (FMIQ)	Mol. wt. 595 F-*N*-(4-methylcyclohexyl)piperidine	
$C_8F_{17}Br$	$C_4F_9CH{=}CHC_4F_9$	$C_6F_{13}CH{=}CHC_6F_{13}$
Mol. wt. 499 F-octylbromide (PFOB)	Mol. wt. 464 bis-(F-butyl)ethene (F-44E)	Mol. wt. 664 bis(F-hexyl)ethene (F-66E)

Molecular weight (code name) \|trade name\|
Source: Ref. 42. Reproduced by permission of Ellis Horwood Ltd.

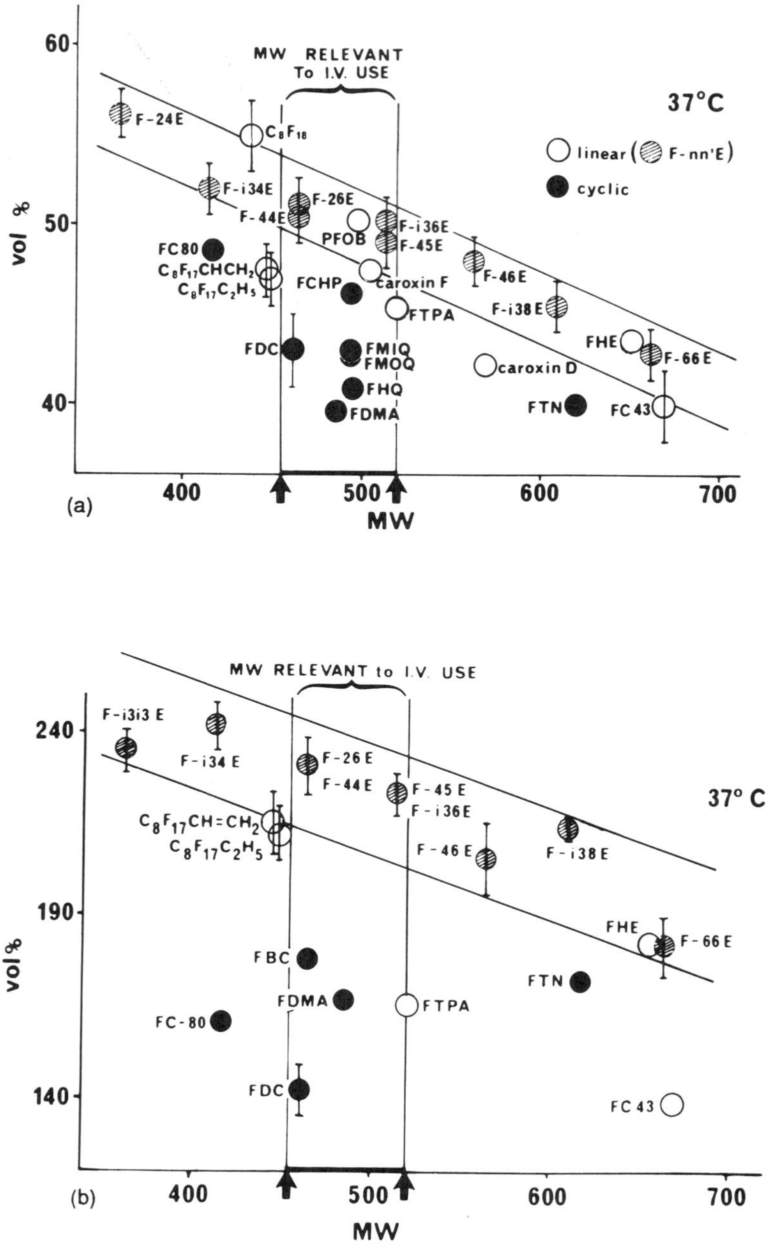

FIG. 10.9 Solubiliy of oxygen (a) and carbon dioxide (b) in various fluorocarbons tested for biomedical use. (From Ref. 42. Reproduced by permission of Ellis Horwood Ltd.)

[64,65]. The concentration of oxygen in a perfluorocarbon emulsion is much lower than the oxygen concentration in saturated arterial blood [66]. The saturation concentration of oxygen in blood with a normal hemoglobin content is about 20 mL O_2/100 mL at about 120 torr. At this pressure, Fluosol-DA, a fluorocarbon-water emulsion, can dissolve about 1.2 mL $O_{2/100}$ mL. Although Fluosol-DA can transport less oxygen than whole blood, the solubility of oxygen in Fluosol-DA is about three times higher than that in plasma [61]. The delivery of oxygen to tissues depends, in addition to the solubility of oxygen in the carrier, on the release rate of oxygen from the carrier. Perfluorocarbon emulsions release oxygen more readily than blood because fluorocarbon emulsions, unlike blood, dissolve oxygen without a chemical reaction. Hence perfluorocarbon emulsions can deliver substantial amounts of oxygen to tissues if the patient inhales supplementary oxygen [67].

In order to be suitable for in vivo application, fluorochemical oxygen carriers have to meet several stringent specifications. The solubility of oxygen, chemical and biochemical inertness, consistent availability in high purity, nontoxicity, a sufficient retention time in the circulating blood, a reasonably fast excretion from the body, and a structure favorable for emulsion stability are the most important and desirable properties. The excretion of the fluorochemical from the body occurs mainly by expiration through the lung [68,69]. The retention time of the perfluorocarbon in blood depends on the in vivo stability of the perfluorocarbon particles in circulation [70]. Tsuda et al. [71] have proposed that particles of a perfluorocarbon emulsion in the vascular system are deposited in tissues of the reticuloendothelial system, such as in liver and spleen, where they are stripped of their surfactant and moved through the cell membranes to blood vessels and adipose tissues. The diffusion rate of the perfluorocarbon across cell membranes is determined by the lipophilicity of the perfluorocarbon [71] and the stability of the emulsion [70]. Lipoproteins in blood transport the perfluorocarbons to the lung, where they are excreted into the expired air. The volatility of the fluorochemical is therefore important. The logarithm of the rate constant for expiratory elimination is inversely related to the boiling point of the fluorocarbon [35] (Fig. 10.10) and increases with increasing vapor pressure of the fluorocarbon (Fig. 10.11) [42,72]. Since the volatility of a compound depends not only on the size of the molecule but also on the shape, it was originally believed that cyclic perfluorochemicals would be most suitable for blood substitutes. Riess and Le Blanc [42] have argued, however, that the molecular weight, and not the molecular shape, is the dominant factor in determining the retention time of the fluorochemical in the body. Unfortunately, the retention time of the fluorochemical oxygen carrier in the body and the stability of its emulsion are conversely related [42]. Both the excretion rate and emulsion stability depend on the vapor pressure of the fluorocarbon. Excretion by exhalation through the lungs is facilitated by the

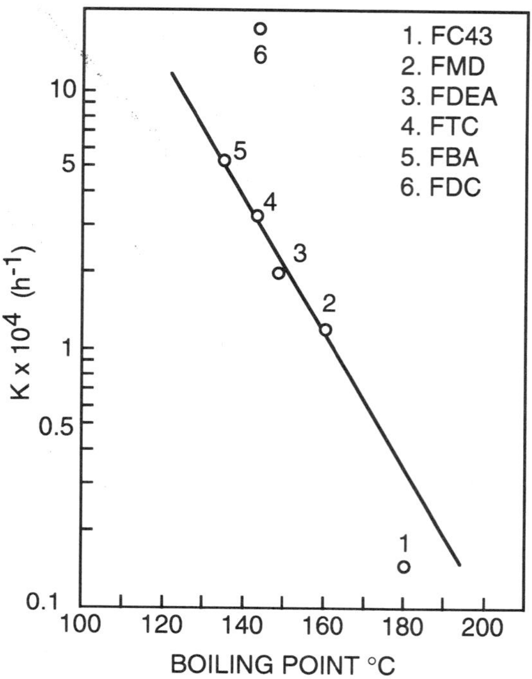

FIG. 10.10 The logarithm of rate constant of the expiratory elimination rate versus the boiling point of the fluorocarbon in rats given 4 g/kg of fluorocarbon intravenously. (From Ref. 72.)

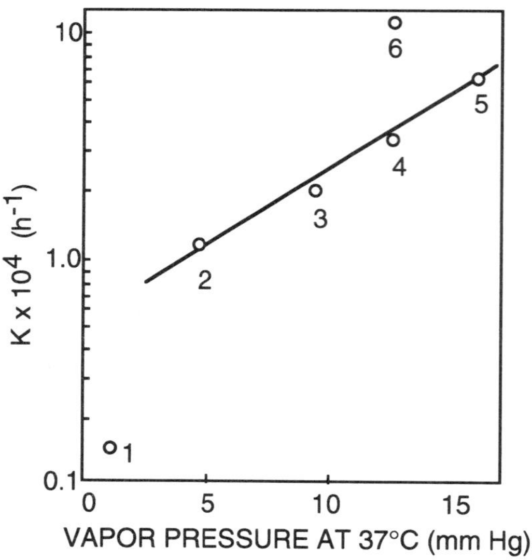

FIG. 10.11 The logarithm of rate constant of the expiratory elimination rate versus the vapor pressure of the fluorocarbon. Symbols are the same as in Fig. 10.10. (From Ref. 72.)

volatility of the fluorocarbon but a higher vapor pressure enhances molecular diffusion and increases emulsion instability.

Volatility and excretion require a low molecular weight but for emulsion stability a larger molecule is needed [42]. Increasing molecular weight decreases the volatility of the fluorochemical but increases emulsion stability. The effect of a fluorochemical on the stability of its emulsion is believed to depend largely on its molecular weight [42]. The volatility and stability requirements limit the molecular weight of fluorochemicals suitable for intravascular use to a range between 460 and 520 [42]. Fluorochemicals of molecular weight 460 are too volatile and may cause embolism, whereas fluorochemicals of molecular weight above 520 have too long retention times in organs. Examples of fluorochemicals used for preparing oxygen-carrying emulsions are shown in Table 10.5.

Perfluorodecalin (FDC), used in the first generation of fluorochemical blood substitutes, has a half-retention time in organs of only 6 days but the stability of its emulsions is insufficient [5]. Perfluorotripropylamine is added to perfluorodecalin in Fluosol-DA to increase emulsion stability. However, the retention time of perfluorotripropylamine in the body is longer than that of perfluorodecalin. Fluosol-DA is stored frozen and diluted with two aqueous solutions containing electrolytes and additives.

The emulsions of perfluorotributylamine (Fluosol 43) are more stable and can be stored refrigerated at 1 to 10°C [42,43] but the retention time in organs is very long [69,73]. Perfluorotributylamine is eliminated from blood probably by storage in tissues. At a dose of 4 g/kg in rats, the half-life of perfluorotri-*n*-butylamine in the body has been estimated to be 900 days.

1-Bromoperfluorooctane (perfluorooctyl bromide, PFOB) has several advantages over other oxygen carriers in a blood substitute [56,62]: (1) The residence time in tissues is relatively short [74,75]. The residence time of 1-bromoperfluorooctane is shorter than expected from the relationship between retention times and molecular weights [42,76]. (2) The bromine substitution increases the solubility of oxygen in a perfluoroalkane [62,77]. (3) The bromine atom permits a convenient determination of 1-bromoperfluorooctane in tissues by neutron activation. (4) 1-Bromoperfluorooctane is readily available in pure form via 1-iodoperfluorooctane by telomerization of tetrafluoroethylene, or as a mixture of normal and isoperfluorooctyl bromide by electrochemical fluorination. (5) 1-Bromoperfluorooctane forms stable highly concentrated emulsions [78]. (6) 1-Bromoperfluorooctane is radiopaque [52–56] and is useful for diagnostic applications [55,56,78–84].

1-Bromoperfluorooctane emulsified with egg yolk phospholipids as a surfactant did not exhibit any toxic effects in the liver, spleen, and bone marrow [85].

Bis(perfluoro-*n*-butyl)ethene (F-44E) can be obtained from pure products, *n*-butyl iodide and ethylene, by a two-step process [6,42,86]. The double bond of the molecule has raised the question of chemical stability, but in biological conditions no evidence of chemical reactions has been found. Animal tests have indicated satisfactory biocompatibility. Bis(perfluorohexyl)ethane (F-66E), however, has caused some toxic effects [85].

The intravascular use of fluorochemicals demands the absence of toxicity and carcinogenic, mutagenic, or teratogenic effects. The toxicity of fluorocarbons is largely affected by their purity. Most fluorocarbons are toxic, unless carefully purified. Tissue culture assays are needed to test for toxicity before a fluorochemical can be considered for biomedical application. Perfluorodecalin and perfluorotripropylamine have been used clinically [35,41,87–89].

Fluorochemical Emulsions

Perfluorochemical oxygen carriers are not soluble in water. Therefore, perfluorochemicals cannot be administered in the pure form but have to be converted to an aqueous emulsion. A surfactant, selected for its effectiveness and biochemical compatibility, serves as an emulsifier. Osmolarity and oncotic pressures are adjusted by adding electrolytes and oncotic agents, such as hydroxyethyl starch. Nutrients, thrombolytic agents, therapeutic agents, and other additives may be included in the emulsion, depending on the particular clinical application of the emulsion. Fluosol-DA and Fluosol 43 are fluorochemical emulsions manufactured by the Green Cross Corporation in Japan, Ftorosan has been made in the Soviet Union and Emulsion No. 2 in China.

Fluosol-DA 20%, the first of the commercial perfluorochemical blood substitutes , contains (% w/v) perfluorodecalin 14.0, perfluorotripropylamine 6.0, Pluronic F-68 2.7, yolk phospholipids (lecithin) 0.4, potassium oleate 0.032, glycerol 0.8, hydroxyethyl starch 3.0, NaCl 0.60, KCl 0.034, $MgCl_2$ 0.020, $CaCl_2$ 0.028, $NaHCO_3$ 0.210, and glucose 0.180 [41]. Fluosol-DA has been tested clinically as an oxygen-carrying blood substitute [89] and has been approved by the Food and Drug Administration for supplying the myocardium with oxygen during percutaneous transluminal coronary angioplasty.

Fluosol 43 [73] contains 20% w/v perfluoro-tri-*n*-butylamine as the oxygen carrier and 2.56% Pluronic F-68 as the emulsifier; the other additives have the same composition as in Fluosol-DA. Riess and Le Blanc [90] have advised against clinical use of Fluosol 43 because of the long retention time of perfluorotri-*n*-butylamine in the body.

The F-44E emulsion, Therox [91,92], contains 24% v/v (42% w/v) F-44E, egg yolk lecithin as the emulsifier, and an isotonic buffer. The half-life of this emulsion in blood is about 2.5 h and the dwell time of F-44E in tissues is about 7 days [90], compared to a blood half-life of 12 h for Fluosol-DA.

The energy needed for emulsification of the perfluorochemical oxygen carrier can be provided by high-pressure homogenization or by sonication. The first step in the high-pressure emulsification process is a dispersion, prepared by high-speed agitation of the perfluorochemical in a surfactant solution. The coarse dispersion is passed through a high-pressure homogenizer until the particle size is reduced to the desired range, about 0.1–0.2 μm. Oxidative degradation of some components, such as lecithin and nonionic hydrocarbon-type surfactants, is prevented by blanketing with nitrogen.

Emulsification by sonication needs a low temperature, close to 0°C. The emulsion is blanketed with carbon dioxide to prevent oxidation and fluoride formation. Sonication permits working on a smaller scale than high-pressure homogenization but the particle size is usually coarser and the size distribution wider [5].

The toxicity of perfluorochemical emulsions is to some extent related to an increase in particle size [35]. Kabal'nov et al. [93–95] have attributed the progressive increase in particle size to Ostwald ripening. The diameter profiles for the emulsified perfluorocarbon droplets in Fluosol-DA and Fluosol 43 have been determined by sedimentation field-flow fractionation [96].

Sterilization of the emulsion by heat can affect the stability of the emulsion. Sterilization of the prepared emulsion by autoclaving can cause coarsening of the emulsion and fluoride formation. Coarsening of the emulsion during sterilization can be avoided by sterilizing all components prior to emulsification but this is a less practical process.

The fluorochemical content of the emulsion can be determined by the density of the emulsion or by analyzing the emulsion for fluorine. The quantitative analysis of fluorochemicals in organs is usually done by gas chromatography [97]. The organs or tissues are homogenized in water. Ethanol is added to break the emulsion. The mixture is centrifuged, the upper layer removed, and the lower layer extracted with F-113. The extract is dried, an internal standard added, and a sample injected into a gas chromatograph.

The determination of the fluorochemical oxygen carrier in the emulsion or organs by gas chromatography assumes that the extraction of the fluorochemical is quantitative. Alternatively, the flurorochemical content of the emulsion can be determined by combustion in an oxygen Parr bomb or, preferably, in a Wickbold oxyhydrogen torch [20,23]. A combustion method can determine fluorinated surfactants which are not sufficiently volatile for gas chromatography. An oxyhydrogen flame combustion method [32] has been used successfully for the analysis of fluorochemical emulsions and fluorochemicals in blood and organs.

The intravascular persistence of perfluorotributylamine in blood has been measured by ^{19}F NMR [73]. The detection limit of 10^{-4}M can be reached with ^{19}F NMR Fourier transform spectrometry.

Surfactants

The properties of a perfluorochemical emulsion depend critically on the surfactant used for emulsification. A surfactant used in fluorochemical blood substitutes has to meet several criteria: (1) provide a fine stable emulsion; (2) be nontoxic, nonmutagenic, and nonhemolytic; (3) be compatible with blood and endothelial cells; (4) be pharmacologically, physiologically, or biochemically inactive; and (5) be excreted either unchanged or in the form of harmless metabolites [41].

The surfactants used in the first-generation fluorochemical blood substitutes were ordinary soaps (potassium oleate), egg yolk phospholipids (lecithin), or Pluronic F-68. These emulsifiers are used individually or as their mixtures (Table 10.6)

Pluronic F-68, a nonionic block polymer, consists of a polyoxypropylene segment located between polyoxyethylene segments. Pluronic F-68 lowers the interfacial tension between perfluorodecalin and water to only 31 mN/m. The stability of a perfluorodecalin–Pluronic F-68 emulsion is therefore not based on a low interfacial tension but probably on a steric stabilization mechanism. The coarsening of perfluorocarbon emulsions, apparently by the Ostwald ripening mechanism [93–95], has been inhibited by the addition of high-boiling polycyclic perfluorocarbons [98,99]. The third component is believed to form a retaining film at the fluorocarbon-water interface and hinder the transmission of the fluorocarbon into water [100]. Meinert et al. [100] explained the stabilizing effect of perfluorodimorpholinoalkanes and semifluorinated alkanes by the enrichment at the interface between the fluorochemical oxygen carrier and a hydrocarbon-type nonionic surfactant, such as Pluronic F-68. Meinert et al. designated such partially fluorinated compounds "interfacial active compounds." The additives have a lower vapor pressure and a higher molecular weight than the fluorochemical oxygen carrier.

The biocompatibility of Pluronic F-68 is questionable, if not inadequate [67,101,102]. Although intravenously infused Fluosol-DA has been reported to be innocuous in rats [103], adverse effects of Fluosol-DA in vivo have been observed and related to Pluronic F-68 or its impurities [67,89,104,105]. Transitory anaphylactoid reactions observed in clinical studies have indicated that more biocompatible emulsifiers are needed [89,106].

Lecithins, found in egg yolk, are natural products biocompatible with blood. The storage stability of emulsions that lecithin has formed with some fluorochemicals is considerable [78,107]. Lecithin-PFOB emulsions, once sterilized, can be stored at room temperature for several months [53,78]. It has been speculated that the bromine atom of PFOB contributes to emulsion stabilization by lecithin. Mukherji and Sloviter [108] prepared a stable emulsion of perfluorodecalin with egg yolk lecithin by sonication in Tyrode's buffer (pH 7.4). The viscosity of the emulsion did not change when stored for 60 weeks at 5°C. At 21°C, viscosity increased after 20 weeks of storage. The viscosity increase was related to slow oxidation of lecithin on storage of the emulsion at either 5 or 21°C.

Lecithins are sensitive to light and oxygen. Some emulsions made with lecithin were not sufficiently stable to be sterilized [42]. Mukherji and Sloviter

TABLE 10.6 Examples of Surfactants Used in Fluorochemical Emulsions for Biomedical Use

Compound	Structure	Reference
Pluronic F-68	$HO(CH_2CH_2O)_n(\underset{\displaystyle CH_3}{\underset{\vert}{C}}HCH_2O)_p(CH_2CH_2O)_nOH$ Average MW 28,300	Wyandotte-ICI
Egg yolk	R_1COOCH_2 $\vert$ $R_2COOCH \quad O\ominus$ $\vert \quad\quad \vert$ $H_2C{-}O{-}P(O){-}OCH_2CH_2R\oplus$ R_1, R_2 — various fatty acid gropus $R\oplus = N(CH_3)_3\oplus$ — lecithin $R\oplus = NH_3$ — phosphatidylethanolamine etc. + lysophospholipids etc.	42,64,78
Potassium oleate	cis $- CH_3(CH_2)_7CH{=}CH(CH_2)_7COO^- K^+$	
Perfluorinated amine oxide	$(CF_3)_2CFO(CF_2)_3\underset{\displaystyle O}{\underset{\Vert}{C}}NH(CH_2)_3\underset{\displaystyle O}{\underset{\Vert}{N}}(CH_3)_2$	109
Fluorinated surfactant	$C_nF_{2n+1}CH_2(OCH_2CH_2)_mOH$ $n = 5\text{–}7, m = 3\text{–}6$	110
Fluorinated surfactant	$C_7F_{15}COO(CH_2CH_2O)_n(\underset{\displaystyle CH_3}{\underset{\vert}{C}}HCH_2O)_p(CH_2CH_2O_2)_nCOC_7F_{15}$	111
Perfluoroalkylated polyols	$C_nF_{2n+1}RX(CHOH)_mH$ $n = 4\text{–}8, m = 5\text{–}12$; R—hydrocarbon segment X—junction unit	112
Perfluoroalkylated fatty acid esters of carbohydrates		42,126

Source: Ref. 42.

[108] measured oxidation of lecithin by analyzing the emulsion for malondialdehyde, a lipid oxidation product. The formation of malondialdehyde during storage and consequently the oxidation of lecithin, at either 5 or 21°C, was markedly reduced by the antioxidant tocopherol. The egg yolk lecithins are used in second-generation fluorochemical emulsions, for example, Therox and

Oxygent, but synthetic surfactants specifically designed for emulsification of fluorochemicals are being developed.

Fluorinated surfactants can lower the interfacial tension between a fluorocarbon, such as perfluorodecalin, and water to as low as 1 mN/m. This suggests that fluorinated surfactants should be effective emulsifiers for fluorochemicals [42]. However, the toxicity of the fluorinated surfactants has hindered their intravascular use. Examples of some synthetic fluorinated surfactants used in fluorochemical emulsions for biomedical use are listed in Table 10.6 [109–113].

Nonionic fluorinated surfactants derived from a telomer mixture and tris(hydroxymethyl)aminomethane by Pavia et al. [114] exhibit a higher surface activity than Pluronic F-68 and are more effective emulsifiers:

$$\begin{array}{l} H-(\underset{|}{C}HCH_2)_nSCH_2C_mF_{2m+1} \\ \quad\;\; C=O \\ \quad\;\; | \\ \;\; HNC(CH_2OH)_3 \end{array}$$

Preliminary cell culture, hemolysis, and in vivo tests indicated that the toxicity of these fluorinated surfactants is low, in spite of their high surface activity. A 0.1 g/L solution of the telomer with $m = 8$ had a surface tension of 32 mN/m and an interfacial tension between water and perfluorodecalin of 12.9 mN/m, compared to 49 and 30 mN/m, respectively, for a 0.1 g/L solution of Pluronic F-68.

The search for stable perfluorochemical emulsions for biomedical applications has included microemulsions formed by fluorinated surfactants (see Sec. 4.8). Microemulsions form spontaneously, are thermodynamically stable, and have a small particle size. However, microemulsions have shared with coarser emulsions the toxicity problems associated with fluorinated surfactants. The early microemulsions made with fluorinated surfactants were toxic and viscous. Chabert et al. [115] used an oxyethylated perfluoroalkanol as the fluorinated surfactant. Yiv [116] has claimed that the toxicity of this fluorinated surfactant is greatly reduced, although not eliminated, if a tertiary carbon atom is inserted between the perfluorinated alkyl group and the oxyethylene hydrophile. Serratrice et al. [117] prepared microemulsions using nonionic fluorinated surfactants of the structures $C_6F_{13}C_2H_4SC_2H_4(OC_2H_4)_nOH$ and $C_6F_{13}C_2H_4SC_2H_4(OC_2H_4)_pSC_2H_4(OC_2H_4)_qOH$. Viscosity and conductivity measurements showed that, in order to form microemulsions of perfluorodecalin or $C_8H_{17}CH{=}CH_2$, the HLB value of the surfactant must be about 7.5, corresponding to a total number of 5–6 units of $-OC_2H_4-$.

The use of fluorinated surfactants in blood substitutes therefore hinges on the question, are all fluorinated surfactants intrinsically toxic or is it possible to synthesize fluorinated surfactants which are not toxic? Meussdoerffer and Niderprüm [7] have explained the toxicity of fluorinated surfactants by their extremely low surface tension and high surface activity. However, Riess and co-workers [118] have reported that perfluoroalkylated xylitol derivatives are biocompatible, in spite of their strong surface activity. Riess and Le Blanc [42] have argued that there is no reason to believe that fluorinated surfactants should necessarily be toxic. Most commercial fluorinated surfactants are mixtures of undefined purity and their toxicity in intravascular use is therefore not surprising. Attempts to reduce their toxicity by various purification methods to a biomedically acceptable level have been unsuccessful. Research on fluorochemical emulsions is therefore focusing on fluorinated surfactants prepared from nontoxic biocompatible intermediates [42,110,119–121]. The search for nontoxic fluorinated surfactants is shifting from fully synthetic surfactants to perfluoroalkylated natural products.

Tsuda et al. [70] have suggested that the efficacy of a perfluorochemical and its retention time in blood depend on the in vivo stability of the emulsion in circulating blood. Hence the efficiency and stability of emulsions have to be increased and the toxicity reduced by lowering the surfactant/fluorocarbon ratios [122].

Fundamentally, the stability of fluorochemical emulsions can be increased by one of the two approaches: by making the surfactant more "fluorophilic" [118,123] or the oxygen carrier less "fluorophilic" [124]. In accord with the latter approach, attempts have been made to prepare stable fluorochemical emulsions using a coventional nonfluorinated surfactant, such as Pluronic F-68 or lecithin, but replacing the perfluorocarbon with a partially fluorinated oxygen carrier [42,124]. Cecutti et al. [124] synthesized a microemulsion consisting of a partially fluorinated alkane, $C_8F_{17}CH_2CH{=}CHC_4H_9$, and a biocompatible hydrocarbon-type surfactant Montanox 80. The solubility of oxygen in the partially fluorinated hydrocarbon was similar to that in blood but lower than the solubility of oxygen in perfluorodecalin. Preliminary tests in the rat suggested that these microemulsions may be nontoxic.

On the other hand, the interfacial tension between the fluorochemical and water can be lowered and the stability of emulsions can be increased by using a fluorinated surfactant as a single emulsifier or as a cosurfactant with a hydrocarbon-type surfactant [100,125].

Riess and his co-workers [126] at the University of Nice, France, have made perfluoroalkylated surfactants with a modular structure:

Hydrophile	Lipophile	Fluorophile
Sugar		
Polyol		
Amine, Amino acid	$(CH_2CH_2)_n-$	C_mF_{2m+1}
Phosphates		
Phosphatidylcholines		

Perfluoroalkyl groups of various lengths were combined with hydrocarbon segments, which in turn are attached to a hydrophilic polar group. To minimize toxic effects, the hydrophilic head groups were derived from atoxic natural products, such as sugars and sugar derivatives [42,118,123,126–133], polyols [118], amines and amino acids [134], phosphocholines and phosphatidylcholines [135–137]. The lengths of the perfluoroalkyl and the hydrocarbon segments affect the solubility and the hemolytic activity of the surfactant [9,133] and have to be selected carefully.

Perfluoroalkylated phosphatidylcholines [135,136] are more effective emulsifiers for fluorocarbons than the natural egg yolk phospholipids. Concentrated (50% w/v) perfluorodecalin emulsions containing only 1% surfactant were stable at 50°C for over 1 month. However, the in vivo tests using mice as test animals gave LD_{50} values in the range of 25–50 mg/kg for a surfactant with the $C_8H_{17}(CH_2)_2-$ group, and 125 mg/kg for the surfactant with the $C_8H_{17}(CH_2)_{11}-$ group, too low for intravascular use.

Riess et al. [118,123,127,138] have shown that perfluoroalkylated xylitol and maltose derivatives exhibit a synergistic stabilizing effect in perfluorocarbon emulsions made with Pluronic F-68. The synergistic effect resulted from an interfacial tension reduction by the fluoroalkyl group and hydrogen bonds formed between the hydroxyl groups of the polyol and the ether groups of Pluronic F-68. The polyhydroxyperfluoroalkylated surfactants are derived from natural products such as mono- and disaccharides and linear pentitols and hexitols, which are not toxic [42,118,123,126–133]. Examples of fluorinated surfactants derived from xylitol or maltose are shown in Fig. 10.12 [123]. Their solubility in water depends, as expected, on the length of the perfluorocarbon chain and the number of hydroxyl groups. The maltose derivative (surfactant 2) is the most soluble and the xylitol derivative (Surfactant 1) the least soluble in water. These semisynthetic fluorinated surfactants can lower the surface tension of water to 18–23 mN/m and the water-perfluorodecalin interfacial tension to 1–5 mN/m, a large improvement over the interfacial tension of 31 mN/m obtained with Pluronic F-68. Small amounts of such fluorinated surfactants have significantly increased the stability of fluorocarbon emulsion prepared with Pluronic F-68. Zarif et al. [123] assumed that the fluorinated surfactants they have prepared are "fluorophilic" and therefore have a higher affinity for the fluorocarbon chain. They have proposed that the hydroxyl groups of the fluorinated surfactant can bond to Pluronic F-68 by forming hydrogen bonds with the ether groups of Pluronic F-68. The effect of surfactant 1 on the surface tension of Pluronic F-68 is shown in Fig. 10.13.

The stabilizing effect of surfactants 1, 2, and 3 on perfluorodecalin–Pluronic F-68 emulsions was evaluated by measuring the average droplet size as a function of time. The emulsions were prepared with a fixed total surfactant concentration and stored at 25°C. Neither the Pluronic F-68 nor fluorinated surfactant 1, 2, or 3 alone can produce a stable perfluorodecalin emulsion. The U-shaped

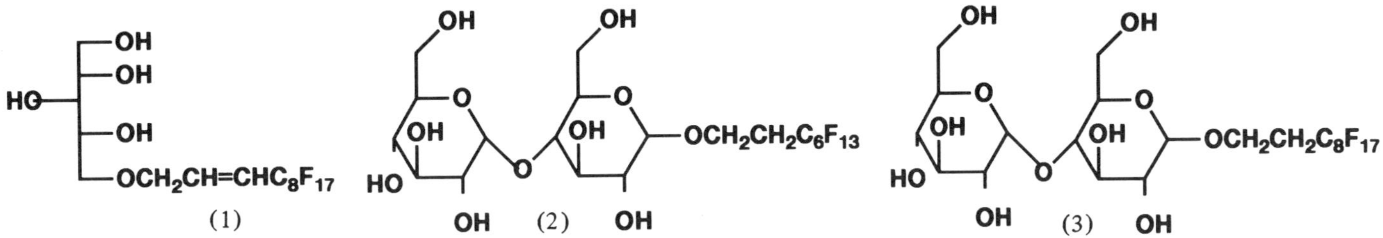

(1) (2) (3)

FIG. 10.12 Structure of fluorinated surfactants derived from xyletol (1) and maltose (2, 3). (From Ref. 123. Reproduced by permission of American Oil Chemists Society.)

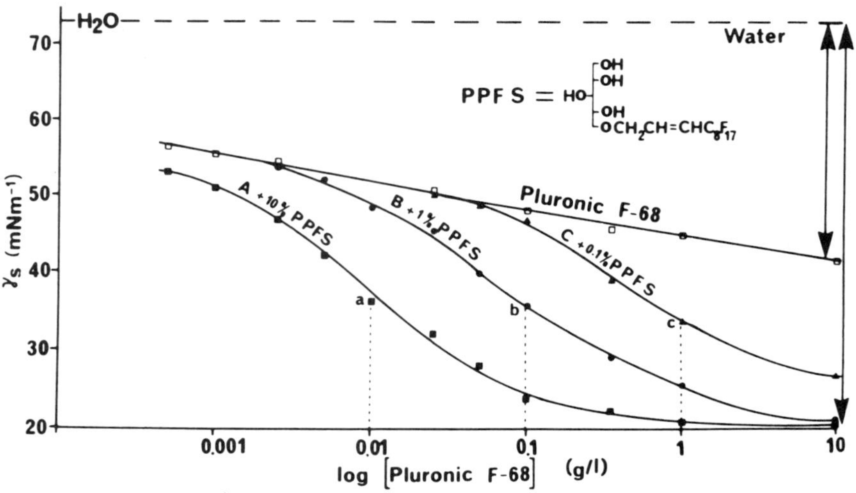

FIG. 10.13 Synergistic effect of the fluorinated surfactant (1) on the surface tension of Pluronic F-68. The percentages of (1) shown are relative to the concentration of Pluronic F-68. (From Ref. 123. Reproduced by permission of American Oil Chemists Society.)

curves indicate that the synergism between the two surfactants, Pluronic F-68 and a fluorinated surfactant, is essential for achieving storage stability.

Fluorinated surfactant 1 was a most effective stabilizer of the three fluorinated surfactants tested. The particle sizes remained the same, regardless of the storage temperature (1, 25, or 50°C) (Fig. 10.14). The higher viscosity of the 75% emulsion may have contributed to the storage stability of the emulsion. However, the viscosity effect is not a dominant one, because the reference emulsions (Pluronic F-68 only) had a higher viscosity than the emulsions with the cosurfactant.

Initial biocompatibility tests of emulsions, prepared with a combination of Pluronic F-68 and a perfluoroalkylated surfactant derived from maltose or xylitol, have been encouraging [118,123,127,138]. Preliminary evaluation has indicated that these fluorinated surfactants are biocompatible, in spite of their high surface activity. The survival ratio, growth, and behavior of mice were normal throughout the 80-day observation period following an injection of about 0.5 mL of dispersion containing the fluorinated surfactant, Pluronic F-68, and sodium chloride. The hemolytic activity of the perfluoroalkyl derivatives of xylitol decreases when the perfluoroalkyl chain length and subsequently the surface activity increase [123]. The hemolytic activity of perfluoroalkylated carbohydrates decreases with increasing length of the perfluoroalkyl group, decreases with increasing length of the hydrocarbon connective link, and

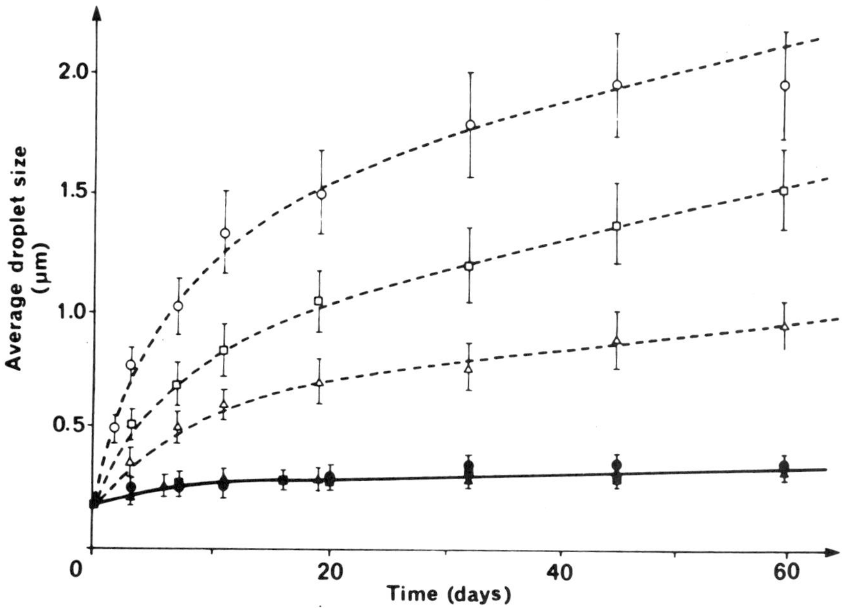

FIG. 10.14 Aging of 75% w/v perfluorodecalin emulsions prepared with 2.5% Pluronic F-68 + 2.5% fluorinated surfactant (1) (solid lines) and with 5% Pluronic F-68 alone (dashed lines). (△) 4°C; (□) 25°C; (○) 50°C. (From Ref. 123. Reproduced by permission of American Oil Chemists Society.)

decreases with increasing size of the hydrophile [9]. This observation suggests that hemolytic activity increases with increasing lipophilicity of the surfactant.

Since Pluronic F-68 has caused adverse reactions in some patients, fluorinated surfactants are being developed to replace Pluronic F-68 as the sole emulsifier. Attempts to use perfluoroalkylated surfactants derived from galactose, glucose, maltose, and related polyols as single surfactants were unsuccessful [118,123,128,131,138]. However, a study of perfluoroalkylated fatty acid monoesters of trehalose and sucrose (Fig. 10.15) [132,133] showed that 6-O-[3′-(perfluorooctyl)propanoyl)-α,α-trehalose can emulsify perfluorodecalin and form stable emulsions in the absence of Pluronic F-68 (Fig. 10.16). Its stabilizing effect is remarkable because other perfluoroalkylated fatty acid esters of trehalose and sucrose were ineffective as single emulsifiers. The effectiveness of the trehalose ester with the $C_8F_{17}(CH_2)_2C(O)-$ acyl group is not solely related to a low cmc value (0.007 mM) because the trehalose ester with the $C_8F_{17}(CH_2)_4C(O)-$ group does not form stable emulsions, in spite of the low cmc (0.004 mM) of the latter.

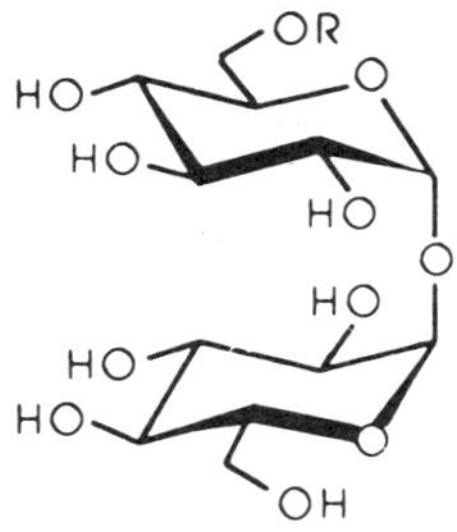

(1)

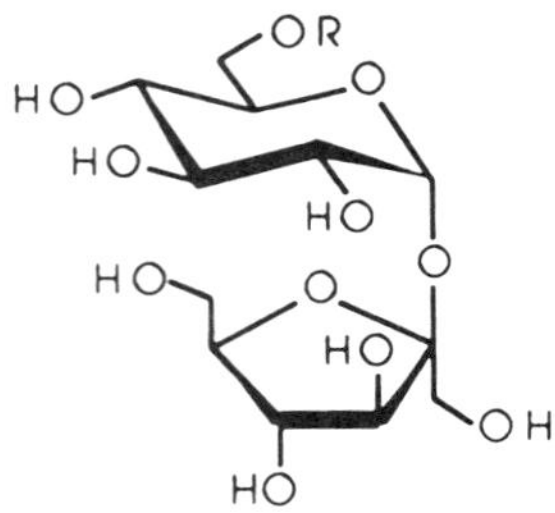

(2)

$R = R_F(CH_2)_nC(O)$

Compound	a	b	c	d	e
n	2	4	4	10	10
R_F	C_8F_{17}	C_6F_{13}	C_8F_{17}	C_4F_9	C_6F_{13}

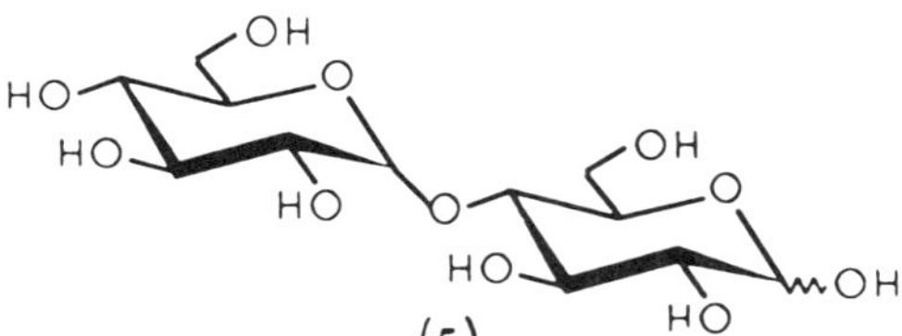

(5)

FIG. 10.15 Structures of perfluoroalkylated fatty acid esters of trehalose (1) and sucrose (2) and the structure of maltose (5). (From Ref. 133. Reproduced by permission of American Oil Chemists Society.)

The trehalose esters were also tested as cosurfactants with Pluronic F-68. For a given hydrophilic group and a given length of the hydrophobic chain the stability of the 50% w/v perfluorodecalin, 5% w/v Pluronic F-68 emulsion increased with increasing fluorination of the hydrophobe (Fig. 10.17). For a given perfluoroalkyl group, the stability of the emulsion appeared to increase when the length (*n* in Fig. 10.15) of the hydrocarbon segment increased and the cmc of the trehalose ester decreased. The authors did not relate the emulsion stability to the HLB of the surfactant. However, since the cmc and HLB values are related (see Chapter 4), the effectiveness of the surfactants evaluated by Abouhilale et al. [133] may have depended on the HLB values.

The biological compatibility of perfluoroalkylated fatty acid monoesters derived from trehalose or sucrose was evaluated by an in vitro cell culture test,

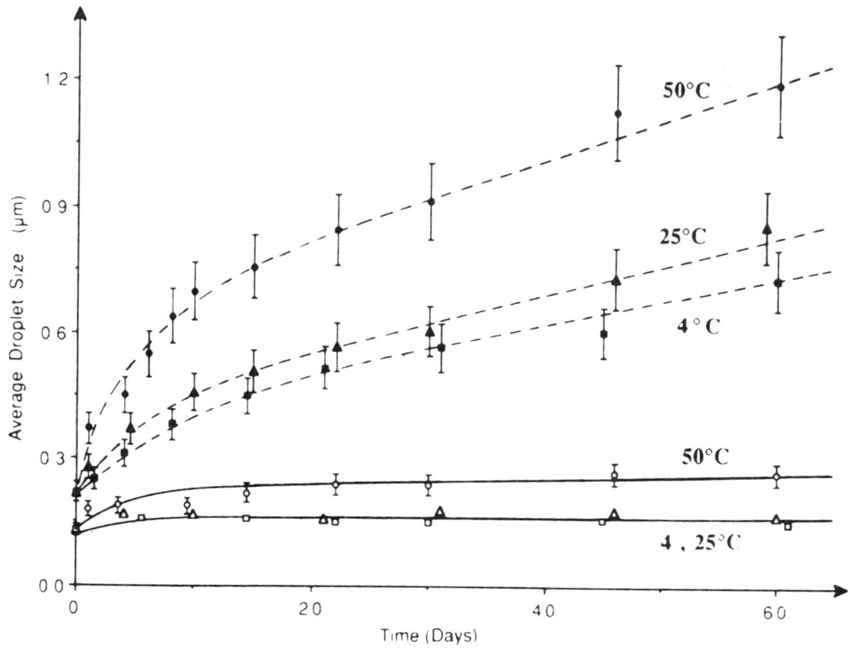

FIG. 10.16 Aging of 50% w/v perfluorodecalin emulsions prepared with 3% w/v of 6-O-[3′-(perflurooctyl)propanoyl]-α,α-trehalose (1a) (solid lines) and with 3% w/v of Pluronic F-68 (dashed lines) at 4, 25, and 50°C. (From Ref. 133. Reproduced by permission of American Oil Chemists Society.)

a hemolytic activity test, and in vivo tests performed in mice. In the cell culture test, the growth and viability of Namalva lymphoblastoid cells in the presence of the surfactant tested were compared to those of control cells (Table 10.7). At a concentration of 0.1 g/L none of the fluorinated surfactants affected the cell growth and viability significantly, in spite of their high surface activity.

The hemolytic activity (Table 10.8) decreased with increasing length of the perfluoroalkyl chain and decreased with increasing hydrocarbon segment. Abouhilale et al. [133] concluded that the most surface-active compounds are the least hemolytic. However, their data appear to suggest also that hemolytic activity is related to lipophilicity, which is reduced by fluorination.

The in vivo tests (Table 10.9) showed that none of the fluorinated surfactants perturbed the normal growth of mice at doses of 250 mg/kg of body weight. The LD_{50} for surfactants 1a and 1b was 250–375 mg/kg. Their hydrocarbon analogue was more toxic (LD_{50} about 125 mg/kg).

Fluorinated surfactants are effective emulsifiers for fluorocarbons, as exemplified by emulsion polymerization of tetrafluoroethylene dispersed with a

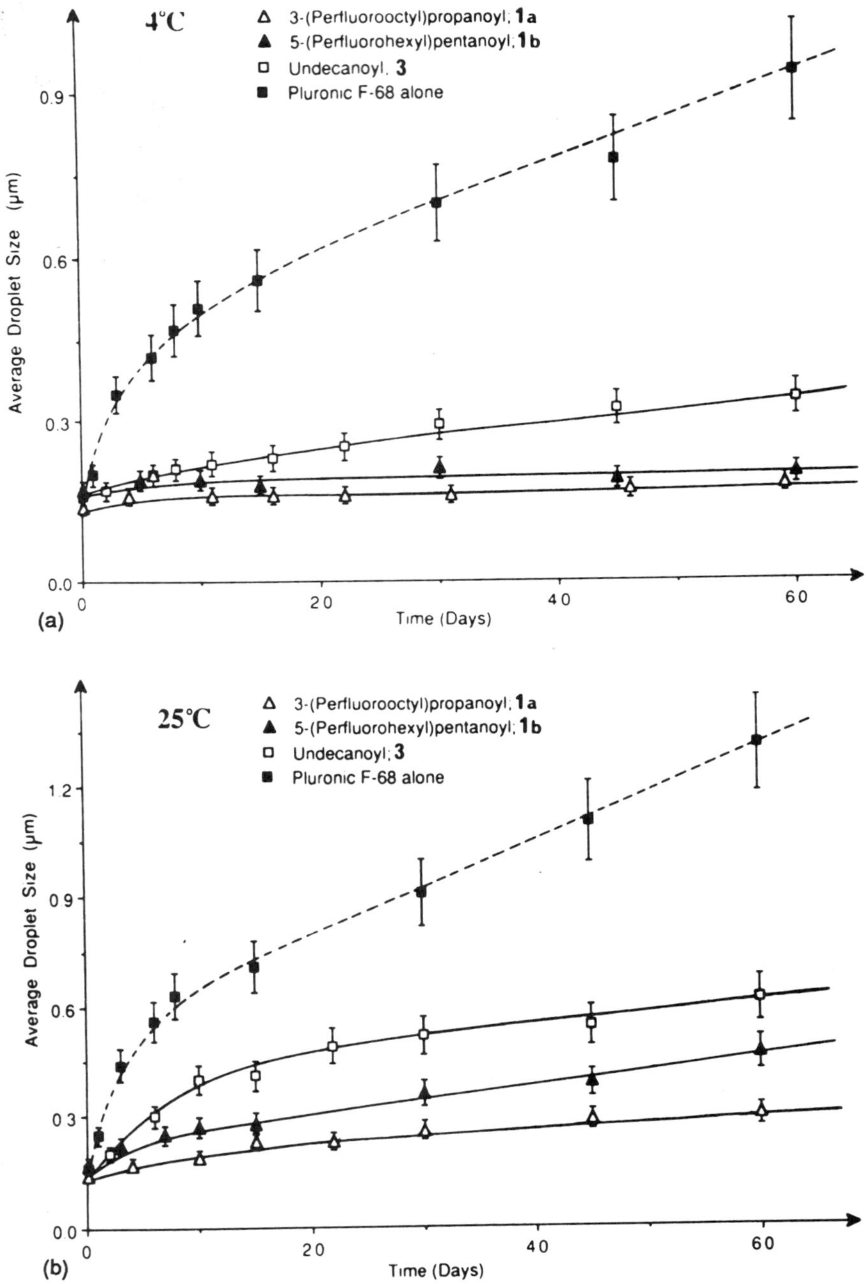

FIG. 10.17 The effect of the prefluoroalkyl chain length at 25°C on aging of 50% w/v perfluorodecalin emulsions prepared with a mixture of 2.5% w/v of Pluronic F-68 and 2.5% w/v of the trehalose esters (solid lines) and with 5% Pluronic F-68 alone (dashed line). (a), 4°C; (b), 25°C. (From Ref. 133. Reproduced by permission of American Oil Chemists Society.)

TABLE 10.7 Toxicity of the Perfluoroalkylated Fatty Acid Monoesters of Trehalose and Sucrose Toward Namalva Lymphoblastoid Cell Cultures

Compound	R_F	n	Concentration (g/L)	γ (mN/m)	Growth/viability (% vs. control)
1a	8	2	1.0	21.7	33/41
			0.1	21.6	99/97
1b	6	4	1.0	—	93/103
			0.1	21.6	108/101
1c	8	4	1.0	—	36/44
			0.1	21.4	108/105
1d	4	10	1.0	25.3	Cells died
			0.1	25.3	100/103
1e	6	10	1.0	—	70/87
			0.1	24.5	95/106
2e	6	10	1.0	24.6	Cells died
			0.1	24.6	100/101
3a	—	10	1.0	38.3	Cells died
			0.1	43.3	78/94

Source: Ref. 133.

fluorinated surfactant (see Sec. 8.3). However, the the potential toxicity of fluorinated surfactants is a concern. Fluorinated surfactants derived by Riess and co-workers from natural products [132,133] appear to be essentially non-toxic in intravascular use, but more testing is needed to remove any doubts about their biocompatibility. The purification of fluorinated natural products is difficult and hence another practical concern. Other emulsification systems are therefore being developed.

The fundamental difficulty with the emulsification of fluorochemical oxygen carriers in water is the interface between the fluorochemical and water. The use of ordinary surfactants as emulsifiers results in a fluorophilic-lipophilic interface between the droplet and the surfactant and consequently in a high interfacial tension. The stability of emulsions prepared with lipophilic-hydrophilic type surfactants is therefore only marginal. Emulsifier systems consisting of two components have therefore been proposed [98–100,125b].

Davis et al. [98,139] were concerned mostly with the coarsening of fluorochemical emulsions by the Ostwald ripening mechanism [93–95]. They added to the emulsion a component which had a lower vapor pressure than the fluorochemical oxygen carrier. The stabilizing effect of these perfluorinated polycyclic oils, such as perfluoroperhydrophenanthrene, was related to the boiling point and consequently to the molecular weight of the additive [98,140].

TABLE 10.8 Hemolytic Activity of the Trehalose and Sucrose Perfluoroalkylated Fatty Acid Monoesters

Compound	Concentration		Hemolysis
	g/L	mmM	%[a]
1a	30/8[b]	36.7	0
1b	50	67.1	0
1c	20/8[b]	23.7	0
1d	0.1	0.14	128
	0.01	0.014	11
	0.005	0.007	0
1e	0.1	0.12	128
	0.01	0.012	10
	0.005	0.006	0
2d	1.0	1.37	117
	0.1	0.14	0
2e	1.0	1.2	117
	0.5	0.6	0
	0.1	0.12	0
3[c]	1.0	1.96	146
	0.1	0.19	0
NaCl 0.9%	Control (−)		0
Water	Control (+)		100

[a] Hemolysis % $= 100(OD_{TEST} - OD_{NaCl})/(OD_{WATER} - OD_{NaCl})$. Optical densities were measured at 540 nm.
[b] *A/B*—Dispersion of *A* g/L of the compound in *B* g/L of Pluronic F-68 in water.
[c] Compound 3 is the hydrocarbon analogue of 1a and 1b.
Source: Ref. 133. Reproduced by permission of American Oil Chemists Society.

Meinert et al. [100] measured the average particle size and the interfacial tensions of perfluorodecalin-water emulsions containing various fluorinated dimorpholines. Their data indicated that the additive was enriched at the perfluorodecalin-water interface and retarded Ostwald ripening by hindering the transmission of perfluorodecalin from the emulsified droplet into water. The stabilizing effect increased with decreasing time needed by the additive to form the retaining film. F-dimorpholinopropane and F-dimorpholinobutane were most effective of the F-dimorpholinoalkanes tested.

Riess et al. [125b] suggested separating the surfactant interactions at the fluorochemical-water interface into two contributions: a fluorophilic-lipophilic interface and a lipophilic-hydrophilic interface. In accord with this concept two

TABLE 10.9 Intravascular Toxicity Tests in Mice of the Trehalose and Sucrose Perfluoroalkylated Fatty Acid Monoesters

Compound	Concentration (g/L)	Injected dose (mg/kg)	Survival ratio	LD_{50} (mg/kg)
1a	10/4[a]	250	10/10	250–375
	15/6	375	2/10	
1b	5	125	10/10	250–375
	10	250	9/10	
	15	375	0/5	
1c	5	125	10/10	>250
	10/4[a]	250	10/10	
1d	10/4[a]	250	10/10	≈375
	15/6[a]	375	6/10	
1e	1	25	10/10	>125
	5/2[a]	125	10/10	
2e	1	25	10/10	250–375
	10/4[a]	250	10/10	
	15/6[a]	375	2/10	
	20/8[a]	500	2/10	
3[b]	5	125	6/10	≈125

[a] A/BN—Dispersion of *A* g/L of the compound in *B* g/L of Pluronic F-68 in water.
[b] Compound 3 is the hydrocarbon analogue of 1a and 1b.
Source: Ref. 133. Reproduced by permission of American Oil Chemists Society.

amphiphiles, a fluorophilic-lipophilic amphiphile (see Sec. 1.8) and a conventional lipophilic-hydrophilic amphiphile, are used to bridge the fluorocarbon and water surfaces. A similar theory was developed by Meinert et al. [100]. A partially fluorinated alkane [100] or alkene [125b] adsorbs on the fluorochemical oxygen carrier with its fluorophilic segment in the surface of the droplet. Its hydrocarbon group oriented away from the droplet surface forms a lipophilic interface. The nonfluorinated surfactant adsorbs on this interface with its lipophilic segment. The hydrophilic groups of the surfactant are oriented toward water and form a hydrophilic outer shell (Fig. 10.18).

The partially fluorinated alkane or alkene is surface active in hydrocarbon media (see Sec. 1.8) and can be visualized to function as a cosurfactant in the emulsion.

Meinert et al. [100] found that small quantities (1 or 2% w/v) of partially fluorinated alkanes, $C_mF_{2m+1}C_nH_{2n+1}$, stabilize perfluorodecalin–Pluronic F-68 emulsions.

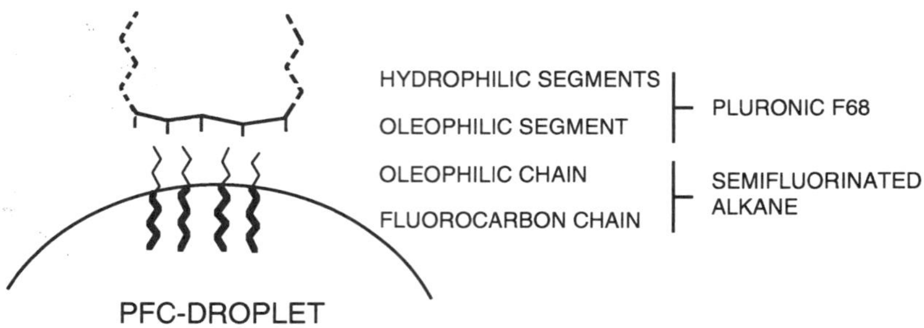

FIG. 10.18 Concept of R_FR_H forming a hydrocarbon sphere around the fluorocarbon droplet. (From Ref. 100b.)

Riess et al. [125b] described a binary emulsifier system consisting of a nonfluorinated surfactant in conjunction with a partially fluorinated alkene. They called the fluorinated amphiphile a "dowel," suggesting that its fluorophilic end adsorbs in the fluorocarbon surface and its lipophilic end penetrates the lipophilic part of the egg yolk phospholipid. The dowel, $C_8F_{17}CH{=}CHC_8H_{17}$, increased the stability of a perfluorooctyl bromide (PFOB)–egg yolk lipid emulsion considerably. The droplet size (0.25 μm) remained constant over 9 months, even at 40°C. In the absence of the dowel the droplet size more than doubled at 25°C to 0.49 μm.

The theories of Meinert et al. [100] and Riess et al. [125b] are similar, except for the interaction between the two amphiphiles forming the surfactant system. Riess et al. [125b] have suggested that the lipophilic group of the fluorinated cosurfactant penetrates the lipophilic segments of the surfactant and is anchored there—hence the term dowel. Meinert et al. [100] visualize the attraction between the two surfactant components as a result of a decrease in interfacial tension.

Future

The history of fluorochemical blood substitutes started with the dramatic demonstration that a mouse can stay alive while submerged in a fluorochemical saturated with oxygen [140]. Since the early days of great optimism, the limitations of fluorochemical emulsions as blood substitutes have been recognized [5]. Difficulties with the preparation of stable emulsions, a low oxygen content at atmospheric pressures, and in vivo accumulation of fluorochemicals in tissues have restricted their intravascular use as blood substitutes [38]. The main emphasis is now on diagnostic applications and oxygen transport into ischemic tissues.

First-generation fluorochemical blood substitutes had a low concentration of the fluorochemical and a marginal storage stability, requiring storage in a freezer or refrigerator. The second-generation emulsions, such as Oxygent, have a higher concentration (90 to 100% w/v) of fluorochemicals [141,142]. These emulsions are made with egg yolk lecithins as the emulsifier, and do not need to be frozen for storage. The side-effects of these emulsions appear to be a normal response of phagocytes to injected particles. These effects are physiological rather than pathological and controllable pharmacologically [142]. Since the molecular weight range of fluorochemicals suitable for intravascular oxygen carriers is limited [42], further progress hinges on the development of more effective, biocompatible surfactants [42,100,125,143]. The fluorochemical oxygen carrier is lipophobic and repels lipophilic groups of conventional surfactants. This phobicity results in a high interfacial tension between the fluorochemical and the surfactant and a low emulsion stability. A fluorinated surfactant is the logical choice, if nontoxic biocompatible fluorinated surfactants can be found. Fluorinated surfactants derived from natural products, such as carbohydrates, appear to be effective, although the toxicity of the least toxic surfactants is marginally acceptable at best. Binary surfactant systems consisting of a fluorophilic-lipophilic fluorinated amphiphile and a lipophilic-hydrophilic surfactant appear promising if adequate biocompatibility can be achieved.

The fear of HIV has heightened the demand for safe blood transfusion procedures and blood substitutes. The safest procedure, autologous blood transfusion, uses blood that patient has donated before the surgery. However, emergency and trauma cases require homologous blood transfusions. The search for safe blood substitutes has therefore intensified. Two avenues are now being explored using either modified hemoglobin or a fluorochemical as the oxygen carrier. The development of blood substitutes based on hemoglobin has been frustrated by the toxicity of hemoglobin deprived from its red blood cell membrane. Free hemoglobin molecule breaks down into two nonfunctional dimers which are filtered by the kidney and cause renal toxicity. The toxicity problem of hemoglobin has lead to modifications of hemoglobin, including crosslinking, genetic engineering, and attachment of a polyoxyethylene chain to the hemoglobin molecule. In spite of earlier difficulties with side effects, blood substitutes based on modified hemoglobin have entered the stage of human trials.

One of the fluorochemical oxygen carriers, Fluosol, has been approved by the FDA for limited use in humans, for oxygenation of myocardium during percutaneous transluminal coronary angioplasty [141]. Further progress in stabilizing fluorochemical emulsions and minimizing side effects will allow the use of fluorochemical oxygen carriers as temporary red blood cell substitutes and for a variety of biomedical uses, such as treatment of cerebral ischemia, organ preservation, diagnostic procedures, drug delivery, and others [141,142].

REFERENCES

1. H. Liebig and K. Ulm, Chem. Z. 100, 3 (1976).
2. F. L. M. Pattison and D. A. V. Peters, Biochem. J. 98, 689 (1966).
3. J. A. Young and T. M. Reed, J. Org. Chem. 32, 1871 (1967).
4. K. Ulm and W. Weigand, Angew. Chem. 87, 171 (1975).
5. J. G. Riess and M. Le Blanc, Angew. Chem. 90, 654 (1978).
6. M. Le Blanc and J. G. Riess, in "Preparation, Properties, and Industrial Applications of Organofluorine Compounds," R. E. Banks, ed., John Wiley & Sons, New York (1982).
7. (a) J. N. Meussdoerffer and H. Niderprüm, Chem. Z. 104, 45 (1980). (b) Mobay Product Information. Fluoro Chemical Surfactant FT 248 (July 1981).
8. M. J. Schwuger and F. G. Bartnik, in "Anionic Surfactants. Biochemistry, Toxicology, Dermatology," Surfactant Science Series, Vol 10, C. Gloxhuber, ed., Marcel Dekker, New York (1980).
9. J. G. Riess, S. Pace, and L. Zarif, Adv. Mater. 3, 249 (1991).
10. Zonyl Fluorosurfactants, Du Pont Performance Products Bulletin (1987).
11. Du Pont Co., Haskell Laboratory unpublished reports.
12. E. Kissa, J. Flurorine chem., accepted for publication.
13. (a) F. Bartnik and K. Künstler, in "Surfactants in Consumer Products," J. Falbe, ed., Springer-Verlag, Berlin (1987). (b) A. Sivak, M. Goyer, J. Perwak, and P. Thayer, in "Solution Behavior of Surfactants," Vol. 1, K. L. Mittal and E. J. Fendler, eds., Plenum, New York (1982).
14. F. D. Griffits and J. E. Long, Am. Ind. Hyg. Assoc. J. 41, 576 (1980). Fluorad Fluorochemical Surfactants, 3M (1992).
15. (a) M. S. Potokar, in "Anionic Surfactants. Biochemistry, Toxicology, Dermatology," C. Gloxhuber, ed., Surfactant Science Series, Vol. 10, Marcel Dekker, New York (1980). (b) Chr. Gloxhuber, Fette Seifen Anstrichm. 74, 49 (1972). (c) Chr. Gloxhuber, Arch. Toxicol. 32, 245 (1974).
16. D. Knaack and H.J. Walther, Acta Hydrochim. Hydrobiol. 11, 245 (1983).
17. D. Prescher, U. Gross, J. Wotzka, M. Tscheu-Schlueter, and W. Starke, Acta Hydrochim. Hydrobiol. 12, 595 (1984).
18. S. Sakakibara, R. Tatematsu, M. Kita, and H. Kimoto, Nagoya Kogyo Gijutsu Shikensho Hokoku 40(1), 20 (1991).
19. D. Prescher, U. Gross, J. Wotzka, M. Tscheu-Schlueter, and W. Starke, Acta Hydrochim. Hydrobiol. 13, 17 (1985).
20. (a) H. Fr. Schröder, Vom Wasser, 77, 277 (1991). (b) H. Fr. Schröder, Vom Wasser, 78, 211 (1992). (c) H. Fr. Schröder, Vom Wasser, 73, 111 (1989).
21. Fluorad FC171, 3M Product Environmental Data (1986).
22. D. Prescher, U. Gross, J. Wotzka, M. Tscheu-Schlueter, and W. Starke, Acta Hydrochim. Hydrobiol. 14, 293 (1986).
23. E. Staude, W. Assenmacher, and H. Hoffmann, Desalination 49, 321 (1984).
24. D. R. Taves, Nature, 217, 1050 (1968).
25. W. S. Guy, D. R. Taves, and W. S. Brey, Am. Chem. Symp. Ser. 28, 117 (1976).
26. R. H. Ophaug and L. Singer, Proc. Soc. Exp. Biol. Med. 163, 19 (1980).

27. J. Belisle and D. F. Hagen, Anal. Biochem. 101, 369 (1980).
28. F. A. Ubel, S. D. Sorenson, and D. E. Roach, Am. Ind. Hyg. Assoc. J. 41, 584 (1980).
29. E. Kissa and L. A. Kinney, Tenside 26, 372 (1989).
30. R. Wickbold, Angew. Chem. 64, 133 (1952).
31. P. B. Sweetser, Anal. Chem. 28, 1766 (1956).
32. E. Kissa, to be published.
33. E. Kissa, Anal. Chem. 55, 1445 (1983).
34. E. Kissa, Clin. Chem. 33, 253 (1987).
35. K. Yokoyama, K. Yamanouchi, M. Watanabe, T. Matsumoto, R. Murashima, T. Daimoto, T. Hamano, H. Okamoto, T. Suyama, R. Watanabe, and R. Naito, Fed. Proc. 34, 1478 (1975).
36. D. M. Long, C. B. Higgins, R. F. Mattrey, R. M. Mitten, and F. K. Multer, in "Preparation, Properties, and Industrial Applications of Organofluorine Compounds," R. F. Banks, ed., pp. 139–156, John Wiley & Sons, New York (1982).
37. D. D. MacNicol and C. D. Robertson, Nature 332, 59 (1988).
38. G. W. Odling-Smee and B. G. Wilson, in "Blood Substitutes," K. C. Lowe, ed., p. 71, VCH Publishers Cambridge, (1988).
39. H. A. Sloviter and T. Kamimoto, Nature 216, 458 (1967).
40. R. B. Bolin, R. P. Geyer, and G. J. Nemo, eds., Advances in Blood Substitute Research, Prog. Clin. Biol. Res. 122 (1983).
41. R. E. Geyer, Int. Anesthesiol. Clin. 23, 25 (1985).
42. J. G. Riess and M. Le Blanc, in "Blood Substitutes," K. C. Lowe, ed., p. 94, VCH Publishers, Cambridge (1988).
43. R. P. Geyer, in "Blood Substitutes," T. M. S. Chang and R. P. Geyer, eds., Vol. 16, pp. 31–49, Marcel Dekker, New York (1989).
44. J. R. Riess, R. P. Greyner, et al., in "Blood Substitutes," T. M. S. Chang and R. P. Geyer, eds., pp. 421–430, Marcel Dekker, New York (1989).
45. T. M. S. Chang, in "Blood Substitutes," K. C. Lowe, ed., p. 87, VCH Publishers, Cambridge (1988).
46. S. Kagawa, K. Koshu, T. Yoshimoto, and J. Suzuki, Surg. Neurol. 17, 66 (1981).
47. J. Suzuki, T. Yoshimoto, and A. Ogawa, Prog. Clin. Biol. Res. 122, 321 (1983).
48. S. J. Peerless, Prog. Clin. Biol. 122, 353 (1983).
49. P. Menasche, M. Fauchet, A. Lavergne, P. Commin, C. Masquet, P. Birkui, P. Lortente, R. P. Geyer, and A. Piwnica, Prog. Clin. Biol. Res. 122, 363 (1983).
50. M. D. Parrish, R. D. Olson, P. S. Mushlin, M. Artman, and R. J. Boucek, Jr., J. Cardiovasc. Pharmacol. 6, 159 (1984).
51. J. M. Yuhas, R. L. Goodman, and R. E. Moore, Int. Anesth. Clin., 23, 199 (1985).
52. D. M. Long, M. S. Liu, P. S. Szanto, and P. Alrenga, Rev. Surg. 29, 71 (1972).
53. D. M. Long, Jr., U.S. Patent 3,975,512 (1976).
54. D. M. Long, Jr., U.S. Patent 4,073,879 (1978).
55. D. M. Long, C. B. Higgins, R. F. Mattrey, R. M. Mitten, and F. K. Multer, in "Preparation, Properties and Industrial Applications of Organofluorine Compounds," R. E. Banks, ed., pp. 139–156, Ellis Horwood, Chichester (1982).

56. D. M. Long, D. C. Long, R. F. Mattrey, et al., in "Blood Substitutes," T. M. S. Chang and R. P. Geyer, eds., pp. 411–420, Marcel Dekker, New York (1989).
57. E. Lowenstein, J. C. Lincoln, J. H. Modell, W. G. Austen, and M. B. Laver, Fed. Proc. 29, 1775 (1970).
58. H. D. Berkowitz, P. McCombs, S. Sheety, L. D. Miller, and H. Sloviter, J. Surg. Res. 20, 595 (1976).
59. K. Honda, Prog. Clin. Biol. Res. 122, 327 (1983).
60. E. M. Levine and K. K. Tremper, in "Perfluorochemical Oxygen Transport," K. K. Tremper, ed., pp. 211–230, Little, Brown, Boston (1985).
61. N. S. Faithfull, in "Blood Substitutes," K. C. Lowe, ed., p. 130, VCH Publishers, Cambridge (1988).
62. C. M. Sharts, A. A. Malik, J. C. Easdon, and L. A. Khawli, J. Fluorine Chem. 34, 365 (1987).
63. C. M. Sharts and H. R. Reese, J. Fluorine Chem. 11, 637 (1978).
64. R. Naito and K. Yokoyama, Technical Information Ser. No 5, Green Cross, Osaka (1978).
65. A. L. Rosen, L. R. Seghal, S. A. Gould, H. Seghal, L. Dalton, C. L. Rice, and G. S. Moss, Crit. Care Med. 10, 149 (1982).
66. R. Zander and H. V. Makowski, in "Oxygen-Carrying Colloidal Blood Substitutes," R. Frey, H. Beisbarth, and K. Stosseck, eds., p. 133, Zuchschwerdt, Munich (1982).
67. K. C. Lowe, in "Blood Substitutes," K. C. Lowe, ed., p. 149, VCH Publishers, Cambridge (1988).
68. H. Okamoto, K. Yamanouchi, and K. Yokoyama, Chem. Pharm. Bull. 23, 1452 (1975).
69. K. Yokoyama, K. Yamanouchi, and R. Murashima, Chem. Pharm. Bull. 23, 1368 (1975).
70. Y. Tsuda, K. Yamanouchi, H. Okamoto, K. Yokoyama, and C. Heldebrant, J. Pharmacobio. Dyn. 13, 165 (1990).
71. Y. Tsuda, K. Yamanouchi, K. Yokoyama, and T. Suyama, Biomater. Artif. Cells Artif. Org. 16, 473 (1988).
72. R. Naito and K. Yokoyama, Proc. Xth International Congress for Nutrition: Symposium on Perfluorochemical Artificial Blood, Kyoto, 1975 (1976).
73. M. C. Malet-Martino, D. Betbeder, A. Lattes, A. Lopez, R. Martino, G. Francois, and S. Cros, J. Pharm. Pharmacol. 36, 556 (1984).
74. D. M. Long, M. S. Liu, G. D. Dobben, P. S. Szanto, and A. S. Arambulo, in "Biochemistry Involving Carbon-Fluorine Bonds," R. Filler, ed., ACS Symp. Ser. 28, Washington, DC (1976).
75. D. M. Long, E. C. Lasser, C. M. Sharts, F. K. Multer, and M. Nielsen, Invest. Radiol. 15, 242 (1980).
76. J. G. Riess, Artif. Organs 8, 44 (1984),
77. E. P. Wesseler, R. Iltis, and L. C. Clark, Jr., J. Fluorine Chem. 9, 137 (1977).
78. D. C. Long, D. M. Long, J. G. Riess, R. Follana, A. Burgan, and R. F. Mattrey, in "Blood Substitutes," T. M. S. Chang and R. P. Geyer, eds., pp. 441–442, Marcel Dekker, New York (1989).
79. R. F. Mattrey, W. W. Peck, R. A. Slutsky, and C. B. Higgins, J. Comput. Assist. Tomogr. 8, 739 (1984).

80. J. N. Bruneton, M. N. Falewee, E. Franois, P. Cambon, C. Philip, J. G. Riess, C. Balumaestro, and A. Rogopoulos, Radiology 170, 179 (1989).
81. R. F. Mattrey, Am. J. Radiol. 152, 247 (1989).
82. R. F. Mattrey, Radiology 170, 18 (1989).
83. M. S. Liu and D. M. Long, Radiology 122, 71 (1977).
84. R. F. Mattrey, Invest. Radiol. 25, 158 (1990).
85. C. Arlen, R. Follana, M. Le Blanc, C. Long, D. Long, J. G. Riess, and A. Valla, Biomater. Artif. Cells Artif. Organs 16(1–3), 455 (1988).
86. S. Pace, M. Le Blanc, R. Follana, A. M. Mahé, J. Manoux, A. Valla, and J. G. Riess, New J. Chem. 8, 251 (1984).
87. T. Mitsuno, H. Ohyanagi, and R. Naito, Ann. Surg. 195, 60 (1982).
88. S. A. Gould, A. L. Rosen, L. R. Sehgal, H. L. Sehgal, L. A. Langdale, L. M. Krause, C. L. Rice, W. H. Chamberlin, and G. S. Moss, N. Engl. J. Med. 314, 1653 (1986).
89. G. M. Vercellotti and D. E. Hammerschmidt, in "Blood Substitutes," K. C. Lowe, ed., p. 173, VCH Publishers, Cambridge (1988).
90. J. G. Riess and M. Le Blanc, Pure Appl. Chem. 54, 2383 (1982).
91. E. I. Du Pont de Nemours. Inc., Chemicals, Jackson Laboratory, Deepwater, NJ.
92. B. Telcher, T. S. Herman, R. E. Hopkins, and K. Menon, Int. J. Radiat. Oncol. Biol. Phys. 21, 969 (1991).
93. A. S. Kabal'nov, A. V. Pertsov, Yu. D. Aprosin, and E. D. Shchukin, Kolloid. Zh. 47, 1048 (1985).
94. A. S. Kabal'nov, Yu. D. Aprosin, O. B. Pavlova-Verevkina, A. V. Pertsov, and E. D. Shchukin, Kolloid. Zh. 48, 27 (1986).
95. (a) A. S. Kabal'nov, K. N. Makarov, L. L. Gervits, A. V. Pertsov, and E. D. Shchukin, Kolloid. Zh. 48, 393 (1986). (b) A. S. Kabalnov and E. D. Shchukin, Adv. Colloid and Interface Sci. 38, 69 (1992). (c) A. S. Kabalnov, K. N. Makarov, and E. D. Shchukin, Colloids Surf. 62, 101 (1992).
96. F.-S. Yang, K. D. Caldwell, J. C. Giddings, and L. Astle, Anal. Biochem. 138, 488 (1984).
97. K. Yamanouchi, R. Murashima, and K. Yoryama, Chem. Pharm. Bull. 23, 1363 (1975).
98. S. S. Davis, K. C. Lowe, and S. K. Sharma, Br. J. Pharmacol. 89(Proc. Suppl.), 66 (1986).
99. S. K. Sharma, K. C. Lowe, and S. S. Davis, Biomater. Artif. Cells Artif. Organs 18, 447 (1988).
100. (a) H. Meinert and J. Mader, J. Fluorine Chem. 51, 335 (1991). (b) H. Meinert, R. Fackler, A. Knoblich, J. Mader, P. Reuter, and W. Röhlke, Biomater. Artif. Cells Immob. Biotech. 20(1), 95 (1992). (c) H. Meinert, P. Reuter, and J. Mader, Biomater. Artif. Cells Immob. Biotech. 20(1), 115 (1992).
101. R. L. Goodman, R. E. Moore, M. E. Davis, D. Stokers, and J. M. Yuhas, Int. J. Radiat. Oncol. Biol. Phys. 10, 1421 (1984).
102. K. C. Lowe and A. D. Bollands, Med. Lab. Sci. 42, 367 (1985).
103. L. West, N. McIntosh, S. Gendler, C. Seymour, and C. Wisdom, Int. J. Radiat. Oncol. Biol. Phys. 12, 1319 (1986).
104. G. Magnusson, T. Olsson, and J.-A. Nyberg, Toxicol. Lett. 30, 203 (1986).
105. A. M. Police, K. Waxman, and G. Tominage, Crit. Care Med. 13, 96 (1985).

106. G. M. Vercellotti and D. E. Hammerschmidt, Int. Anest. Clin. 23, 47 (1985).
107. D. E. Hammerschmidt and G. M. Vercellotti, in "Blood Substitutes," T. M. S. Chang and R. P. Geyer, eds., pp. 431–438, Marcel Dekker, New York (1989).
108. B. Mukherji and H. A. Sloviter, Transfusion 31, 325 (1991).
109. (a) L. C. Clark, E. W. Clark, R. E. Moore, and D. G. I. Kinnett, Prog. Clin. Biol. Res. 122, 169 (1983). (b) R. E. Moore, U.S. Patent 4,461,717 (1984).
110. J-J. Delpuech, G. Mathis, J. C. Ravey, C. Selve, G. Serratrice, and M. J. Stebe, Bull. Soc. Chim. Fr. 578 (1985).
111. U. Gross and H. Meinert, 11th Symposium on Fluorine Chemistry, Berlin, DDR (1985).
112. G. Charpiot, J. Greiner, M. Le Blanc, A. Manfredi, J. G. Riess, and L. Zarif, Fr. Pat. Dem. 86/11, 084 (1986).
113. J. G. Riess, C. Arlen, J. Greiner, M. Le Blanc, A. Manfredi, S. Pace, C. Varescon, and L. Zarif, Biomater. Artif. Cells Artif. Organs 16, 421 (1988).
114. A. A. Pavia, B. Pucci, J. G. Riess, and L. Zarif, Bioorgan. Med. Chem. Lett. 1, 103 (1991).
115. P. Chabert, L. Foulletier, and A. Lantz, Ger. Offen 2452513 (1975); FR. Appl. 73 39533 (1973).
116. H. S. Yiv, U.S. Patent 4,975,468 (1990).
117. G. Serratrice, L. Matos, J. J. Delpuech, and A. Cambon, J. Chim. Phys. 87, 1969 (1990).
118. L. Zarif, J. Greiner, S. Pace, and J. G. Riess, J. Med. Chem. 33, 1262 (1990).
119. A. Faradji, M. Giunta, Y. Dayan, L. Foulletier, and F. Oberling, Rev. Fr. Transf. Immunohematol. 22, 119 (1979).
120. G. Mathis and J. J. Delpuech, Br. Eur. No. 0,051,526 (1981).
121. G. Mathis, P. Leempoel, J.-C. Ravey, C. Selve, and J.-J. Delpuech, J. Am. Chem. Soc. 106, 6162 (1985).
122. J. G. Riess, Curr. Surg. 45, 365 (1988).
123. L. Zarif, A. Manfredi, C. Varescon, M. Le Blanc, and J. G. Riess, J. Am. Oil Chem. Soc. 66, 1515 (1989).
124. C. Cecutti, A. Novelli, I. Rico, and A. Lattes, J. Dispersion Sci. and Technol. 11, 115 (1990).
125. (a) J. G. Riess, Artif. Organs 15, 408 (1991). (b) J. G. Riess, L. Sole-Violan, and M. Postel, J. Dispers. Sci. Technol. 13, 349 (1991).
126. (a) A. Milius, J. Greiner, and J. G. Riess, New J. Chem. 15, 337 (1991). (b) A. Milius, J. Greiner, and J. G. Riess, Colloids Surf. 63, 281 (1992). (c) A. Milius, J. Greiner, and J. G. Riess, New J. Chem. 16, 771 (1992). (d) J. G. Riess, J. Greiner, S. Abouhilale, and A. Milius, Progr. Colloid Polym. Sci. 88, 123 (1992).
127. J. G. Riess, C. Arlen, J. Greiner, M. Le Blanc, A. Manfredi, S. Pace, C. Varescon, and L. Zarif, Biomater. Artif. Cells Artif. Organs 16, 421 (1988).
128. J. Greiner, A. Manfredi, and J. G. Riess, New J. Chem. 13, 247 (1989).
129. A. Manfredi, S. Abouhilale, J. Greiner, and J. G. Riess, Bull. Soc. Chim. Fr. 872 (1989). (b) A. Milius, F. Guillod, E. Myrtil, and J. G. Riess, J. Fluorine Chem. 58, 205 (1992).
130. L. Zarif, J. Greiner, and J. G. Riess, J. Fluorine Chem. 44, 73 (1989).

131. C. Varescon, A. Manfredi, M. Le Blanc, and J. G. Riess, J. Colloid Interface Sci. 137, 373 (1990).
132. S. J. Abouhilale, J. Greiner, and J. G. Riess, Carbohydr. Res. 212, 55 (1991).
133. S. Abouhilale, J. Greiner, and J. G. Riess, J. Am. Oil Chem. Soc. 69, 1 (1992).
134. (a) C. Blaignon, M. Le Blanc, and J. G. Riess, Pat. Appl. France FR 244 803 (1988). (b) C. Blaignon, M. Le Blanc, and J. G. Riess, Proceed. 2nd World Surfactant Congress (Paris, May 1988), Vol. 2, pp 137–142 (1988). (c) J. B. Nivet, M. Le Blanc, and J. G. Riess, J. Dispersion Science and Technology 13, 627 (1992). (d) J. B. Nivet, M. Le Blanc, and J. G. Riess, Eur. J. Med. Chem. 26, 953 (1991).
135. M. P. Krafft, J. P. Rolland, P. Vierling, and J. G. Riess, New J. Chem. 14, 869 (1990).
136. (a) C. Santaella, P. Vierling, and J. G. Riess, J. Colloid Interface Sci. 148, 288 (1991). (b) C. Santaella, P. Vierling, and J. G. Riess, New. J. Chem. 15, 685 (1991). (c) C. Santaella, P. Vierling, and J. G. Riess, Angew. Chem. Int. Ed. Engl. 30, 567 (1991).
137. J. G. Riess, F. Jeanneaux, M. P. Krafft, and P. Vierling, U.S. Patent Appl. 07/542 227 (1990).
138. J. G. Riess, Proc. 2nd World Surfactant Congress, Vol. 4, p. 256, A.S.P.A., Paris (1988).
139. S. S. Davis, H. P. Round, and T. S. Purewal, J. Colloid Interface Sci. 80, 508 (1981).
140. L. C. Clark and F. Gollan, Science 152, 1755 (1966).
141. J. G. Riess, Vox Sang 61, 225 (1991).
142. J. G. Riess, Biomat., Art. Cells, Immob. Biotech. 20, 183 (1992).
143. J. Greiner, J. G. Riess, and P. Vierling, in "Organofluorine Compounds in Medicinal Chemistry and Biomedical Applications," eds. R. Filler, Y. Kobayashi, and L. M. Yagupolski, Elsevier (in press).

Index